Graduate Texts in Mathematics 217

Springer
New York
Berlin
Heidelberg
Hong Kong
London
Milan
Paris
Tokyo

Graduate Texts in Mathematics

(continued after index)

David Marker

Model Theory:

An Introduction

Springer

David Marker
Department of Mathematics
University of Illinois
351 S. Morgan Street
Chicago, IL 60607-7045
USA
marker@math.uic.edu

Mathematics Subject Classification (2000): 03-01, 03Cxx

Library of Congress Cataloging-in-Publication Data
Marker, D. (David), 1958–
    Model theory : an introduction / David Marker
        p. cm. — (Graduate texts in mathematics ; 217)
    Includes bibliographical references and index.

    1. Model theory.   I. Title.   II. Series.
QA9.7 .M367 2002
511.3—dc21   2002024184

ISBN 978-1-4419-3157-3   e-ISBN 978-0-387-22734-4

Printed in the United States of America.

9 8 7 6 5 4 3 2

Springer-Verlag is a part of *Springer Science+Business Media*

springeronline.com

In memory of Laura

Contents

Introduction

Model theory is a branch of mathematical logic where we study mathematical structures by considering the first-order sentences true in those structures and the sets definable by first-order formulas. Traditionally there have been two principal themes in the subject:

- starting with a concrete mathematical structure, such as the field of real numbers, and using model-theoretic techniques to obtain new information about the structure and the sets definable in the structure;
- looking at theories that have some interesting property and proving general structure theorems about their models.

A good example of the first theme is Tarski's work on the field of real numbers. Tarski showed that the theory of the real field is decidable. This is a sharp contrast to Gödel's Incompleteness Theorem, which showed that the theory of the seemingly simpler ring of integers is undecidable. For his proof, Tarski developed the method of quantifier elimination which can be used to show that all subsets of $\mathbb{R}^n$ definable in the real field are geometrically well-behaved. More recently, Wilkie [103] extended these ideas to prove that sets definable in the real exponential field are also well-behaved.

The second theme is illustrated by Morley's Categoricity Theorem, which says that if T is a theory in a countable language and there is an uncountable cardinal κ such that, up to isomorphism, T has a unique model of cardinality κ, then T has a unique model of cardinality λ for every uncountable κ. This line has been extended by Shelah [92], who has developed deep general classification results.

For some time, these two themes seemed like opposing directions in the subject, but over the last decade or so we have come to realize that there are fascinating connections between these two lines. Classical mathematical structures, such as groups and fields, arise in surprising ways when we study general classification problems, and ideas developed in abstract settings have surprising applications to concrete mathematical structures. The most striking example of this synthesis is Hrushovski's [43] application of very general model-theoretic methods to prove the Mordell–Lang Conjecture for function fields.

My goal was to write an introductory text in model theory that, in addition to developing the basic material, illustrates the abstract and applied directions of the subject and the interaction of these two programs.

Chapter 1 begins with the basic definitions and examples of languages, structures, and theories. Most of this chapter is routine, but, because studying definability and interpretability is one of the main themes of the subject, I have included some nontrivial examples. Section 1.3 ends with a quick introduction to M^{eq}. This is a rather technical idea that will not be needed until Chapter 6 and can be omitted on first reading.

The first results of the subject, the Compactness Theorem and the Löwenheim–Skolem Theorem, are introduced in Chapter 2. In Section 2.2 we show that even these basic results have interesting mathematical consequences by proving the decidability of the theory of the complex field. Section 2.4 discusses the back-and-forth method beginning with Cantor's analysis of countable dense linear orders and moving on to Ehrenfeucht–Fraïssé Games and Scott's result that countable structures are determined up to isomorphism by a single infinitary sentence.

Chapter 3 shows how the ideas from Chapter 2 can be used to develop a model-theoretic test for quantifier elimination. We then prove quantifier elimination for the fields of real and complex numbers and use these results to study definable sets.

Chapters 4 and 5 are devoted to the main model-building tools of classical model theory. We begin by introducing types and then study structures built by either realizing or omitting types. In particular, we study prime, saturated, and homogeneous models. In Section 4.3, we show that even these abstract constructions have algebraic applications by giving a new quantifier elimination criterion and applying it to differentially closed fields. The methods of Sections 4.2 and 4.3 are used to study countable models in Section 4.4, where we examine $\aleph_0$-categorical theories and prove Morley's result on the number of countable models. The first two sections of Chapter 5 are devoted to basic results on indiscernibles. We then illustrate the usefulness of indiscernibles with two important applications—a special case of Shelah's Many-Models Theorem in Section 5.3 and the Paris–Harrington independence result in Section 5.4. Indiscernibles also later play an important role in Section 6.5.

Chapter 6 begins with a proof of Morley's Categoricity Theorem in the spirit of Baldwin and Lachlan. The Categoricity Theorem can be thought of as the beginning of modern model theory and the rest of the book is devoted to giving the flavor of the subject. I have made a conscious pedagogical choice to focus on ω-stable theories and avoid the generality of stability, superstability, or simplicity. In this context, forking has a concrete explanation in terms of Morley rank. One can quickly develop some general tools and then move on to see their applications. Sections 6.2 and 6.3 are rather technical developments of the machinery of Morley rank and the basic results on forking and independence. These ideas are applied in Sections 6.4 and 6.5 to study prime model extensions and saturated models of ω-stable theories.

Chapters 7 and 8 are intended to give a quick but, I hope, seductive glimpse at some current directions in the subject. It is often interesting to study algebraic objects with additional model-theoretic hypotheses. In Chapter 7 we study ω-stable groups and show that they share many properties with algebraic groups over algebraically closed fields. We also include Hrushovski's theorem about recovering a group from a generically associative operation which is a generalization of Weil's theorem on group chunks. Chapter 8 begins with a seemingly abstract discussion of the combinatorial geometry of algebraic closure on strongly minimal sets, but we see in Section 8.3 that this geometry has a great deal of influence on what algebraic objects are interpretable in a structure. We conclude with an outline of Hrushovski's proof of the Mordell–Lang Conjecture in one special case.

Because I was trying to write an introductory text rather than an encyclopedic treatment, I have had to make a number of ruthless decisions about what to include and what to omit. Some interesting topics, such as ultraproducts, recursive saturation, and models of arithmetic, are relegated to the exercises. Others, such as modules, the p-adic field, or finite model theory, are omitted entirely. I have also frequently chosen to present theorems in special cases when, in fact, we know much more general results. Not everyone would agree with these choices.

The Reader

While writing this book I had in mind three types of readers:
- graduate students considering doing research in model theory;
- graduate students in logic outside of model theory;
- mathematicians in areas outside of logic where model theory has had interesting applications.

For the graduate student in model theory, this book should provide a firm foundation in the basic results of the subject while whetting the appetite for further exploration. My hope is that the applications given in Chapters 7 and 8 will excite students and lead them to read the advanced texts [7], [18], [76], and [86] written by my friends.

The graduate student in logic outside of model theory should, in addition to learning the basics, get an idea of some of the main directions of the modern subject. I have also included a number of special topics that I think every logician should see at some point, namely the random graph, Ehrenfeucht–Fraïssé Games, Scott's Isomorphism Theorem, Morley's result on the number of countable models, Shelah's Many-Models Theorem, and the Paris–Harrington Theorem.

For the mathematician interested in applications, I have tried to illustrate several of the ways that model theory can be a useful tool in analyzing classical mathematical structures. In Chapter 3, we develop the method of quantifier elimination and show how it can be used to prove results about algebraically closed fields and real closed fields. One of the areas where model-theoretic ideas have had the most fruitful impact is differential algebra. In Chapter 4, we introduce differentially closed fields. Differentially closed fields are very interesting ω-stable structures. Chapters 6, 7, and 8 contain a number of illustrations of the impact of stability-theoretic ideas on differential algebra. In particular, in Section 7.4 we give Poizat's proof of Kolchin's theorem on differential Galois groups of strongly normal extensions. In Chapter 7, we look at classical mathematical objects—groups—under additional model-theoretic assumptions—ω-stability. We also use these ideas to give more information about algebraically closed fields. In Section 8.3, we give an idea of how ideas from geometric model theory can be used to answer questions in Diophantine geometry.

Prerequisites

Chapter 1 begins with the basic definitions of languages and structures. Although a mathematically sophisticated reader with little background in mathematical logic should be able to read this book, I expect that most readers will have seen this material before. The ideal reader will have already taken one graduate or undergraduate course in logic and be acquainted with mathematical structures, formal proofs, Gödel's Completeness and Incompleteness Theorems, and the basics about computability. Shoenfield's *Mathematical Logic* [94] or Ebbinghaus, Flum, and Thomas' *Mathematical Logic* [31] are good references.

I will assume that the reader has some familiarity with very basic set theory, including Zorn's Lemma, ordinals, and cardinals. Appendix A summarizes all of this material. More sophisticated ideas from combinatorial set theory are needed in Chapter 5 but are developed completely in the text.

Many of the applications and examples that we will investigate come from algebra. The ideal reader will have had a year-long graduate algebra course and be comfortable with the basics about groups, commutative rings, and fields. Because I suspect that many readers will not have encountered the algebra of formally real fields that is essential in Section 3.3, I have included this material in Appendix B. Lang's *Algebra* [58] is a good reference for most of the material we will need. Ideally the reader will have also seen some elementary algebraic geometry, but we introduce this material as needed.

Using This Book as a Text

I suspect that in most courses where this book is used as a text, the students will have already seen most of the material in Sections 1.1, 1.2, and 2.1. A reasonable one-semester course would cover Sections 2.2, 2.3, the beginning of 2.4, 3.1, 3.2, 4.1–4.3, the beginning of 4.4, 5.1, 5.2, and 6.1. In a year-long course, one has the luxury of picking and choosing extra topics from the remaining text. My own choices would certainly include Sections 3.3, 6.2–6.4, 7.1, and 7.2.

Exercises and Remarks

Each chapter ends with a section of exercises and remarks. The exercises range from quite easy to quite challenging. Some of the exercises develop important ideas that I would have included in a longer text. I have left some important results as exercises because I think students will benefit by working them out. Occasionally, I refer to a result or example from the exercises later in the text. Some exercises will require more comfort with algebra, computability, or set theory than I assume in the rest of the book. I mark those exercises with a dagger.[†]

The Remarks sections have two purposes. I make some historical remarks and attributions. With a few exceptions, I tend to give references to secondary sources with good presentations rather than the original source. I also use the Remarks section to describe further results and give suggestions for further reading.

Notation

Most of my notation is standard. I use $A \subseteq B$ to mean that A is a subset of B, and $A \subset B$ means A is a proper subset (i.e., $A \subseteq B$ but $A \neq B$).

If A is a set,

$$A^{<\omega} = \bigcup_{n=1}^{\infty} A^n$$

is the set of all finite sequences from A. I write $\bar{a}$ to indicate a finite sequence $(a_1, \ldots, a_n)$. When I write $\bar{a} \in A$, I really mean $\bar{a} \in A^{<\omega}$.

If A is a set, then $|A|$ is the cardinality of A. The *power set* of A is $\mathcal{P}(A) = \{X : X \subseteq A\}$.

In displays, I sometimes use $\Leftarrow$, $\Rightarrow$ as abbreviations for "implies" and $\Leftrightarrow$ as an abbreviation for "if and only if".

Acknowledgments

My approach to model theory has been greatly influenced by many discussions with my teachers, colleagues, collaborators, students, and friends.

My thesis advisor and good friend, Angus Macintyre, has been the greatest influence, but I would also like to thank John Baldwin, Elisabeth Bouscaren, Steve Buechler, Zoé Chatzidakis, Lou van den Dries, Bradd Hart, Leo Harrington, Kitty Holland, Udi Hrushovski, Masanori Itai, Julia Knight, Chris Laskwoski, Dugald Macpherson, Ken McAloon, Margit Messmer, Ali Nesin, Kobi Peterzil, Anand Pillay, Wai Yan Pong, Charlie Steinhorn, Alex Wilkie, Carol Wood, and Boris Zil'ber for many enlightening conversations and Alan Taylor and Bill Zwicker, who first interested me in mathematical logic.

I would also like to thank John Baldwin, Amador Martin Pizarro, Dale Radin, Kathryn Vozoris, Carol Wood, and particularly Eric Rosen for extensive comments on preliminary versions of this book.

Finally, I, like every model theorist of my generation, learned model theory from two wonderful books, C. C. Chang and H. J. Keisler's *Model Theory* and Gerald Sacks *Saturated Model Theory*. My debt to them for their elegant presentations of the subject will be clear to anyone who reads this book.

1

Structures and Theories

1.1 Languages and Structures

In mathematical logic, we use first-order languages to describe mathematical structures. Intuitively, a structure is a set that we wish to study equipped with a collection of distinguished functions, relations, and elements. We then choose a language where we can talk about the distinguished functions, relations, and elements and nothing more. For example, when we study the ordered field of real numbers with the exponential function, we study the structure $(\mathbb{R}, +, \cdot, \exp, <, 0, 1)$, where the underlying set is the set of real numbers, and we distinguish the binary functions addition and multiplication, the unary function $x \mapsto e^x$, the binary order relation, and the real numbers 0 and 1. To describe this structure, we would use a language where we have symbols for $+, \cdot, \exp, <, 0, 1$ and can write statements such as $\forall x \forall y \ \exp(x) \cdot \exp(y) = \exp(x+y)$ and $\forall x \ (x > 0 \rightarrow \exists y \ \exp(y) = x)$. We interpret these statements as the assertions "$e^x e^y = e^{x+y}$ for all x and y" and "for all positive x, there is a y such that $e^y = x$."

For another example, we might consider the structure $(\mathbb{N}, +, 0, 1)$ of the natural numbers with addition and distinguished elements 0 and 1. The natural language for studying this structure is the language where we have a binary function symbol for addition and constant symbols for 0 and 1. We would write sentences such as $\forall x \exists y \ (x = y + y \ \lor \ x = y + y + 1)$, which we interpret as the assertion that "every number is either even or 1 plus an even number."

Definition 1.1.1 A *language* $\mathcal{L}$ is given by specifying the following data:
 i) a set of function symbols $\mathcal{F}$ and positive integers n_f for each $f \in \mathcal{F}$;
 ii) a set of relation symbols $\mathcal{R}$ and positive integers n_R for each $R \in \mathcal{R}$;
 iii) a set of constant symbols $\mathcal{C}$.

The numbers n_f and n_R tell us that f is a function of n_f variables and R is an n_R-ary relation.

Any or all of the sets $\mathcal{F}$, $\mathcal{R}$, and $\mathcal{C}$ may be empty. Examples of languages include:
 i) the language of rings $\mathcal{L}_r = \{+, -, \cdot, 0, 1\}$, where $+, -$ and $\cdot$ are binary function symbols and 0 and 1 are constants;
 ii) the language of ordered rings $\mathcal{L}_{or} = \mathcal{L}_r \cup \{<\}$, where $<$ is a binary relation symbol;
 iii) the language of pure sets $\mathcal{L} = \emptyset$;
 iv) the language of graphs is $\mathcal{L} = \{R\}$ where R is a binary relation symbol.

Next, we describe the structures where $\mathcal{L}$ is the appropriate language.

Definition 1.1.2 An *$\mathcal{L}$-structure* $\mathcal{M}$ is given by the following data:
 i) a nonempty set M called the *universe*, *domain*, or *underlying set* of $\mathcal{M}$;
 ii) a function $f^{\mathcal{M}} : M^{n_f} \to M$ for each $f \in \mathcal{F}$;
 iii) a set $R^{\mathcal{M}} \subseteq M^{n_R}$ for each $R \in \mathcal{R}$;
 iv) an element $c^{\mathcal{M}} \in M$ for each $c \in \mathcal{C}$.

We refer to $f^{\mathcal{M}}$, $R^{\mathcal{M}}$, and $c^{\mathcal{M}}$ as the *interpretations* of the symbols f, R, and c. We often write the structure as $\mathcal{M} = (M, f^{\mathcal{M}}, R^{\mathcal{M}}, c^{\mathcal{M}} : f \in \mathcal{F}, R \in \mathcal{R}$, and $c \in \mathcal{C})$. We will use the notation $A, B, M, N, \ldots$ to refer to the underlying sets of the structures $\mathcal{A}, \mathcal{B}, \mathcal{M}, \mathcal{N}, \ldots$.

For example, suppose that we are studying groups. We might use the language $\mathcal{L}_g = \{\cdot, e\}$, where $\cdot$ is a binary function symbol and e is a constant symbol. An $\mathcal{L}_g$-structure $\mathcal{G} = (G, \cdot^{\mathcal{G}}, e^{\mathcal{G}})$ will be a set G equipped with a binary relation $\cdot^{\mathcal{G}}$ and a distinguished element $e^{\mathcal{G}}$. For example, $\mathcal{G} = (\mathbb{R}, \cdot, 1)$ is an $\mathcal{L}_g$-structure where we interpret $\cdot$ as multiplication and e as 1; that is, $\cdot^{\mathcal{G}} = \cdot$ and $e^{\mathcal{G}} = 1$. Also, $\mathcal{N} = (\mathbb{N}, +, 0)$ is an $\mathcal{L}_g$-structure where $\cdot^{\mathcal{N}} = +$ and $e^{\mathcal{G}} = 1$. Of course, $\mathcal{N}$ is not a group, but it is an $\mathcal{L}_g$-structure.

Usually, we will choose languages that closely correspond to the structure that we wish to study. For example, if we want to study the real numbers as an ordered field, we would use the language of ordered rings $\mathcal{L}_{or}$ and give each symbol its natural interpretation.

We will study maps that preserve the interpretation of $\mathcal{L}$.

Definition 1.1.3 Suppose that $\mathcal{M}$ and $\mathcal{N}$ are $\mathcal{L}$-structures with universes M and N, respectively. An *$\mathcal{L}$-embedding* $\eta : \mathcal{M} \to \mathcal{N}$ is a one-to-one map $\eta : M \to N$ that preserves the interpretation of all of the symbols of $\mathcal{L}$. More precisely:

i) $\eta(f^{\mathcal{M}}(a_1,\ldots,a_{n_f})) = f^{\mathcal{N}}(\eta(a_1),\ldots,\eta(a_{n_f}))$ for all $f \in \mathcal{F}$ and $a_1,\ldots,a_n \in M$;

ii) $(a_1,\ldots,a_{m_R}) \in R^{\mathcal{M}}$ if and only if $(\eta(a_1),\ldots,\eta(a_{m_R})) \in R^{\mathcal{N}}$ for all $R \in \mathcal{R}$ and $a_1,\ldots,a_{m_j} \in M$;

iii) $\eta(c^{\mathcal{M}}) = c^{\mathcal{N}}$ for $c \in \mathcal{C}$.

A bijective $\mathcal{L}$-embedding is called an $\mathcal{L}$-*isomorphism*. If $M \subseteq N$ and the inclusion map is an $\mathcal{L}$-embedding, we say either that $\mathcal{M}$ is a *substructure* of $\mathcal{N}$ or that $\mathcal{N}$ is an *extension* of $\mathcal{M}$.

For example:

i) $(\mathbb{Z},+,0)$ is a substructure of $(\mathbb{R},+,0)$.

ii) If $\eta : \mathbb{Z} \to \mathbb{R}$ is the function $\eta(x) = e^x$, then η is an $\mathcal{L}_g$-embedding of $(\mathbb{Z},+,0)$ into $(\mathbb{R},\cdot,1)$.

The *cardinality of* $\mathcal{M}$ is $|M|$, the cardinality of the universe of $\mathcal{M}$. If $\eta : \mathcal{M} \to \mathcal{N}$ is an embedding then the cardinality of $\mathcal{N}$ is at least the cardinality of $\mathcal{M}$.

We use the language $\mathcal{L}$ to create formulas describing properties of $\mathcal{L}$-structures. Formulas will be strings of symbols built using the symbols of $\mathcal{L}$, variable symbols $v_1, v_2,\ldots$, the equality symbol $=$, the Boolean connectives $\wedge$, $\vee$, and $\neg$, which we read as "and," "or," and "not", the quantifiers $\exists$ and $\forall$, which we read as "there exists" and "for all", and parentheses (,).

Definition 1.1.4 The set of $\mathcal{L}$-*terms* is the smallest set $\mathcal{T}$ such that

i) $c \in \mathcal{T}$ for each constant symbol $c \in \mathcal{C}$,

ii) each variable symbol $v_i \in \mathcal{T}$ for $i = 1, 2,\ldots$, and

iii) if $t_1,\ldots,t_{n_f} \in \mathcal{T}$ and $f \in \mathcal{F}$, then $f(t_1,\ldots,t_{n_f}) \in \mathcal{T}$.

For example, $\cdot(v_1, -(v_3, 1))$, $\cdot(+(v_1, v_2), +(v_3, 1))$ and $+(1, +(1, +(1, 1)))$ are $\mathcal{L}_r$-terms. For simplicity, we will usually write these terms in the more standard notation $v_1(v_3 - 1)$, $(v_1 + v_2)(v_3 + 1)$, and $1 + (1 + (1 + 1))$ when no confusion arises. In the $\mathcal{L}_r$-structure $(\mathbb{Z}, +, \cdot, 0, 1)$, we think of the term $1 + (1 + (1 + 1))$ as a name for the element 4, while $(v_1 + v_2)(v_3 + 1)$ is a name for the function $(x, y, z) \mapsto (x + y)(z + 1)$. This can be done in any $\mathcal{L}$-structure.

Suppose that $\mathcal{M}$ is an $\mathcal{L}$-structure and that t is a term built using variables from $\overline{v} = (v_{i_1},\ldots,v_{i_m})$. We want to interpret t as a function $t^{\mathcal{M}} : M^m \to M$. For s a subterm of t and $\overline{a} = (a_{i_1},\ldots,a_{i_m}) \in M$, we inductively define $s^{\mathcal{M}}(\overline{a})$ as follows.

i) If s is a constant symbol c, then $s^{\mathcal{M}}(\overline{a}) = c^{\mathcal{M}}$.

ii) If s is the variable v_{i_j}, then $s^{\mathcal{M}}(\overline{a}) = a_{i_j}$.

iii) If s is the term $f(t_1,\ldots,t_{n_f})$, where f is a function symbol of $\mathcal{L}$ and $t_1,\ldots,t_{n_f}$ are terms, then $s^{\mathcal{M}}(\overline{a}) = f^{\mathcal{M}}(t_1^{\mathcal{M}}(\overline{a}),\ldots,t_{n_f}^{\mathcal{M}}(\overline{a}))$.

The function $t^{\mathcal{M}}$ is defined by $\overline{a} \mapsto t^{\mathcal{M}}(\overline{a})$.

For example, let $\mathcal{L} = \{f, g, c\}$, where f is a unary function symbol, g is a binary function symbol, and c is a constant symbol. We will consider the $\mathcal{L}$-terms $t_1 = g(v_1, c)$, $t_2 = f(g(c, f(v_1)))$, and $t_3 =$

$g(f(g(v_1, v_2)), g(v_1, f(v_2)))$. Let $\mathcal{M}$ be the $\mathcal{L}$-structure $(\mathbb{R}, \exp, +, 1)$; that is, $f^{\mathcal{M}} = \exp$, $g^{\mathcal{M}} = +$, and $c^{\mathcal{M}} = 1$.

Then

$$t_1^{\mathcal{M}}(a_1) = a_1 + 1,$$

$$t_2^{\mathcal{M}}(a_1) = e^{1+e^{a_1}}, \text{ and}$$

$$t_3^{\mathcal{M}}(a_1, a_2) = e^{a_1 + a_2} + (a_1 + e^{a_2}).$$

We are now ready to define $\mathcal{L}$-formulas.

Definition 1.1.5 We say that ϕ is an *atomic $\mathcal{L}$-formula* if ϕ is either
i) $t_1 = t_2$, where t_1 and t_2 are terms, or
ii) $R(t_1, \ldots, t_{n_R})$, where $R \in \mathcal{R}$ and $t_1, \ldots, t_{n_R}$ are terms.

The set of *$\mathcal{L}$-formulas* is the smallest set $\mathcal{W}$ containing the atomic formulas such that
i) if ϕ is in $\mathcal{W}$, then $\neg\phi$ is in $\mathcal{W}$,
ii) if ϕ and ψ are in $\mathcal{W}$, then $(\phi \wedge \psi)$ and $(\phi \vee \psi)$ are in $\mathcal{W}$, and
iii) if ϕ is in $\mathcal{W}$, then $\exists v_i \ \phi$ and $\forall v_i \ \phi$ are in $\mathcal{W}$.

Here are three examples of $\mathcal{L}_{or}$-formulas.
- $v_1 = 0 \vee v_1 > 0$.
- $\exists v_2 \ v_2 \cdot v_2 = v_1$.
- $\forall v_1 \ (v_1 = 0 \vee \exists v_2 \ v_2 \cdot v_1 = 1)$.

Intuitively, the first formula asserts that $v_1 \geq 0$, the second asserts that v_1 is a square, and the third asserts that every nonzero element has a multiplicative inverse. We would like to define what it means for a formula to be true in a structure, but these examples already show one difficulty. While in any $\mathcal{L}_{or}$-structure the third formula will either be true or false, the first two formulas express a property that may or may not be true of particular elements of the structure. In the $\mathcal{L}_{or}$-structure $(\mathbb{Z}, +, -, \cdot, <, 0, 1)$, the second formula would be true of 9 but false of 8.

We say that a variable v *occurs freely* in a formula ϕ if it is not inside a $\exists v$ or $\forall v$ quantifier; otherwise, we say that it is *bound*.[1] For example v_1 is free in the first two formulas and bound in the third, whereas v_2 is bound in both formulas. We call a formula a *sentence* if it has no free variables.

Let $\mathcal{M}$ be an $\mathcal{L}$-structure. We will see that each $\mathcal{L}$-sentence is either true or false in $\mathcal{M}$. On the other hand, if ϕ is a formula with free variables

---

[1] To simplify some bookkeeping we will tacitly restrict our attention to formulas where in each subformula no variable v_i has both free and bound occurrences. For example we will not consider formulas such as $(v_1 > 0 \vee \exists v_1 \ v_1 \cdot v_1 = v_2)$, because this formula could be replaced by the clearer formula $v_1 > 0 \vee \exists v_3 \ v_3 \cdot v_3 = v_2$ with the same meaning. There are some areas of mathematical logic where one wants to be frugal with variables, but we will not consider such issues here. See [94] for a definition of satisfaction for arbitrary formulas.

$v_1, \ldots, v_n$, we will think of ϕ as expressing a property of elements of M^n. We often write $\phi(v_1, \ldots, v_n)$ to make explicit the free variables in ϕ. We must define what it means for $\phi(v_1, \ldots, v_n)$ to hold of $(a_1, \ldots, a_n) \in M^n$.

Definition 1.1.6 Let ϕ be a formula with free variables from $\bar{v} = (v_{i_1}, \ldots, v_{i_m})$, and let $\bar{a} = (a_{i_1}, \ldots, a_{i_m}) \in M^m$. We inductively define $\mathcal{M} \models \phi(\bar{a})$ as follows.

  i) If ϕ is $t_1 = t_2$, then $\mathcal{M} \models \phi(\bar{a})$ if $t_1^{\mathcal{M}}(\bar{a}) = t_2^{\mathcal{M}}(\bar{a})$.

  ii) If ϕ is $R(t_1, \ldots, t_{n_R})$, then $\mathcal{M} \models \phi(\bar{a})$ if $(t_1^{\mathcal{M}}(\bar{a}), \ldots, t_{n_R}^{\mathcal{M}}(\bar{a})) \in R^{\mathcal{M}}$.

  iii) If ϕ is $\neg\psi$, then $\mathcal{M} \models \phi(\bar{a})$ if $\mathcal{M} \not\models \psi(\bar{a})$.

  iv) If ϕ is $(\psi \wedge \theta)$, then $\mathcal{M} \models \phi(\bar{a})$ if $\mathcal{M} \models \psi(\bar{a})$ and $\mathcal{M} \models \theta(\bar{a})$.

  v) If ϕ is $(\psi \vee \theta)$, then $\mathcal{M} \models \phi(\bar{a})$ if $\mathcal{M} \models \psi(\bar{a})$ or $\mathcal{M} \models \theta(\bar{a})$.

  vi) If ϕ is $\exists v_j \psi(\bar{v}, v_j)$, then $\mathcal{M} \models \phi(\bar{a})$ if there is $b \in M$ such that $\mathcal{M} \models \psi(\bar{a}, b)$.

  vii) If ϕ is $\forall v_j \psi(\bar{v}, v_j)$, then $\mathcal{M} \models \phi(\bar{a})$ if $\mathcal{M} \models \psi(\bar{a}, b)$ for all $b \in M$.

If $\mathcal{M} \models \phi(\bar{a})$ we say that $\mathcal{M}$ *satisfies* $\phi(\bar{a})$ or $\phi(\bar{a})$ is *true* in $\mathcal{M}$.

Remarks 1.1.7 • There are a number of useful abbreviations that we will use: $\phi \rightarrow \psi$ is an abbreviation for $\neg\phi \vee \psi$, and $\phi \leftrightarrow \psi$ is an abbreviation for $(\phi \rightarrow \psi) \wedge (\psi \rightarrow \phi)$. In fact, we did not really need to include the symbols $\vee$ and $\forall$. We could have considered $\phi \vee \psi$ as an abbreviation for $\neg(\neg\phi \wedge \neg\psi)$ and $\forall v\phi$ as an abbreviation for $\neg(\exists v\neg\phi)$. Viewing these as abbreviations will be an advantage when we are proving theorems by induction on formulas because it eliminates the $\vee$ and $\forall$ cases.

We also will use the abbreviations $\bigwedge_{i=1}^n \psi_i$ and $\bigvee_{i=1}^n \psi_i$ for $\psi_1 \wedge \ldots \wedge \psi_n$ and $\psi_1 \vee \ldots \vee \psi_n$, respectively.

• In addition to $v_1, v_2, \ldots$, we will use $w, x, y, z, \ldots$ as variable symbols.

• It is important to note that the quantifiers $\exists$ and $\forall$ range only over elements of the model. For example the statement that an ordering is complete (i.e., every bounded subset has a least upper bound) cannot be expressed as a formula because we cannot quantify over subsets. The fact that we are limited to quantification over elements of the structure is what makes it "first-order" logic.

When proving results about satisfaction in models, we often must do an induction on the construction of formulas. The next proposition asserts that if a formula without quantifiers is true in some structure, then it is true in every extension. It is proved by induction on quantifier-free formulas.

Proposition 1.1.8 *Suppose that $\mathcal{M}$ is a substructure of $\mathcal{N}$, $\bar{a} \in M$, and $\phi(\bar{v})$ is a quantifier-free formula. Then, $\mathcal{M} \models \phi(\bar{a})$ if and only if $\mathcal{N} \models \phi(\bar{a})$.*

Proof

Claim If $t(\bar{v})$ is a term and $\bar{b} \in M$, then $t^{\mathcal{M}}(\bar{b}) = t^{\mathcal{N}}(\bar{b})$. This is proved by induction on terms.

If t is the constant symbol c, then $c^{\mathcal{M}} = c^{\mathcal{N}}$.

If t is the variable v_i, then $t^{\mathcal{M}}(\bar{b}) = b_i = t^{\mathcal{N}}(\bar{b})$.

Suppose that $t = f(t_1, \ldots, t_n)$, where f is an n-ary function symbol, $t_1, \ldots, t_n$ are terms, and $t_i^{\mathcal{M}}(\bar{b}) = t_i^{\mathcal{N}}(\bar{b})$ for $i = 1, \ldots, n$. Because $\mathcal{M} \subseteq \mathcal{N}$, $f^{\mathcal{M}} = f^{\mathcal{N}}|M^n$. Thus,

$$
\begin{aligned}
t^{\mathcal{M}}(\bar{b}) &= f^{\mathcal{M}}(t_1^{\mathcal{M}}(\bar{b}), \ldots, t_n^{\mathcal{M}}(\bar{b})) \\
&= f^{\mathcal{N}}(t_1^{\mathcal{M}}(\bar{b}), \ldots, t_n^{\mathcal{M}}(\bar{b})) \\
&= f^{\mathcal{N}}(t_1^{\mathcal{N}}(\bar{b}), \ldots, t_n^{\mathcal{N}}(\bar{b})) \\
&= t^{\mathcal{N}}(\bar{b}).
\end{aligned}
$$

We now prove the proposition by induction on formulas.

If ϕ is $t_1 = t_2$, then

$$
\mathcal{M} \models \phi(\bar{a}) \Leftrightarrow t_1^{\mathcal{M}}(\bar{a}) = t_2^{\mathcal{M}}(\bar{a}) \Leftrightarrow t_1^{\mathcal{N}}(\bar{a}) = t_2^{\mathcal{N}}(\bar{a}) \Leftrightarrow \mathcal{N} \models \phi(\bar{a}).
$$

If ϕ is $R(t_1, \ldots, t_n)$, where R is an n-ary relation symbol, then

$$
\begin{aligned}
\mathcal{M} \models \phi(\bar{a}) \quad &\Leftrightarrow \quad (t_1^{\mathcal{M}}(\bar{a}), \ldots, t_n^{\mathcal{M}}(\bar{a})) \in R^{\mathcal{M}} \\
&\Leftrightarrow \quad (t_1^{\mathcal{M}}(\bar{a}), \ldots, t_n^{\mathcal{M}}(\bar{a})) \in R^{\mathcal{N}} \\
&\Leftrightarrow \quad (t_1^{\mathcal{N}}(\bar{a}), \ldots, t_n^{\mathcal{N}}(\bar{a})) \in R^{\mathcal{N}} \\
&\Leftrightarrow \quad \mathcal{N} \models \phi(\bar{a}).
\end{aligned}
$$

Thus, the proposition is true for all atomic formulas.

Suppose that the proposition is true for ψ and that ϕ is $\neg\psi$. Then,

$$
\mathcal{M} \models \neg\phi(\bar{a}) \Leftrightarrow \mathcal{M} \not\models \psi(\bar{a}) \Leftrightarrow \mathcal{N} \not\models \psi(\bar{a}) \Leftrightarrow \mathcal{N} \models \phi(\bar{a}).
$$

Finally, suppose that the proposition is true for ψ_0 and ψ_1 and that ϕ is $\psi_0 \wedge \psi_1$. Then,

$$
\begin{aligned}
\mathcal{M} \models \phi(\bar{a}) \quad &\Leftrightarrow \quad \mathcal{M} \models \psi_0(\bar{a}) \text{ and } \mathcal{M} \models \psi_1(\bar{a}) \\
&\Leftrightarrow \quad \mathcal{N} \models \psi_0(\bar{a}) \text{ and } \mathcal{M} \models \psi_1(\bar{a}) \\
&\Leftrightarrow \quad \mathcal{N} \models \phi(\bar{a}).
\end{aligned}
$$

We have shown that the proposition holds for all atomic formulas and that if it holds for ϕ and ψ, then it also holds for $\neg\phi$ and $\phi \wedge \psi$. Because the set of quantifier-free formulas is the smallest set of formulas containing the atomic formulas and closed under negation and conjunction, the proposition is true for all quantifier-free formulas.

Elementary Equivalence and Isomorphism

We next consider structures that satisfy the same sentences.

Definition 1.1.9 We say that two $\mathcal{L}$-structures $\mathcal{M}$ and $\mathcal{N}$ are *elementarily equivalent* and write $\mathcal{M} \equiv \mathcal{N}$ if

$$\mathcal{M} \models \phi \text{ if and only if } \mathcal{N} \models \phi$$

for all $\mathcal{L}$-sentences ϕ.

We let $\mathrm{Th}(\mathcal{M})$, the *full theory of* $\mathcal{M}$, be the set of $\mathcal{L}$-sentences ϕ such that $\mathcal{M} \models \phi$. It is easy to see that $\mathcal{M} \equiv \mathcal{N}$ if and only if $\mathrm{Th}(\mathcal{M}) = \mathrm{Th}(\mathcal{N})$. Our next result shows that $\mathrm{Th}(\mathcal{M})$ is an isomorphism invariant of $\mathcal{M}$. The proof uses the important technique of "induction on formulas."

Theorem 1.1.10 *Suppose that* $j : \mathcal{M} \rightarrow \mathcal{N}$ *is an isomorphism. Then,* $\mathcal{M} \equiv \mathcal{N}$.

Proof We show by induction on formulas that $\mathcal{M} \models \phi(a_1, \ldots, a_n)$ if and only if $\mathcal{N} \models \phi(j(a_1), \ldots, j(a_n))$ for all formulas ϕ.

We first must show that terms behave well.

Claim Suppose that t is a term and the free variables in t are from $\bar{v} = (v_1, \ldots, v_n)$. For $\bar{a} = (a_1, \ldots, a_n) \in M$, we let $j(\bar{a})$ denote $(j(a_1), \ldots, j(a_n))$. Then $j(t^{\mathcal{M}}(\bar{a})) = t^{\mathcal{N}}(j(\bar{a}))$.

We prove this by induction on terms.

i) If $t = c$, then $j(t^{\mathcal{M}}(\bar{a})) = j(c^{\mathcal{M}}) = c^{\mathcal{N}} = t^{\mathcal{N}}(j(\bar{a}))$.

ii) If $t = v_i$, then $j(t^{\mathcal{M}}(\bar{a})) = j(a_i) = t^{\mathcal{N}}(j(a_i))$.

iii) If $t = f(t_1, \ldots, t_m)$, then

$$
\begin{aligned}
j(t^{\mathcal{M}}(\bar{a})) &= j(f^{\mathcal{M}}(t_1^{\mathcal{M}}(\bar{a}), \ldots, t_m^{\mathcal{M}}(\bar{a}))) \\
&= f^{\mathcal{N}}(j(t_1^{\mathcal{M}}(\bar{a})), \ldots, j(t_m^{\mathcal{M}}(\bar{a}))) \\
&= f^{\mathcal{N}}(t_1^{\mathcal{N}}(j(\bar{a})), \ldots, t_m^{\mathcal{N}}(j(\bar{a}))) \\
&= t^{\mathcal{N}}(j(\bar{a})).
\end{aligned}
$$

We proceed by induction on formulas.

i) If $\phi(\bar{v})$ is $t_1 = t_2$, then

$$
\begin{aligned}
\mathcal{M} \models \phi(\bar{a}) &\Leftrightarrow t_1^{\mathcal{M}}(\bar{a}) = t_2^{\mathcal{M}}(\bar{a}) \\
&\Leftrightarrow j(t_1^{\mathcal{M}}(\bar{a})) = j(t_2^{\mathcal{M}}(\bar{a})) \text{ because } j \text{ is injective} \\
&\Leftrightarrow t_1^{\mathcal{N}}(j(\bar{a})) = t_2^{\mathcal{N}}(j(\bar{a})) \\
&\Leftrightarrow \mathcal{N} \models \phi(j(\bar{a})).
\end{aligned}
$$

ii) If $\phi(\bar{v})$ is $R(t_1, \ldots, t_n)$, then

$$
\begin{aligned}
\mathcal{M} \models \phi(\bar{a}) &\Leftrightarrow (t_1^{\mathcal{M}}(\bar{a}), \ldots, t_n^{\mathcal{M}}(\bar{a})) \in R^{\mathcal{M}} \\
&\Leftrightarrow (j(t_1^{\mathcal{M}}(\bar{a})), \ldots, j(t_n^{\mathcal{M}}(\bar{a}))) \in R^{\mathcal{N}} \\
&\Leftrightarrow (t_1^{\mathcal{N}}(j(\bar{a})), \ldots, t_n^{\mathcal{N}}(j(\bar{a}))) \in R^{\mathcal{N}} \\
&\Leftrightarrow \mathcal{N} \models \phi(j(\bar{a})).
\end{aligned}
$$

iii) If ϕ is $\neg\psi$, then by induction

$$\mathcal{M} \models \phi(\bar{a}) \Leftrightarrow \mathcal{M} \not\models \psi(\bar{a}) \Leftrightarrow \mathcal{N} \not\models \psi(j(\bar{a})) \Leftrightarrow \mathcal{N} \models \phi(j(\bar{a})).$$

iv) If ϕ is $\psi \wedge \theta$, then

$$\begin{aligned}\mathcal{M} \models \phi(\bar{a}) \quad &\Leftrightarrow \quad \mathcal{M} \models \psi(\bar{a}) \text{ and } \mathcal{M} \models \theta(\bar{a}) \\ &\Leftrightarrow \quad \mathcal{N} \models \psi(j(\bar{a})) \text{ and } \mathcal{N} \models \theta(j(\bar{a})) \Leftrightarrow \mathcal{N} \models \phi(j(\bar{a})).\end{aligned}$$

v) If $\phi(\bar{v})$ is $\exists w\ \psi(\bar{v}, w)$, then

$$\begin{aligned}\mathcal{M} \models \phi(\bar{a}) \quad &\Leftrightarrow \quad \mathcal{M} \models \psi(\bar{a}, b) \text{ for some } b \in M \\ &\Leftrightarrow \quad \mathcal{N} \models \psi(j(\bar{a}), c) \text{ for some } c \in N \text{ because } j \text{ is onto} \\ &\Leftrightarrow \quad \mathcal{N} \models \phi(j(\bar{a})).\end{aligned}$$

1.2   Theories

Let $\mathcal{L}$ be a language. An $\mathcal{L}$-*theory* T is simply a set of $\mathcal{L}$-sentences. We say that $\mathcal{M}$ is a *model* of T and write $\mathcal{M} \models T$ if $\mathcal{M} \models \phi$ for all sentences $\phi \in T$.

The set $T = \{\forall x\ x = 0, \exists x\ x \neq 0\}$ is a theory. Because the two sentences in T are contradictory, there are no models of T. We say that a theory is *satisfiable* if it has a model.

We say that a class of $\mathcal{L}$-structures $\mathcal{K}$ is an *elementary class* if there is an $\mathcal{L}$-theory T such that $\mathcal{K} = \{\mathcal{M} : \mathcal{M} \models T\}$.

One way to get a theory is to take $\mathrm{Th}(\mathcal{M})$, the full theory of an $\mathcal{L}$-structure $\mathcal{M}$. In this case, the elementary class of models of $\mathrm{Th}(\mathcal{M})$ is exactly the class of $\mathcal{L}$-structures elementarily equivalent to $\mathcal{M}$. More typically, we have a class of structures in mind and try to write a set of properties T describing these structures. We call these sentences *axioms* for the elementary class.

We give a few basic examples of theories and elementary classes that we will return to frequently.

Example 1.2.1 *Infinite Sets*

Let $\mathcal{L} = \emptyset$.

Consider the $\mathcal{L}$-theory where we have, for each n, the sentence ϕ_n given by

$$\exists x_1 \exists x_2 \ldots \exists x_n \bigwedge_{i < j \leq n} x_i \neq x_j.$$

The sentence ϕ_n asserts that there are at least n distinct elements, and an $\mathcal{L}$-structure $\mathcal{M}$ with universe M is a model of T if and only if M is infinite.

Example 1.2.2 *Linear Orders*

Let $\mathcal{L} = \{<\}$, where $<$ is a binary relation symbol. The class of linear orders is axiomatized by the $\mathcal{L}$-sentences
$$\forall x \, \neg(x < x),$$
$$\forall x \forall y \forall z \, ((x < y \wedge y < z) \rightarrow x < z),$$
$$\forall x \forall y \, (x < y \vee x = y \vee y < x).$$

There are a number of interesting extensions of the theory of linear orders. For example, we could add the sentence

$$\forall x \forall y \, (x < y \rightarrow \exists z \, (x < z \wedge z < y))$$

to get the theory of dense linear orders, or we could instead add the sentence

$$\forall x \exists y \, (x < y \wedge \forall z(x < z \rightarrow (z = y \vee y < z)))$$

to get the theory of linear orders where every element has a unique successor. We could also add sentences that either assert or deny the existence of top or bottom elements.

Example 1.2.3 *Equivalence Relations*

Let $\mathcal{L} = \{E\}$, where E is a binary relation symbol. The theory of equivalence relations is given by the sentences
$$\forall x \, E(x, x),$$
$$\forall x \forall y(E(x, y) \rightarrow E(y, x)),$$
$$\forall x \forall y \forall z((E(x, y) \wedge E(y, z)) \rightarrow E(x, z)).$$

If we added the sentence

$$\forall x \exists y(x \neq y \wedge E(x, y) \wedge \forall z \, (E(x, z) \rightarrow (z = x \vee z = y)))$$

we would have the theory of equivalence relations where every equivalence class has exactly two elements. If instead we added the sentence

$$\exists x \exists y(\neg E(x, y) \wedge \forall z(E(x, z) \vee E(y, z)))$$

and the infinitely many sentences

$$\forall x \exists x_1 \exists x_2 \ldots \exists x_n \left(\bigwedge_{i < j \leq n} x_i \neq x_j \wedge \bigwedge_{i=1}^{n} E(x, x_i) \right)$$

we would axiomatize the class of equivalence relations with exactly two classes, both of which are infinite.

Example 1.2.4 *Graphs*

Let $\mathcal{L} = \{R\}$ where R is a binary relation. We restrict our attention to irreflexive graphs. These are axiomatized by the two sentences
$\forall x \, \neg R(x, x),$
$\forall x \forall y \, (R(x, y) \to R(y, x)).$

Example 1.2.5 *Groups*

Let $\mathcal{L} = \{\cdot, e\}$, where $\cdot$ is a binary function symbol and e is a constant symbol. We will write $x \cdot y$ rather than $\cdot(x, y)$. The class of groups is axiomatized by
$\forall x \, e \cdot x = x \cdot e = x,$
$\forall x \forall y \forall z \, x \cdot (y \cdot z) = (x \cdot y) \cdot z,$
$\forall x \exists y \, x \cdot y = y \cdot x = e.$

We could also axiomatize the class of Abelian groups by adding $\forall x \forall y \, x \cdot y = y \cdot x$.

Let $\phi_n(x)$ be the $\mathcal{L}$-formula

$$\underbrace{x \cdot x \cdots x}_{n-\text{times}} = e;$$

which asserts that $nx = e$.

We could axiomatize the class of torsion-free groups by adding $\{\forall x \, (x = e \lor \neg \phi_n(x)) : n \geq 2\}$ to the axioms for groups. Alternatively, we could axiomatize the class of groups where every element has order at most N by adding to the axioms for groups the sentence

$$\forall x \bigvee_{n \leq N} \phi_n(x).$$

Note that the same idea will not work to axiomatize the class of torsion groups because the corresponding sentence would be infinitely long. In the next chapter, we will see that the class of torsion groups is not elementary.

Let $\psi_n(x, y)$ be the formula

$$\underbrace{x \cdot x \cdots x}_{n-\text{times}} = y;$$

which asserts that $x^n = y$. We can axiomatize the class of divisible groups by adding the axioms $\{\forall y \exists x \, \psi_n(x, y) : n \geq 2\}$.

It will often be useful to deal with additive groups instead of multiplicative groups. The class of additive groups is the collection structures in the language $\mathcal{L} = \{+, 0\}$, axiomatized as above replacing $\cdot$ by $+$ and e by 0.

Example 1.2.6 *Ordered Abelian Groups*

Let $\mathcal{L} = \{+, <, 0\}$, where $+$ is a binary function symbol, $<$ is a binary relation symbol, and 0 is a constant symbol. The axioms for ordered groups are

the axioms for additive groups,
the axioms for linear orders, and
$\forall x \forall y \forall z (x < y \rightarrow x + z < y + z)$.

Example 1.2.7 *Left R-modules*

Let R be a ring with multiplicative identity 1. Let $\mathcal{L} = \{+, 0\} \cup \{r : r \in R\}$ where $+$ is a binary function symbol, 0 is a constant, and r is a unary function symbol for $r \in R$. In an R-module, we will interpret r as scalar multiplication by R. The axioms for left R-modules are

the axioms for additive commutative groups,
$\forall x \; r(x + y) = r(x) + r(y) \quad$ for each $r \in R$,
$\forall x \; (r + s)(x) = r(x) + s(x) \quad$ for each $r, s \in R$,
$\forall x \; r(s(x)) = rs(x) \quad$ for $r, s \in R$,
$\forall x \; 1(x) = x$.

Example 1.2.8 *Rings and Fields*

Let $\mathcal{L}_r$ be the language of rings $\{+, -, \cdot, 0, 1\}$, where $+$, $-$, and $\cdot$ are binary function symbols and 0 and 1 are constants. The axioms for rings are given by

the axioms for additive commutative groups,
$\forall x \forall y \forall z \; (x - y = z \leftrightarrow x = y + z)$,
$\forall x \; x \cdot 0 = 0$,
$\forall x \forall y \forall z \; (x \cdot (y \cdot z) = (x \cdot y) \cdot z)$,
$\forall x \; x \cdot 1 = 1 \cdot x = x$,
$\forall x \forall y \forall z \; x \cdot (y + z) = (x \cdot y) + (x \cdot z)$,
$\forall x \forall y \forall z \; (x + y) \cdot z = (x \cdot z) + (y \cdot z)$.

The second axiom is only necessary because we include $-$ in the language (this will be useful later). We axiomatize the class of fields by adding the axioms

$\forall x \forall y \; x \cdot y = y \cdot x$,
$\forall x \; (x \neq 0 \rightarrow \exists y \; x \cdot y = 1)$.

We axiomatize the class of algebraically closed fields by adding to the field axioms the sentences

$$\forall a_0 \ldots \forall a_{n-1} \exists x \; x^n + \sum_{i=0}^{n-1} a_i x^i = 0$$

for $n = 1, 2, \ldots$. Let ACF be the axioms for algebraically closed fields.

Let ψ_p be the $\mathcal{L}_r$-sentence $\forall x \underbrace{x + \ldots + x}_{p-\text{times}} = 0$, which asserts that a field

has characteristic p. For $p > 0$ a prime, let $\text{ACF}_p = \text{ACF} \cup \{\psi_p\}$ and $\text{ACF}_0 = \text{ACF} \cup \{\neg\psi_p : p > 0\}$, be the theories of algebraically closed fields of characteristic p and characteristic zero, respectively.

Example 1.2.9 *Ordered Fields*

Let $\mathcal{L}_{\text{or}} = \mathcal{L}_r \cup \{<\}$. The class of ordered fields is axiomatized by the axioms for fields,

the axioms for linear orders,
$$\forall x \forall y \forall z \ (x < y \rightarrow x + z < y + z),$$
$$\forall x \forall y \forall z \ ((x < y \wedge z > 0) \rightarrow x \cdot z < y \cdot z).$$

Example 1.2.10 *Differential Fields*

Let $\mathcal{L} = \mathcal{L}_r \cup \{\delta\}$, where δ is a unary function symbol. The class of differential fields is axiomatized by

the axioms of fields,
$$\forall x \forall y \ \delta(x + y) = \delta(x) + \delta(y),$$
$$\forall x \forall y \ \delta(x \cdot y) = x \cdot \delta(y) + y \cdot \delta(x).$$

Example 1.2.11 *Peano Arithmetic*

Let $\mathcal{L} = \{+, \cdot, s, 0\}$, where $+$ and $\cdot$ are binary functions, s is a unary function, and 0 is a constant. We think of s as the successor function $x \mapsto x + 1$. The Peano axioms for arithmetic are the sentences
$$\forall x \ s(x) \neq 0,$$
$$\forall x \ (x \neq 0 \rightarrow \exists y \ s(y) = x),$$
$$\forall x \ x + 0 = x,$$
$$\forall x \ \forall y \ x + (s(y)) = s(x + y),$$
$$\forall x \ \ x \cdot 0 = 0,$$
$$\forall x \forall y \ x \cdot s(y) = (x \cdot y) + x,$$
and the axioms $\text{Ind}(\phi)$ for each formula $\phi(v, \overline{w})$, where $\text{Ind}(\phi)$ is the sentence
$$\forall \overline{w} \ [(\phi(0, \overline{w}) \wedge \forall v \ (\phi(v, \overline{w}) \rightarrow \phi(s(v), \overline{w}))) \rightarrow \forall x \ \phi(x, \overline{w})].$$
    The axiom $\text{Ind}(\phi)$ formalizes an instance of induction. It asserts that if $\overline{a} \in M$, $X = \{m \in M : M \models \phi(m, \overline{a})\}$, $0 \in X$, and $s(m) \in X$ whenever $m \in X$, then $X = M$.

Logical Consequence

Definition 1.2.12 Let T be an $\mathcal{L}$-theory and ϕ an $\mathcal{L}$-sentence. We say that ϕ is a *logical consequence* of T and write $T \models \phi$ if $M \models \phi$ whenever $M \models T$.

We give two examples.

Proposition 1.2.13 *a) Let $\mathcal{L} = \{+, <, 0\}$ and let T be the theory of or-dered Abelian groups. Then, $\forall x(x \neq 0 \to x + x \neq 0)$ is a logical consequence of T.*

b) Let T be the theory of groups where every element has order 2. Then, $T \not\models \exists x_1 \exists x_2 \exists x_3 (x_1 \neq x_2 \wedge x_2 \neq x_3 \wedge x_a 1 \neq x_3)$.

Proof

a) Suppose that $\mathcal{M} = (M, +, <, 0)$ is an ordered Abelian group. Let $a \in M \setminus \{0\}$. We must show that $a + a \neq 0$. Because $(M, <)$ is a linear order $a < 0$ or $0 < a$. If $a < 0$, then $a + a < 0 + a = a < 0$. Because $\neg(0 < 0)$, $a + a \neq 0$. If $0 < a$, then $0 < a = 0 + a < a + a$ and again $a + a \neq 0$.

b) Clearly, $\mathbb{Z}/2\mathbb{Z} \models T \wedge \neg\exists x_1 \exists x_2 \exists x_3(x_1 \neq x_2 \wedge x_2 \neq x_3 \wedge x_1 \neq x_3)$.

In general, to show that $T \models \phi$, we give an informal mathematical proof as above that $\mathcal{M} \models \phi$ whenever $\mathcal{M} \models T$. To show that $T \not\models \phi$, we usually construct a counterexample.

1.3   Definable Sets and Interpretability

Definable Sets

Definition 1.3.1 Let $\mathcal{M} = (M, \ldots)$ be an $\mathcal{L}$-structure. We say that $X \subseteq M^n$ is *definable* if and only if there is an $\mathcal{L}$-formula $\phi(v_1, \ldots, v_n, w_1, \ldots, w_m)$ and $\bar{b} \in M^m$ such that $X = \{\bar{a} \in M^n : \mathcal{M} \models \phi(\bar{a}, \bar{b})\}$. We say that $\phi(\bar{v}, \bar{b})$ *defines* X. We say that X is *A-definable* or *definable over A* if there is a formula $\psi(\bar{v}, w_1, \ldots, w_l)$ and $\bar{b} \in A^l$ such that $\psi(\bar{v}, \bar{b})$ defines X.

We give a number of examples using $\mathcal{L}_r$, the language of rings.

• Let $\mathcal{M} = (R, +, -, \cdot, 0, 1)$ be a ring. Let $p(X) \in R[X]$. Then, $Y = \{x \in R : p(x) = 0\}$ is definable. Suppose that $p(X) = \sum_{i=0}^{m} a_i X^i$. Let $\phi(v, w_0, \ldots, w_n)$ be the formula

$$w_n \cdot \underbrace{v \cdots v}_{n-\text{times}} + \ldots + w_1 \cdot v + w_0 = 0$$

(in the future, when no confusion arises, we will abbreviate such a formula as "$w_n v^n + \ldots + w_1 v + w_0 = 0$"). Then, $\phi(v, a_0, \ldots, a_n)$ defines Y. Indeed, Y is A-definable for any $A \supseteq \{a_0, \ldots, a_n\}$.

• Let $\mathcal{M} = (\mathbb{R}, +, -, \cdot, 0, 1)$ be the field of real numbers. Let $\phi(x, y)$ be the formula

$$\exists z(z \neq 0 \wedge y = x + z^2).$$

Because $a < b$ if and only if $\mathcal{M} \models \phi(a, b)$, the ordering is $\emptyset$-definable.

• Let $\mathcal{M} = (\mathbb{Z}, +, -, \cdot, 0, 1)$ be the ring of integers. Let $X = \{(m, n) \in \mathbb{Z}^2 : m < n\}$. Then, X is definable (indeed $\emptyset$-definable). By Lagrange's Theorem, every nonnegative integer is the sum of four squares. Thus, if we let $\phi(x, y)$ be the formula

$$\exists z_1 \exists z_2 \exists z_3 \exists z_4 (z_1 \neq 0 \wedge y = x + z_1^2 + z_2^2 + z_3^2 + z_4^2),$$

then $X = \{(m, n) \in \mathbb{Z}^2 : \mathcal{M} \models \phi(m, n)\}$.

• Let F be a field and $\mathcal{M} = (F[X], +, -, \cdot, 0, 1)$ be the ring of polynomials over F. Then F is definable in $\mathcal{M}$. Indeed, F is the set of units of $F[X]$ and is defined by the formula $x = 0 \vee \exists y \; xy = 1$.

• Let $\mathcal{M} = (\mathbb{C}(X), +, -, \cdot, 0, 1)$ be the field of complex rational functions in one variable. We claim that $\mathbb{C}$ is defined in $\mathbb{C}(X)$ by the formula

$$\exists x \exists y \; y^2 = v \wedge x^3 + 1 = v.$$

For any $z \in \mathbb{C}$ we can find x and y such that $y^2 = x^3 + 1 = z$. Suppose that h is a nonconstant rational function and that there are nonconstant rational functions f and g such that $h = g^2 = f^3 + 1$. Then $t \mapsto (f(t), g(t))$ is a nonconstant rational function from an open subset of $\mathbb{C}$ into the curve E given by the equation $y^2 = x^3 + 1$. But E is an elliptic curve and it is known (see for example [95]) that there are no such functions.

A similar argument shows that $\mathbb{C}$ is the set of rational functions f such that f and $f + 1$ are both fourth powers. These ideas generalize to show that $\mathbb{C}$ is definable in any finite algebraic extension of $\mathbb{C}(X)$.

• Let $\mathcal{M} = (\mathbb{Q}_p, +, -, \cdot, 0, 1)$ be the field of p-adic numbers. Then $\mathbb{Z}_p$ the ring of p-adic integers is definable. Suppose $p \neq 2$ (we leave $\mathbb{Q}_2$ for Exercise 1.4.13) and $\phi(x)$ is the formula $\exists y \; y^2 = px^2 + 1$. We claim that $\phi(x)$ defines $\mathbb{Z}_p$.

First, suppose that $y^2 = pa^2 + 1$. Let v denote the p-adic valuation. Because $v(pa^2) = 2v(a) + 1$, if $v(a) < 0$, then $v(pa^2)$ is an odd negative integer and $v(y^2) = v(pa^2 + 1) = v(pa^2)$. On the other hand, $v(y^2) = 2v(y)$, an even integer. Thus, if $\mathcal{M} \models \phi(a)$, then $v(a) \geq 0$ so $a \in \mathbb{Z}_p$.

On the other hand, suppose that $a \in \mathbb{Z}_p$. Let $F(X) = X^2 - (pa^2 + 1)$. Let $\overline{F}$ be the reduction of F mod p. Because $v(a) \geq 0$, $v(pa) > 0$ and $\overline{F}(X) = X^2 - 1$ and $\overline{F}' = 2X$. Thus, $\overline{F}(1) = 0$ and $\overline{F}'(1) \neq 0$ so, by Hensel's Lemma, there is $b \in \mathbb{Z}_p$ such that $F(b) = 0$. Hence $\mathcal{M} \models \phi(a)$.

• Let $\mathcal{M} = (\mathbb{Q}, +, -, \cdot, 0, 1)$ be the field of rational numbers. Let $\phi(x, y, z)$ be the formula

$$\exists a \exists b \exists c \; xyz^2 + 2 = a^2 + xy^2 - yc^2$$

and let $\psi(x)$ be the formula

$$\forall y \forall z \; ([\phi(y, z, 0) \wedge (\forall w(\phi(y, z, w) \rightarrow \phi(y, z, w + 1)))] \rightarrow \phi(y, z, x)).$$

A remarkable result of Julia Robinson (see [34]) shows that $\psi(x)$ defines the integers in $\mathbb{Q}$.

• Consider the natural numbers $\mathbb{N}$ as an $\mathcal{L} = \{+, \cdot, 0, 1\}$ structure. The definable sets are quite complex. For example, there is an $\mathcal{L}$-formula $T(e, x, s)$ such that $\mathbb{N} \models T(e, x, s)$ if and only if the Turing machine with program coded by e halts on input x in at most s steps (see, for example, [51]). Thus, the Turing machine with program e halts on input x if and only if $\mathbb{N} \models \exists s\, T(e, x, s)$, so the set of halting computations is definable. It is well known that this set is not computable (see, for example, [94]). This leads to an interesting conclusion.

Proposition 1.3.2 *The full $\mathcal{L}$-theory of the natural numbers is undecidable (i.e., there is no algorithm that when given an $\mathcal{L}$-sentence ψ as input will always halt answering "yes" if $\mathbb{N} \models \psi$ and "no" if $\mathbb{N} \models \neg\psi$).*

Proof For each e and x, let $\phi_{e,x}$ be the $\mathcal{L}$-sentence

$$\exists s\, T(\underbrace{1 + \ldots + 1}_{e-\text{times}}, \underbrace{1 + \ldots + 1}_{x-\text{times}}, s).$$

If there were such an algorithm we could decide whether the program coded by e halts on input x by asking whether $\mathbb{N} \models \phi_{e,x}$.

Recursively enumerable sets have simple mathematical definitions. By the Matijasevič–Robinson–Davis–Putnam solution to Hilbert's 10th Problem (see [24]).

for any recursively enumerable set $A \subseteq \mathbb{N}^n$ there is a polynomial $p(X_1, \ldots, X_n, Y_1, \ldots, Y_m) \in \mathbb{Z}[\overline{X}, \overline{Y}]$ such that

$$A = \{\overline{x} \in \mathbb{N}^n : \mathbb{N} \models \exists y_1 \ldots \exists y_m\, p(\overline{x}, \overline{y}) = 0\}.$$

The following example will be useful later.

Lemma 1.3.3 *Let $\mathcal{L}_r$ be the language of ordered rings and $(\mathbb{R}, +, -, \cdot, <, 0, 1)$ be the ordered field of real numbers. Suppose that $X \subseteq \mathbb{R}^n$ is A-definable. Then, the topological closure of X is also A-definable.*

Proof Let $\phi(v_1, \ldots, v_n, \overline{a})$ define X. Let $\psi(v_1, \ldots, v_n, \overline{w})$ be the formula

$$\forall \epsilon \left[\epsilon > 0 \rightarrow \exists y_1, \ldots, y_n \left(\phi(\overline{y}, \overline{w}) \wedge \sum_{i=1}^{n} (x_i - y_i)^2 < \epsilon \right) \right].$$

Then, $\overline{b}$ is in the closure of X if and only if $\mathcal{M} \models \phi(\overline{b}, \overline{a})$.

We can give a more concrete characterization of the definable sets.

Proposition 1.3.4 *Let $\mathcal{M}$ be an $\mathcal{L}$-structure. Suppose that D_n is a collection of subsets of M^n for all $n \geq 1$ and $\mathcal{D} = (D_n : n \geq 1)$ is the smallest collection such that:*
i) $M^n \in D_n$;

ii) for all n-ary function symbols f of $\mathcal{L}$, the graph of $f^{\mathcal{M}}$ is in D_{n+1};
iii) for all n-ary relation symbols R of $\mathcal{L}$, $R^{\mathcal{M}} \in D_n$;
iv) for all $i, j \leq n$, $\{(x_1, \ldots, x_n) \in M^n : x_i = x_j\} \in D_n$;
v) if $X \in D_n$, then $M \times X \in D_{n+1}$;
vi) each D_n is closed under complement, union, and intersection;
vii) if $X \in D_{n+1}$ and $\pi : M^{n+1} \rightarrow M^n$ is the projection map $(x_1, \ldots, x_{n+1}) \mapsto (x_1, \ldots, x_n)$, then $\pi(X) \in D_n$;
viii) if $X \in D_{n+m}$ and $b \in M^m$, then $\{a \in M^n : (a, b) \in X\} \in D_n$.

Then, $X \subseteq M^n$ is definable if and only if $X \in D_n$.

Proof We first show that the definable sets satisfy the closure properties
i)–viii). Because $\mathcal{D}$ is the smallest collection with these closure properties,
every $X \in D_n$ is definable.

i) M^n is definable by $v_1 = v_1$.

ii) The graph of $f^{\mathcal{M}}$ is definable by $f(x_1, \ldots, x_{n_f}) = y$.

iii) The relation $R^{\mathcal{M}}$ is defined by R.

iv) The set $\{x \in M^n : x_i = x_j\}$ is defined by $v_i = v_j$.

v) If $X \subseteq M^n$ is defined by $\phi(v_1, \ldots, v_n, \bar{a})$, then $M \times X$ is defined by
$\phi(v_2, \ldots, v_{n+1}, \bar{a})$.

vi) If $X \subseteq M^n$ is defined by $\phi(\bar{v}, \bar{a})$ and $Y \subseteq M^n$ is defined by $\psi(\bar{v}, \bar{b})$,
then $M \setminus X$ is defined by $\neg\phi(\bar{v}, \bar{a})$, $X \cap Y$ is defined by $\phi(\bar{v}, \bar{a}) \wedge \psi(\bar{v}, \bar{b})$ and
$X \cup Y$ is defined by $\phi(\bar{v}, \bar{a}) \vee \psi(\bar{v}, \bar{b})$.

vii) If $X \subseteq M^{n+1}$ is defined by $\phi(v_1, \ldots, v_{n+1}, \bar{a})$, then $\pi(X)$ is defined
by $\exists v_{n+1}\, \phi(\bar{v}, \bar{a})$.

viii) If $X \subset M^{n+m}$ is defined by $\phi(v_1, \ldots, v_{n+m}, \bar{c})$ and $\bar{b} \in M^m$, then
$\{\bar{a} \in M^n : (\bar{a}, \bar{b}) \in X\}$ is defined by $\phi(v_1, \ldots, v_n, \bar{b}, \bar{c})$.

Thus, if $X \in D_n$, then X is definable.

Next we show that if $X \subseteq M^n$ is definable, then $X \in D_n$. We first show
by induction that if $t(v_1, \ldots, v_n)$ is a term, then $\{(\bar{x}, y) \in M^{n+1} : t^{\mathcal{M}}(\bar{x}) = y\} \in D_{n+1}$.

If t is a constant term c, then we must show that $\{(\bar{x}, c^{\mathcal{M}}) : \bar{x} \in M^n\} \in D_{n+1}$. By iv) and viii), $\{c^{\mathcal{M}}\} \in D_1$. Thus, by n applications of v), $\{(\bar{x}, c^{\mathcal{M}}) : \bar{x} \in M^n\} \in D_{n+1}$.

If t is v_i, then we must show that $\{(\bar{x}, x_i) : \bar{x} \in M^n\} \in D_{n+1}$, but this
follows easily from i) and iv).

Suppose that $t = f(t_1, \ldots, t_m)$. By induction we suppose that $G_i \in D_{n_i}$,
where G_i is the graph of $t_i^{\mathcal{M}} : M^n \rightarrow M$. Let $G \in D_{m+1}$ be the graph of
$f^{\mathcal{M}}$. Then, the graph of $t^{\mathcal{M}}$ is

$$\left\{ (\bar{x}, y) : \exists z_1 \ldots \exists z_m \left(\bigwedge_{i=1}^{m} (\bar{x}, z_i) \in G_i \wedge (\bar{z}, y) \in G \right) \right\}.$$

Using the closure properties of $\mathcal{D}$, we see this is in D_{n+1}.

We now proceed by induction on formulas to show that every $\emptyset$-definable
$X \subseteq M^n$ is in D_n.

Let ϕ be $t_1 = t_2$. Then $\{\bar{x} \in M^n : t_1^{\mathcal{M}}(\bar{x}) = t_2^{\mathcal{M}}(\bar{x})\}=\{\bar{x} \in M^n : \exists y \exists z \, (t_1^{\mathcal{M}}(\bar{x}) = y \wedge t_2^{\mathcal{M}}(\bar{x}) = z \wedge y = z)\} \in D_n$.

Let ϕ be $R(t_1, \ldots, t_m)$. Then $\{\bar{x} \in M^n : \mathcal{M} \models \phi(\bar{x})\} =$

$$\left\{ \bar{x} \in M^n : \exists z_1 \ldots \exists z_m \bigwedge_{i=1}^{m} t_i^{\mathcal{M}}(\bar{x}) = z_i \wedge \bar{z} \in R^{\mathcal{M}} \right\}.$$

Thus, all sets defined over $\emptyset$ by atomic formulas are in $\mathcal{D}$. Because $\mathcal{D}$ is closed under Boolean combinations and projections, all $\emptyset$-definable sets are in $\mathcal{D}$. Property viii) ensures that all definable sets are in $\mathcal{D}$.

How do we show that $X \subset M^n$ is not definable? The following proposition will often be useful.

Proposition 1.3.5 *Let $\mathcal{M}$ be an $\mathcal{L}$-structure. If $X \subset M^n$ is A-definable, then every $\mathcal{L}$-automorphism of $\mathcal{M}$ that fixes A pointwise fixes X setwise (that is, if σ is an automorphism of M and $\sigma(a) = a$ for all $a \in A$, then $\sigma(X) = X$).*

Proof Let $\psi(\bar{v}, \bar{a})$ be the $\mathcal{L}$-formula defining X where $\bar{a} \in A$. Let σ be an automorphism of $\mathcal{M}$ with $\sigma(\bar{a}) = \bar{a}$, and let $\bar{b} \in M^n$.

In the proof of Theorem 1.1.10, we showed that if $j : \mathcal{M} \to \mathcal{N}$ is an isomorphism, then $\mathcal{M} \models \phi(\bar{a})$ if and only if $\mathcal{N} \models \phi(j(\bar{a}))$. Thus

$$\mathcal{M} \models \psi(\bar{b}, \bar{a}) \leftrightarrow \mathcal{M} \models (\sigma(\bar{b}), \sigma(\bar{a})) \Leftrightarrow \mathcal{M} \models (\sigma(\bar{b}), \bar{a}).$$

In other words, $\bar{b} \in X$ if and only if $\sigma(\bar{b}) \in X$ as desired.

We give a sample application.

Corollary 1.3.6 *The set of real numbers is not definable in the field of complex numbers.*

Proof If $\mathbb{R}$ were definable, then it would be definable over a finite $A \subset \mathbb{C}$. Let $r, s \in \mathbb{C}$ be algebraically independent over A with $r \in \mathbb{R}$ and $s \notin \mathbb{R}$. There is an automorphism σ of $\mathbb{C}$ such that $\sigma|A$ is the identity and $\sigma(r) = s$. Thus, $\sigma(\mathbb{R}) \neq \mathbb{R}$ and $\mathbb{R}$ is not definable over A.

This proof worked because $\mathbb{C}$ has many automorphisms. The situation is much different for $\mathbb{R}$. Any automorphism of the real field must fix the rational numbers. Because the ordering is definable it must be preserved by any automorphism. Because the rationals are dense in $\mathbb{R}$, the only automorphism of the real field is the identity. Most subsets of $\mathbb{R}$ are undefinable (there are $2^{2^{\aleph_0}}$ subsets of $\mathbb{R}$ and only $2^{\aleph_0}$ possible definitions), but we cannot use Proposition 1.3.5 to show any particular set is undefinable. In Proposition 4.3.25 we show that the converse to Proposition 1.3.5 holds for sufficiently rich models.

Interpretability

It is often very useful to study the structures that can be defined inside a given structure. For example, let K be a field and G be $\mathrm{GL}_2(K)$, the group of invertible 2×2 matrices over K. Let $X = \{(a, b, c, d) \in K^4 : ad - bc \neq 0\}$. Let $f : X^2 \to X$ by

$$f((a_1, b_1, c_1, d_1), (a_2, b_2, c_2, d_2)) = \\ (a_1 a_2 + b_1 c_2, a_1 b_2 + b_1 d_2, c_1 a_2 + d_1 c_2, c_1 b_2 + d_1 d_2).$$

Clearly, X and f are definable in $(K, +, \cdot)$ and the set X with the operation f is isomorphic to $\mathrm{GL}_2(K)$, where the identity element of X is $(1, 0, 0, 1)$.

We say that an $\mathcal{L}_0$-structure $\mathcal{N}$ is *definably interpreted* in an $\mathcal{L}$-structure $\mathcal{M}$ if and only if we can find a definable $X \subseteq M^n$ for some n and we can interpret the symbols of $\mathcal{L}_0$ as definable subsets and functions on X (where "definable" means definable using $\mathcal{L}$) so that the resulting $\mathcal{L}_0$-structure is isomorphic to $\mathcal{N}$.

The example above shows that $(\mathrm{GL}_n(K), \cdot, e)$ is definably interpreted in $(K, +, \cdot, 0, 1)$. A *linear algebraic group* over K is a subgroup of $\mathrm{GL}_n(K)$ defined by polynomial equations over K. It is easy to see that any linear algebraic group over K is definably interpreted in K.

For another example, let $\mathcal{L} = \emptyset$ and let $\mathcal{M}$ be an $\mathcal{L}$-structure where $|M| = \aleph_0$. Let $\mathcal{L}_0 = \{E\}$, where E is a binary relation. Let $\mathcal{N} = (N, E)$ be the $\mathcal{L}_0$-structure where E is an equivalence relation with $\aleph_0$-classes each of size $\aleph_0$. Let

$$R = \{((x_1, x_2), (y_1, y_2)) \in M^2 \times M^2 : x_1 = y_1\}.$$

Then, $\mathcal{N} \cong (M^2, R)$, so $\mathcal{N}$ is definably interpreted in $\mathcal{M}$.

Interpreting a Field in the Affine Group

We give a more interesting example. Let F be an infinite field and let G be the group of matrices of the form

$$\begin{pmatrix} a & b \\ 0 & 1 \end{pmatrix},$$

where $a, b \in F, a \neq 0$. This group is isomorphic to the group of affine transformations $x \mapsto ax + b$, where $a, b \in F$ and $a \neq 0$.

We will show that F is definably interpreted in the group G. Let

$$\alpha = \begin{pmatrix} 1 & 1 \\ 0 & 1 \end{pmatrix} \text{ and } \beta = \begin{pmatrix} \tau & 0 \\ 0 & 1 \end{pmatrix},$$

where $\tau \neq 0, 1$. Let

$$A = \{g \in G : g\alpha = \alpha g\} = \left\{ \begin{pmatrix} 1 & x \\ 0 & 1 \end{pmatrix} : x \in F \right\}$$

and

$$B = \{g \in G : g\beta = \beta g\} = \left\{ \begin{pmatrix} x & 0 \\ 0 & 1 \end{pmatrix} : x \neq 0 \right\}.$$

Clearly, A and B are definable using parameters α and β.
  B acts on A by conjugation

$$\begin{pmatrix} x & 0 \\ 0 & 1 \end{pmatrix}^{-1} \begin{pmatrix} 1 & y \\ 0 & 1 \end{pmatrix} \begin{pmatrix} x & 0 \\ 0 & 1 \end{pmatrix} = \begin{pmatrix} 1 & \frac{y}{x} \\ 0 & 1 \end{pmatrix}.$$

The action $(a, b) \mapsto b^{-1}ab$ is definable. We can define the map $i : A \backslash \{1\} \rightarrow$
B by $i(a) = b$ if and only if $b^{-1}ab = \alpha$, that is,

$$i \begin{pmatrix} 1 & x \\ 0 & 1 \end{pmatrix} = \begin{pmatrix} x & 0 \\ 0 & 1 \end{pmatrix}.$$

Define an operation $*$ on A by

$$a * b = \begin{cases} i(b)a(i(b))^{-1} & \text{if } b \neq I \\ 1 & \text{if } b = I \end{cases},$$

where I is the identity matrix. It is now easy to see that $(F, +, \cdot, 0, 1)$ is
isomorphic to $(A, \cdot, *, 1, \alpha)$.

Interpreting Orders in Graphs

Very complicated structures can often be interpreted in seemingly simpler
ones. For example, any structure in a countable language can be interpreted
in a graph. We give one simple example of this construction where we
interpret linear orders in graphs. Let $(A, <)$ be a linear order. We will build
a graph G_A as follows. For each $a \in A$, G_A will have vertices a, x_1^a, x_2^a, x_3^a
and contain the subgraph

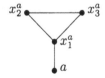

  No edge aside from the ones above will have any x_i^a as a vertex. If $a < b$,
G_A will have vertices $y_1^{a,b}$, $y_2^{a,b}$, and $y_3^{a,b}$ and contain the subgraph

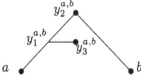

  No edge other than the ones shown will contain any vertex $y_i^{a,b}$, and
G_A will have no edges other than the ones we have described. Let $V_A =$

$A \cup \{x_1^a, x_2^a, x_3^a : a \in A\} \cup \{y_1^{a,b}, y_2^{a,b}, y_3^{a,b} : a, b \in A$ and $a < b\}$ and let R_A be the smallest symmetric relation containing all edges (a, x_1^a), (x_1^a, x_2^a), (x_1^a, x_3^a), (x_2^a, x_3^a), for $a \in A$ and $(a, y_1^{a,b})$, $(y_1^{a,b}, y_2^{a,b})$, $(y_1^{a,b}, y_3^{a,b})$, $(y_2^{a,b}, b)$ for $a < b$.

For example, if A is the three-element linear order $a < b < c$, then G_A is the graph

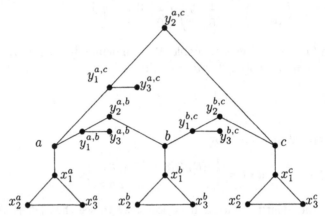

Let $\mathcal{L} = \{R\}$ where R is a binary relation. We will describe an $\mathcal{L}$-theory T of graphs such that every model of T is G_A for a unique linear order A. We begin by giving two formulas in the language of graphs that describe the first two diagrams.

Let $\phi(x, u, v, w)$ be the formula asserting that x, u, v, w are distinct, there are edges $(x, u), (u, v), (u, w)$, and (v, w), and these are the only edges involving vertices u, v, and w. Note that $G_A \models \phi(a, x_1^a, x_2^a, x_3^a)$ for all $a \in A$.

Let $\psi(x, y, u, v, w)$ be the formula asserting that x, y, u, v, w are distinct, there are edges $(x, u), (u, v), (u, w)$ and (v, y), and these are the only edges involving vertices u, v and w. Note that $G_A \models \phi(a, b, y_1^{a,b}, y_2^{a,b}, y_3^{a,b})$ whenever $a < b$ in A.

We define formulas $\theta_0(z), \ldots, \theta_5(z)$ as follows:
$\theta_0(z)$ is $\exists u \exists v \exists w \; \phi(z, u, v, w)$,
$\theta_1(z)$ is $\exists x \exists v \exists w \; \phi(x, z, u, w)$,
$\theta_2(z)$ is $\exists x \exists u \exists w \; \phi(x, u, z, w)$,
$\theta_3(z)$ is $\exists x \exists y \exists v \exists w \; \psi(x, y, z, v, w)$,
$\theta_4(z)$ is $\exists x \exists y \exists u \exists w \; \psi(x, y, u, z, w)$,
$\theta_5(z)$ is $\exists x \exists y \exists u \exists v \; \psi(x, y, u, v, z)$.

If $a, b \in A$ with $a < b$, then

$$G_A \models \theta_0(a) \wedge \theta_1(x_1^a) \wedge \theta_2(x_2^a) \wedge \theta_2(x_3^a)$$

and

$$G_A \models \theta_3(y_1^{a,b}) \wedge \theta_4(y_2^{a,b}) \wedge \theta_5(y_3^{a,b}).$$

Thus, for each vertex x in G_A, $G_A \models \theta_i(x)$ for some $i = 0, \ldots, 5$.

Lemma 1.3.7 *If $(A, <)$ is a linear order, then for all vertices x in G_A, there is a unique $i \leq 5$ such that $G_A \models \theta_i(x)$.*

Proof Let x be a vertex in G_A. If x is part of a 3-cycle, then $x = x_i^a$ for some $a \in A$. Recall that the *valence* of x is the number of edges with x as a vertex. If x has valence 3, then $\theta_1(x)$ is the unique formula that holds. If x has valence 2, then $\theta_2(x)$ is the unique formula that holds.

Suppose that x is not part of a 3-cycle. Then, either x is in A or $x = y_i^{a,b}$, where $a, b \in A$, $a < b$, and $i \leq 3$. If there is an edge (x, v) such that v is part of a 3-cycle, then $\theta_0(x)$ is the unique formula that holds. If x has valence 1, then $\theta_5(x)$. If there is an edge (x, v) such that v has valence 1, then $\theta_3(x)$ is the unique formula that holds of x. Otherwise, $\theta_4(x)$ is the unique formula that holds of x.

Let T be the $\mathcal{L}$-theory with the following axioms:
i) R is symmetric and irreflexive;
ii) for all x, exactly one θ_i holds;
iii) if $\theta_0(x)$ and $\theta_0(y)$, then $\neg R(x, y)$;
iv) if $\exists u \exists v \exists w \, \psi(x, y, u, v, w)$, then $\forall u_1 \forall v_1 \forall w_1 \neg \psi(y, x, u_1, v_1, w_1)$;
v) if $\exists u \exists v \exists w \, \psi(x, y, u, v, w)$ and $\exists u_1 \exists v_1 \exists w_1 \psi(y, z, u_1, v_1, w_1)$, then $\exists u_2 \exists v_2 \exists w_2 \, \psi(x, z, u_2, v_2, w_2)$;
vi) if $\theta_0(x)$ and $\theta_0(y)$, then either $x = y$ or $\exists u \exists v \exists w \, \psi(x, y, u, v, w)$ or $\exists u \exists v \exists w \, \psi(y, x, u, v, w)$;
vii) if $\phi(x, u, v, w) \wedge \phi(x, u', v', w')$, then $u = u'$, $v = v'$ and $w = w'$;
viii) if $\psi(x, y, u, v, w) \wedge \psi(x, y, u', v', w')$, then $u = u'$, $v = v'$ and $w = w'$.

If $(A, <)$ is a linear order, then $G_A \models T$. Thus, every linear order can be interpreted in a model of T.

Suppose that $G \models T$. Let $X_G = \{x \in G : G \models \theta_0(x)\}$. Because $G \models T$, axioms iv)–vi) ensure that we can linearly order X_G by $x <_G y$ if and only if $G \models \exists u \exists v \exists w \, \psi(x, y, u, v, w)$. Thus, we can interpret linear orders into every model of T.

The next lemma shows that in fact $(A, <) \mapsto G_A$ is a bijection between linear orders and models of T. We leave the proof as an exercise.

Lemma 1.3.8 *If $(A, <)$ is a linear order, then $(X_{G_A}, <_{G_A}) \cong (A, <)$. Moreover, $G_{X_G} \cong G$ for any $G \models T$.*

Quotients

Often we want to do more general constructions. For example, suppose that we have a definable group G and a definable normal subgroup H. We might want to look at the group G/H. It is possible that G/H does not correspond to a definable group in our structure. But it does correspond to the cosets of a definable equivalence relation.

Definition 1.3.9 We say that an $\mathcal{L}_0$-structure $\mathcal{N}$ is *interpretable* in an $\mathcal{L}$-structure M if there is a definable $X \subseteq M^n$, a definable equivalence relation

E on X, and for each symbol of $\mathcal{L}_0$ we can find definable E-invariant sets on X (where "definable" means definable in $\mathcal{L}$) such that X/E with the induced structure is isomorphic to $\mathcal{N}$.

Let K be a field. Let

$$X = \left\{ (a_0, \ldots, a_n) \in K^{n+1} : \bigvee_{i=0}^{n} a_i \neq 0 \right\}.$$

Define an equivalence relation $\sim$ on X by $\overline{a} \sim \overline{b}$ if and only if there is a non-zero $\lambda \in K$ such that $\lambda \overline{a} = \overline{b}$. Clearly, X and $\sim$ are definable, and the quotient $X/\sim$ is $\mathbb{P}^n(K)$, projective n-space over K. Let $f(X_0, \ldots, X_n)$ be a homogeneous polynomial over K (i.e., there is a d such that $f(\lambda(\overline{x})) = \lambda^d f(\overline{x})$ for any λ and $\overline{x}$). Let $V = \{\overline{x} \in X : f(\overline{x}) = 0\}$. Because f is homogeneous, V is $\sim$-invariant. Thus, we can interpret $(\mathbb{P}^n(K), V/\sim)$ in $(K, +, \cdot, 0, 1)$.

Let us show that we can interpret the ordered additive group of integers in the field $\mathbb{Q}_p$. We saw above that $\mathbb{Z}_p$ is a definable subset of $\mathbb{Q}_p$. Let $U = \{x \in \mathbb{Z}_p : \exists y \in \mathbb{Z}_p : xy = 1\}$ be the units of $\mathbb{Z}_p$. Then, $(\mathbb{Z}, +)$ is isomorphic to the multiplicative group $\mathbb{Q}_p^\times / U$. We can define the ordering on $\mathbb{Q}_p^\times / U$ by $x/U \geq y/U \Leftrightarrow \frac{x}{y} \in \mathbb{Z}_p$.

Many-sorted Structures and $\mathcal{M}^{\mathrm{eq}}$

Quotient constructions are so useful that we often enrich our structure so that we can deal with all quotients as elements of the structure. Although this material will be useful in Chapters 6–8, the reader can safely skip this material for the time being.

We need to generalize our definition of structures to include *many-sorted structures*.

Let S be a set. The universe of a many-sorted structure $\mathcal{N}$ with sorts S is a set N that is partitioned into disjoint sets $\{N_i : i \in S\}$. For each n-ary relation symbol R, there are $s_1, \ldots, s_n \in S$ such that $R^\mathcal{N} \subset N^{s_1} \times \ldots \times N^{s_n}$. For each n-ary function symbol f, there are $s_0, \ldots, s_n$ such that $f^\mathcal{N} : N^{s_1} \times \ldots \times N^{s_n} \to N^{s_0}$.

Let $\mathcal{M}$ be an $\mathcal{L}$-structure. We consider the set of sorts $S = \{S_E : E$ an $\emptyset$-definable equivalence relation on M^n for some $n\}$. In the many-sorted structure $\mathcal{M}^{\mathrm{eq}}$ we interpret the sort S_E as M^n/E for E an $\emptyset$-definable equivalence relation on M^n. Because $=$ is a definable equivalence relation on M, we can identify M with the sort $S_=$. All relations and functions of $\mathcal{L}$ are relations and functions on M^k. For each $\emptyset$-definable equivalence relation E on M^n we have in $\mathcal{M}^{\mathrm{eq}}$ an n-ary function $f_E : M^n \to S_E$ given by $f_E(\overline{x}) = \overline{x}/E$.

The next lemma summarizes some basic facts we will need about $\mathcal{M}^{\mathrm{eq}}$. We leave the proof as an exercise.

Lemma 1.3.10 *i)* If $X \subseteq M^n$ is definable in $\mathcal{M}^{eq}$, then X is definable in $\mathcal{M}$.

ii) If $\mathcal{M} \equiv \mathcal{N}$, then $\mathcal{M}^{eq} \equiv \mathcal{N}^{eq}$.

iii) If $\mathcal{N}$ is interpretable in $\mathcal{M}$, then $\mathcal{N}$ is definably interpretable in $\mathcal{M}^{eq}$.

iv) If σ is an automorphism of $\mathcal{M}^{eq}$, then $\sigma | M$ is an automorphism of $\mathcal{M}$.

v) If σ is an automorphism of $\mathcal{M}$, there is $\hat{\sigma}$ an automorphism of $\mathcal{M}^{eq}$ such that $\sigma = \hat{\sigma} | M$.

1.4 Exercises and Remarks

Exercise 1.4.1 a) Suppose that $\phi_1, \ldots, \phi_n$ are $\mathcal{L}$-formulas and ψ is a Boolean combination of $\phi_1, \ldots, \phi_n$. Then there is $S \subseteq \mathcal{P}(\{1, \ldots, n\})$ such that

$$\models \psi \Leftrightarrow \bigvee_{X \in S} \left(\bigwedge_{i \in X} \phi_i \wedge \bigwedge_{i \notin X} \neg \phi_i \right).$$

b) Show that every formula is equivalent to one of the form

$$Q_1 v_1 \ldots Q_m v_m \; \psi,$$

where ψ is quantifier-free and each Q_i is either $\forall$ or $\exists$.

Exercise 1.4.2 a) Let $\mathcal{L} = \{\cdot, e\}$ be the language of groups. Show that there is a sentence ϕ such that $\mathcal{M} \models \phi$ if and only if $\mathcal{M} \cong \mathbb{Z}/2\mathbb{Z} \times \mathbb{Z}/2\mathbb{Z}$.

b) Let $\mathcal{L}$ be any finite language and let $\mathcal{M}$ be a finite $\mathcal{L}$-structure. Show that there is an $\mathcal{L}$-sentence ϕ such that $\mathcal{N} \models \phi$ if and only if $\mathcal{N} \cong \mathcal{M}$.

Exercise 1.4.3 Let $\mathcal{L}$ be any countable language. Show that for any infinite cardinal κ there are at most 2^κ nonisomorphic $\mathcal{L}$-structures of cardinality κ.

Exercise 1.4.4 Let T be an $\mathcal{L}$-theory. We say that T' is an *axiomatization* of T if $\mathcal{M} \models T$ if and only if $\mathcal{M} \models T'$ for any $\mathcal{L}$-structure $\mathcal{M}$. Suppose that T' is an axiomatization of T. Show that $T \models \phi$ if and only if $T' \models \phi$ for all $\mathcal{L}$-sentences ϕ.

Exercise 1.4.5 Show that the following classes are elementary. In each case, you should first pick an appropriate language.

a) Partial orders

b) Lattices

c) Boolean algebras

d) Integral domains

e) Trees

Exercise 1.4.6 Show that if T is an unsatisfiable, $\mathcal{L}$-theory then every $\mathcal{L}$-sentence ϕ is a logical consequence of T.

Exercise 1.4.7 Let ϕ be an $\mathcal{L}$-sentence. The *finite spectrum* of ϕ is the set $\{n \in \mathbb{N}^+ :$ there is $\mathcal{M} \models \phi$ with $|M| = n\}$, where $\mathbb{N}^+$ is the set of positive natural numbers.

a) Let $\mathcal{L} = \{E\}$ where E is a binary relation, and let ϕ be the sentence that asserts that E is an equivalence relation where every equivalence class has exactly two elements. Show that the finite spectrum of ϕ is the set of positive even numbers.

b) For each of the following subsets X of $\mathbb{N}^+$, show that X occurs as the finite spectrum of an $\mathcal{L}$-sentence for some language $\mathcal{L}$:

    i) $\{2^n 3^m : n, m > 0\}$;

    ii) $\{m > 0 : m$ is composite$\}$ (i.e. $m = ab$ where $a \neq 1$ and $b \neq 1$);

    iii) $\{p^n : p$ is prime and $n > 0\}$;

    iv) $\{p : p$ is prime$\}$;

c)[†] Show that $X \subseteq \mathbb{N}^+$ is a finite spectrum if and only if there is a nondeterministic Turing machine M running in exponential time such that given a string of n 1's as input M halts accepting if and only if $n \in X$. [Remark: An interesting open problem is whether the complement of a finite spectrum is a finite spectrum. This problem shows that it is equivalent to the question of whether the collection of sets recognizable in nondeterministic exponential time is closed under complement.]

Exercise 1.4.8 Let $\mathcal{L} = \{+, 0\}$. Show that $\mathbb{Z} \oplus \mathbb{Z} \not\equiv \mathbb{Z}$.

Exercise 1.4.9 Let $\mathcal{M}$ be an $\mathcal{L}$-structure. We say that $f : M^n \to M^m$ is definable if the graph of f is a definable set in M^{n+m}.

a) Show that if $f : M^n \to M^m$ and $g : M^m \to M^l$ are definable, then so is $g \circ f$.

b) Suppose that $f : M^n \to M$ is definable. Show that the image of f is definable.

c) Suppose that $f : M^n \to M$ is definable and one-to-one. Show that f^{-1} is definable.

Exercise 1.4.10 Let $\mathcal{M}$ be an $\mathcal{L}$-structure and $A \subseteq M$. We say that $b \in M$ is definable over A if there is a formula $\phi(v, \overline{w})$ and $\overline{a} \in A$ such that

$$\mathcal{M} \models \phi(b, \overline{a}) \wedge \forall y \, (\phi(y, \overline{a}) \to y = b).$$

In other words, $\{b\}$ is A-definable.

a) Show that x is definable over A if and only if for some n there is an A-definable function $f : M^n \to M$ and $\overline{a} \in M$ such that $f(\overline{a}) = x$.

b) Suppose that x is definable from A and σ is an automorphism of $\mathcal{M}$ such that $\sigma(a) = a$ for all $a \in A$. Show that $\sigma(x) = x$.

Let $\mathrm{dcl}(A) = \{x \in M : x$ is definable from $A\}$.

c) Show that $\mathrm{dcl}(\mathrm{dcl}(A)) = A$.

Exercise 1.4.11 Let $\mathcal{M}$ be an $\mathcal{L}$-structure and $A \subseteq M$. We say that $b \in M$ is *algebraic* over A if there is an $\mathcal{L}$-formula $\phi(v, \overline{w})$ and $\overline{a} \in A$ such that

$\mathcal{M} \models \phi(b, \bar{a})$ and $\{y \in M : \mathcal{M} \models \phi(y, \bar{a})\}$ is finite. We let $\mathrm{acl}(A) = \{x : x$ is algebraic over $A\}$.

a) Suppose that $x \in \mathrm{acl}(A)$. Show that there are $x_1, \ldots, x_m$ such that if σ is an automorphism of $\mathcal{M}$ with $\sigma(a) = a$ for all $a \in A$, then $\sigma(x) = x_i$ for some i. In other words, there are only finitely many conjugates of x under automorphisms of $\mathcal{M}$ fixing a.

b) Show that $\mathrm{acl}(\mathrm{acl}(A)) = \mathrm{acl}(A)$.

c) Show that if $x \in \mathrm{acl}(A)$, then $x \in \mathrm{acl}(A_0)$ for some finite $A_0 \subseteq A$.

d) Show that if $A \subseteq B$, then $\mathrm{acl}(A) \subseteq \mathrm{acl}(B)$.

Exercise 1.4.12 Let K be a field and let F be a finite algebraic extension of K. Show that the field F is interpretable in the field K. [Hint: F is a finite-dimensional K-vector space.]

Exercise 1.4.13 [†] Show that the ring of 2-adic integers $\mathbb{Z}_2$ is definable in the field of 2-adic numbers $\mathbb{Q}_2$.

Exercise 1.4.14 a) Prove Lemma 1.3.8.

b) Let T_0 be a theory of linear orders. Show that there is T_1 a theory of graphs such that every model of T_0 is interpretable in a model of T_1, every model of T_1 interprets a model of T_0, and the number of nonisomorphic models of T_0 of cardinality κ is equal to the number of nonisomorphic models of T_1 of cardinality κ for all infinite cardinals κ.

c) Find a way to interpret groups into graphs such that the analog of Lemma 1.3.8 holds.

Exercise 1.4.15 (Expansions and Reducts) Let $\mathcal{L}_1 \supset \mathcal{L}$. If $\mathcal{M}_1$ is an $\mathcal{L}_1$-structure, then by ignoring the interpretations of the symbols in $\mathcal{L}_1 \setminus \mathcal{L}$ we get an $\mathcal{L}$-structure $\mathcal{M}$. We call $\mathcal{M}$ a *reduct* of $\mathcal{M}_1$ and $\mathcal{M}_1$ an *expansion* of $\mathcal{M}$.

a) Show that if $X \subseteq M^n$ is definable in $\mathcal{M}$, then it is definable in $\mathcal{M}_1$.

b) Give an example showing that there may be sets definable in the expansion that are not definable in the original structure.

c) Suppose that $\mathcal{M}_1$ is an expansion of $\mathcal{M}$ where for each symbol f or R in $\mathcal{L}_1 \setminus \mathcal{L}$ the interpretation $f^{\mathcal{M}_1}$ or $R^{\mathcal{M}_1}$ is definable in M. In this case we call $\mathcal{M}_1$ an *expansion by definitions*. For example because we can define the ordering of $\mathbb{R}$ in the field language, $(\mathbb{R}, +, \cdot, <, 0, 1)$ is an expansion by definitions of $(\mathbb{R}, +, \cdot, 0, 1)$. Show that if $\mathcal{M}_1$ is an expansion by definitions of $\mathcal{M}$, then every subset of M^n definable in the structure $\mathcal{M}_1$ is definable in the structure $\mathcal{M}$.

Exercise 1.4.16 Suppose that $\mathcal{N}$ is interpretable in $\mathcal{M}$. Say $\mathcal{N} \cong (X/E, \ldots)$, where X and E are definable in $\mathcal{M}$. We say that the interpretation is *pure* if any subset of $(X/E)^n$ definable in $\mathcal{M}$ is definable in $\mathcal{N}$. Show that the interpretation of the complex field $\mathbb{C}$ in the real field $\mathbb{R}$ is not pure.

Exercise 1.4.17 Let $\mathcal{L}$ be a language Let $\mathcal{L}_0$ be the language containing all relation symbols of $\mathcal{L}$, an $(n+1)$-ary relation symbol R_f for each n-ary function symbol of $\mathcal{L}$, and a unary relation symbol R_c for each constant symbol of $\mathcal{L}$. Let $\mathcal{M}$ be an $\mathcal{L}$-structure. Let $\mathcal{M}^*$ be the $\mathcal{L} \cup \mathcal{L}_0$ expansion of $\mathcal{M}$ where we interpret R_f as the graph of $f^{\mathcal{M}}$ and R_c as $\{c^{\mathcal{M}}\}$. Let $\mathcal{M}_0$ be the reduct of $\mathcal{M}^*$ to $\mathcal{L}_0$.

a) Show that $X \subseteq M^n$ is definable in $\mathcal{M}$ if and only if it is definable in $\mathcal{M}^*$ if and only if it is definable in $\mathcal{M}_0$.

b) Show that for any $\mathcal{L}$-theory T, there is an $\mathcal{L}_0$-theory T_0 such that the map $\mathcal{M} \mapsto \mathcal{M}_0$ above is a bijection between models of T and models of T_0 that preserves the collection of sets definable in structures.

These results show that for many purposes we can assume that out language is relational (i.e., has only relation symbols).

Exercise 1.4.18 Prove Lemma 1.3.10.

Exercise 1.4.19 Show that if $\mathcal{N}$ is interpretable in $\mathcal{M}^{\mathrm{eq}}$, then $\mathcal{N}$ is definably interpretable in $\mathcal{M}^{\mathrm{eq}}$.

Remarks

Our treatment of languages, structures, and theories has been very terse. For a more detailed and leisurely discussion, we refer the reader to [8], [94], or [31].

If $\mathcal{L}$ is a finite language and T is an $\mathcal{L}$-theory, then we can find a theory T' of graphs such that the analog of Lemma 1.3.8 holds. There are also generalizations for countable languages. See [40] for details. Mekler [70] proved an analogous result showing that any structure in a countable language can be interpreted in a nilpotent group.

Exercise 1.4.7 c) is due to Bennett. The relationship between finite spectra and computational complexity is discussed in Section 7.26 of [46].

2
Basic Techniques

2.1 The Compactness Theorem

Let T be an $\mathcal{L}$-theory and ϕ an $\mathcal{L}$-sentence. To show that $T \models \phi$, we must show that ϕ holds in every model of T. Checking all models of T sounds like a daunting task, but in practice we usually show that $T \models \phi$ by giving an informal mathematical proof that ϕ is true in every model of T. One of the first great achievements of mathematical logic was giving a rigorous definition of "proof" that completely captures the notion of "logical consequence."

A proof of ϕ from T is a finite sequence of $\mathcal{L}$-formulas $\psi_1, \ldots, \psi_m$ such that $\psi_m = \phi$ and $\psi_i \in T$ or ψ_i follows from $\psi_1, \ldots, \psi_{i-1}$ by a simple logical rule for each i. We write $T \vdash \phi$ if there is a proof of ϕ from T. Examples of a "simple" logical rules are:

"from ϕ and ψ conclude $\phi \wedge \psi$," or

"from $\phi \wedge \psi$ conclude ϕ."

It will not be important for our purposes to go into the details of the proof system, but we stress the following points. (See [94], for example, for complete details of one possible proof system.)

- Proofs are finite.
- (Soundness) If $T \vdash \phi$, then $T \models \phi$.
- If T is a finite set of sentences, then there is an algorithm that, when given a sequence of $\mathcal{L}$-formulas σ and an $\mathcal{L}$-sentence ϕ, will decide whether σ is a proof of ϕ from T.

Note that the last point does not say that there is an algorithm that will decide if $T \vdash \phi$. It only says that there is an algorithm that can check each purported proof.

We say that a language $\mathcal{L}$ is *recursive* if there is an algorithm that decides whether a sequence of symbols is an $\mathcal{L}$-formula. We say that an $\mathcal{L}$-theory T is recursive if there is an algorithm that, when given an $\mathcal{L}$-sentence ϕ as input, decides whether $\phi \in T$.

Proposition 2.1.1 *If $\mathcal{L}$ is a recursive language and T is a recursive $\mathcal{L}$-theory, then $\{\phi : T \vdash \phi\}$ is recursively enumerable; that is, there is an algorithm, that when given ϕ as input will halt accepting if $T \vdash \phi$ and not halt if $T \not\vdash \phi$.*

Proof There is $\sigma_0, \sigma_1, \sigma_2, \ldots$, a computable listing of all finite sequences of $\mathcal{L}$-formulas. At stage i of our algorithm, we check to see whether σ_i is a proof of ψ from T. This involves checking that each formula either is in T (which we can check because T is recursive) or follows by a logical rule from earlier formulas in the sequence σ_i and that the last formula is ϕ. If σ_i is a proof of ϕ from T, then we halt accepting; otherwise we go on to stage $i + 1$.

Remarkably, the finitistic syntactic notion of "proof" completely captures the semantic notion of "logical consequence."

Theorem 2.1.2 (Gödel's Completeness Theorem) *Let T be an $\mathcal{L}$-theory and ϕ an $\mathcal{L}$-sentence, then $T \models \phi$ if and only if $T \vdash \phi$.*

The Completeness Theorem gives a criterion for testing whether an $\mathcal{L}$-theory is satisfiable. We say that an $\mathcal{L}$-theory T is *inconsistent* if $T \vdash (\phi \wedge \neg\phi)$ for some sentence ϕ; otherwise we say that T is *consistent*. Because our proof system is sound, any satisfiable theory is consistent. The Completeness Theorem implies that the converse is true.

Corollary 2.1.3 *T is consistent if and only if T is satisfiable.*

Proof Suppose that T is not satisfiable. Because there are no models of T, every model of T is a model of $(\phi \wedge \neg\phi)$. Thus, $T \models (\phi \wedge \neg\phi)$ and by the Completeness Theorem $T \vdash (\phi \wedge \neg\phi)$.

This has a deceptively simple consequence.

Theorem 2.1.4 (Compactness Theorem) *T is satisfiable if and only if every finite subset of T is satisfiable.*

Proof Clearly, if T is satisfiable, then every subset of T is satisfiable. On the other hand, if T is not satisfiable, then T is inconsistent. Let σ be a proof of a contradiction from T. Because σ is finite, only finitely many assumptions from T are used in the proof. Thus, there is a finite $T_0 \subseteq T$

such that σ is a proof of a contradiction from T_0. But then T_0 is a finite unsatisfiable subset of T.

Although it is a simple consequence of the Completeness Theorem and the finite nature of proof, the Compactness Theorem is the cornerstone of model theory. Because it will not be useful for us to understand the exact nature of our proof system, we will not prove the Completeness Theorem. Instead we will give a second proof of the Compactness Theorem that does not appeal directly to the Completeness Theorem. This proof can be adapted to prove the Completeness Theorem as well.

Henkin Constructions

We say that a theory T is *finitely satisfiable* if every finite subset of T is satisfiable. We will show that every finitely satisfiable theory T is satisfiable. To do this, we must build a model of T. The main idea of the construction is that we will add enough constants to the language so that every element of our model will be named by a constant symbol. The following definition will give us sufficient conditions to construct a model from the constants.

Definition 2.1.5 We say that an $\mathcal{L}$-theory T has the *witness property* if whenever $\phi(v)$ is an $\mathcal{L}$-formula with one free variable v, then there is a constant symbol $c \in \mathcal{L}$ such that $T \models (\exists v\ \phi(v)) \rightarrow \phi(c)$.
    An $\mathcal{L}$-theory T is *maximal* if for all ϕ either $\phi \in T$ or $\neg\phi \in T$.

Our proof will frequently use the following simple lemma.

Lemma 2.1.6 *Suppose T is a maximal and finitely satisfiable $\mathcal{L}$-theory. If $\Delta \subseteq T$ is finite and $\Delta \models \psi$, then $\psi \in T$.*

Proof If $\psi \notin T$, then, because T is maximal, $\neg\psi \in T$. But then $\Delta \cup \{\neg\psi\}$ is a finite unsatisfiable subset of T, a contradiction.

Lemma 2.1.7 *Suppose that T is a maximal and finitely satisfiable $\mathcal{L}$-theory with the witness property. Then, T has a model. In fact, if κ is a cardinal and $\mathcal{L}$ has at most κ constant symbols, then there is $\mathcal{M} \models T$ with $|\mathcal{M}| \leq \kappa$.*

Proof Let $\mathcal{C}$ be the set of constant symbols of $\mathcal{L}$. For $c, d \in \mathcal{C}$, we say $c \sim d$ if $T \models c = d$.

Claim 1 $\sim$ is an equivalence relation.
    Clearly, $c = c$ is in T. Suppose that $c = d$ and $d = e$ are in T. By Lemma 2.1.6, $d = c$ and $c = e$ are in T.

The universe of our model will be $M = \mathcal{C}/\sim$, the equivalence classes of $\mathcal{C}$ mod $\sim$. Clearly, $|M| \leq \kappa$. We let c^* denote the equivalence class of c and interpret c as its equivalence class, that is, $c^{\mathcal{M}} = c^*$. Next we show how to interpret the relation and function symbols of $\mathcal{L}$.

Suppose that R is an n-ary relation symbol of $\mathcal{L}$.

Claim 2 Suppose that $c_1, \ldots, c_n, d_1, \ldots, d_n \in \mathcal{C}$, and $c_i \sim d_i$ for $i = 1, \ldots, n$, then, $R(\bar{c}) \in T$ if and only if $R(\bar{d}) \in T$.

Because $c_i = d_i \in T$ for $i = 1, \ldots, n$, by Lemma 2.1.6, if one of $R(\bar{c})$ and $R(\bar{d})$ is in T, then both are in T.

We will interpret R as

$$R^{\mathcal{M}} = \{(c_1^*, \ldots, c_n^*) : R(c_1, \ldots, c_n) \in T\}.$$

By Claim 2, $R^{\mathcal{M}}$ is well-defined.

Suppose that f is an n-ary function symbol of $\mathcal{L}$ and $c_1, \ldots, c_n \in \mathcal{C}$. Because $\emptyset \models \exists v \, f(c_1, \ldots, c_n) = v$ and T has the witness property, by Lemma 2.1.6, there is $c_{n+1} \in \mathcal{C}$ such that $f(c_1, \ldots, c_n) = c_{n+1} \in T$. As above, if $d_i \sim c_i$ for $i = 1, \ldots, n+1$, then $f(d_1, \ldots, d_n) = d_{n+1} \in T$. Moreover, because f is a function symbol, if $e_i \sim c_i$ for $i = 1, \ldots n$ and $f(e_1, \ldots, e_n) = e_{n+1} \in T$, then $e_{n+1} \sim c_{n+1}$. Thus, we get a well-defined function $f^{\mathcal{M}} : M^n \to M$ by

$$f^{\mathcal{M}}(c_1^*, \ldots, c_n^*) = d^* \text{ if and only if } f(c_1, \ldots, c_n) = d \in T.$$

This completes the description of the structure $\mathcal{M}$. Before showing that $\mathcal{M} \models T$, we must show that terms behave correctly.

Claim 3 Suppose that t is a term using free variables from $v_1, \ldots, v_n$. If $c_1, \ldots, c_n, d \in \mathcal{C}$, then $t(c_1, \ldots, c_n) = d \in T$ if and only if $t^{\mathcal{M}}(c_1^*, \ldots, c_n^*) = d^*$.

($\Rightarrow$) We first prove, by induction on terms, that if $t(c_1, \ldots, c_n) = d \in T$, then $t^{\mathcal{M}}(c_1^*, \ldots, c_n^*) = d^*$. If t is a constant symbol c, then $c = d \in T$ and $c^{\mathcal{M}} = c^* = d^*$.

If t is the variable v_i, then $c_i = d \in T$ and $t^{\mathcal{M}}(c_1^*, \ldots, c_n^*) = c_i^* = d^*$.

Suppose that the claim is true for $t_1, \ldots, t_m$ and t is $f(t_1, \ldots, t_m)$. Using the witness property and Lemma 2.1.6, we can find $d, d_1, \ldots, d_n \in \mathcal{C}$ such that $t_i(c_1, \ldots, c_n) = d_i \in T$ for $i \leq m$ and $f(d_1, \ldots, d_m) = d \in T$. By our induction hypothesis, $t_i^{\mathcal{M}}(c_1^*, \ldots, c_n^*) = d_i^*$ and $f^{\mathcal{M}}(d_1^*, \ldots, d_m^*) = d^*$. Thus $t^{\mathcal{M}}(c_1^*, \ldots, c_n^*) = d^*$.

($\Leftarrow$) Suppose, on the other hand, than $t^{\mathcal{M}}(c_1^*, \ldots, c_n^*) = d^*$. By the witness property and Lemma 2.1.6, there is $e \in \mathcal{C}$ such that $t(c_1, \ldots, c_n) = e \in T$. Using the ($\Rightarrow$) direction of the proof, $t^{\mathcal{M}}(c_1^*, \ldots, c_n^*) = e^*$. Thus, $e^* = d^*$ and $e = d \in T$. By Lemma 2.1.6, $t(c_1, \ldots, c_n) = d \in T$.

Claim 4 For all $\mathcal{L}$-formulas $\phi(v_1, \ldots, v_n)$ and $c_1, \ldots, c_n \in \mathcal{C}$, $\mathcal{M} \models \phi(\bar{c}^*)$ if and only if $\phi(\bar{c}) \in T$.

We prove this claim by induction on formulas.

Suppose that ϕ is $t_1 = t_2$. By Lemma 2.1.6 and the witness property, we can find d_1 and d_2 such that $t_1(\bar{c}) = d_1$ and $t_2(\bar{c}) = d_2$ are in T. By Claim 3, $t_i^{\mathcal{M}}(\bar{c}^*) = d_i^*$ for $i = 1, 2$. Then

$$\mathcal{M} \models \phi(\bar{c}^*) \quad \Leftrightarrow \quad d_1^* = d_2^*$$
$$\Leftrightarrow \quad d_1 = d_2 \in T$$
$$\Leftrightarrow \quad t_1(\bar{c}) = t_2(\bar{c}) \in T \text{ by Lemma 2.1.6.}$$

Suppose that ϕ is $R(t_1, \ldots, t_m)$. Because T has the witness property, by Lemma 2.1.6 there are $d_1, \ldots, d_m \in \mathcal{C}$ such that $t_i(\bar{c}) = d_i \in T$ and, Claim 4, $t_i^{\mathcal{M}}(\bar{c}^*) = d_i^*$ for $i = 1, \ldots, m$. Thus,

$$\mathcal{M} \models \phi(\bar{c}^*) \quad \Leftrightarrow \quad \bar{d}^* \in R^{\mathcal{M}}$$
$$\Leftrightarrow \quad R(\bar{d}) \in T$$
$$\Leftrightarrow \quad \phi(\bar{c}) \in T \text{ by Lemma 2.1.6.}$$

Suppose that the claim is true for ϕ. If $\mathcal{M} \models \neg\phi(\bar{c}^*)$, then $\mathcal{M} \not\models \phi(\bar{c}^*)$. By the induction hypothesis, $\phi(\bar{c}) \notin T$. Thus by maximality, $\neg\phi(\bar{c}) \in T$. On the other hand, if $\neg\phi(\bar{c}) \in T$, then, because T is finitely satisfiable, $\phi(\bar{c}) \notin T$. Thus, by induction, $\mathcal{M} \not\models \phi(\bar{c}^*)$ and $\mathcal{M} \models \neg\phi(\bar{c}^*)$.

Suppose that the claim is true for ϕ and ψ. Then

$$\mathcal{M} \models (\phi \wedge \psi)(\bar{c}^*) \quad \Leftrightarrow \quad \phi(\bar{c}) \in T \text{ and } \psi(\bar{c}) \in T$$
$$\Leftrightarrow \quad (\phi \wedge \psi)(\bar{c}) \in T \text{ by Lemma 2.1.6.}$$

Suppose that ϕ is $\exists v \ \psi(v)$ and the claim is true for ψ. If $\mathcal{M} \models \psi(d^*, \bar{c}^*)$, then, by the inductive assumption, $\psi(d, \bar{c}) \in T$ and $\exists v \ \psi(v, \bar{c}) \in T$, by 2.1.6. On the other hand if $\exists v \ \psi(v, \bar{c}) \in T$, then by the witness property and Lemma 2.1.6, $\psi(d, \bar{c}) \in T$ for some c. By induction, $\mathcal{M} \models \psi(d^*, \bar{c}^*)$ and $\mathcal{M} \models \exists v \ \psi(v, \bar{c}^*)$.

This completes the induction. In particular, we have $\mathcal{M} \models T$, as desired.

The following lemmas show that any finitely satisfiable theory can be extended to a maximal finitely satisfiable theory with the witness property.

Lemma 2.1.8 *Let T be a finitely satisfiable $\mathcal{L}$-theory. There is a language $\mathcal{L}^* \supseteq \mathcal{L}$ and $T^* \supseteq T$ a finitely satisfiable $\mathcal{L}^*$-theory such that any $\mathcal{L}^*$-theory extending T^* has the witness property. We can choose $\mathcal{L}^*$ such that $|\mathcal{L}^*| = |\mathcal{L}| + \aleph_0$.*

Proof We first show that there is a language $\mathcal{L}_1 \supseteq \mathcal{L}$ and a finitely satisfiable $\mathcal{L}_1$-theory $T_1 \supseteq T$ such that for any $\mathcal{L}$-formula $\phi(v)$ there is an $\mathcal{L}_1$-constant symbol c such that $T_1 \models (\exists v \ \phi(v)) \rightarrow \phi(c)$. For each $\mathcal{L}$-formula $\phi(v)$, let c_ϕ be a new constant symbol and let $\mathcal{L}_1 = \mathcal{L} \cup \{c_\phi : \phi(v) \text{ an } \mathcal{L}\text{-formula}\}$. For each $\mathcal{L}$-formula $\phi(v)$, let Θ_ϕ be the $\mathcal{L}_1$-sentence $(\exists v \ \phi(v)) \rightarrow \phi(c_\phi)$. Let $T_1 = T \cup \{\Theta_\phi : \phi(v) \text{ an } \mathcal{L}\text{-formula}\}$.

Claim T_1 is finitely satisfiable.

Suppose that Δ is a finite subset of T_1. Then, $\Delta = \Delta_0 \cup \{\Theta_{\phi_1}, \ldots, \Theta_{\phi_n}\}$, where Δ_0 is a finite subset of T. Because T is finitely satisfiable, there is

$\mathcal{M} \models \Delta_0$. We will make $\mathcal{M}$ into an $\mathcal{L} \cup \{c_{\phi_1}, \ldots, c_{\phi_n}\}$-structure $\mathcal{M}'$. Because we will not change the interpretation of the symbols of $\mathcal{L}$, we will have $\mathcal{M}' \models \Delta_0$. To do this, we must show how to interpret the symbols c_{ϕ_i} in $\mathcal{M}'$. If $\mathcal{M} \models \exists v \ \phi(v)$, choose a_i some element of M such that $\mathcal{M} \models \phi(a_i)$ and let $c_{\phi_i}^{\mathcal{M}'} = a_i$. Otherwise, let $c_{\phi_i}^{\mathcal{M}'}$ be any element of $\mathcal{M}$. Clearly, $\mathcal{M}' \models \Theta_{\phi_i}$ for $i \leq n$. Thus, T_1 is finitely satisfiable.

We now iterate the construction above to build a sequence of languages $\mathcal{L} \subseteq \mathcal{L}_1 \subseteq \mathcal{L}_2 \subseteq \ldots$ and a sequence of finitely satisfiable $\mathcal{L}_i$-theories $T \subseteq T_1 \subseteq T_2 \subseteq \ldots$ such that if $\phi(v)$ is an $\mathcal{L}_i$-formula, then there is a constant symbol $c \in \mathcal{L}_{i+1}$ such that $T_{i+1} \models (\exists v \phi(v)) \rightarrow \phi(c)$.

Let $\mathcal{L}^* = \bigcup \mathcal{L}_i$ and $T^* = \bigcup T_i$. By construction, T^* has the witness property. If Δ is a finite subset of T^*, then $\Delta \subseteq T_i$ for some i. Thus, Δ is satisfiable and T^* is finitely satisfiable.

If $|\mathcal{L}_i|$ is the number of relation, function and constant symbols in $\mathcal{L}_i$, then there are at most $|\mathcal{L}_i| + \aleph_0$ formulas in $\mathcal{L}_i$. Thus, by induction, $|\mathcal{L}^*| = |\mathcal{L}| + \aleph_0$.

Lemma 2.1.9 *Suppose that T is a finitely satisfiable $\mathcal{L}$-theory and ϕ is an $\mathcal{L}$-sentence, then, either $T \cup \{\phi\}$ or $T \cup \{\neg\phi\}$ is finitely satisfiable.*

Proof Suppose that $T \cup \{\phi\}$ is not finitely satisfiable. Then, there is a finite $\Delta \subseteq T$ such that $\Delta \models \neg\phi$. We claim that $T \cup \{\neg\phi\}$ is finitely satisfiable. Let Σ be a finite subset of T. Because $\Delta \cup \Sigma$ is satisfiable and $\Delta \cup \Sigma \models \neg\phi$, $\Sigma \cup \{\neg\phi\}$ is satisfiable. Thus, $T \cup \{\neg\phi\}$ is finitely satisfiable.

Corollary 2.1.10 *If T is a finitely satisfiable $\mathcal{L}$-theory, then there is a maximal finitely satisfiable $\mathcal{L}$-theory $T' \supseteq T$.*

Proof Let I be the set of all finitely satisfiable $\mathcal{L}$-theories containing T. We partially order I by inclusion. If $C \subseteq I$ is a chain, let $T_C = \bigcup \{\Sigma : \Sigma \in C\}$. If Δ is a finite subset of T_C, then there is $\Sigma \in C$ such that $\Delta \subseteq \Sigma$, so T_C is finitely satisfiable and $T_C \supseteq \Sigma$ for all $\Sigma \in C$. Thus, every chain in I has an upper bound, and we can apply Zorn's Lemma (see Appendix A) to find a $T' \in I$ maximal with respect to the partial order. By Lemma 2.1.9, either $T' \cup \{\phi\}$ or $T' \cup \{\neg\phi\}$ is finitely satisfiable. Because T' is maximal in the partial order, one of ϕ or $\neg\phi$ is in T'. Thus, T' is a maximal theory.

We can now prove the following strengthening of the Compactness Theorem.

Theorem 2.1.11 *If T is a finitely satisfiable $\mathcal{L}$-theory and κ is an infinite cardinal with $\kappa \geq |\mathcal{L}|$, then there is a model of T of cardinality at most κ.*

Proof By Lemma 2.1.8, we can find $\mathcal{L}^* \supseteq \mathcal{L}$ and $T^* \supseteq T$ a finitely satisfiable $\mathcal{L}^*$-theory such that any $\mathcal{L}^*$-theory extending T^* has the witness property and the cardinality of $\mathcal{L}^*$ is at most κ. By Corollary 2.1.10, we can find a maximal finitely satisfiable $\mathcal{L}^*$-theory $T' \supseteq T^*$. Because T' has the witness property, Lemma 2.1.7 ensures that there is $\mathcal{M} \models T$ with $|M| \leq \kappa$.

This proof of the Compactness Theorem is based on Henkin's proof on the Completeness Theorem. The method of constructing a model where the universe is built from the constant symbols is referred to as a "Henkin construction" and the theories with the witness property are sometimes called "Henkinized".

We conclude this section with several standard applications of the Compactness Theorem. We will give a few more in the exercises.

Proposition 2.1.12 *Let $\mathcal{L}$ be a language containing $\{\cdot, e\}$, the language of groups, let T be an $\mathcal{L}$-theory extending the theory of groups, and let $\phi(v)$ be an $\mathcal{L}$-formula. Suppose that for all n there is $G_n \models T$ and $g_n \in G_n$ with finite order greater than n such that $G_n \models \phi(g_n)$. Then, there is $G \models T$ and $g \in G$ such that $G \models \phi(g)$ and g has infinite order. In particular, there is no formula that defines the torsion points in all models of T.*

Proof Let $\mathcal{L}^* = \mathcal{L} \cup \{c\}$, where c is a new constant symbol. Let T^* be the $\mathcal{L}$-theory

$$T \cup \{\phi(c)\} \cup \{\underbrace{c \cdot c \cdots c}_{n-\text{times}} \neq e : n = 1, 2, \ldots\}.$$

If G is a model of T^* and g is the interpretation of c in G then $G \models \phi(g)$ and g has infinite order. Hence, it suffices to show that T^* is satisfiable.

Let $\Delta \subseteq T^*$ be finite. Then

$$\Delta \subseteq T \cup \{\phi(c)\} \cup \{\underbrace{c \cdot c \cdots c}_{n-\text{times}} \neq e : n = 1, 2, \ldots, m\}$$

for some m. View G_m as an $\mathcal{L}^*$ structure by interpreting c as the element g_m. Because $G_m \models T \cup \{\phi(g_m)\}$ and g_m has order greater than m, $G_m \models \Delta$. Thus, T^* is finitely satisfiable and hence, by the Compactness Theorem, satisfiable.

Proposition 2.1.13 *Let $\mathcal{L} = \{\cdot, +, <, 0, 1\}$ and let $\mathrm{Th}(\mathbb{N})$ be the full $\mathcal{L}$-theory of the natural numbers. There is $\mathcal{M} \models \mathrm{Th}(\mathbb{N})$ and $a \in M$ such that a is larger than every natural number.*

Proof Let $\mathcal{L}^* = \mathcal{L} \cup \{c\}$ where c is a new constant symbol and let

$$T = \mathrm{Th}(\mathbb{N}) \cup \{\underbrace{1 + 1 + \ldots + 1}_{n-\text{times}} < c : \text{for } n = 1, 2, \ldots\}.$$

If Δ is a finite subset of T, we can make $\mathbb{N}$ a model of Δ by interpreting c as a suitably large natural number. Thus, T is finitely satisfiable and there is $\mathcal{M} \models T$. If $a \in M$ is the interpretation of c, then a is larger than every natural number.

The next lemma is an easy consequence of the Completeness Theorem, but it also can be deduced from the Compactness Theorem.

Lemma 2.1.14 *If $T \models \phi$, then $\Delta \models T$ for some finite $\Delta \subseteq T$.*

Proof Suppose not. Let $\Delta \subseteq T$ be finite. Because $\Delta \not\models \phi$, $\Delta \cup \{\neg\phi\}$ is satisfiable. Thus, $T \cup \{\neg\phi\}$ is finitely satisfiable and, by the Compactness Theorem, $T \not\models \phi$.

2.2 Complete Theories

Definition 2.2.1 An $\mathcal{L}$-theory T is called *complete* if for any $\mathcal{L}$-sentence ϕ either $T \models \phi$ or $T \models \neg\phi$.

For $\mathcal{M}$ an $\mathcal{L}$-structure, then the full theory

$$\mathrm{Th}(\mathcal{M}) = \{\phi : \phi \text{ is an } \mathcal{L}\text{-sentence and } \mathcal{M} \models \phi\}$$

is a complete theory. Usually, it is difficult to determine exactly what sentences are in $\mathrm{Th}(\mathcal{M})$. In many cases, the key to understanding $\mathrm{Th}(\mathcal{M})$ is finding a simpler $\mathcal{L}$-theory T such that $\mathcal{M} \models T$ and T is complete. In this case, $\mathcal{M} \models \phi$ if and only if $T \models \phi$, because if $T \not\models \phi$, then $T \models \neg\phi$ and $\mathcal{M} \models \neg\phi$.

The Compactness Theorem yields a very useful completeness test that we will use in this section to show that several natural theories are complete. This will have striking consequences for algebraically closed fields.

Proposition 2.2.2 *Let T be an $\mathcal{L}$-theory with infinite models. If κ is an infinite cardinal and $\kappa \geq |\mathcal{L}|$, then there is a model of T of cardinality κ.*

Proof Let $\mathcal{L}^* = \mathcal{L} \cup \{c_\alpha : \alpha < \kappa\}$, where each c_α is a new constant symbol, and let T^* be the $\mathcal{L}^*$-theory $T \cup \{c_\alpha \neq c_\beta : \alpha, \beta < \kappa, \alpha \neq \beta\}$. Clearly if $\mathcal{M} \models T^*$, then $\mathcal{M}$ is a model of T of cardinality at least κ. Thus, by Theorem 2.1.11, it suffices to show that T^* is finitely satisfiable. But if $\Delta \subset T^*$ is finite, then $\Delta \subseteq T \cup \{c_\alpha \neq c_\beta : \alpha \neq \beta, \alpha, \beta \in I\}$, where I is a finite subset of κ. Let $\mathcal{M}$ be an infinite model of T. We can interpret the symbols $\{c_\alpha : \alpha \in I\}$ as $|I|$ distinct elements of M. Because $\mathcal{M} \models \Delta$, T^* is finitely satisfiable.

Definition 2.2.3 Let κ be an infinite cardinal and let T be a theory with models of size κ. We say that T is κ-*categorical* if any two models of T of cardinality κ are isomorphic.

Let $\mathcal{L} = \{+, 0\}$ be the language of additive groups and let T be the $\mathcal{L}$-theory of torsion-free divisible Abelian groups. The axioms of T are the axioms for Abelian groups together with the axioms

$$\forall x (x \neq 0 \rightarrow \underbrace{x + \ldots + x}_{n-\text{times}} \neq 0)$$

and

$$\forall y \exists x \underbrace{x + \ldots + x}_{n-\text{times}} = y$$

for $n = 1, 2, \ldots$.

Proposition 2.2.4 *The theory of torsion-free divisible Abelian groups is κ-categorical for all $\kappa > \aleph_0$.*

Proof We first argue that models of T are essentially vector spaces over the field of rational numbers $\mathbb{Q}$. Clearly, if V is any vector space over $\mathbb{Q}$, then the underlying additive group of V is a model of T. On the other hand, if $G \models T$, $g \in G$, and $n \in \mathbb{N}$ with $n > 0$, we can find $h \in G$ such that $nh = g$. If $nk = g$, then $n(h - k) = 0$. Because G is torsion-free there is a unique $h \in G$ such that $nh = g$. We call this element g/n. We can view G as a $\mathbb{Q}$-vector space under the action $\frac{m}{n}g = m(g/n)$.

Two $\mathbb{Q}$-vector spaces are isomorphic if and only if they have the same dimension. Thus, models of T are determined up to isomorphism by their dimension. If G has dimension λ, then $|G| = \lambda + \aleph_0$. If κ is uncountable and G has cardinality κ, then G has dimension κ. Thus, for $\kappa > \aleph_0$ any two models of T of cardinality κ are isomorphic.

Note that T is not $\aleph_0$-categorical. Indeed, there are $\aleph_0$ nonisomorphic models corresponding to vector spaces of dimension $1, 2, 3, \ldots$ and $\aleph_0$.

A similar argument applies to the theory of algebraically closed fields. Let ACF_p be the theory of algebraically closed fields of characteristic p, where p is either 0 or a prime number.

Proposition 2.2.5 ACF_p *is κ-categorical for all uncountable cardinals κ.*

Proof Two algebraically closed fields are isomorphic if and only if they have the same characteristic and transcendence degree (see, for example [58] X §1). An algebraically closed field of transcendence degree λ has cardinality $\lambda + \aleph_0$. If $\kappa > \aleph_0$, an algebraically closed field of cardinality κ also has transcendence degree κ. Thus, any two algebraically closed fields of the same characteristic and same uncountable cardinality are isomorphic.

We give two simpler examples.

• Let $\mathcal{L}$ be the empty language. Then the theory of an infinite set is κ-categorical for all cardinals κ.

• Let $\mathcal{L} = \{E\}$, where E is a binary relation, and let T be the theory of an equivalence relation with exactly two classes, both of which are infinite. It is easy to see that any two countable models of T are isomorphic. On the other hand, T is not κ-categorical for $\kappa > \aleph_0$. To see this, let $\mathcal{M}_0$ be a model where both classes have cardinality κ, and let $\mathcal{M}_1$ be a model where one class has cardinality κ and the other has cardinality $\aleph_0$. Clearly, $\mathcal{M}_0$ and $\mathcal{M}_1$ are not isomorphic.

Theorem 2.2.6 (Vaught's Test) *Let T be a satisfiable theory with no finite models that is κ-categorical for some infinite cardinal $\kappa \geq |\mathcal{L}|$. Then T is complete.*

Proof Suppose that T is not complete. Then there is a sentence ϕ such that $T \not\models \phi$ and $T \not\models \neg\phi$. Because $T \not\models \psi$ if and only if $T \cup \{\neg\psi\}$ is satisfiable, the theories $T_0 = T \cup \{\phi\}$ and $T_1 = T \cup \{\neg\phi\}$ are satisfiable. Because T has no finite models, both T_0 and T_1 have infinite models. By Proposition 2.2.2 we can find $\mathcal{M}_0$ and $\mathcal{M}_1$ of cardinality κ with $\mathcal{M}_i \models T_i$. Because $\mathcal{M}_0$ and $\mathcal{M}_1$ disagree about ϕ, there are not elementarily equivalent and hence, by Theorem 1.1.10, nonisomorphic. This contradicts the κ-categoricity of T.

The assumption that T has no finite models is necessary. Suppose that T is the $\{+, 0\}$-theory of groups, where every element has order 2. In the exercises, we will show that T is κ-categorical for all $\kappa \geq \aleph_0$. However, T is not complete. The sentence $\exists x \exists y \exists z \, (x \neq y \wedge y \neq z \wedge z \neq x)$ is false in the two-element group but true in every other model of T.

Vaught's test implies that all of the categorical theories discussed above are complete. In particular, algebraically closed fields are complete. This result of Tarski has several immediate interesting consequences.

Definition 2.2.7 We say that an $\mathcal{L}$-theory T is *decidable* if there is an algorithm that when given an $\mathcal{L}$-sentence ϕ as input decides whether $T \models \phi$.

Lemma 2.2.8 *Let T be a recursive complete satisfiable theory in a recursive language $\mathcal{L}$. Then T is decidable.*

Proof Because T is satisfiable $A = \{\phi : T \models \phi\}$ and $B = \{\phi : T \models \neg\phi\}$ are disjoint. Because T is consistent $A \cup B$ is the set of all $\mathcal{L}$-sentences. By the Completeness Theorem, $A = \{\phi : T \vdash \phi\}$ and $B = \{\phi : T \vdash \neg\phi\}$. By Proposition 2.1.1 A and B are recursively enumerable. But any recursively enumerable set with a recursively enumerable complement is recursive.

Informally, to decide whether ϕ is a logical consequence of a complete satisfiable recursive theory T, we begin searching through possible proofs from T until we find either a proof of ϕ or a proof of $\neg\phi$. Because T is satisfiable, we will not find proofs of both. Because T is complete, we will eventually find a proof of one or the other.

Corollary 2.2.9 *For $p = 0$ or p prime, ACF_p is decidable. In particular, $\mathrm{Th}(\mathbb{C})$, the first-order theory of the field of complex numbers, is decidable.*

The completeness of ACF_p can also be thought of as a first-order version of the Lefschetz Principle from algebraic geometry.

Corollary 2.2.10 *Let ϕ be a sentence in the language of rings. The following are equivalent.*
i) ϕ is true in the complex numbers.

ii) ϕ *is true in every algebraically closed field of characteristic zero.*

iii) ϕ *is true in some algebraically closed field of characteristic zero.*

iv) *There are arbitrarily large primes p such that ϕ is true in some algebraically closed field of characteristic p.*

v) *There is an m such that for all $p > m$, ϕ is true in all algebraically closed fields of characteristic p.*

Proof The equivalence of i)–iii) is just the completeness of ACF_0 and v)$\Rightarrow$ iv) is obvious.

For ii) $\Rightarrow$ v) suppose that $\text{ACF}_0 \models \phi$. By Lemma 2.1.14, there is a finite $\Delta \subset \text{ACF}_0$ such that $\Delta \models \phi$. Thus, if we choose p large enough, then $\text{ACF}_p \models \Delta$. Thus, $\text{ACF}_p \models \phi$ for all sufficiently large primes p.

For iv) $\Rightarrow$ ii) suppose $\text{ACF}_0 \not\models \phi$. Because ACF_0 is complete, $\text{ACF}_0 \models \neg\phi$. By the argument above, $\text{ACF}_p \models \neg\phi$ for sufficiently large p; thus, iv) fails.

Ax found the following striking application of Corollary 2.2.10.

Theorem 2.2.11 *Every injective polynomial map from $\mathbb{C}^n$ to $\mathbb{C}^n$ is surjective.*

Proof Remarkably, the key to the proof is the simple observation that if k is a finite field, then every injective function $f : k^n \to k^n$ is surjective. From this observation it is easy to show that the same is true for $\mathbb{F}_p^{\text{alg}}$, the algebraic closure of the p-element field.

Claim Every injective polynomial map $f : (\mathbb{F}_p^{\text{alg}})^n \to (\mathbb{F}_p^{\text{alg}})^n$ is surjective.

Suppose not. Let $\bar{a} \in \mathbb{F}_p^{\text{alg}}$ be the coefficients of f and let $\bar{b} \in (\mathbb{F}_p^{\text{alg}})^n$ such that $\bar{b}$ is not in the range of f. Let k be the subfield of $\mathbb{F}_p^{\text{alg}}$ generated by $\bar{a}, \bar{b}$. Then $f|k^n$ is an injective but not surjective polynomial map from k^n into itself. But $\mathbb{F}_p^{\text{alg}} = \bigcup_{n=1}^{\infty} \mathbb{F}_{p^n}$ is a locally finite field. Thus k is finite, a contradiction.

Suppose that the theorem is false. Let $X = (X_1, \ldots, X_n)$. Let $f(X) = (f_1(X), \ldots, f_n(X))$ be a counterexample where each $f_i \in \mathbb{C}[X]$ has degree at most d. There is an $\mathcal{L}$-sentence $\Phi_{n,d}$ such that for K a field, $K \models \Phi_{n,d}$ if and only if every injective polynomial map from K^n to K^n where each coordinate function has degree at most d is surjective. We can quantify over polynomials of degree at most d by quantifying over the coefficients. For example, $\Phi_{2,2}$ is the sentence

$$\forall a_{0,0} \forall a_{0,1} \forall a_{0,2} \forall a_{1,0} \forall a_{1,1} \forall a_{2,0} \forall b_{0,0} \forall b_{0,1} \forall b_{0,2} \forall b_{1,0} \forall b_{1,1} \forall b_{2,0}$$

$$\Big[\big(\forall x_1 \forall y_1 \forall x_2 \forall y_2 ((\textstyle\sum a_{i,j} x_1^i y_1^j = \sum a_{i,j} x_2^i y_2^j \wedge \sum b_{i,j} x_1^i y_1^j = \sum b_{i,j} x_2^i y_2^j) \to$$

$$(x_1 = x_2 \wedge y_1 = y_2))\big) \to \forall u \forall v \exists x \exists y \textstyle\sum a_{i,j} x^i y^j = u \wedge \sum b_{i,j} x^i y^j = v \Big].$$

By the claim $\mathbb{F}_p^{\text{alg}} \models \Phi_{n,d}$ for all primes p. By Corollary 2.2.10, $\mathbb{C} \models \Phi_{n,d}$, a contradiction.

2.3   Up and Down

In Chapter 1, we looked at the category of $\mathcal{L}$-structures and introduced the notion of $\mathcal{L}$-embeddings between structures. These are the maps that preserve the relations, constants, and functions of $\mathcal{L}$. Often we will want to consider the more restrictive class of maps that preserve all first-order properties.

Definition 2.3.1 If $\mathcal{M}$ and $\mathcal{N}$ are $\mathcal{L}$-structures, then an $\mathcal{L}$-embedding $j : \mathcal{M} \to \mathcal{N}$ is called an *elementary embedding* if

$$\mathcal{M} \models \phi(a_1, \ldots, a_n) \Leftrightarrow \mathcal{N} \models \phi(j(a_1), \ldots, j(a_n))$$

for all $\mathcal{L}$-formulas $\phi(v_1, \ldots, v_n)$ and all $a_1, \ldots, a_n \in M$.

If $\mathcal{M}$ is a substructure of $\mathcal{N}$, we say that it is an *elementary substructure* and write $\mathcal{M} \prec \mathcal{N}$ if the inclusion map is elementary. We also say that $\mathcal{N}$ is an *elementary extension* of $\mathcal{M}$.

The proof of Theorem 1.1.10 shows that isomorphisms are elementary maps.

Next we give a way to construct embeddings and elementary embeddings.

Definition 2.3.2 Suppose that $\mathcal{M}$ is an $\mathcal{L}$-structure. Let $\mathcal{L}_M$ be the language where we add to $\mathcal{L}$ constant symbols m for each element of M. The *atomic diagram* of $\mathcal{M}$ is $\{\phi(m_1, \ldots, m_n) : \phi$ is either an atomic $\mathcal{L}$-formula or the negation of an atomic $\mathcal{L}$-formula and $\mathcal{M} \models \phi(m_1, \ldots, m_n)\}$. The *elementary diagram* of $\mathcal{M}$ is

$$\{\phi(m_1, \ldots, m_n) : \mathcal{M} \models \phi(m_1, \ldots, m_n), \phi \text{ an } \mathcal{L}\text{-formula}\}.$$

We let $\mathrm{Diag}(\mathcal{M})$ and $\mathrm{Diag}_{\mathrm{el}}(\mathcal{M})$ denote the atomic and elementary diagrams of $\mathcal{M}$, respectively.

Lemma 2.3.3 *i) Suppose that $\mathcal{N}$ is an $\mathcal{L}_M$-structure and $\mathcal{N} \models \mathrm{Diag}(\mathcal{M})$; then, viewing $\mathcal{N}$ as an $\mathcal{L}$-structure, there is an $\mathcal{L}$-embedding of $\mathcal{M}$ into $\mathcal{N}$.*
ii) If $\mathcal{N} \models \mathrm{Diag}_{\mathrm{el}}(\mathcal{M})$, then there is an elementary embedding of $\mathcal{M}$ into $\mathcal{N}$.

Proof i) Let $j : M \to N$ by $j(m) = m^{\mathcal{N}}$; that is, $j(m)$ is the interpretation of this constant symbol m in $\mathcal{N}$. If m_1, m_2 are distinct elements of M, then $m_1 \neq m_2 \in \mathrm{Diag}(\mathcal{M})$; thus, $j(m_1) \neq j(m_2)$ so j is an embedding. If f is a function symbol of $\mathcal{L}$ and $f^{\mathcal{M}}(m_1, \ldots, m_n) = m_{n+1}$, then $f(m_1, \ldots, m_n) = m_{n+1}$ is a formula in $\mathrm{Diag}(\mathcal{M})$ and $f^{\mathcal{N}}(j(m_1), \ldots, j(m_n)) = j(m_{n+1})$. If R is a relation symbol and $\overline{m} \in R^{\mathcal{M}}$, then $R(m_1, \ldots, m_n) \in \mathrm{Diag}(\mathcal{M})$ and $(j(m_1), \ldots, j(m_n)) \in R^{\mathcal{N}}$. Hence, j is an $\mathcal{L}$-embedding.
ii) If $\mathcal{N} \models \mathrm{Diag}_{\mathrm{el}}(\mathcal{M})$, then the map j above is elementary.

Combining these observations with the Compactness Theorem allows us to build elementary extensions.

Theorem 2.3.4 (Upward Löwenheim–Skolem Theorem) *Let M be an infinite $\mathcal{L}$-structure and κ be an infinite cardinal $\kappa \geq |M| + |\mathcal{L}|$. Then, there is N an $\mathcal{L}$-structure of cardinality κ and $j : M \to N$ is elementary.*

Proof Because $M \models \mathrm{Diag}_{el}(M)$, $\mathrm{Diag}_{el}(M)$ is satisfiable. By Theorem 2.1.11, there is $N \models \mathrm{Diag}_{el}(M)$ of cardinality κ. By Lemma 2.3.3, there is an elementary $j : M \to N$.

The Downward Löwenheim–Skolem Theorem will give us a method for building small elementary submodels. Its proof will use the following test for elementary extensions due to Tarski and Vaught.

Proposition 2.3.5 (Tarski–Vaught Test) *Suppose that M is a substructure of N. Then, M is an elementary substructure if and only if, for any formula $\phi(v, \overline{w})$ and $\overline{a} \in M$, if there is $b \in N$ such that $N \models \phi(b, \overline{a})$, then there is $c \in M$ such that $N \models \phi(c, \overline{a})$.*

Proof If N is an elementary extension of M, the condition is clearly true. To prove the converse, we must show that for all $\overline{a} \in M$ and all $\mathcal{L}$-formulas $\psi(\overline{v})$

$$M \models \psi(\overline{a}) \Leftrightarrow N \models \psi(\overline{a}).$$

We prove this by induction on formulas.

In Proposition 1.1.8, we showed that if $\phi(\overline{v})$ is quantifier free then $M \models \phi(\overline{a})$ if and only if $N \models \phi(\overline{a})$. Thus, the claim is true for all atomic formulas.

If the claim is true for ψ, then

$$M \models \neg\psi(\overline{a}) \Leftrightarrow M \not\models \psi(\overline{a}) \Leftrightarrow N \not\models \psi(\overline{a}) \Leftrightarrow N \models \neg\psi(\overline{a}).$$

Similarly, if the claim is true for ψ and θ, then

$$
\begin{aligned}
M \models (\psi \wedge \theta)(\overline{a}) \quad &\Leftrightarrow \quad M \models \psi(\overline{a}) \text{ and } M \models \theta(\overline{a}) \\
&\Leftrightarrow \quad N \models \psi(\overline{a}) \text{ and } N \models \theta(\overline{a}) \\
&\Leftrightarrow \quad N \models (\psi \wedge \theta)(\overline{a}).
\end{aligned}
$$

Suppose that the claim is true for $\psi(v, \overline{w})$. Let $\overline{a} \in M$. If $M \models \exists v\, \psi(v, \overline{a})$, then there is $b \in M$ such that $M \models \psi(b, \overline{a})$. By our inductive assumption, $N \models \psi(b, \overline{a})$ and hence $N \models \exists v\, \psi(v, \overline{a})$.

If, on the other hand, $N \models \exists v\, \psi(v, \overline{a})$, then, by our assumptions, there is $c \in M$ such that $N \models \psi(c, \overline{a})$. By induction, $M \models \psi(c, \overline{a})$ and $M \models \exists v\, \psi(v, \overline{a})$.

This proposition tells us that to find elementary substructures we must be able to witness quantifiers. We now examine one systematic way of doing that.

We say that an $\mathcal{L}$-theory T has *built-in Skolem functions* if for all $\mathcal{L}$-formulas $\phi(v, w_1, \ldots, w_n)$ there is a function symbol f such that $T \models \forall \overline{w}\, ((\exists v \phi(v, \overline{w})) \to \phi(f(\overline{w}), \overline{w}))$. In other words, there are enough function symbols in the language to witness all existential statements.

Lemma 2.3.6 *Let T be an $\mathcal{L}$-theory. There are $\mathcal{L}^* \supseteq \mathcal{L}$ and $T^* \supseteq T$ an $\mathcal{L}^*$-theory such that T^* has built-in Skolem functions, and if $\mathcal{M} \models T$, then we can expand $\mathcal{M}$ to $\mathcal{M}^* \models T^*$. We can choose $\mathcal{L}^*$ such that $|\mathcal{L}^*| = |\mathcal{L}| + \aleph_0$. We call T^* a skolemization of T.*

Proof Recall from Exercise 1.4.15 that to expand $\mathcal{M}$ we must interpret all of the symbols in $\mathcal{L}^* \setminus \mathcal{L}$ to make $\mathcal{M}$ a model of T^*.

We build a sequence of languages $\mathcal{L} = \mathcal{L}_0 \subseteq \mathcal{L}_1 \subseteq \ldots$ and $\mathcal{L}_i$-theories T_i such that $T = T_0 \subseteq T_1 \subseteq \ldots$.

Given $\mathcal{L}_i$, let $\mathcal{L}_{i+1} = \mathcal{L} \cup \{f_\phi : \phi(v, w_1, \ldots, w_n)$ an $\mathcal{L}_i$-formula, $n = 1, 2, \ldots\}$, where f_ϕ is an n-ary function symbol. For $\phi(v, \overline{w})$ an $\mathcal{L}_i$-formula, let Ψ_ϕ be the sentence

$$\forall \overline{w} \left((\exists v \ \phi(v, \overline{w})) \to \phi(f_\phi(\overline{w}), \overline{w}) \right)$$

and let $T_{i+1} = T_i \cup \{\Psi_\phi : \phi$ an $\mathcal{L}_i$-formula$\}$.

Claim If $\mathcal{M} \models T_i$, then we can interpret the function symbols of $\mathcal{L}_{i+1} \setminus \mathcal{L}_i$ so that $\mathcal{M} \models T_{i+1}$.

Let c be some fixed element of M. If $\phi(v, w_1, \ldots, w_n)$ is an $\mathcal{L}_i$-formula, we find a function $g : M^n \to M$ such that if $\overline{a} \in M^n$ and $X_{\overline{a}} = \{b \in M : \mathcal{M} \models \phi(b, \overline{a})\}$ is nonempty, then $g(\overline{a}) \in X_{\overline{a}}$, and if $X_{\overline{a}} = \emptyset$, then $g(\overline{a}) = c$ (the choice in this case is irrelevant). Thus, if $\mathcal{M} \models \exists v \ \phi(v, \overline{a})$, then $\mathcal{M} \models \phi(g(\overline{a}), \overline{a})$. If we interpret f_ϕ as g, then $\mathcal{M} \models \Psi_\phi$.

Let $\mathcal{L}^* = \bigcup \mathcal{L}_i$ and $T^* = \bigcup T_i$. If $\phi(v, \overline{w})$ is an $\mathcal{L}^*$-formula, then $\phi \in \mathcal{L}_i$ for some i and $\Psi_\phi \in T_{i+1} \subseteq T^*$, so T^* has built-in Skolem functions. By iterating the claim, we see that for any $\mathcal{M} \models T$ we can interpret the symbols of $\mathcal{L}^* \setminus \mathcal{L}$ to make $\mathcal{M} \models T^*$.

Because we have added one function symbol to $\mathcal{L}_{i+1}$ for each $\mathcal{L}_i$-formula, $|\mathcal{L}_{i+1}| = |\mathcal{L}_i| + \aleph_0$ so $|\mathcal{L}^*|$ has the desired cardinality.

Theorem 2.3.7 (Löwenheim–Skolem Theorem) *Suppose that $\mathcal{M}$ is an $\mathcal{L}$-structure and $X \subseteq M$, there is an elementary submodel $\mathcal{N}$ of $\mathcal{M}$ such that $X \subseteq N$ and $|\mathcal{N}| \leq |X| + |\mathcal{L}| + \aleph_0$.*

Proof By Lemma 2.3.6, we may assume that $\mathrm{Th}(\mathcal{M})$ has built in Skolem functions. Let $X_0 = X$. Given X_i, let $X_{i+1} = X_i \cup \{f^{\mathcal{M}}(\overline{a}) : f$ an n-ary function symbol, $\overline{a} \in X_i^n$ and $n = 1, 2, \ldots\}$. Let $N = \bigcup X_i$, then $N \leq |X| + |\mathcal{L}| + \aleph_0$ (see Corollary A.15).

If f is an n-ary function symbol of $\mathcal{L}$ and $\overline{a} \in N^n$, then $\overline{a} \in X_i^n$ for some i and $f^{\mathcal{M}}(\overline{a}) \in X_{i+1} \subseteq N$. Thus $f^{\mathcal{M}} | N : N^n \to N$. Thus, we can interpret f as $f^{\mathcal{N}} = f^{\mathcal{M}} | N^n$. If R is an n-ary relation symbol, let $R^{\mathcal{N}} = R^{\mathcal{M}} \cap N^n$. If c is a constant symbol of $\mathcal{L}$, there is a Skolem function $f \in \mathcal{L}$ such that $f(x) = c^{\mathcal{M}}$ for all $x \in M$ (for example, f is the Skolem function for the formula $v = c$). Thus $c^{\mathcal{M}} \in N$. Let $c^{\mathcal{N}} = c^{\mathcal{M}}$. This makes N into an $\mathcal{L}$-structure $\mathcal{N}$ which is a substructure of $\mathcal{M}$.

If $\phi(v, \overline{w})$ is any $\mathcal{L}$-formula, $\overline{a}, b \in M$, and $\mathcal{M} \models \phi(b, \overline{a})$, then $\mathcal{M} \models \phi(f(\overline{a}), \overline{a})$ for some function symbol f of $\mathcal{L}$. By construction, $f^{\mathcal{M}}(\overline{a}) \in N$. Thus, by Proposition 2.3.5, $\mathcal{N} \prec \mathcal{M}$.

Lemma 2.3.3 i) has a useful application in the following preservation theorem.

Definition 2.3.8 A *universal sentence* is one of the form $\forall \overline{v} \phi(\overline{v})$, where ϕ is quantifier-free. We say that an $\mathcal{L}$-theory T has a *universal axiomatization* if there is a set of universal $\mathcal{L}$-sentences Γ such that $\mathcal{M} \models \Gamma$ if and only if $\mathcal{M} \models T$ for all $\mathcal{L}$-structures $\mathcal{M}$.

Theorem 2.3.9 *An $\mathcal{L}$-theory T has a universal axiomatization if and only if whenever $\mathcal{M} \models T$ and $\mathcal{N}$ is a substructure of $\mathcal{M}$, then $\mathcal{N} \models T$. In other words, a theory is preserved under substructure if and only if it has a universal axiomatization.*

Proof Suppose that $\mathcal{N} \subseteq \mathcal{M}$. We showed in Proposition 1.1.8 that if $\phi(\overline{v})$ is a quantifier-free formula and $\overline{a} \in N$, then $\mathcal{N} \models \phi(\overline{a})$ if and only if $\mathcal{M} \models \phi(\overline{a})$. Thus, if $\mathcal{M} \models \forall \overline{v} \, \phi(\overline{v})$, then so does $\mathcal{N}$.

Suppose that T is preserved under substructures. Let $\Gamma = \{\phi : \phi \text{ is universal and } T \models \phi\}$. Clearly, if $\mathcal{N} \models T$, then $\mathcal{N} \models \Gamma$. For the other direction, suppose that $\mathcal{N} \models \Gamma$. We claim that $\mathcal{N} \models T$.

Claim $T \cup \mathrm{Diag}(\mathcal{N})$ is satisfiable.

Suppose not. Then, by the Compactness Theorem, there is a finite $\Delta \subseteq \mathrm{Diag}(\mathcal{N})$ such that $T \cup \Delta$ is not satisfiable. Let $\Delta = \{\psi_1, \ldots, \psi_n\}$. Let $\overline{c}$ be the new constant symbols from N used in $\psi_1, \ldots, \psi_n$ and say $\psi_i = \phi_i(\overline{c})$, where ϕ_i is a quantifier-free $\mathcal{L}$-formula. Because the constants in $\overline{c}$ do not occur in T, if there is a model of $T \cup \{\exists \overline{v} \bigwedge \phi_i(\overline{v})\}$, then by interpreting $\overline{c}$ as witnesses to the existential formula, $T \cup \Delta$ would be satisfiable. Thus $T \models \forall \overline{v} \bigvee \neg \phi_i(\overline{v})$. As the latter formula is universal, $\forall \overline{v} \bigvee \neg \phi_i(\overline{v}) \in \Gamma$, contradicting $\mathcal{N} \models \Gamma$.

By Lemma 2.3.3, there is $\mathcal{M} \models T$ with $\mathcal{M} \supseteq \mathcal{N}$. Because T is preserved under substructure, $\mathcal{N} \models T$ and Γ is a universal axiomatization of T.

We conclude this section with one useful observation about elementary chains.

Definition 2.3.10 Suppose that $(I, <)$ is a linear order. Suppose that $\mathcal{M}_i$ is an $\mathcal{L}$-structure for $i \in I$. We say that $(\mathcal{M}_i : i \in I)$ is a *chain* of $\mathcal{L}$-structures if $\mathcal{M}_i \subseteq \mathcal{M}_j$ for $i < j$. If $\mathcal{M}_i \prec \mathcal{M}_j$ for $i < j$, we call $(\mathcal{M}_i : i \in I)$ an *elementary chain*.

If $(\mathcal{M}_i : i \in I)$ is a nonempty chain of structures, then we can define $\mathcal{M} = \bigcup_{i \in I} \mathcal{M}_i$, the union of the chain, as follows. The universe of $\mathcal{M}$ will be $M = \bigcup_{i \in I} M_i$. If c is a constant in the language, then $c^{\mathcal{M}_i} = c^{\mathcal{M}_j}$ for all $i, j \in I$. Let $c^{\mathcal{M}} = c^{\mathcal{M}_i}$.

Suppose that $\bar{a} \in M$. Because I is linearly ordered, we can find $i \in I$ such that $\bar{a} \in M_i$. If f is a function symbol of $\mathcal{L}$ and $i < j$, then $f^{M_i}(\bar{a}) = f^{M_j}(\bar{a})$. Thus, $f^M = \bigcup_{i \in I} f^{M_i}$ is a well-defined function. Similarly, if f is a relation symbol of $\mathcal{L}$ and $i < j$, then $\bar{a} \in R^{M_i}$ if and only if $\bar{a} \in R^{M_j}$. Let $R^M = \bigcup_{i \in I} R^{M_i}$. It is now easy to see that $M_i \subseteq M$ for all $i \in I$.

Proposition 2.3.11 *Suppose that $(I, <)$ is a linear order and $(M_i : i \in I)$ is an elementary chain. Then, $M = \bigcup_{i \in I} M_i$ is an elementary extension of each M_i.*

Proof We prove by induction on formulas that

$$M \models \phi(\bar{a}) \Leftrightarrow M_i \models \phi(\bar{a})$$

for all $i \in I$, all formulas $\phi(\bar{v})$, and all $\bar{a} \in M_i^n$.

Because M_i is a substructure of M, by Proposition 1.1.8 this is true for all atomic ϕ. It is easy to see that if it is true for ϕ and ψ, then it is true for $\neg \phi$ and $\phi \wedge \psi$.

Suppose that ϕ is $\exists v\, \psi(v, \bar{w})$ and the claim holds for ψ. If $M_i \models \psi(b, \bar{a})$, then so does M. Thus if $M_i \models \phi(\bar{a})$, then so does M. On the other hand if $M \models \psi(b, \bar{a})$, there is $j \geq i$ such that $b \in M_j$. By induction, $M_j \models \psi(b, \bar{a})$, so $M_j \models \phi(\bar{a})$. Because $M_i \prec M_j$, $M_i \models \phi(\bar{a})$, as desired.

2.4   Back and Forth

The "back-and-forth" method is a style of argument that we will encounter several times. In this section, we will examine several manifestations of the method, starting with Cantor's proof that any two countable dense linear orders are isomorphic and leading up to Scott's use of infinitary logic to characterize isomorphism of countable models.

Dense Linear Orders

Let $\mathcal{L} = \{<\}$ and let DLO be the theory of dense linear orders without endpoints. DLO is axiomatized by the axioms for linear orders plus the axioms

$$\forall x \forall y\ (x < y \rightarrow \exists z\, x < z < y)$$

and

$$\forall x \exists y \exists z\ y < x < z.$$

Theorem 2.4.1 *The theory DLO is $\aleph_0$-categorical and complete.*

Proof Let $(A, <)$ and $(B, <)$ be two countable models of DLO. Let $a_0, a_1, a_2, \ldots$ and $b_0, b_1, b_2, \ldots$ be one-to-one enumerations of A and B. We will build a sequence of partial bijections $f_i : A_i \to B_i$ where $A_i \subset A$ and $B_i \subset B$ are finite such that $f_0 \subseteq f_1 \subseteq \ldots$ and if $x, y \in A_i$ and $x < y$, then $f_i(x) < f_i(y)$. We call f_i a *partial embedding*. We will build these sequences such that $A = \bigcup A_i$ and $B = \bigcup B_i$. In this case, $f = \bigcup f_i$ is the desired isomorphism from $(A, <)$ to $(B, <)$.

At odd stages of the construction we will ensure that $\bigcup A_i = A$, and at even stages we will ensure that $\bigcup B_i = B$.

stage 0: Let $A_0 = B_0 = f_0 = \emptyset$.

stage $n + 1 = 2m + 1$: We will ensure that $a_m \in A_{n+1}$.

If $a_m \in A_n$, then let $A_{n+1} = A_n, B_{n+1} = B_n$ and $f_{n+1} = f_n$. Suppose that $a_m \notin A_n$. To add a_m to the domain of our partial embedding, we must find $b \in B \setminus B_n$ such that

$$\alpha < a_m \Leftrightarrow f_n(\alpha) < b$$

for all $\alpha \in A_n$. In other words, we must find $b \in B$, which is in the image under f_n of the cut of a_m in A_n. Exactly one of the following holds:

i) a_m is greater than every element of A_n, or

ii) a_m is less than every element of A_n, or

iii) there are α and $\beta \in A_n$ such that $\alpha < \beta$, $\gamma \leq \alpha$ or $\gamma \geq \beta$ for all $\gamma \in A_n$ and $\alpha < a_m < \beta$.

In case i) because B_n is finite and $B \models \text{DLO}$, we can find $b \in B$ greater than every element of B_n. Similarly in case ii) we can find $b \in B$ less than every element of B_n. In case iii), because f_n is a partial embedding, $f_n(\alpha) < f_n(\beta)$ and we can choose $b \in B \setminus B_n$ such that $f_n(\alpha) < b < f_n(\beta)$. Note that

$$\alpha < a_m \Leftrightarrow f_n(\alpha) < b$$

for all $\alpha \in A_n$.

In any case, we let $A_{n+1} = A_n \cup \{a_m\}$, $B_{n+1} = B_n \cup \{b\}$, and extend f_n to $f_{n+1} : A_{n+1} \to B_{n+1}$ by sending a_m to b. This concludes stage n.

stage $n + 1 = 2m + 2$: We will ensure that $b_m \in B_{n+1}$.

Again, if b_m is already in B_n, then we make no changes and let $A_{n+1} = A_n, B_{n+1} = B_n$ and $f_{n+1} = f_n$. Otherwise, we must find $a \in A$ such that the image of the cut of a in A_n is the cut of b_m in B_n. This is done as in the odd case.

Clearly, at odd stages we have ensured that $\bigcup A_n = A$ and at even stages we have ensured that $\bigcup B_n = B$. Because each f_n is a partial embedding, $f = \bigcup f_n$ is an isomorphism from A onto B.

Because there are no finite dense linear orders, Vaught's test implies that DLO is complete.

The proof of Theorem 2.4.1 is an example of a *back-and-forth* construction. At odd stages, we go forth trying to extend the domain, whereas at even stages we go back trying to extend the range. We give another example of this method.

The Random Graph

Let $\mathcal{L} = \{R\}$, where R is a binary relation symbol. We will consider an $\mathcal{L}$-theory containing the graph axioms $\forall x \; \neg R(x,x)$ and $\forall x \forall y \; R(x,y) \to R(y,x)$. Let ψ_n be the "extension axiom"

$$\forall x_1 \ldots \forall x_n \forall y_1 \ldots \forall y_n \left(\bigwedge_{i=1}^{n} \bigwedge_{j=1}^{n} x_i \neq y_j \to \exists z \bigwedge_{i=1}^{n} (R(x_i, z) \wedge \neg R(y_i, z)) \right).$$

We let T be the theory of graphs where we add $\{\exists x \exists y x \neq y\} \cup \{\psi_n : n = 1, 2, \ldots\}$ to the graph axioms. A model of T is a graph where for any finite disjoint sets X and Y we can find a vertex with edges going to every vertex in X and no vertex in Y.

Theorem 2.4.2 T *is satisfiable and* $\aleph_0$-*categorical. In particular,* T *is complete and decidable.*

Proof We first build a countable model of T. Let G_0 be any countable graph.

Claim There is a graph $G_1 \supset G_0$ such that G_1 is countable and if X and Y are disjoint finite subsets of G_0 then there is $z \in G_1$ such that $R(x, z)$ for $x \in X$ and $\neg R(y, z)$ for $y \in Y$.

Let the vertices of G_1 be the vertices of G_0 plus new vertices z_X for each finite $X \subseteq G_0$. The edges of G_1 are the edges of G together with new edges between x and z_X whenever $X \subseteq G_0$ is finite and $x \in X$. Clearly, G_1 is countable and has the desired property.

We iterate this construction to build a sequence of countable graphs $G_0 \subset G_1 \subset \ldots$ such that if X and Y are disjoint finite subsets of G_i, then there is $z \in G_{i+1}$ such that $R(x, z)$ for $x \in X$ and $\neg R(y, z)$ for $y \in Y$. Then, $G = \bigcup G_n$ is a countable model of T.

Next we show that T is $\aleph_0$-categorical. Let G_1 and G_2 be countable models of T. Let $a_0, a_1, \ldots$ list G_1, and let $b_0, b_1, \ldots$ list G_2. We will build a sequence of finite partial one-to-one maps $f_0 \subseteq f_1 \subseteq f_2 \subseteq \ldots$ such that for all x, y in the domain of f_s,

$$G_1 \models R(x,y) \quad \text{if and only if} \quad G_2 \models R(f_s(x), f_s(y)). \qquad (*)$$

Let $f_0 = \emptyset$.

stage $s + 1 = 2i + 1$: We make sure that a_i is in the domain.

If a_i is in the domain of f_s, let $f_{s+1} = f_s$. If not, let $\alpha_1, \ldots, \alpha_m$ list the domain of f_s and let $X = \{j \leq m : R(\alpha_j, a_i)\}$ and let $Y = \{j \leq m : \neg R(\alpha_j, a_i)\}$. Because $G_2 \models T$, we can find $b \in G_2$ such that $G_2 \models R(f_s(\alpha_j), b)$ for $j \in X$ and $G_2 \models \neg R(f_s(\alpha_j), b)$ for $j \in Y$. Let $f_{s+1} = f_s \cup \{(a_i, b)\}$. By choice of b and induction, f_{s+1} satisfies $(*)$.

stage $s + 1 = 2i + 2$: By a similar argument, we can ensure that f_{s+1} satisfies $(*)$ and b_i is in the image of f_{s+1}.

Let $f = \bigcup f_s$. We have ensured that f maps G_1 onto G_2. By $(*)$, f is a graph isomorphism. Thus, $G_1 \cong G_2$ and T is $\aleph_0$-categorical.

Because all models of T are infinite, T is complete. Because T is recursively axiomatized, T is decidable.

The theory T is very interesting because it gives us insights into random finite graphs. Let $\mathcal{G}_N$ be the set of all graphs with vertices $\{1, 2, \ldots, N\}$. We consider a probability measure on $\mathcal{G}_N$ where we make all graphs equally likely. This is the same as constructing a random graph where we independently decide whether there is an edge between i and j with probability $\frac{1}{2}$. For any $\mathcal{L}$-sentence ϕ,

$$p_N(\phi) = \frac{|\{G \in \mathcal{G}_N : G \models \phi\}|}{|\mathcal{G}_N|}$$

is the probability that a random element of $\mathcal{G}_N$ satisfies ϕ.

We argue that large graphs are likely to satisfy the extension axioms.

Lemma 2.4.3 $\lim\limits_{N \to \infty} p_N(\psi_n) = 1$ for $n = 1, 2, \ldots$.

Proof Fix n. Let G be a random graph in $\mathcal{G}_N$ where $N > 2n$. Fix $x_1, \ldots, x_n, y_1, \ldots, y_n, z \in G$ distinct. Let q be the probability that

$$\neg \left(\bigwedge_{i=1}^{n} (R(x_i, z) \wedge \neg R(y_i, z)) \right).$$

Then $q = 1 - 2^{-2n}$. Because these probabilities are independent, the probability that

$$G \models \neg \exists z \neg \left(\bigwedge_{i=1}^{n} (R(x_i, z) \wedge \neg R(y_i, z)) \right)$$

is q^{N-2n}. Let M be the number of pairs of disjoint subsets of G of size n. Thus

$$p_N(\neg \psi_n) \leq M q^{N-2n} < N^{2n} q^{N-2n}.$$

Because $q < 1$,

$$\lim_{N \to \infty} p_N(\neg \psi_n) = \lim_{N \to \infty} N^{2n} q^N = 0,$$

as desired.

We can now use the fact that T is complete to get a good understanding of the asymptotic properties of random graphs.

Theorem 2.4.4 (Zero-One Law for Graphs) *For any $\mathcal{L}$-sentence ϕ either $\lim\limits_{N \to \infty} p_N(\phi) = 0$ or $\lim\limits_{N \to \infty} p_N(\phi) = 1$. Moreover, T axiomatizes $\{\phi : \lim\limits_{N \to \infty} p_N(\phi) = 1\}$, the almost sure theory of graphs. The almost sure theory of graphs is decidable and complete.*

Proof If $T \models \phi$, then there is n such that if G is a graph and $G \models \psi_n$, then $G \models \phi$. Thus, $p_N(\phi) \geq p_N(\psi_n)$ and by Lemma 2.4.3, $\lim\limits_{N \to \infty} p_N(\phi) = 1$. On the other hand, if $T \not\models \phi$, then, because T is complete, $T \models \neg\phi$ and $\lim\limits_{N \to \infty} p_N(\neg\phi) = 1$ so $\lim\limits_{N \to \infty} p_N(\phi) = 0$.

Ehrenfeucht–Fraïssé Games

The type of back-and-forth constructions we did in Theorems 2.4.1 and 2.4.2 will appear several times in the book. It is useful to recast constructions as games. We will do this in a bit more generality. Let $\mathcal{L}$ be a language and $\mathcal{M} = (M, \ldots)$ and $\mathcal{N} = (N, \ldots)$ be two $\mathcal{L}$-structures with $M \cap N = \emptyset$. If $A \subseteq M$, $B \subseteq N$ and $f : A \to B$, we say that f is a *partial embedding* if $f \cup \{(c^{\mathcal{M}}, c^{\mathcal{N}}) : c$ a constant of $\mathcal{L}\}$ is a bijection preserving all relations and functions of $\mathcal{L}$.

We will define an infinite two-player game $G_\omega(\mathcal{M}, \mathcal{N})$. We will call the two players player I and player II; together they will build a partial embedding f from M to N. A play of the game will consist of ω stages. At the ith-stage, player I moves first and either plays $m_i \in M$, challenging player II to put m_i into the domain of f, or $n_i \in N$, challenging player II to put n_i into the range. If player I plays $m_i \in M$, then player II must play $n_i \in N$, whereas if player I plays $n_i \in M$, then player II must play $m_i \in M$. Player II wins the play of the game if $f = \{(m_i, n_i) : i = 1, 2, \ldots\}$ is the graph of a partial embedding.

A *strategy* for player II in $G_\omega(\mathcal{M}, \mathcal{N})$ is a function τ such that if player I's first n moves are $c_1, \ldots, c_n$, then player II's nth move will be $\tau(c_1, \ldots, c_n)$. We say that player II uses the strategy τ in the play of the game if the play looks like:

Player I	Player II
c_1	
	$\tau(c_1)$
c_2	
	$\tau(c_1, c_2)$
c_3	
	$\tau(c_1, c_2, c_3)$
$\vdots$	$\vdots$

We say that τ is a *winning strategy* for player II, if for any sequence of plays $c_1, c_2, \ldots$ player I makes, player II will win by following τ. We define strategies for player I analogously.

For example, suppose that $\mathcal{M}, \mathcal{N} \models$ DLO. Then, player II has a winning strategy. Suppose that up to stage n they have built a partial embedding $g : A \to B$. If player I plays $a \in M$, then player II plays $b \in N$ such that the cut b makes in B is the image of the cut of a in A under g. Similarly, if player I plays $b \in N$, player II plays $a \in M$ such that the cut of a in A is the image under g^{-1} of the cut of b in B. This can be done as in the proof of Theorem 2.4.1.

Proposition 2.4.5 *If $\mathcal{M}$ and $\mathcal{N}$ are countable, then the second player has a winning strategy in $G_\omega(\mathcal{M}, \mathcal{N})$ if and only if $\mathcal{M} \cong \mathcal{N}$.*

Proof If $\mathcal{M}$ and $\mathcal{N}$ are isomorphic, then player II can win by playing according to the isomorphism.

Suppose that player II has a winning strategy. Let $m_0, m_1, \ldots$ list M and $n_0, n_1, \ldots$ list N. Consider a play of the game where the second player uses the winning strategy and the first player plays $m_0, n_0, m_1, n_1, m_2, n_2, \ldots$. If f is the partial embedding built during this play of the game then the domain of f is M and the range of f is N. Thus, f is an isomorphism.

By weakening the game, we can, for suitable languages, give a characterization of elementary equivalence. Fix $\mathcal{L}$ a finite language with no function symbols, and let $\mathcal{M}$ and $\mathcal{N}$ be $\mathcal{L}$-structures. We define a game $G_n(\mathcal{M}, \mathcal{N})$ for $n = 1, 2, \ldots$. The game will have n rounds. On the ith round player I plays first and either plays $a_i \in M$ or $b_i \in N$. On player II's turn, if player I played $a_i \in M$, then player II must play $b_i \in N$, and if player I plays $b_i \in N$, then player II must play $a_i \in M$. The game stops after the nth round. Player II wins if $\{(a_i, b_i) : i = 1, \ldots, n\}$ is the graph of a partial embedding from $\mathcal{M}$ into $\mathcal{N}$. We call $G_n(\mathcal{M}, \mathcal{N})$ an *Ehrenfeucht–Fraïssé game*.

Our goal is to prove the following theorem.

Theorem 2.4.6 *Let $\mathcal{L}$ be a finite language without function symbols and let $\mathcal{M}$ and $\mathcal{N}$ be $\mathcal{L}$-structures. Then, $\mathcal{M} \equiv \mathcal{N}$ if and only if the second player has a winning strategy in $G_n(\mathcal{M}, \mathcal{N})$ for all n.*

Before proving this, we will need several lemmas.

Lemma 2.4.7 *One of the players has a winning strategy in $G_n(\mathcal{M}, \mathcal{N})$.*

Proof (sketch) This follows from Zermelo's theorem that in any two-person finite length game of perfect information without ties one of the players has a winning strategy (see [10] 1.7.1). It also follows from the determinacy of closed games (see [52]). We outline the proof. Suppose that player II does not have a winning strategy. Then, there is some move player I can make in round one so that player II has no move available to force a win. Player I makes that move. Now, whatever player II does, there is still a move that if made by player I means that player II cannot force a win. Player I makes that move and continues in this way. On the last round, there is still a move possible so that player II has no winning move. Player I makes that move and wins. This informally describes a winning strategy for player I (the strategy can be summarized as "avoid losing positions").

We inductively define depth(ϕ), the *quantifier depth* of an $\mathcal{L}$-formula ϕ, as follows:

depth(ϕ) = 0 if and only if ϕ is quantifier-free;

depth($\neg\phi$) = depth(ϕ);

depth($\phi \wedge \psi$) = depth($\phi \vee \psi$) = max{depth(ϕ), depth(ψ)};

depth($\exists v\, \phi$) = depth(ϕ) + 1.

We say that $\mathcal{M} \equiv_n \mathcal{N}$ if $\mathcal{M} \models \phi \Leftrightarrow \mathcal{N} \models \phi$ for all sentences of depth at most n. We will show that player II has a winning strategy in $G_n(\mathcal{M}, \mathcal{N})$ if and only if $\mathcal{M} \equiv_n \mathcal{N}$. We first argue that there are only finitely many inequivalent formulas of a fixed quantifier depth.

Lemma 2.4.8 *For each n and l, there is a finite list of formulas $\phi_1, \ldots, \phi_k$ of depth at most n in free variables $x_1, \ldots, x_l$ such that every formula of depth at most n in free variables $x_1, \ldots, x_l$ is equivalent to some ϕ_i.*

Proof We first prove this for quantifier-free formulas. Because $\mathcal{L}$ is finite and has no constant symbols, there are only finitely many atomic $\mathcal{L}$-formulas in free variables $x_1, \ldots, x_l$. Let $\sigma_1, \ldots, \sigma_s$ list all such formulas.

If ϕ is a Boolean combination of formulas $\tau_1, \ldots, \tau_s$, then there is S a collection of subsets of $\{1, \ldots, s\}$ such that

$$\models \phi \leftrightarrow \bigvee_{X \in S} \left(\bigwedge_{i \in X} \tau_i \wedge \bigwedge_{i \notin X} \neg\tau_i \right)$$

(see Exercise 1.4.1). This gives a list of 2^{2^s} formulas such that every Boolean combination of $\tau_1, \ldots, \tau_s$ is equivalent to a formula in this list. In particular, because quantifier free formulas are Boolean combinations of atomic formulas, there is a finite list of depth-zero formulas such that every depth-zero formula is equivalent to one in the list.

Because formulas of depth $n + 1$ are Boolean combinations of $\exists v\phi$ and $\forall v\phi$ where ϕ has depth at most n, the lemma follows by induction.

We can give a characterization of $\equiv_n$ using Ehrenfeucht–Fraïssé games. Theorem 2.4.6 will follow immediately.

Lemma 2.4.9 *Let $\mathcal{L}$ be a finite language without function symbols and $\mathcal{M}$ and $\mathcal{N}$ be $\mathcal{L}$-structures. The second player has a winning strategy in $G_n(\mathcal{M}, \mathcal{N})$ if and only if $\mathcal{M} \equiv_n \mathcal{N}$.*

Proof We prove this by induction on n.

Suppose that $\mathcal{M} \equiv_n \mathcal{N}$. Consider a play of the game where in round one player I plays $a \in M$. (The case where player I plays $b \in N$ is similar.) We claim that there is $b \in \mathcal{N}$ such that $\mathcal{M} \models \phi(a) \Leftrightarrow \mathcal{N} \models \phi(b)$ whenever depth(ϕ) $< n$. Let $\phi_0(v), \ldots, \phi_m(v)$ list, up to equivalence, all formulas of depth less than n. Let $X = \{i \leq m : \mathcal{M} \models \phi_i(a)\}$, and let $\Phi(v)$ be the formula

$$\bigwedge_{i \in X} \phi_i(v) \wedge \bigwedge_{i \notin X} \neg\phi_i(v).$$

Then, depth($\exists v\ \Phi(v)$) $\leq n$ and $\mathcal{M} \models \Phi(a)$; thus, there is $b \in N$ such that $\mathcal{N} \models \Phi(b)$. Player II plays b in round one.

If $n = 1$, the game has now concluded and $a \mapsto b$ is a partial embedding so player II wins. Suppose that $n > 1$.

Let $\mathcal{L}^* = \mathcal{L} \cup \{c\}$, where c is a new constant symbol. View $\mathcal{M}$ and $\mathcal{N}$ as $\mathcal{L}^*$-structures $(\mathcal{M}, a)$ and $(\mathcal{N}, b)$ where we interpret the new constant as a and b, respectively. Because

$$\mathcal{M} \models \phi(a) \Leftrightarrow \mathcal{N} \models \phi(b)$$

for $\phi(v)$ an $\mathcal{L}$-formula with depth$(\phi) < n$, $(\mathcal{M}, a) \equiv_{n-1} (\mathcal{N}, b)$. By induction, player II has a winning strategy in $G_{n-1}((\mathcal{M}, a), (\mathcal{N}, b))$. If player I's second play is d, player II responds as if d was player I's first play in $G_{n-1}((\mathcal{M}, a), (\mathcal{N}, b))$ and continues playing using this strategy, that is, in round i player I has plays $a, d_2, \ldots, d_i$, then player II plays $\tau(d_2, \ldots, d_i)$, where τ is his winning strategy in $G((\mathcal{M}, a), (\mathcal{N}, b))$. Let $f : X \to N$ be the function built by this play of the game. Because τ is a winning strategy, f^* is partial $\mathcal{L}^*$-embedding. Extend f^* to $f : X \cup \{a\} \to N$ by $f(a) = b$. Because f^* preserves $\mathcal{L}$-formulas with an additional constant denoting a in $\mathcal{M}$ and b in $\mathcal{N}$, f is a partial $\mathcal{L}$-embedding. Thus a winning strategy for player II can be summarized as: given player I's first play a, find b such that $(\mathcal{M}, a) \equiv_{n-1} (\mathcal{N}, b)$ and follow the winning strategy of $G_{n-1}((\mathcal{M}, a), (\mathcal{N}, b))$.

On the other hand, suppose that $\mathcal{M} \not\equiv_n \mathcal{N}$. Because formulas of depth at most n are Boolean combinations of formulas of the form $\exists v\ \phi(v)$ where depth$(\phi) < n$, $\mathcal{M}$ and $\mathcal{N}$ must disagree about a formula of this type. Without loss of generality, we may assume that $\mathcal{M} \models \exists v\ \phi(v)$ and $\mathcal{N} \models \forall v \neg \phi(v)$ where depth$(\phi) < n$. We claim that player I has a winning strategy. In round one player I plays $a \in M$ such that $\mathcal{M} \models \phi(a)$. Suppose that player II responds with $b \in N$. Let $(\mathcal{M}, a)$ and $(\mathcal{N}, b)$ be as above. Then $(\mathcal{M}, a) \not\equiv_{n-1} (\mathcal{N}, b)$ and, by induction, player I has a winning strategy in $G_{n-1}((\mathcal{M}, a), (\mathcal{N}, b))$. Player I continues playing as if just starting a game of $G_{n-1}((\mathcal{M}, a), (\mathcal{N}, b))$. The function f^* played starting at the second move will not be a partial $\mathcal{L}^*$-embedding so the whole function played is not a partial $\mathcal{L}$-embedding.

We give one application of Theorem 2.4.6. Let $\mathcal{L} = \{<\}$. Let T be the $\mathcal{L}$-theory that asserts $<$ is a linear order and $\forall x \exists y \exists z\ (y < x < z \land \forall w\ (w \leq y \lor w = x \lor w \geq z))$. T is the theory of discrete orderings with no top or bottom element.

Suppose that $\mathcal{N} \models T$. For $a, b \in N$ say aEb if b is the nth successor or predecessor of a for some natural number n. Then, E is an equivalence relation. Each E-class is a linear order that looks like $(\mathbb{Z}, <)$. If aEb, $\neg(aEc)$, and $a < c$, then $b < c$. Thus, the E-classes are linearly ordered and every model of T is of the form $(L \times \mathbb{Z}, <)$, where L is a linear order and $<$ is the lexicographic order on $L \times \mathbb{Z}$ (i.e., $(a, n) < (b, m)$ if $a < b$ or $a = b$ and $n < m$). Also, every linear order of this form is a model of T.

Proposition 2.4.10 *The theory of discrete linear orders with no top or bottom element is a complete theory. In particular,* $(\mathbb{Z}, <) \models \phi$ *if and only if* $T \models \phi$ *for all* $\mathcal{L}$*-sentences* ϕ.

Proof Let $\mathcal{M}$ be the ordered set of integers $(\mathbb{Z}, <)$, and let $\mathcal{N}$ be $L \times \mathbb{Z}$ with the lexicographic order where L is any linearly ordered set.

We claim that $\mathcal{M} \equiv \mathcal{N}$. We must show that player II has a winning strategy in $G_n(\mathcal{M}, \mathcal{N})$ for all n.

If $a, b \in \mathbb{Z}$, we define the distance between a and b to be $\text{dist}(a, b) = |b - a|$, and if $x = (i, a), y = (j, b) \in L \times \mathbb{Z}$, we define the distance to be $\text{dist}(x, y) = |b - a|$ if $i = j$ and $\text{dist}(a, b) = \infty$ if $i \neq j$. The problem for player II is that player I can play elements that are infinitely far apart in $\mathcal{N}$ and force player II to play elements that are finitely far apart in $\mathcal{M}$. Because player II knows how long the game will last, player II can play elements sufficiently far apart to avoid conflicts. Player II will try to ensure:

$(*)$ after m rounds of $G_n(\mathcal{M}, \mathcal{N})$, if $a_1 < a_2 < \ldots < a_l$ are the element of $\mathbb{N}$ that have been played, where $l \leq m$ (player I might—for no good reason—play an element more than once, so possibly $l < m$) and $b_1 < b_2 < \ldots < b_l$ are the elements of $\mathbb{Z}$ that have been played, then $a_i \mapsto b_i$ is a partial embedding corresponding to the play of the game, and if $\text{dist}(a_i, a_{i+1}) > 3^{n-m}$, then $\text{dist}(b_i, b_{i+1}) > 3^{n-m}$ and if $\text{dist}(a_i, a_{i+1}) \leq 3^{n-m}$ then $\text{dist}(a_i, a_{i+1}) = \text{dist}(b_i, b_{i+1})$ for $i = 1, \ldots, m - 1$.

By doing this, player II will win because after n rounds there will be a partial embedding.

We argue that player II can always choose a move to preserve $(*)$. In round 1, player II chooses an arbitrary element and $(*)$ holds. Suppose that we have played m rounds and $(*)$ holds. Let $a_1 < \ldots < a_l$ and $b_1 < \ldots < b_l$ be as above. Suppose that player I plays $b \in L \times \mathbb{Z}$. There are several cases to consider.

case 1: $b < b_1$.

If $\text{dist}(b, b_1) = k < \infty$, then player II plays $a = a_1 - k$. If $\text{dist}(b, b_1) = \infty$, then player II plays $a_1 - 3^n$. In either case, $(*)$ holds.

case 2: $b_i < b < b_{i+1}$ and $\text{dist}(b_i, b_{i+1}) \leq 3^{n-m}$.

Then $\text{dist}(a_i, a_{i+1}) = \text{dist}(b_i, b_{i+1})$. Play $a = a_i + \text{dist}(b, b_i)$. Then, $\text{dist}(a, a_{i+1}) = \text{dist}(b, b_{i+1})$, as desired.

case 3: $b_i < b < b_{i+1}$, $\text{dist}(b_i, b_{i+1}) > 3^{n-m}$, and $\text{dist}(b, b_i) < 3^{n-m-1}$.

In this case $\text{dist}(a_i, a_{i+1}) > 3^{n-m}$. Play $a = a_i + \text{dist}(b, b_i)$. Then, $\text{dist}(a, a_{i+1})$ and $\text{dist}(b, b_{i+1})$ are greater than 3^{n-m-1}, as desired.

case 4: $b_i < b < b_{i+1}$, $\text{dist}(b_i, b_{i+1}) > 3^{n-m}$, and $\text{dist}(b, b_{i+1}) < 3^{n-m-1}$.

As in case 3, let $a = a_{i+1} - \text{dist}(b, b_{i+1})$.

<u>case 5</u>: $b_i < b < b_{i+1}$, $\text{dist}(b_i, b_{i+1}) > 3^{n-m}$, $\text{dist}(b, b_i) > 3^{n-m-1}$, and $\text{dist}(b, b_{i+1}) > 3^{n-m-1}$.

In this case, $\text{dist}(a_i, a_{i+1})$ is greater than 3^{n-m}. Choose a such that $a_i < a < a_{i+1}$, and $\text{dist}(a_i, a)$ and $\text{dist}(a, a_{i+1})$ are both greater than 3^{n-m-1}. If player II plays a, then $(*)$ holds.

<u>case 6</u>: $b > b_l$.

Similar to case 1.

This explains the strategy if player I plays $b \in L \times \mathbb{Z}$. The case where player I plays $a \in \mathbb{Z}$ is analogous and left to the reader.

Scott–Karp Analysis

We give an extension of these ideas. Let $\mathcal{L}$ be an arbitrary language and $\mathcal{M}$ and $\mathcal{N}$ be $\mathcal{L}$-structures. For each ordinal α, we will have a relation $(\mathcal{M}, \bar{a}) \sim_\alpha (\mathcal{N}, \bar{b})$ where $\bar{a} \in M^n$ and $\bar{b} \in M^n$ and $n = 0, 1, 2, \ldots$.

$(\mathcal{M}, \bar{a}) \sim_0 (\mathcal{N}, \bar{b})$ if $\mathcal{M} \models \phi(\bar{a})$ if and only if $\mathcal{N} \models \phi(\bar{b})$ for all atomic $\mathcal{L}$-formulas ϕ.

For all ordinals α, $(\mathcal{M}, \bar{a}) \sim_{\alpha+1} (\mathcal{N}, \bar{b})$ if for all $c \in M$ there is $d \in N$ such that $(\mathcal{M}, \bar{a}, c) \sim_\alpha (\mathcal{N}, \bar{b}, d)$ and for all $d \in N$ there is $c \in M$ such that $(\mathcal{M}, \bar{a}, c) \sim_\alpha (\mathcal{N}, \bar{b}, d)$.

For all limit ordinals β, $(\mathcal{M}, \bar{a}) \sim_\beta (\mathcal{N}, \bar{b})$ if and only if $(\mathcal{M}, \bar{a}) \sim_\alpha (\mathcal{N}, \bar{b})$ for all $\alpha < \beta$. We will leave the following lemma for the exercises.

Lemma 2.4.11 *Let $\mathcal{L}$ be a finite language without function symbols. If $\bar{a} \in M^l, \bar{b} \in N^l$, then $(\mathcal{M}, \bar{a}) \sim_n (\mathcal{N}, \bar{b})$ if and only if player II has a winning strategy in $G_n((\mathcal{M}, \bar{a}), (\mathcal{N}, \bar{b}))$ if and only if $(\mathcal{M}, \bar{a}) \equiv_n (\mathcal{N}, \bar{b})$ for all $n = 1, 2, \ldots$. In particular, $\mathcal{M} \equiv \mathcal{N}$ if and only if $\mathcal{M} \sim_\omega \mathcal{N}$.*

In particular, note that even for finite languages without function symbols $\mathcal{M} \sim_\omega \mathcal{N}$ is much weaker than player II having a winning strategy in $G_\omega(\mathcal{M}, \mathcal{N})$.

We would like to prove a variant of Theorem 2.4.11 characterizing $\sim_\alpha$. To do this, we must leave first-order logic for infinitary languages.

Definition 2.4.12 Let $\mathcal{L}$ be a language and κ be an infinite cardinal. The formulas of the infinitary logic $\mathcal{L}_{\kappa,\omega}$ are defined inductively as follows:

i) Every atomic $\mathcal{L}$-formula is a formula of $\mathcal{L}_{\kappa,\omega}$.

ii) If X is a set of formulas of $\mathcal{L}_{\kappa,\omega}$ such that all of the free variables come from a fixed finite set and $|X| < \kappa$, then

$$\bigwedge_{\phi \in X} \phi \quad \text{and} \quad \bigvee_{\phi \in X} \phi$$

are formulas of $\mathcal{L}_{\kappa,\omega}$.

iii) If ϕ is a formula of $\mathcal{L}_{\kappa,\omega}$, then so are $\neg\phi$, $\forall v \, \phi$, and $\exists v \, \phi$.

We say that ϕ is a formula of $\mathcal{L}_{\infty,\omega}$ if it is an $\mathcal{L}_{\kappa,\omega}$-formula for some infinite cardinal κ. When $\kappa = \aleph_1$, it is traditional to write $\mathcal{L}_{\omega_1,\omega}$. Intuitively, $\mathcal{L}_{\omega_1,\omega}$ is the language where we allow countable conjunctions and countable disjunctions.

As in Definition 1.1.6, we can define satisfaction for formulas of $\mathcal{L}_{\infty,\omega}$. The only difference is that $\bigwedge_{\phi \in X} \phi$ is true if all of the $\phi \in X$ are true and $\bigvee_{\phi \in X} \phi$ is true if at least one of the formulas $\phi \in X$ is true.

If $\mathcal{L}$ is any first-order language and $\mathcal{M}$ is an $\mathcal{L}$-structure we define a sequence of $\mathcal{L}_{\infty,\omega}$-formulas $\phi^{\mathcal{M}}_{\bar{a},\alpha}(\bar{v})$, where $\bar{a} \in M^l$ and α is an ordinal as follows:

$$\phi^{\mathcal{M}}_{\bar{a},0}(\bar{v}) = \bigwedge_{\psi \in X} \psi(\bar{v}),$$

where $X = \{\psi : \mathcal{M} \models \psi(\bar{a})$ and ψ is atomic or the negation of an atomic $\mathcal{L}$-formula$\}$. If α is a limit ordinal, then

$$\phi^{\mathcal{M}}_{\bar{a},\alpha}(\bar{v}) = \bigwedge_{\beta < \alpha} \phi^{\mathcal{M}}_{\bar{a},\beta}(\bar{v}).$$

If $\alpha = \beta + 1$, then

$$\phi^{\mathcal{M}}_{\bar{a},\alpha}(\bar{v}) = \bigwedge_{b \in M} \exists w \, \phi^{\mathcal{M}}_{\bar{a}b,\beta}(\bar{v}, w) \wedge \forall w \bigvee_{b \in M} \phi^{\mathcal{M}}_{\bar{a}b,\beta}(\bar{v}, w).$$

Lemma 2.4.13 Let $\mathcal{M}$ and $\mathcal{N}$ be $\mathcal{L}$-structures, $\bar{a} \in M^l$, and $\bar{b} \in N^l$. Then, $(\mathcal{M}, \bar{a}) \sim_\alpha (\mathcal{N}, \bar{b})$ if and only if $\mathcal{N} \models \phi^{\mathcal{M}}_{\bar{a},\alpha}(\bar{b})$.

Proof We prove this by induction on α (see Appendix A). Because $(\mathcal{M}, \bar{a}) \sim_0 (\mathcal{N}, \bar{b})$ if and only if they satisfy the same atomic formulas, the lemma holds for $\alpha = 0$.

Suppose that γ is a limit ordinal and the lemma is true for all $\alpha < \gamma$. Then

$$
\begin{aligned}
(\mathcal{M}, \bar{a}) \sim_\gamma (\mathcal{N}, \bar{b}) \quad &\Leftrightarrow \quad (\mathcal{M}, \bar{a}) \sim_\alpha (\mathcal{N}, \bar{b}) \text{ for all } \alpha < \gamma \\
&\Leftrightarrow \quad \mathcal{N} \models \phi^{\mathcal{M}}_{\bar{a},\alpha}(\bar{b}) \text{ for all } \alpha < \gamma \\
&\Leftrightarrow \quad \mathcal{N} \models \phi^{\mathcal{M}}_{\bar{a},\gamma}(\bar{b}).
\end{aligned}
$$

Suppose that the lemma is true for α. First, suppose that $\mathcal{N} \models \phi^{\mathcal{M}}_{\bar{a},\alpha+1}(\bar{b})$. Let $c \in M$. Because

$$\mathcal{N} \models \bigwedge_{x \in M} \exists w \, \phi^{\mathcal{M}}_{\bar{a}x,\alpha}(\bar{b}, w),$$

there is $d \in N$ such that $\mathcal{N} \models \phi^{\mathcal{M}}_{\bar{a}c,\alpha}(\bar{b}, d)$. By induction, $(\mathcal{M}, \bar{a}, c) \sim_\alpha (\mathcal{N}, \bar{b}, d)$. If $d \in N$, then because

$$\mathcal{N} \models \forall w \bigvee_{c \in M} \phi^{\mathcal{M}}_{\bar{a}c,\alpha}(\bar{b}, w)$$

there is $c \in M$ such that $\mathcal{N} \models \phi^{\mathcal{M}}_{\bar{a}c,\alpha}(\bar{b},d)$ and $(\mathcal{M},\bar{a},c) \sim_{\alpha} (\mathcal{N},\bar{b},d)$. Thus $(\mathcal{M},\bar{a}) \sim_{\alpha+1} (\mathcal{N},\bar{b})$.

Suppose, on the other hand, that $(\mathcal{M},\bar{a}) \sim_{\alpha+1} (\mathcal{N},\bar{b})$. Suppose that $c \in M$, then there is $d \in N$ such that $(\mathcal{M},\bar{a},c) \sim_{\alpha} (\mathcal{N},\bar{b},d)$ and $\mathcal{N} \models \phi^{\mathcal{M}}_{\bar{a}c,\alpha}(\bar{b},d)$. Similarly, if $d \in N$, then there is $c \in M$ such that $\mathcal{N} \models \phi^{\mathcal{M}}_{\bar{a}c,\alpha}(\bar{b},d)$. Thus, $\mathcal{N} \models \phi^{\mathcal{M}}_{\bar{a},\alpha+1}(\bar{b})$, as desired.

Lemma 2.4.14 *For any infinite $\mathcal{L}$-structure $\mathcal{M}$, there is an ordinal $\alpha < |M|^{+}$ such that if $\bar{a},\bar{b} \in M^{l}$ and $(\mathcal{M},\bar{a}) \sim_{\alpha} (\mathcal{M},\bar{b})$, then $(\mathcal{M},\bar{a}) \sim_{\beta} (\mathcal{M},\bar{b})$ for all β. We call the least such α the Scott rank of $\mathcal{M}$.*

Proof Let $\Gamma_{\alpha} = \{(\bar{a},\bar{b}) : \bar{a},\bar{b} \in M^{l}$ for some $l = 0, 1, \ldots$ and $(\mathcal{M},\bar{a}) \not\sim_{\alpha} (\mathcal{M},\bar{b})\}$. Clearly, $\Gamma_{\alpha} \subseteq \Gamma_{\beta}$ for $\alpha < \beta$.

Claim 1 If $\Gamma_{\alpha} = \Gamma_{\alpha+1}$, then $\Gamma_{\alpha} = \Gamma_{\beta}$ for all $\beta > \alpha$.

We prove this by induction on β. If β is a limit ordinal and the claim holds for all $\gamma < \beta$, then it also holds for β. Suppose that the claim is true for $\beta > \alpha$ and we want to show that it holds for $\beta+1$. Suppose that $(\mathcal{M},\bar{a}) \sim_{\beta} (\mathcal{M},\bar{b})$ and $c \in M$. Because $(\mathcal{M},\bar{a}) \sim_{\alpha+1} (\mathcal{M},\bar{b})$, there is $d \in N$ such that $(\mathcal{M},\bar{a},c) \sim_{\alpha} (\mathcal{M},\bar{b},d)$. By our inductive assumption, $(\mathcal{M},\bar{a},c) \sim_{\beta} (\mathcal{M},\bar{b},d)$. Similarly, if $d \in M$, then $c \in M$ such that $(\mathcal{M},\bar{a},c) \sim_{\beta} (\mathcal{M},\bar{b},d)$. Thus, $(\mathcal{M},\bar{a}) \sim_{\beta+1} (\mathcal{M},\bar{b})$ as desired.

Claim 2 There is an ordinal $\alpha < |M|^{+}$ such that $\Gamma_{\alpha} = \Gamma_{\alpha+1}$.

Suppose not. Then, for each $\alpha < |M|^{+}$, choose $(\bar{a}_{\alpha},\bar{b}_{\alpha}) \in \Gamma_{\alpha+1} \setminus \Gamma_{\alpha}$. Because $\Gamma_{\alpha} \subseteq \Gamma_{\beta}$ for $\alpha < \beta$, the function $\alpha \mapsto (\bar{a}_{\alpha},\bar{b}_{\alpha})$ is one-to-one. Because there are only $|M|$ finite sequences from M this is impossible.

We conclude this section with Scott's Isomorphism Theorem that every countable $\mathcal{L}$-structure is described up to isomorphism by a single $\mathcal{L}_{\omega_1,\omega}$-sentence.

Let $\mathcal{M}$ be an infinite $\mathcal{L}$-structure of cardinality κ, and let α be the Scott rank of $\mathcal{M}$. Let $\Phi^{\mathcal{M}}$ be the sentence

$$\phi^{\mathcal{M}}_{\emptyset,\alpha} \wedge \bigwedge_{l=0}^{\infty} \bigwedge_{\bar{a} \in M^{l}} \forall \bar{v}(\phi^{\mathcal{M}}_{\bar{a},\alpha}(\bar{v}) \rightarrow \phi^{\mathcal{M}}_{\bar{a},\alpha+1}(\bar{v})).$$

Because all of the conjunctions and disjunctions in $\phi^{\mathcal{M}}_{\bar{a},\beta}$ are of size κ, $\phi^{\mathcal{M}}_{\bar{a},\beta} \in \mathcal{L}_{\kappa^{+},\omega}$ for all ordinals $\beta < \kappa^{+}$. Thus $\Phi^{\mathcal{M}}$ is an $\mathcal{L}_{\kappa^{+},\omega}$-sentence. We call $\Phi^{\mathcal{M}}$ the *Scott sentence* of $\mathcal{M}$. If $\mathcal{M}$ is countable, then $\Phi^{\mathcal{M}} \in \mathcal{L}_{\omega_1,\omega}$.

Theorem 2.4.15 (Scott's Isomorphism Theorem) *Let $\mathcal{M}$ be a countable $\mathcal{L}$-structure, and let $\Phi^{\mathcal{M}} \in \mathcal{L}_{\omega_1,\omega}$ be the Scott sentence of $\mathcal{M}$. Then, $\mathcal{N} \cong \mathcal{M}$ if and only if $\mathcal{N} \models \Phi^{\mathcal{M}}$.*

Proof Because α is the Scott rank of $\mathcal{M}$, $\mathcal{M} \models \Phi^{\mathcal{M}}$. An easy induction left to the exercises shows that if $\mathcal{N} \cong \mathcal{M}$, then $\mathcal{M}$ and $\mathcal{N}$ model the same $\mathcal{L}_{\infty,\omega}$-sentences.

On the other hand, suppose that $\mathcal{N}$ models $\Phi^{\mathcal{M}}$. We do a back-and-forth argument to build a sequence of finite partial embeddings $f_0 \subseteq f_1 \subseteq \dots$ from $\mathcal{M}$ to $\mathcal{N}$ such that if $\bar{a}$ is the domain of f_i, then

$$(\mathcal{M}, \bar{a}) \sim_\alpha (\mathcal{N}, f_i(\bar{a})). \tag{$*$}$$

Let $m_0, m_1, \dots$ list M and $n_0, \dots, n_1, \dots$ list N.

At stage 0, we let $f_0 = \emptyset$. Because $\mathcal{N} \models \phi_{\emptyset, \alpha}^{\mathcal{M}}$, $\mathcal{M} \sim_\alpha \mathcal{N}$ and ($*$) holds.

Suppose we are at stage $n + 1$. Let $\bar{a}$ be the domain of f_n. Because $(\mathcal{M}, \bar{a}) \sim_\alpha (\mathcal{N}, f(\bar{a}))$, $\mathcal{N} \models \phi_{\bar{a}, \alpha}^{\mathcal{M}}(f(\bar{a}))$. Because $\mathcal{N} \models \Phi^{\mathcal{M}}$, $\mathcal{N} \models \phi_{\bar{a}, \alpha+1}^{\mathcal{M}}(f(\bar{a}))$ and $(\mathcal{M}, \bar{a}) \sim_{\alpha+1} (\mathcal{N}, f(\bar{a}))$.

If $n + 1 = 2i + 1$, we want to ensure that m_i is in the domain of f_{n+1}. If m_i is in the domain of f_n, then $f_n = f_{n+1}$. If not, choose $b \in N$ such that $(\mathcal{M}, \bar{a}, m_i) \sim_\alpha (\mathcal{N}, f(\bar{a}), b)$ and extend f_n to f_{n+1} by sending m_i to b.

If $n = 2i + 2$, we want to ensure that n_i is in the image of f_{n+1}. If it is already in the image of f_n, let $f_{n+1} = f_n$. Otherwise, we can find $m \in M$ such that $(\mathcal{M}, \bar{a}, m) \sim_\alpha (\mathcal{N}, f(\bar{a}), n_i)$ and extend f_n to f_{n+1} by simply sending m to n_i.

2.5   Exercises and Remarks

Exercise 2.5.1 We say that an ordered group $(G, +, <)$ is *Archimedian* if for all $x, y \in G$ with $x, y > 0$ there is an integer m such that $|x| < m|y|$. Show that there are non-Archimedian fields elementarily equivalent to the field of real numbers.

Exercise 2.5.2 Suppose that T has arbitrarily large finite models. Show that T has an infinite model.

Exercise 2.5.3 Let $\mathcal{L}$ be the language with one binary relation symbol $<$. Let T be an $\mathcal{L}$-theory extending the theory of linear orders such that T has infinite models. Show that there is $\mathcal{M} \models T$ and an order-preserving embedding $\sigma : \mathbb{Q} \to \mathcal{M}$ of the rational numbers into M.

For example, if T is the full theory of the $(\mathbb{Z}, <)$, there is $\mathcal{M} \equiv (\mathbb{Z}, <)$ in which the rational order embeds.

Exercise 2.5.4 Show that every torsion-free Abelian group $(G, +)$ can be linearly ordered such that $(a < b \wedge c \leq d) \to a + c < b + d$. [Hint: First show this for finitely generated groups. Then use compactness.]

Exercise 2.5.5 Let $\mathcal{L} = \{E\}$ where E is a binary relation symbol. Let T be the $\mathcal{L}$-theory of an equivalence relation with infinitely many infinite classes.

a) Write axioms for T.

b) How many models of T are there of cardinality $\aleph_0$? $\aleph_1$? $\aleph_2$? $\aleph_{\omega_1}$?

c) Is T complete?

Exercise 2.5.6 (Skolem's Paradox) Let ZFC be the Zermelo–Frankel axioms for set theory with the Axiom of Choice. Show that there is a countable model $\mathcal{M}$ of ZFC. How do you explain the fact that $\mathcal{M} \models$ "there is an uncountable set"?[1]

Exercise 2.5.7 (Overspill) Let $\mathcal{M}$ be a nonstandard model of Peano arithmetic, $\phi(v, \overline{w})$ a formula in the language of arithmetic, and $\overline{a} \in M$. Suppose that $\mathcal{M} \models \phi(n, \overline{a})$ for all $n < \omega$. Then, there is an infinite $c \in M$ such that $\mathcal{M} \models \phi(c, \overline{a})$.

Exercise 2.5.8 Suppose that $\mathcal{M} \prec \mathcal{N}$ and $A \subseteq M$.
a) Show that the definable (algebraic) closure of A in $\mathcal{N}$ is equal to the definable (algebraic) closure of A in $\mathcal{M}$. (See Exercises 1.4.10 and 1.4.11.) Thus, algebraic closure and definable closure are preserved under elementary extension.
b) Give examples showing that this is not true if we only have $\mathcal{M} \equiv \mathcal{N}$ and $\mathcal{M} \subseteq \mathcal{N}$.

Exercise 2.5.9 Suppose that $\mathcal{M}_0 \subset \mathcal{M}_1 \subset \mathcal{M}_2$, $\mathcal{M}_0 \prec \mathcal{M}_2$, and $\mathcal{M}_1 \prec \mathcal{M}_2$. Show that $\mathcal{M}_0 \prec \mathcal{M}_1$.

Exercise 2.5.10 Let T be an $\mathcal{L}$-theory and $T_\forall$ be all of the universal sentences ϕ such that $T \models \phi$. Show that $\mathcal{A} \models T_\forall$ if and only if there is $\mathcal{M} \models T$ with $\mathcal{A} \subseteq \mathcal{M}$.

Exercise 2.5.11 (Amalgamation) Suppose that $\mathcal{M}_0, \mathcal{M}_1$ and $\mathcal{M}_2$ are $\mathcal{L}$-structures and $j_i : \mathcal{M}_0 \to \mathcal{M}_i$ is an elementary embedding for $i = 1, 2$. Show that there is an $\mathcal{L}$-structure $\mathcal{N}$ and elementary embeddings $f_i : \mathcal{M}_i \to \mathcal{N}$ such that $f_1 \circ j_1 = f_2 \circ j_2$.

Exercise 2.5.12 Show that the following are equivalent.
i) There is a universal formula $\psi(\overline{v})$ such that $T \models \forall \overline{v}(\phi(\overline{v}) \leftrightarrow \psi(\overline{v}))$.
ii) If $\mathcal{M}$ and $\mathcal{N}$ are models of T with $\mathcal{M} \subset \mathcal{N}$, $\overline{a} \in M$, and $\mathcal{N} \models \phi(\overline{a})$, then $\mathcal{M} \models \phi(\overline{a})$.

Exercise 2.5.13 Let $\mathcal{L} = \{s\}$, where s is a unary function symbol. Let T be the $\mathcal{L}$-theory that asserts that s is a bijection with no cycles (i.e., $s^{(n)}(x) \neq x$ for $n = 1, 2, \ldots$). For what cardinals κ is T κ-categorical?

Exercise 2.5.14 Let T be the theory of Abelian groups where every element has order 2. Show that T is κ-categorical for all infinite cardinals κ but not complete. Find $T' \supset T$ a complete theory with the same infinite models as T.

Exercise 2.5.15 We say that T has a $\forall\exists$-axiomatization if it can be axiomatized by sentences of the form $\forall v_1 \ldots \forall v_n \exists w_1 \ldots \exists w_m \, \phi(\overline{v}, \overline{w})$ where ϕ is a quantifier-free formula.

---

[1] Some philosophers have found Skolem's Paradox very interesting (see [88]).

a) Suppose that T has a $\forall\exists$-axiomatization, $(I, <)$ is a linear order, and $(\mathcal{M}_i : i \in I)$ is a chain of models of T. Show that $\bigcup \mathcal{M}_i$ is a model of T.

We will show that the converse also holds. Suppose that whenever $(\mathcal{M}_i : i \in I)$ is a chain of models of T, then $\bigcup \mathcal{M}_i \models T$. Let $\Gamma = \{\phi : \phi$ is a $\forall\exists$-sentence and $T \models \phi\}$. Let $\mathcal{M} \models \Gamma$. We will show that $\mathcal{M} \models T$.

b) Show that there is $\mathcal{N} \models T$ such that if ψ is an $\exists\forall$-sentence and $\mathcal{M} \models \psi$, then $\mathcal{N} \models \psi$.

c) Show that there is $\mathcal{N}' \supseteq \mathcal{M}$ with $\mathcal{N}' \equiv \mathcal{N}$.

d) Show that there is $\mathcal{M}' \supseteq \mathcal{N}'$ such that $\mathcal{M} \prec \mathcal{M}'$.

e) Iterate the constructions from c) and d) to build a chain of structures

$$\mathcal{M} = \mathcal{M}_0 \subseteq \mathcal{N}_1 \subseteq \mathcal{M}_1 \subseteq \mathcal{N}_2 \ldots$$

such that $\mathcal{M}_i \prec \mathcal{M}_{i+1}$ for $i = 0, 1, \ldots$ and each $\mathcal{N}_i \prec \mathcal{N}_{i+1}$. Let $\mathcal{M}^* = \bigcup \mathcal{M}_i = \bigcup \mathcal{N}_i$. Show that $\mathcal{M}^* \models T$ and $\mathcal{M} \prec \mathcal{M}^*$.

f) Conclude that T is $\forall\exists$-axiomatizable.

Exercise 2.5.16 (Finitely Axiomatizable $\aleph_1$-categorical Theory) Let $\mathcal{L} = \{U, V, E, R, s, \pi_1, \pi_2, \pi_3, s_1, s_2, s_3\}$, where U, V, E, and R are unary predicates and s, π_i, and s_i are unary function symbols for $i = 1, 2, 3$.

We consider the structure $\mathcal{M}$, where $M = U \cup V$, $U = \mathbb{Z}$, $V = \mathbb{Z}^3 \times \{0, 1\}$, $E \subset U$ is the even integers, $s : U \to U$ by $s(x) = x + 1$, $\pi_i : V \to U$ by $\pi_i(x_1, x_2, x_3, x_4) = x_i$, $s_i : V \to U$ by

$$s_i(x_1, x_2, x_3, x_4) = \begin{cases} (x_1 + 1, x_2, x_3, x_4) & i = 1 \\ (x_1, x_2 + 1, x_3, x_4) & i = 2 \\ (x_1, x_2, x_3 + 1, x_4) & i = 3 \end{cases},$$

and if $D = \{(x, y, z, w) \in V : x = y \lor y = z \lor z = w\}$, then $R = \{(x, y, z, w) \in D : (x \leq y \lor y = z) \leftrightarrow w = 0\}$.

Let T be the following $\mathcal{L}$-theory:

i) $\forall x \; U(x) \leftrightarrow \neg V(x)$;

ii) $E(x) \to U(x)$;

iii) $s(x) = x$ for $x \in V$ and $\pi_i(x) = s_i(x) = x$ for $x \in U$;

iv) $\pi_i : V \to U$ and if $x_1, x_2, x_3 \in U$, there are exactly two $y \in V$ such that $\pi_i(y) = x_i$ for $i = 1, 2, 3$;

v) s is a bijection between E and $U \backslash E$, $\pi_i(s_i(y)) = s(\pi_i(y))$, $\pi_i(s_j(y))\pi_i(y)$ for $i = 1, 2, 3$ and $j \neq i$;

vi) $s_i s_j(x) = s_j s_i(x)$ for $1 \leq i, j \leq 3$;

vii) Let $D = \{y \in V : \pi_i(x) = \pi_j(y) \text{ for some } i \neq j\}$. Then $R(y) \to y \in D$ and for all $x \in D$ there is a unique $y \in V$ such that $R(y)$ and $\pi_i(x) = \pi_i(y)$ for $i = 1, 2, 3$;

viii) $\pi_1(x) = \pi_2(x) \to [(R(x) \leftrightarrow R(s_3(x))) \land (R(x) \leftrightarrow R(s_1 s_2(x)))]$ for $x \in V$;

ix) $\pi_2(x) = \pi_3(x) \to [(R(x) \leftrightarrow R(s_1(x))) \land (R(x) \leftrightarrow R(s_2 s_3(x)))]$ for $x \in V$;

x) $\pi_1(x) = \pi_3(x) \rightarrow [(R(x) \leftrightarrow (R(s_2(x)) \leftrightarrow \pi_1(x) \neq s(\pi_2(x))) \wedge (R(x) \leftrightarrow (R(s_1 s_3(x)) \leftrightarrow \pi_1(x) \neq \pi_2(x)))]$.

a) Show that $\mathcal{M} \models T$.

Suppose that $\mathcal{N} \models T$, $x \in U^{\mathcal{N}}$, and $n > 0$. We want to show that $s^n(x) \neq x$. Suppose, for purposes of contradiction, that $s^n(x) = x$.

b) Show that n is even and there is $y \in R$ such that $\pi_i(y) = x$ for $i = 1, 2, 3$.

c) Let $\sigma = s_1 s_3 s_2^{-1}$ and $\tau = s_1 s_2^{-1} s_3^{-1}$. Show that $\neg R(\sigma^i(y))$ for $1 \leq i \leq \frac{n}{2}$, $R(s_3^n(y))$, and $R(\tau^{\frac{n}{2}}(s_3^n(y)))$. Use the fact that $\sigma^{\frac{n}{2}} = \tau^{\frac{n}{2}} s_3^n$ to derive a contradiction.

d) Argue that $(U^{\mathcal{N}}, s^{\mathcal{N}} | U^{\mathcal{N}}) \equiv (\mathbb{Z}, s)$.

e) Show that T is κ-categorical for all uncountable κ but not $\aleph_0$-categorical.

Exercise 2.5.17 We say that $\mathcal{M} \models T$ is *existentially closed* if whenever $\mathcal{N} \models T, \mathcal{N} \supseteq \mathcal{M}$, and $\mathcal{N} \models \exists \overline{v} \, \phi(\overline{v}, \overline{a})$, where $\overline{a} \in M$ and ϕ is quantifier-free, then $\mathcal{M} \models \exists \overline{v} \, \phi(\overline{v}, \overline{a})$.

a) Show that if T is $\forall \exists$-axiomatizable, then T has an existentially closed model. Indeed, if $\mathcal{M} \models T$, there is $\mathcal{N} \supseteq \mathcal{M}$ existentially closed with $|N| = |M| + |\mathcal{L}| + \aleph_0$.

b) Suppose that T has an infinite nonexistentially closed model. Prove that T has a nonexistentially closed model of cardinal κ for any infinite cardinal $\kappa \geq |\mathcal{L}|$. [Hint: Suppose that $\mathcal{M} \subset \mathcal{N}$ are models of T and $\mathcal{N}$ satisfies an existential formula not satisfied in $\mathcal{M}$. Consider models of the theory of $\mathcal{N}$ where we add a unary predicate for $\mathcal{M}$.]

c) Suppose that T is κ-categorical for some infinite $\kappa \geq |\mathcal{L}|$ and axiomatized by $\forall \exists$-sentences. Prove that all models of T are existentially closed. Conclude that every algebraically closed field is existentially closed.

Exercise 2.5.18 (Ultrafilters) Let I be a set and let $\mathcal{P}(I) = \{X : X \subset I\}$ denote the power set of I. A *filter* on I is a collection $D \subset \mathcal{P}(I)$ such that:

i) $I \in D, \emptyset \notin D$;

ii) if $A, B \in D$, then $A \cap B \in D$;

iii) if $A \in D$ and $A \subseteq B \subseteq I$, then $B \in D$.

Intuitively a filter is a collection of "big" subsets of I.

a) Suppose that $I = \mathbb{R}$. Show that $D = \{X \subseteq \mathbb{R} : \mathbb{R} \setminus X$ has Lebesgue measure zero$\}$ is a filter.

b) Let κ be an infinite cardinal with $\kappa \leq |I|$. Show that $D = \{X \subseteq I : |I \setminus X| < \kappa\}$ is a filter. If $\kappa = \aleph_0$, we call D the *Frechet filter*.

c) Show that for $x \in I$, the *principal filter* $D = \{X \subseteq I : x \in X\}$ is a filter on I.

d) Suppose that D is a filter on I and $X \notin D$. Let $E = \{Y \subseteq I : Z \setminus X \subseteq Y$ for some $Z \in D\}$. Show that E is a filter, $D \subset E$, and $I \setminus X \in E$.

We say that a filter D on I is an *ultrafilter* if $X \in D$ or $I \setminus X \in D$ for all $X \subseteq I$. We can think of an ultrafilter as a finitely additive two-valued measure on the subsets of I.

e) Show that every principal filter is an ultrafilter.

f) Show that for all filters D on I there is an ultrafilter U on I with $D \subseteq U$. [Hint: Suppose that D is a filter on I. Let $\mathcal{I} = \{F \subseteq P(I) : F \supseteq D$ and F is a filter$\}$. Use Zorn's Lemma to show that there is a maximal element of $\mathcal{I}$. Use d) to show that a maximal filter is an ultrafilter.] In particular if D is the Frechet filter on I, then U is a nonprincipal ultrafilter.

Exercise 2.5.19 (Ultraproducts) Let $\mathcal{L}$ be a language and suppose that I is an infinite set. Suppose that $\mathcal{M}_i$ is an $\mathcal{L}$-structure for each $i \in I$. Let D be an ultrafilter on I. We define a new structure $\mathcal{M} = \prod \mathcal{M}_i / D$, which we call the *ultraproduct* of the $\mathcal{M}_i$ using D. Define a relation $\sim$ on

$$X = \prod_{i \in I} M_i = \left\{ f : I \to \bigcup_{i \in I} M_i : f(i) \in M_i \text{ for all } i \right\}$$

by $f \sim g$ if and only if $\{i \in I : f(i) = g(i)\} \in D$.

a) Show that $\sim$ is an equivalence relation.

The universe of $\mathcal{M}$ will be $M = X / \sim$, the $\sim$ equivalence classes. We must show how to view $\mathcal{M}$ as an $\mathcal{L}$-structure. If c is a constant symbol of $\mathcal{L}$, let $c^{\mathcal{M}}$ be the $\sim$ class of $f_c \in X$ where $f_c(i) = c^{\mathcal{M}_i}$ for all $i \in I$.

Let f be an n-ary function symbol of $\mathcal{L}$.

b) Suppose that $g_1, \ldots, g_n, h_1, \ldots, h_n \in X$, and $g_i \sim h_i$ for $i = 1, \ldots, n$. Define $g_{n+1}(i) = f^{\mathcal{M}_i}(g_1(i), \ldots, g_n(i))$ and $h_{n+1}(i) = f^{\mathcal{M}_i}(h_1(i), \ldots, h_n(i))$ for $i \in I$. Show that $g_{n+1} \sim h_{n+1}$. Thus $f^{\mathcal{M}}(g_1/\sim, \ldots, g_n/\sim) = g_{n+1}/\sim$ determines a well-defined function on $\mathcal{M}$.

c) Suppose that R is a relation symbol of $\mathcal{L}$ and $g_1, \ldots, g_n, h_1, \ldots, h_n$ are as above. Show that $\{i \in I : (g_1(i), \ldots, g_n(i)) \in R^{\mathcal{M}_i}\} \in D$ if and only if $\{i \in I : (h_1(i), \ldots, h_n(i)) \in R^{\mathcal{M}_i}\} \in D$.

Thus, we can interpret

$$R^{\mathcal{M}} = \{(g_1/\sim, \ldots, g_m/\sim) : \{i \in I : (g_1(i), \ldots, g_m(i)) \in R^{\mathcal{M}_i}\} \in D\}.$$

d) (Łoś's Theorem) Let $\phi(v_1, \ldots, v_n)$ be an $\mathcal{L}$-formula. Then, $\mathcal{M} \models \phi(g_1/\sim, \ldots, g_n/\sim)$ if and only if $\{i \in I : \mathcal{M}_i \models \phi(g_1(i), \ldots, g_n(i))\} \in D$.

e) What goes wrong in the proof of d) if D is a filter but not an ultrafilter?

Exercise 2.5.20 (Ultraproduct and the Compactness Theorem) We show how ultraproducts can be used to give a different proof of the Compactness Theorem. Suppose that T is a finitely satisfiable theory. If T is finite, then T is satisfiable, so we may assume that T is infinite. Let $I = \{\Delta \subseteq T : \Delta$ is finite$\}$.

a) For $\phi \in T$, let $X_\phi = \{\Delta \in I : \phi \in \Delta\}$. Let

$$D = \{Y \subseteq I : X_\phi \subseteq Y \text{ for some } \phi \in T\}.$$

Show that D is a filter on I.

b) For $\Delta \in I$, let $\mathcal{M}_\Delta \models \Delta$. Let U be an ultrafilter on I with $U \supseteq D$. Show that $\prod_{\Delta \in I} M_\Delta / U \models T$.

Exercise 2.5.21 [†] For each prime p, let $\mathbb{F}_p$ be the field with p elements. Let D be a nonprincipal ultrafilter on the set of primes, and let $K = \prod \mathbb{F}_p / D$.

i) What can you say about the characteristic of K?

ii) Show that K has a unique algebraic extension of each degree.

iii) Let $a, b \in K \setminus \{0\}$. Show that there are infinitely many points (x, y) in K^2 such that $y^2 = x^3 + x$. [Hint: The equation defines an elliptic curve. Hasse showed that if E is an elliptic curve defined over a finite field $\mathbb{F}_q$ and N_q is the number of points on E with coordinates in $\mathbb{F}_q$, then $|N_q - q| \leq 2\sqrt{q}$ (see [95]).]

Exercise 2.5.22 Let $\mathcal{M}$ be a fixed $\mathcal{L}$ structure, and let $\mathcal{M}_i = \mathcal{M}$ for every $i \in \omega$. Let D be a nonprincipal ultrafilter on ω. Let $\mathcal{M}^* = \prod \mathcal{M}_i / D$. We call $\mathcal{M}^*$ an *ultrapower* of $\mathcal{M}$. Let $d : M \to M^*$ by setting $d(m)$ equal to the $\sim$ class of the constant function $i \mapsto m$. Show that d is an elementary embedding and d is surjective if and only if $\mathcal{M}$ is finite.

Exercise 2.5.23 (Effective Henkin Constructions) Let $\mathcal{L}$ be a recursive language. Consider an $\mathcal{L}$-structure $\mathcal{M}$ with underlying set $\mathbb{N}$. We say that $\mathcal{M}$ is *decidable* if $\{(\phi(n_1, \ldots, n_m)) : \overline{n} \in \mathbb{N}$ and $\mathcal{M} \models \phi(n_1, \ldots, n_m)$, ϕ and $\mathcal{L}$-formula$\}$ is recursive. Show that if T is a complete, recursive $\mathcal{L}$-theory, then T has a decidable model. [Sketch: Let $\mathcal{L}^* = \mathcal{L} \cup \{c_i : i = 0, 1, \ldots\}$, where the c_i are new constant symbols. We do a recursive Henkin construction. Let $\phi_0, \phi_1, \ldots$ list all $\mathcal{L}^*$-sentences, and let $\psi_0, \psi_2, \ldots$ list all $\mathcal{L}^*$-sentences with one free variable. At any stage s of the construction, we will have a sentence θ_s such that $T \cup \{\theta_s : s = 0, 1, 2, \ldots\}$ is a complete satisfiable theory with the witnessing property. Let $\theta_0 = \forall x \; x = x$. At stage $s = 2m$, if $T \cup \{\theta_s, \phi_m\}$ is satisfiable, let $\theta_{s+1} = \theta_s \wedge \phi_m$; otherwise, let $\theta_{s+1} = \theta_s \wedge \neg \phi_m$. Show that we can make this decision recursively and that $T \cup \{\theta_{s+1}\}$ is satisfiable. At stage $s = 2m + 1$, let i be least such that the constant c_i does not occur in $\theta_{s+1} = \theta_s \wedge (\exists v \; \psi(v)) \to \psi(c_i)$. Show that $T \cup \{\theta_{s+1}\}$ is satisfiable. Argue that $T^* = T \cup \{\theta_s : s = 0, 1, \ldots\}$ is a satisfiable complete decidable theory with the witness property. Build $\mathcal{M}$ as in Lemma 2.1.7. Let $\sigma : \mathbb{N} \to M$ by $\sigma(i) = c_j / \sim$, where j is least such that $|\{c_0 / \sim, \ldots, c_j / \sim\}| = i + 1$. Use σ to make $\mathbb{N}$ into an $\mathcal{L}^*$-structure $\mathcal{N}$ so that σ is an isomorphism. Show that $\mathcal{N}$ is decidable.]

Exercise 2.5.24 We can also view a countable Henkin construction as a forcing construction (see Appendix A). Let $\mathcal{L}, \mathcal{L}^*$, and T be as in the previous exercise. Let $P = \{\Sigma : \Sigma$ is a finite set of $\mathcal{L}^*$-formulas and $T \cup \Sigma$ is satisfiable$\}$. We order P by $\Sigma < \Delta$ if and only if $\Delta \subset \Sigma$. For each $\mathcal{L}^*$-sentence ϕ, let $D_\phi = \{\Sigma \in P : \phi \in \Sigma$ or $\neg\phi \in \Sigma\}$. For each $\mathcal{L}^*$-formula

$\psi(v)$ in one free variable v, let $E_\psi = \{\Sigma \in P : \neg \exists v \ \psi(v) \in \Sigma$ or $\psi(c) \in \Sigma$ for some constant $c\}$.

a) Show that each D_ϕ and E_ψ is dense.

b) Let $\mathcal{D}$ be the collection of all D_ϕ and E_ψ. If $G \subseteq P$ is a $\mathcal{D}$-generic filter, then $T^* = \bigcup_{\Sigma \in G} \Sigma$ is a satisfiable complete theory with the witness property.

Exercise 2.5.25 We say that a theory T has *definable Skolem functions* if for any formula $\phi(\overline{w}, v)$ there is a formula $\psi(\overline{w}, v)$ such that

$T \models \forall \overline{w} \exists v \ \psi(\overline{w}, v)$,

$T \models \forall \overline{w} \forall u \forall v \ ((\psi(\overline{w}, u) \wedge \psi(\overline{w}, v)) \rightarrow u = v)$,

$T \models \forall \overline{w} \ (\exists v \phi(\overline{w}, v) \rightarrow \exists u \ (\psi(\overline{w}, u) \wedge \phi(\overline{w}, u)))$.

In other words, for each formula $\phi(\overline{w}, v)$ there is a function f with a definable graph that is a Skolem function for ϕ.

a) Show that Peano arithmetic has definable Skolem functions.

b) Show that if T has definable Skolem functions, then there is an expansion by definitions of T with built-in Skolem functions.

Exercise 2.5.26 Suppose that T has built-in Skolem functions. Show that T has a universal axiomatization.

Exercise 2.5.27 If $(I, <)$ is a linear order and $(A_i, <)$ is a linear order for $i \in I$, we may linearly order $\{(i, x) : i \in I, x \in A_i\}$ by $(i, x) < (j, y)$ if and only if $i < j$ or $i = j$ and $x < y$. We call this order $\sum_{i \in I} A_i$.

Let κ be an infinite cardinal. Let A be the linear order $\mathbb{Q} + 2 + \mathbb{Q}$ (that is, a copy of the rationals, followed by two discrete points, followed by a copy of the rationals), and let B be the linear order $\mathbb{Q} + 3 + \mathbb{Q}$.

a) Let $X \subseteq \kappa$. For $\alpha < \kappa$, let

$$C_\alpha = \begin{cases} A & \text{if } \alpha \in X \\ B & \text{if } \alpha \notin X \end{cases},$$

and let L_X be the linear order $\sum_{\alpha < \kappa} C_\alpha$. Show that if $X \neq Y$, then $L_X \not\cong L_Y$. Conclude that there are 2^κ nonisomorphic linear orders of cardinality κ.

b)[†] Show that if $\kappa \geq \aleph_1$, then there are 2^κ nonisomorphic dense linear orders of cardinality κ.

Exercise 2.5.28 Let $\mathcal{L}_3 = \{<, c_0, c_1, \ldots\}$, where $c_0, c_1, \ldots$ are constant symbols. Let T_3 be the theory of dense linear orders with sentences added asserting $c_0 < c_1 < \ldots$.

a) Show that T_3 has exactly three countable models up to isomorphism. [Hint: Consider the questions: Does $c_0, c_1, c_2, \ldots$ have an upper bound? A least upper bound?]

b) Prove the following two general results and use them to prove that T_3 is complete.

i) For any language $\mathcal{L}$, two $\mathcal{L}$-structures $\mathcal{M}$ and $\mathcal{N}$ are elementarily equivalent if and only if they are elementarily equivalent for every finite sublanguage.

ii) If $\mathcal{L}$ is countable, T is an $\mathcal{L}$-theory with no finite models, and any two countable models of T are elementarily equivalent, then T is complete.

c) Let $\mathcal{L}_4 = \mathcal{L}_3 \cup \{P\}$, where P is a unary predicate. Let T_4 be T_3 with the added sentences

$$\forall x \forall y (x < y \rightarrow \exists z \exists w (x < z < y \wedge x < w < y \wedge P(z) \wedge \neg P(w))).$$

In other words, P is a dense-codense subset. Show that T_4 is a complete theory with exactly four countable models.

d) Generalize c) to give examples of complete theories with exactly n countable models for $n = 5, 6, \ldots$.

Exercise 2.5.29 [†] a) Show that all ordinals $\alpha, \beta < \omega^\omega 2$, $\alpha \not\equiv \beta$. [Here, ω^ω is the ordinal $\sup\{\omega, \omega^2, \ldots\}$ (see Appendix A).]

b) Show that for all ordinals α there is an ordinal $\beta < \omega^\omega 2$ such that $\alpha \equiv \beta$.

Exercise 2.5.30 Prove Lemma 2.4.11.

Exercise 2.5.31 Show that the Compactness Theorem fails for $\mathcal{L}_{\omega_1,\omega}$.

Exercise 2.5.32 Show that if $\mathcal{M} \cong \mathcal{N}$, then $\mathcal{M} \equiv_{\infty,\omega} \mathcal{N}$.

Exercise 2.5.33 The definition of quantifier depth extends to formulas of $\mathcal{L}_{\infty,\omega}$ by defining

$$\text{depth}\left(\bigwedge \phi_i\right) = \text{depth}\left(\bigwedge \psi_i\right) = \sup \text{depth}(\phi_i).$$

We say that $\mathcal{M} \equiv^\alpha_{\infty,\omega}$ if $\mathcal{M} \models \phi$ if and only if $\mathcal{N} \models \phi$ for all $\mathcal{L}_{\infty,\omega}$-sentences of quantifier depth at most α.

a) Show that $(\mathcal{M}, \bar{a}) \equiv^\alpha_{\infty,\omega} (\mathcal{N}, \bar{b})$ if and only if $(\mathcal{M}, \bar{a}) \sim_\alpha (\mathcal{N}, \bar{b})$.

b) Let $\Phi^{\mathcal{M}}$ be the Scott sentence of $\mathcal{M}$. Show that if $\mathcal{N} \models \Phi^{\mathcal{M}}$, then $\mathcal{N} \equiv_{\infty,\omega} \mathcal{M}$.

Exercise 2.5.34 If $\mathcal{M}$ and $\mathcal{N}$ are $\mathcal{L}$-structures, then a *back-and-forth system* for $\mathcal{M}$ and $\mathcal{N}$ is a family F of partial embeddings from $\mathcal{M}$ into $\mathcal{N}$ with finite domain such that:

i) for all $f \in F$ and $a \in M$, there is $g \in F$ such that $g \supseteq f$ and a is in the domain of g, and

ii) for all $f \in F$ and $b \in N$, there is $g \in F$ such that $g \supseteq f$ and b is in the image of g.

a) Show that if $\mathcal{M}$ and $\mathcal{N}$ are countable and there is a back-and-forth system, then $\mathcal{M} \cong \mathcal{N}$.

b) Suppose that F is a back-and-forth system between $\mathcal{M}$ and $\mathcal{N}$, $f \in F$, and $\bar{a}$ is the domain of F. Show that $(\mathcal{M}, \bar{a}) \sim_\alpha (\mathcal{N}, f(\bar{a}))$ for all α.

c) Show that there is a back-and-forth system between $\mathcal{M}$ and $\mathcal{N}$ if and only if $\mathcal{M} \equiv_{\infty,\omega} \mathcal{N}$.

d)† Show that $\mathcal{M} \equiv_{\infty,\omega} \mathcal{N}$ if and only if there is a forcing extension of the universe in which $\mathcal{M}$ is isomorphic to $\mathcal{N}$.

Exercise 2.5.35 † Let $\mathcal{L} = \{R\}$, where R is a binary relation. If $f \in 2^{\mathbb{N}\times\mathbb{N}}$, let $\mathcal{M}_f$ be the $\mathcal{L}$-structure with universe $\mathbb{N}$ and $(x,y) \in R^{\mathcal{M}_f}$ if and only if $f(x,y) = 1$. We topologize $2^{\mathbb{N}\times\mathbb{N}}$ by taking subbasic open sets $\{f : f(m,n) = i\}$ where $m, n \in \mathbb{N}$ and $i \in \{0,1\}$. If σ is a permutation of $\mathbb{N}$ we view σ as acting on $2^{\mathbb{N}\times\mathbb{N}}$ by letting $\sigma(f)(i,j) = f(\sigma(i),\sigma(j))$. We say that $X \subseteq 2^{\mathbb{N}\times\mathbb{N}}$ is *invariant* if

$$f \in X \Leftrightarrow \sigma(f) \in X$$

for all permutations σ of f.

a) Show that for any sentence ϕ of $\mathcal{L}_{\omega_1,\omega}$, $\{f \in 2^{\mathbb{N}\times\mathbb{N}} : \mathcal{M}_f \models \phi\}$ is an invariant Borel set.

b) Show that for any countable $\mathcal{L}$-structure $\mathcal{M}$, $\{f \in 2^{\mathbb{N}\times\mathbb{N}} : \mathcal{M}_f \cong \mathcal{M}\}$ is an invariant Borel set.

c) Show that the equivalence relation fEg if and only if $\mathcal{M}_f \cong \mathcal{M}_g$ is an invariant analytic subset of $2^{\mathbb{N}\times\mathbb{N}} \times 2^{\mathbb{N}\times\mathbb{N}}$. (Recall that an analytic subset of Y is the continuous image of a Borel set in $Y \times \mathbb{N}^{\mathbb{N}}$.)

d) Show that every invariant Borel subset of $2^{\mathbb{N}\times\mathbb{N}}$ is of the form $\{f : \mathcal{M}_f \models \phi\}$ for some $\mathcal{L}_{\omega_1,\omega}$-sentence.

Remarks

The Compactness Theorem was proved for countable languages by Gödel. Malcev proved the general result and was the first who saw its power. Indeed in a review of a paper of Skolem's, Gödel [35] points out that the existence of a nonstandard model of arithmetic follows from the Incompleteness Theorem, but does not mention the simple compactness proof.

The completeness and decidability of ACF_p for $p = 0$ or prime was first proved by Tarski using the method of quantifier elimination from Chapter 3.

Ax's Theorem 2.2.11 was first proved as we described above. Later, Borel [13] gave a topological proof and Rudin [90] gave an algebraic proof.

The method of diagrams from Section 3.2 is due to Abraham Robinson. Theorem 2.3.9 is due to Łoś and Tarski. Theorem 2.3.11 is due to Tarski and Vaught.

The analysis of the random graph is due to Fagin. See [97] for more on zero-one laws for random graphs where we consider other probability measures. These ideas can easily be generalized to study a random n-ary relation. Constructions of this sort are generally referred to as *Fraïssé constructions*. This topic is discussed more carefully in [40]. Hrushovski [41],

[42] has expanded these ideas into a powerful tool for constructing interesting new structures.

Ehrenfeucht–Fraïssé games, and their variants, play an interesting role in finite model theory (see, for example, [32]). The use of games and back-and-forth systems to understand infinitary equivalence is due to Karp. The use of $\mathcal{L}_{\omega_1,\omega}$ to characterize countable structures and Exercise 2.5.35 are due to Scott. The reader is referred to [72] for a survey of the model theory of infinitary logic and its connection to classical model theory.

Exercise 2.5.15 is due to Chang, Łoś, and Suszko. The finitely axiomatizable $\aleph_1$-categorical theory given in Exercise 2.5.16 is Morley's simplification of an example by Peretjat'kin. Exercise 2.5.17 is due to Lindström.

Ultraproducts and ultrapowers are an important tool in model theory, although they will not play a central role in this text. A very thorough development of the model theory of ultraproducts is contained in [22]. A high point of the theory is the Keisler–Shelah Theorem (see [22] 6.1.15).

Theorem 2.5.36 (Keisler–Shelah Theorem) *Two* $\mathcal{L}$-*structures* $\mathcal{M}$ *and* $\mathcal{N}$ *are elementarily equivalent if and only if there is an index set* I *and an ultrafilter* D *on* I *such that* $\prod \mathcal{M}/D \cong \prod \mathcal{N}/D$.

The study of ultraproducts of finite fields mentioned in Exercise 2.5.21 is a fascinating one initiated by Ax [2] and developed further by Chatzidakis, van den Dries, and Macintyre [23].

The examples from Exercise 2.5.28 are due to Ehrenfeucht. The analysis of elementary equivalence of ordinals is due to Mostowski and Tarski.

3
Algebraic Examples

3.1 Quantifier Elimination

The study of definable sets is often made quite complicated by quantifiers. For example, in the structure $(\mathbb{N}, +, \cdot, <)$ the quantifier-free definable sets are defined by polynomial equations and inequalities. Even if we use only existential quantifiers the definable sets become complicated. By the Matijasevič–Robinson–Davis–Putnam solution to Hilbert's 10th problem [24], every recursively enumerable subset of $\mathbb{N}$ is defined by a formula

$$\exists v_1 \ldots \exists v_n \; p(x, v_1, \ldots, v_n) = 0$$

for some polynomial $p \in \mathbb{N}[X, Y_1, \ldots, Y_n]$. As we allow more alternations of quantifiers, we get even more complicated definable sets.

Not surprisingly, it will be easiest to study definable sets that are defined by quantifier-free formulas. Sometimes formulas with quantifiers can be shown to be equivalent to formulas without quantifiers. Here are two well-known examples. Let $\phi(a, b, c)$ be the formula

$$\exists x \; ax^2 + bx + c = 0.$$

By the quadratic formula,

$$\mathbb{R} \models \phi(a, b, c) \leftrightarrow [(a \neq 0 \land b^2 - 4ac \geq 0) \lor (a = 0 \land (b \neq 0 \lor c = 0))],$$

whereas in the complex numbers

$$\mathbb{C} \models \phi(a, b, c) \leftrightarrow (a \neq 0 \lor b \neq 0 \lor c = 0).$$

In either case, ϕ is equivalent to a quantifier-free formula. However, ϕ is not equivalent to a quantifier-free formula over the rational numbers $\mathbb{Q}$.

For a second example, let $\phi(a, b, c, d)$ be the formula

$$\exists x \exists y \exists u \exists v \ (xa + yc = 1 \wedge xb + yd = 0 \wedge ua + vc = 0 \wedge ub + vd = 1).$$

The formula $\phi(a, b, c, d)$ asserts that the matrix

$$\begin{pmatrix} a & b \\ c & d \end{pmatrix}$$

is invertible. By the determinant test,

$$F \models \phi(a, b, c, d) \leftrightarrow ad - bc \neq 0$$

for any field F.

Definition 3.1.1 We say that a theory T has *quantifier elimination* if for every formula ϕ there is a quantifier-free formula ψ such that

$$T \models \phi \leftrightarrow \psi.$$

We will start by showing that DLO, the theory of dense linear orders without endpoints, has quantifier elimination. We need a slight variant of the proof of Theorem 2.4.1.

Lemma 3.1.2 *Let $(A, <)$ and $(B, <)$ be countable dense linear orders, $a_1, \ldots, a_n \in A$, $b_1, \ldots, b_n \in B$, such that $a_1 < \ldots < a_n$ and $b_1 < \ldots < b_n$. Then there is an isomorphism $f : A \rightarrow B$ such that $f(a_i) = b_i$ for $i = 1, \ldots, n$.*

Proof Modify the proof of Theorem 2.4.1 starting with $A_0 = \{a_1, \ldots, a_n\}$, $B_0 = \{b_1, \ldots, b_n\}$, and the partial isomorphism $f_0 : A_0 \rightarrow B_0$, where $f_0(a_i) = b_i$. The rest of the proof works, and we build $f : A \rightarrow B$, an isomorphism extending f_0.

Theorem 3.1.3 DLO *has quantifier elimination.*

Proof First, suppose that ϕ is a sentence. If $\mathbb{Q} \models \phi$, then because DLO is complete, $\text{DLO} \models \phi$ and

$$\text{DLO} \models \phi \leftrightarrow x_1 = x_1,$$

whereas if $\mathbb{Q} \models \neg\phi$,

$$\text{DLO} \models \phi \leftrightarrow x_1 \neq x_1.$$

Next, suppose that ϕ is a formula with free variables $x_1, \ldots, x_n$, where $n \geq 1$. We will show that there is a quantifier-free formula ψ with free variables from among $x_1, \ldots, x_n$ such that

$$\mathbb{Q} \models \forall \overline{x} \ (\phi(\overline{x}) \leftrightarrow \psi(\overline{x})).$$

Because DLO is complete,

$$\text{DLO} \models \forall \overline{x} \; (\phi(\overline{x}) \leftrightarrow \psi(\overline{x})),$$

so this will suffice.

For $\sigma : \{(i,j) : 1 \leq i < j \leq n\} \to 3$, let $\chi_\sigma(x_1, \ldots, x_n)$ be the formula

$$\bigwedge_{\sigma(i,j)=0} x_i = x_j \wedge \bigwedge_{\sigma(i,j)=1} x_i < x_j \wedge \bigwedge_{\sigma(i,j)=2} x_i > x_j.$$

We call χ_σ a *sign condition*. Each sign condition describes a (possibly inconsistent) arrangement of n elements in an ordered set.

Let $\mathcal{L}$ be the language of linear orders and ϕ be an $\mathcal{L}$-formula with $n \geq 1$ free variables. Let Λ_ϕ be the set of sign conditions $\sigma : \{(i,j) : 1 \leq i < j \leq n\} \to 3$ such that there is $\overline{a} \in \mathbb{Q}$ such that $\mathbb{Q} \models \chi_\sigma(\overline{a}) \wedge \phi(\overline{a})$. There are two cases to consider.

<u>case 1</u>: $\Lambda_\phi = \emptyset$.
  Then $\mathbb{Q} \models \forall \overline{x} \; \neg \phi(\overline{x})$ and $\mathbb{Q} \models \phi(\overline{x}) \leftrightarrow x_1 \neq x_1$.
<u>case 2</u>: $\Lambda_\phi \neq \emptyset$.
  Let

$$\psi_\phi(\overline{x}) = \bigvee_{\sigma \in \Lambda_\phi} \chi_\sigma(\overline{x}).$$

By choice of Λ_ϕ,

$$\mathbb{Q} \models \phi(\overline{x}) \to \psi_\phi(\overline{x}).$$

On the other hand, suppose that $\overline{b} \in \mathbb{Q}$ and $\mathbb{Q} \models \psi_\phi(\overline{b})$. Let $\sigma \in \Lambda_\phi$ such that $\mathbb{Q} \models \chi_\sigma(\overline{b})$. There is $\overline{a} \in \mathbb{Q}$ such that $\mathbb{Q} \models \phi(\overline{a}) \wedge \chi_\sigma(\overline{a})$. By Theorem 2.4.1, there is f, an automorphism of $(\mathbb{Q}, <)$, such that $f(\overline{a}) = \overline{b}$. By Theorem 1.1.10, $\mathbb{Q} \models \phi(\overline{b})$. Thus $\phi(\overline{b}) \leftrightarrow \psi_\phi(\overline{b})$.

Note that there is a slight anomaly here. If ϕ is not a sentence, then we can find an equivalent quantifier-free sentence using the same variables. Because there are no quantifier-free $\mathcal{L}$-sentences, to find a quantifier-free formula equivalent to a sentence, we must introduce a new free variable. If our language has constant symbols, this is unnecessary.

DLO is an example where we can give a direct explicit proof of quantifier elimination. In the exercises, we will look at several more simple examples where there is an easy explicit elimination of quantifiers. For more complicated theories explicit proofs of quantifier elimination are often quite difficult. Next we will give a useful model-theoretic criterion for quantifier elimination.

Theorem 3.1.4 *Suppose that $\mathcal{L}$ contains a constant symbol c, T is an $\mathcal{L}$-theory, and $\phi(\overline{v})$ is an $\mathcal{L}$-formula. The following are equivalent:*
  i) There is a quantifier-free $\mathcal{L}$-formula $\psi(\overline{v})$ such that $T \models \forall \overline{v} \; (\phi(\overline{v}) \leftrightarrow \psi(\overline{v}))$.

ii) *If* $\mathcal{M}$ *and* $\mathcal{N}$ *are models of* T, $\mathcal{A}$ *is an* $\mathcal{L}$-*structure,* $\mathcal{A} \subseteq \mathcal{M}$, *and* $\mathcal{A} \subseteq \mathcal{N}$, *then* $\mathcal{M} \models \phi(\bar{a})$ *if and only if* $\mathcal{N} \models \phi(\bar{a})$ *for all* $\bar{a} \in \mathcal{A}$.

Proof i)$\Rightarrow$ ii) Suppose that $T \models \forall \bar{v} \, (\phi(\bar{v}) \leftrightarrow \psi(\bar{v}))$, where ψ is quantifier-free. Let $\bar{a} \in \mathcal{A}$, where $\mathcal{A}$ is a common substructure of $\mathcal{M}$ and $\mathcal{N}$ and the latter two structures are models of T. In Proposition 1.1.8, we saw that quantifier-free formulas are preserved under substructure and extension. Thus

$$
\begin{aligned}
\mathcal{M} \models \phi(\bar{a}) \quad &\Leftrightarrow \quad \mathcal{M} \models \psi(\bar{a}) \\
&\Leftrightarrow \quad \mathcal{A} \models \psi(\bar{a}) \quad (\text{because } \mathcal{A} \subseteq \mathcal{M}) \\
&\Leftrightarrow \quad \mathcal{N} \models \psi(\bar{a}) \quad (\text{because } \mathcal{A} \subseteq \mathcal{N}) \\
&\Leftrightarrow \quad \mathcal{N} \models \phi(\bar{a}).
\end{aligned}
$$

ii) $\Rightarrow$ i) First, if $T \models \forall \bar{v} \, \phi(\bar{v})$, then $T \models \forall \bar{v} \, (\phi(\bar{v}) \leftrightarrow c = c)$. Second, if $T \models \forall \bar{v} \, \neg\phi(\bar{v})$, then $T \models \forall \bar{v} \, (\phi(\bar{v}) \leftrightarrow c \neq c)$.

Thus, we may assume that both $T \cup \{\phi(\bar{v})\}$ and $T \cup \{\neg\phi(\bar{v})\}$ are satisfiable.

Let $\Gamma(\bar{v}) = \{\psi(\bar{v}) : \psi$ is quantifier-free and $T \models \forall \bar{v} \, (\phi(\bar{v}) \rightarrow \psi(\bar{v}))\}$. Let $d_1, \ldots, d_m$ be new constant symbols. We will show that $T \cup \Gamma(\bar{d}) \models \phi(\bar{d})$. Then, by compactness, there are $\psi_1, \ldots, \psi_n \in \Gamma$ such that

$$
T \models \forall \bar{v} \left(\bigwedge_{i=1}^{n} \psi_i(\bar{v}) \; \rightarrow \phi(\bar{v}) \right).
$$

Thus

$$
T \models \forall \bar{v} \left(\bigwedge_{i=1}^{n} \psi_i(\bar{v}) \; \leftrightarrow \phi(\bar{v}) \right)
$$

and $\bigwedge_{i=1}^{n} \psi_i(\bar{v})$ is quantifier-free. We need only prove the following claim.

Claim $T \cup \Gamma(\bar{d}) \models \phi(\bar{d})$.

Suppose not. Let $\mathcal{M} \models T \cup \Gamma(\bar{d}) \cup \{\neg\phi(\bar{d})\}$. Let $\mathcal{A}$ be the substructure of $\mathcal{M}$ generated by $\bar{d}$.

Let $\Sigma = T \cup \mathrm{Diag}(\mathcal{A}) \cup \phi(\bar{d})$. If Σ is unsatisfiable, then there are quantifier-free formulas $\psi_1(\bar{d}), \ldots, \psi_n(\bar{d}) \in \mathrm{Diag}(\mathcal{A})$ such that

$$
T \models \forall \bar{v} \left(\bigwedge_{i=1}^{n} \psi_i(\bar{v}) \rightarrow \neg\phi(\bar{v}) \right).
$$

But then

$$
T \models \forall \bar{v} \left(\phi(\bar{v}) \rightarrow \bigvee_{i=1}^{n} \neg\psi_i(\bar{v}) \right),
$$

so $\bigvee_{i=1}^{n} \neg\psi_i(\bar{v}) \in \Gamma$ and $\mathcal{A} \models \bigvee_{i=1}^{n} \neg\psi_i(\bar{d})$, a contradiction. Thus, Σ is satisfiable.

Let $\mathcal{N} \models \Sigma$. Then $\mathcal{N} \models \phi(\bar{d})$. Because $\Sigma \supseteq \text{Diag}(\mathcal{A})$, $\mathcal{A} \subseteq \mathcal{N}$, by Lemma 2.3.3 i). But $\mathcal{M} \models \neg\phi(\bar{d})$; thus, by ii), $\mathcal{N} \models \neg\phi(\bar{d})$, a contradiction.

The proof above can easily be adapted to the case where $\mathcal{L}$ contains no constant symbols. In this case, there are no quantifier-free sentences, but for each sentence we can find a quantifier-free formula $\psi(v_1)$ such that $T \models \phi \leftrightarrow \psi(v_1)$.

The next lemma shows that we can prove quantifier elimination by getting rid of one existential quantifier at a time.

Lemma 3.1.5 *Let T be an $\mathcal{L}$-theory. Suppose that for every quantifier-free $\mathcal{L}$-formula $\theta(\bar{v}, w)$ there is a quantifier-free formula $\psi(\bar{v})$ such that $T \models \exists w\, \theta(\bar{v}, w) \leftrightarrow \psi(\bar{v})$. Then, T has quantifier elimination.*

Proof Let $\phi(\bar{v})$ be an $\mathcal{L}$-formula. We wish to show that $T \models \forall \bar{v}\, (\phi(\bar{v}) \leftrightarrow \psi(\bar{v}))$ for some quantifier-free formula $\phi(\bar{v})$. We prove this by induction on the complexity of $\phi(\bar{v})$.

If ϕ is quantifier-free, there is nothing to prove. Suppose that for $i = 0, 1$, $T \models \forall \bar{v}\, (\theta_i(\bar{v}) \leftrightarrow \psi_i(\bar{v}))$, where ψ_i is quantifier free.

If $\phi(\bar{v}) = \neg\theta_0(\bar{v})$, then $T \models \forall \bar{v}\, (\phi(\bar{v}) \leftrightarrow \neg\psi_0(\bar{v}))$.

If $\phi(\bar{v}) = \theta_0(\bar{v}) \wedge \theta_1(\bar{v})$, then $T \models \forall v\, (\phi(\bar{v}) \leftrightarrow (\psi_0(\bar{v}) \wedge \psi_1(\bar{v})))$.

In either case, ϕ is equivalent to a quantifier-free formula.

Suppose that $T \models \forall \bar{v}(\theta(\bar{v}, w) \leftrightarrow \psi_0(\bar{v}, w))$, where ψ_0 is quantifier-free and $\phi(\bar{v}) = \exists w\theta(\bar{v}, w)$. Then $T \models \forall \bar{v}\, (\phi(\bar{v}) \leftrightarrow \exists w\, \psi_0(\bar{v}, w))$. By our assumptions, there is a quantifier-free $\psi(\bar{v})$ such that $T \models \forall \bar{v}\, (\exists w\, \psi_0(\bar{v}, w) \leftrightarrow \psi(\bar{v}))$. But then $T \models \forall \bar{v}\, (\phi(\bar{v}) \leftrightarrow \psi(\bar{v}))$.

Combining Theorem 3.1.4 and Lemma 3.1.5 gives us the following simple, yet useful, test for quantifier elimination.

Corollary 3.1.6 *Let T be an $\mathcal{L}$-theory. Suppose that for all quantifier-free formulas $\phi(\bar{v}, w)$, if $\mathcal{M}, \mathcal{N} \models T$, $\mathcal{A}$ is a common substructure of $\mathcal{M}$ and $\mathcal{N}$, $\bar{a} \in A$, and there is $b \in M$ such that $\mathcal{M} \models \phi(\bar{a}, b)$, then there is $c \in N$ such that $\mathcal{N} \models \phi(\bar{a}, c)$. Then, T has quantifier elimination.*

Divisible Abelian Groups

In Proposition 2.2.4 we showed that the theory of nontrivial torsion-free divisible Abelian groups is κ-categorical for uncountable cardinals and hence complete by Vaught's test. To illustrate Corollary 3.1.6, we will show that the theory of divisible ordered Abelian groups has quantifier elimination. It will be convenient (although not essential) to work with the language $\mathcal{L} = \{+, -, 0\}$ because in this language substructures of groups are groups, whereas with $\{+, 0\}$ substructures are semigroups. Let DAG be the $\mathcal{L}$-theory of nontrivial torsion-free divisible Abelian groups. We will show DAG has quantifier elimination.

We start by verifying a special case of the quantifier elimination test.

Lemma 3.1.7 *Suppose G and H are nontrivial torsion free divisible Abelian groups, $G \subseteq H$, $\psi(\overline{v}, w)$ is quantifier-free, $\overline{a} \in G$, $b \in H$, and $H \models \phi(\overline{a}, b)$. Then, there is $c \in G$ such that $G \models \phi(\overline{a}, c)$.*

Proof We first note that ψ can be put in disjunctive normal form, namely there are atomic or negated atomic formulas $\theta_{i,j}(\overline{v}, w)$ such that:

$$\psi(\overline{v}, w) \leftrightarrow \bigvee_{i=1}^{n} \bigwedge_{j=1}^{m} \theta_{i,j}(\overline{v}, w).$$

Because $H \models \psi(\overline{a}, b)$, $H \models \bigwedge_{j=1}^{m} \theta_{i,j}(\overline{a}, b)$ for some i. Thus, without loss of generality, we may assume that ψ is a conjunction of atomic and negated atomic formulas. If $\theta(v_1, \ldots, v_m, w)$ is an atomic formula, then for some integers $n_1, \ldots, n_m, m$, $\theta(\overline{v}, w)$ is $\sum n_i v_i + m w = 0$.

Thus, we may assume that

$$\psi(\overline{a}, w) = \bigwedge_{i=1}^{s} \sum_{j=1}^{m} n_{i,j} a_j + m_i w = 0 \wedge \bigwedge_{i=1}^{s} \sum_{j=1}^{m} n'_{i,j} a_j + m'_i w \neq 0.$$

Let $g_i = \sum n_{i,j} a_j$ and $h_i = \sum n'_{i,j} a_j$. Then, $g_i, h_i \in G$ and

$$\psi(\overline{a}, w) \leftrightarrow \bigwedge g_i + m_i w = 0 \wedge \bigwedge h_i + m'_i w \neq 0.$$

If any $m_i \neq 0$, then $b = \frac{-g_i}{m_i} \in G$ and $G \models \theta(\overline{a}, b)$, so suppose that $\psi(\overline{a}, w) = \bigwedge h_i + m'_i w \neq 0$. Thus, $\psi(\overline{a}, w)$ is satisfied by any element of H that is not equal to any one of $\frac{-h_1}{m'_1}, \ldots, \frac{-h_s}{m'_s}$. Because G is infinite, there is an element of G satisfying $\psi(\overline{a}, w)$.

We will need the following algebraic lemma.

Lemma 3.1.8 *Suppose that G is a torsion-free Abelian group. Then, there is a torsion-free divisible Abelian group H, called the divisible hull of G, and an embedding $i : G \to H$ such that if $j : G \to H'$ is an embedding of G into a torsion-free Abelian group, then there is $h : H \to H'$ such that $j = h \circ i$.*

Proof If G is the trivial group, then we can take $H = \mathbb{Q}$, so suppose that G is non-trivial.

Let $X = \{(g, n) : g \in G, n \in \mathbb{N}, n > 0\}$. We think of (g, n) as g/n.

We define an equivalence relation $\sim$ on X by $(g, n) \sim (h, m)$ if and only if $mg = nh$. Let $H = X/\sim$. For $(g, n) \in X$, let $[(g, n)]$ denote the $\sim$-class of (g, n). We define $+$ on H by $[(g, n)] + [(h, m)] = [(mg + nh, mn)]$. We must show that $+$ is well defined.

Suppose that $(g_0, n_0) \sim (g, n)$. We claim that $(mg_0 + n_0 h, mn_0) \sim (mg + nh, mn)$. We must verify that $mn_0(mg + nh) = mn(mg_0 + n_0 h)$. Because G is Abelian, $mn_0(mg + nh) = m^2 n_0 g + mn_0 nh$. But $ng_0 = n_0 g$. Thus, $mn_0(mg + nh) = m^2 ng_0 + mn_0 nh = mn(mg_0 + n_0 h)$, as desired. Thus, $+$ is well-defined.

Similarly, we can define $-$ by $[(g, n)] - [(h, m)] = [(mg - nh, mn)]$. This is also well-defined. It is easy to show that $(H, +)$ is an Abelian group, where $[(0, 1)]$ is the identity and $[(-g, n)]$ is the inverse of $[(g, n)]$.

If $[(g, m)] \in H$ and $n > 0$, then $n[(g, m)] = [(ng, m)]$. If $(ng, m) \sim (0, k)$, then $kng = 0$. Because $k > 0$, $n > 0$, and G is torsion-free, $g = 0$. But then $[(g, m)] = [(0, 1)]$. Thus, H is torsion-free.

Suppose that $[(g, m)] \in H$ and $n > 0$, then $n[(g, mn)] = [(ng, mn)] = [(g, m)]$. Thus, H is divisible.

We can embed G into H by the map $i(g) = [(g, 1)]$. Clearly, for $g_0 \neq g_1$, $[(g_0, 1)] \neq [(g_1, 1)]$. Also $[(g, 1)] + [(h, 1)] = [(g + h, 1)]$, as desired.

Suppose that H' is a divisible torsion-free Abelian group and $j : G \to H'$ is an embedding. Let $h : H \to H'$ by $h([g, n]) = j(g)/n$. The reader should verify that h is a well-defined embedding and $j = h \circ i$.

Theorem 3.1.9 DAG *has quantifier elimination.*

Proof Suppose that G_0 and G_1 are torsion-free divisible Abelian groups, G is a common subgroup of G_0 and G_1, $\bar{g} \in G$, $h \in G_0$, and $G_0 \models \phi(\bar{g}, h)$, where ϕ is quantifier-free. Let H be the divisible hull of G. Because we can embed H into G_0, by Lemma 3.1.7, $H \models \exists w \, \phi(\bar{g}, w)$. Because we can embed H into G_1, there is $h' \in G_1$ such that $G_1 \models \phi(\bar{g}, h')$. By Corollary 3.1.6, DAG has quantifier elimination.

Quantifier elimination gives us a good picture of the definable sets in a model of DAG. Suppose that $\phi(v_1, \ldots, v_n, w_1, \ldots, w_m)$ is an atomic formula. Then, there are integers $k_1, \ldots, k_n$ and $l_1, \ldots, l_m$ such that $\phi(\bar{v}, \bar{w}) \leftrightarrow \sum k_i x_i + \sum l_i y_i = 0$. If $G \models$ DAG and $a_1, \ldots, a_m \in G$, $\phi(\bar{v}, \bar{a})$ defines $\{\bar{g} \in G^n : \sum k_i g_i + \sum l_i a_i = 0\}$, a hyperplane in G^n. Because any $\mathcal{L}$-formula $\phi(\bar{v}, \bar{w})$ is equivalent in DAG to a Boolean combination of atomic $\mathcal{L}$-formulas, every definable subset of G^n is a Boolean combination of hyperplanes.

In particular, suppose that $\bar{a} \in G^m$ and $\phi(v, \bar{a})$ defines a subset of G. The "hyperplanes" in G are just single points. Thus, $\{g \in G : G \models \phi(g, \bar{a})\}$ is either finite or cofinite. Thus, every definable subset of G was already definable already in the language of equality.[1] This is an example of a very important phenomenon.

---

[1] Of course, in G^2 there are definable sets which are not definable in the pure language of equality.

Definition 3.1.10 We say that an $\mathcal{L}$-theory T is *strongly minimal* if for any $\mathcal{M} \models T$ every definable subset of M is either finite or cofinite.

Corollary 3.1.11 DAG *is strongly minimal.*

In Chapter 6.1, we will see that strongly minimal theories are $\aleph_1$-categorical and that their analysis is crucial in any understanding of $\aleph_1$-categorical theories.

Because several of the proofs of quantifier elimination that we give below will follow the exact pattern of the proof of Theorem 3.1.9, we will isolate the properties highlighted in Lemmas 3.1.7 and 3.1.8 and show that they suffice for quantifier elimination.

If T is a theory then $T_\forall$ is the set of all universal consequences of T. In Exercise 2.5.10 we saw that $\mathcal{A} \models T_\forall$ if and only if there is $\mathcal{M} \models T$ with $\mathcal{A} \subseteq \mathcal{M}$. One consequence of Lemma 3.1.8 is that every torsion-free Abelian group is a substructure of a nontrivial divisible Abelian group. Because the axioms for torsion-free Abelian groups are universal, $DAG_\forall$ is exactly the theory of torsion-free Abelian groups.

We say that a theory T has *algebraically prime models* if for any $\mathcal{A} \models T_\forall$ there is $\mathcal{M} \models T$ and an embedding $i : \mathcal{A} \to \mathcal{M}$ such that for all $\mathcal{N} \models T$ and embeddings $j : \mathcal{A} \to \mathcal{N}$ there is $h : \mathcal{M} \to \mathcal{N}$ such that $j = h \circ i$. Lemma 3.1.8 asserts that DAG has algebraically prime models.

If $\mathcal{M}, \mathcal{N} \models T$ and $\mathcal{M} \subseteq \mathcal{N}$, we say that $\mathcal{M}$ is *simply closed* in $\mathcal{N}$ and write $\mathcal{M} \prec_s \mathcal{N}$ if for any quantifier free formula $\phi(\overline{v}, w)$ and any $\overline{a} \in M$, if $\mathcal{N} \models \exists w\, \phi(\overline{a}, w)$ then so does $\mathcal{M}$. Lemma 3.1.7 says that if G and H are models of DAG and $G \subseteq H$, then $G \prec_s H$.

The proof of Theorem 3.1.9 can be easily modified to yield the following quantifier elimination test.

Corollary 3.1.12 *Suppose that T is an $\mathcal{L}$-theory such that*
i) T has algebraically prime models and
ii) $\mathcal{M} \prec_s \mathcal{N}$ whenever $\mathcal{M} \subseteq \mathcal{N}$ are models of T.
Then, T has quantifier elimination.

Quantifier elimination implies a significant strengthening of ii).

Definition 3.1.13 An $\mathcal{L}$-theory T is *model-complete* $\mathcal{M} \prec \mathcal{N}$ whenever $\mathcal{M} \subseteq \mathcal{N}$ and $\mathcal{M}, \mathcal{N} \models T$.

Stated in terms of embeddings: T is model-complete if and only if all embeddings are elementary.

Proposition 3.1.14 *If T has quantifier elimination, then T is model-complete.*

Proof Suppose that $\mathcal{M} \subseteq \mathcal{N}$ are models of T. We must show that $\mathcal{M}$ is an elementary submodel. Let $\phi(\overline{v})$ be an $\mathcal{L}$-formula, and let $\overline{a} \in M$. There is a quantifier-free formula $\psi(\overline{v})$ such that $\mathcal{M} \models \forall \overline{v}\, (\phi(\overline{v}) \leftrightarrow \psi(\overline{v}))$. Because quantifier-free formulas are preserved under substructures and extensions, $\mathcal{M} \models \psi(\bar{a})$ if and only if $\mathcal{N} \models \psi(\bar{a})$. Thus

$$\mathcal{M} \models \phi(\bar{a}) \Leftrightarrow \mathcal{M} \models \psi(\bar{a}) \Leftrightarrow \mathcal{N} \models \psi(\bar{a}) \Leftrightarrow \mathcal{N} \models \phi(\bar{a}).$$

There are model-complete theories that do not have quantifier elimination. We will investigate model-completeness further in the exercises. For now, let us just point out the following test for completeness of model-complete theories.

Proposition 3.1.15 *Let T be a model-complete theory. Suppose that there is $\mathcal{M}_0 \models T$ such that $\mathcal{M}_0$ embeds into every model of T. Then, T is complete.*

Proof If $\mathcal{M} \models T$, the embedding of $\mathcal{M}_0$ into $\mathcal{M}$ is elementary. In particular $\mathcal{M}_0 \equiv \mathcal{M}$. Thus, any two models of T are elementarily equivalent.

Because $(\mathbb{Q}, +, 0)$ embeds in every model of DAG, this gives another proof of the completeness of DAG. We will use Proposition 3.1.15 below in several cases where Vaught's test does not apply.

Ordered Divisible Abelian Groups

Let us use the tools we have developed to analyze the theory of $(\mathbb{Q}, +, <, 0)$. Let $\mathcal{L} = \{+, -, <, 0\}$ and let ODAG be the theory of nontrivial divisible ordered Abelian groups. We will show that ODAG is a complete theory with quantifier elimination. It follows from completeness that ODAG axiomatizes the theory of the ordered group of rationals.

We start by trying to identify $\text{ODAG}_\forall$. It is easy to see that the axioms for ordered Abelian groups are universal and hence contained in $\text{ODAG}_\forall$. We claim that these axioms suffice. We must show that every ordered Abelian group embeds in an ordered divisible Abelian group. Because ordered groups are torsion-free, it suffices to show that the ordering of the group extends to an ordering of the divisible hull. The next lemma will show this and prove that ODAG has algebraically prime models.

Lemma 3.1.16 *Let G be an ordered Abelian group and H be the divisible hull of G. We can order H such that $i : G \to H$ is order-preserving, $(H, +, <) \models \text{ODAG}$ and if $H' \models \text{ODAG}$ and $j : G \to H'$ is an embedding, then there is an embedding $h : H \to H'$ such that $j = h \circ i$.*

Proof We let $\frac{g}{n}$ denote $[(g, n)]$. We can order H by $\frac{g}{n} < \frac{h}{m}$ if and only if $mg < nh$. If $g < h$, then $\frac{g}{1} < \frac{h}{1}$ so this extends the ordering of G. If $\frac{g_1}{n_1} < \frac{g_2}{n_2}$ and $\frac{h_1}{m_1} \leq \frac{h_2}{m_2}$, then $n_2 g_1 < n_1 g_2$ and $m_2 h_1 \leq m_1 h_2$. Then,

$$m_1 m_2 n_2 g_1 + n_1 n_2 m_2 h_1 < m_1 m_2 n_1 g_2 + n_1 n_2 m_1 h_2$$

and

$$\frac{m_1 g_1 + n_1 h_1}{m_1 n_1} < \frac{m_2 g_2 + n_2 h_2}{m_2 n_2}.$$

Thus, $<$ makes H an ordered group.

If H' is another ordered divisible Abelian group and $j : G \to H'$ is an embedding, let h be as in Lemma 3.1.8. It is easy to see that h is order-preserving.

To prove quantifier elimination, we must show that if G and H are ordered divisible Abelian groups and $G \subseteq H$, then $G \prec_s H$.

Suppose that $\phi(v, \overline{w})$ is a quantifier-free formula, $\overline{a} \in G$, and for some $b \in H$, $H \models \phi(b, \overline{a})$. As above, it suffices to consider the case where ϕ is a conjunction of atomic and negated atomic formulas. If $\theta(v, \overline{w})$ is atomic, then θ is equivalent to either $\sum n_i w_i + mv = 0$ or $\sum n_i w_i + mv > 0$ for some $n_i, m \in \mathbb{Z}$. In particular, there is an element $g \in G$ such that $\theta(v, \overline{a})$ is of the form $mv = g$ or $mv > g$. Also note that any formula $mv \neq g$ is equivalent to $mv > g$ or $-mv > g$. Thus we may assume that

$$\phi(v, \overline{a}) \leftrightarrow \bigwedge m_i v = g_i \wedge \bigwedge n_i v > h_i,$$

where $g_i, h_i \in G$ and $m_i, n_i \in \mathbb{Z}$.

If there is actually a conjunct $m_i v = g_i$, then we must have $b = \frac{g_i}{m_i} \in G$; otherwise $\phi(v, \overline{a}) = \bigwedge m_i v > h_i$. Let $k_0 = \min\{\frac{h_i}{m_i} : m_i < 0\}$ and $k_i = \max\{\frac{h_i}{m_i} : m_i > 0\}$. Then, $c \in H$ satisfies $\phi(v, \overline{a})$ if and only if $k_0 < v < k_1$. Because b satisfies ϕ, we must have $k_0 < k_1$. But it is easy to see that any ordered divisible Abelian group is densely ordered because if $g < h$, then $g < \frac{g+h}{2} < h$, so there is $d \in G$ such that $k_0 < d < k_1$. Thus $G \prec_s H$.

Corollary 3.1.17 ODAG *is a complete decidable theory with quantifier elimination. In particular, every ordered divisible Abelian group is elementarily equivalent to* $(\mathbb{Q}, +, <)$.

Proof By Lemma 3.1.16, ODAG$_\forall$ is the theory of ordered Abelian groups and ODAG has algebraically prime models. From the remarks above and Corollary 3.1.12 we see that ODAG has quantifier elimination. The ordered group of rationals embeds into every ordered divisible Abelian group; thus, by Proposition 3.1.15, ODAG is complete. Because ODAG has a recursive axiomatization, it is decidable by Lemma 2.2.8.

ODAG is not strongly minimal. For example, $\{a \in \mathbb{Q} : a < 0\}$ is infinite and coinfinite. On the other hand, definable subsets are quite well-behaved. Suppose that G is an ordered divisible Abelian group and $X \subseteq G$ is definable. By quantifier elimination, X is a Boolean combination of sets defined by atomic formulas. If $\phi(v, w_1, \ldots, w_n)$ is atomic, then there are integers $k_0, \ldots, k_n$ such that ϕ is equivalent to either

$$k_0 v + \sum k_i w_i = 0$$

or
$$k_0 v + \sum k_i w_i > 0.$$

If $\bar{a} \in G^n$, in the first case $\phi(v, \bar{a})$ defines a finite set whereas in the second case it defines an interval. It follows that X is a finite union of points and intervals with endpoints in $G \cup \{\pm\infty\}$. This is also a very useful property.

Definition 3.1.18 We say that an ordered structure $(M, <, \ldots)$ is o-minimal (where "o" comes from "order") if for any definable $X \subseteq M$ there are finitely many intervals $I_1, \ldots I_m$ with endpoints in $M \cup \{\pm\infty\}$ and a finite set X_0 such that $X = X_0 \cup I_1 \cup \ldots \cup I_m$.

If $\mathcal{M}$ is o-minimal, then the only definable subsets of M are already definable using only the ordering. Although there may be more complicated definable subsets in M^k, these sets will still be quite well behaved. We will say a bit more about this in Section 3.3 (see [29] for a thorough treatment of this important subject).

Presburger Arithmetic

We conclude this chapter by considering a slightly more complicated example. Let $\mathcal{L} = \{+, -, <, 0, 1\}$ and consider the $\mathcal{L}$-theory of the ordered group of integers. In fact this theory will not have quantifier elimination in the language $\mathcal{L}$. Let $\psi_n(v)$ be the formula

$$\exists y \; v = \underbrace{y + \ldots + y}_{n-\text{times}},$$

which asserts that v is divisible by n. We will see in the exercises that ψ_n is not equivalent to a quantifier free formula. It turns out that this is the only obstruction to quantifier elimination. Let $\mathcal{L}^* = \mathcal{L} \cup \{P_n : n = 2, 3, \ldots\}$, where P_n is a unary predicate which we will interpret as the elements divisible by n. We will see that the $\mathcal{L}^*$-theory of $\mathbb{Z}$ has quantifier elimination and is decidable. Because we are only adding predicates for sets that we could define already in the language $\mathcal{L}$, we will not change the definable sets (see Exercise 1.4.15).

There is something slippery going on here that we should be careful about. For any language $\mathcal{L}$ and $\mathcal{L}$-theory T, there is a language $\mathcal{L}' \supseteq \mathcal{L}$ and an $\mathcal{L}'$-theory $T' \supseteq T$ such that for any $\mathcal{M} \models T$ we can interpret the new symbols of $\mathcal{L}'$ to make $\mathcal{M}' \models T'$ such that any subset of M^k definable using $\mathcal{L}'$ is already definable using $\mathcal{L}$, and any $\mathcal{L}'$-formula is equivalent to an atomic $\mathcal{L}'$-formula!

Let $\mathcal{L}' = \mathcal{L} \cup \{R_\phi : \phi \text{ an } \mathcal{L}\text{-formula}\}$, where if ϕ is a formula in n free variables, R_ϕ is an n-ary predicate symbol. Let T' be the theory obtained by adding to T the sentences

$$\forall \bar{v} \; (\phi(\bar{v}) \leftrightarrow R_\phi(\bar{v}))$$

for each $\mathcal{L}$-formula ϕ. As an exercise, show that T' has the desired property. Thus, by adding lots of predicate symbols for sets defined in the original language, we obtain a new language in which we have quantifier elimination.

In general, this is a completely useless construction. Quantifier elimination in the new language is only helpful if the quantifier-free definable sets are easy to understand. If we were having trouble understanding the sets defined using $\mathcal{L}$, we would have the same problems with sets defined using $\mathcal{L}'$.

In our analysis of the ordered group of integers, we will see that adding predicates for subgroups of elements divisible by n suffices to allow us to eliminate quantifiers, but the language is simple enough that we can prove decidability and easily understand definable sets. We will consider the $\mathcal{L}^*$-theory, which we call Pr for *Presburger arithmetic*, with axioms:

i) axioms for ordered Abelian groups;

ii) $0 < 1$;

iii) $\forall x\ (x \leq 0 \vee x \geq 1)$;

iv)$_n$ $\forall x(P_n(x) \leftrightarrow \exists y\ x = \underbrace{y + \ldots + y}_{n-\text{times}})$, for $n = 2, 3, \ldots$;

v)$_n$ $\forall x\ \bigvee_{i=0}^{n-1}[P_n(x + \underbrace{1 + \ldots + 1}_{i\ \text{times}}) \wedge \bigwedge_{j \neq i} \neg P_n(x + \underbrace{1 + \ldots + 1}_{j\ \text{times}})]$

for $n = 2, 3, \ldots$.

Suppose that $(G, +, -, <, 0, 1)$ is a model of Pr. For each n, axiom iv)$_n$ asserts that $P_n^G = nG$. Axiom v)$_n$ asserts that $\frac{G}{nG} \cong \frac{\mathbb{Z}}{n\mathbb{Z}}$.

What is Pr$_\forall$? Clearly the axioms i), ii), iii), and v)$_n$ are universal, whereas axiom iv) is not. Let us define a theory T that we will eventually show is Pr$_\forall$. The axioms for T are:

axioms i), ii), iii), and v)$_n$;

vi)$_n$ P_n is closed under $+$ and $-$;

vii)$_n$ $\forall x, y\ (\underbrace{y + \ldots + y}_{n\ \text{times}} = x) \rightarrow P_n(x)$;

viii)$_{n,m}$ (for m dividing n) $\forall x(P_m(x) \rightarrow P_n(x))$;

ix)$_{n,k}\forall x(P_{kn}(\underbrace{x + \ldots + x}_{k\ \text{times}}) \rightarrow P_n(x))$ for $k, n = 2, 3, \ldots$.

Axiom vi) ensures that the P_n are additive subgroups. Axiom vii)$_n$ asserts that $nG \subset P_n$. Axiom viii)$_{n,m}$ asserts that if $m|n$, then $P_m \subset P_n$. Axiom ix)$_{n,k}$ asserts that if $kx \in P_{kn}$, then $x \in P_n$. Clearly, $T \subseteq$ Pr$_\forall$. The next lemma shows that T axiomatizes Pr$_\forall$ and Pr has algebraically prime models.

Lemma 3.1.19 Let $(G, +, <, P_2, P_3, \ldots) \models T$. There is $H \supseteq G$ such that $H \models$ Pr and if $H' \supseteq G$ and $H' \models$ Pr, then there is $h : H \to H'$ such that $h|G$ is the identity.

Proof We define H a subgroup of the divisible hull of G. Let $H = \{\frac{x}{n} : x \in G$ and $n = 1$ or $P_n(x)\}$. We let $P_n^H = nH$.

We first show that H is a subgroup of the divisible hull of G. Suppose that $\frac{x}{m}, \frac{y}{n} \in H$. We assume that $m \neq 1$ and $n \neq 1$, leaving the other cases to the reader. Then $x \in P_m$ and $y \in P_n$. Thus, by axiom ix), $nx \in P_{mn}$ and $my \in P_{mn}$. Because P_{mn} is closed under addition and subtraction, $nx \pm my \in P_{mn}$. Thus, $\frac{nx \pm my}{mn} \in H$ and H is an ordered subgroup of the divisible hull of G. In particular, H is Abelian.

Clearly, $H \models 0 < 1$. Suppose that $\frac{x}{m} \in H$ and $0 < \frac{x}{m} < 1$. Because G is discretely ordered and 1 is the least positive element, $m \neq 1$ and $0 < x < m$. Thus $x \in \{1, 2, \ldots, m-1\}$. But then, by axiom v), $_n$, $G \models \neg P_m(x)$, a contradiction.

P_n^H is defined so that axiom iv)$_n$ holds. We need only check axiom v)$_n$. Let $\frac{x}{m} \in H$. Thus $P_m(x)$. We would like $\frac{x}{m}$ to be congruent one of $0, 1, \ldots, n-1 \mod n$. By axiom v), there is a unique i such that $0 \leq i < mn$ such that $x + i \in P_{mn}$. By axiom viii), $x + i \in P_m$. Because P_m is a subgroup, $i \in P_m$. Thus, $i = lm$, where $0 \leq l < n$. In H, there is y such that $\underbrace{y + \ldots + y}_{mn \text{ times}} = x + lm$. Then $\underbrace{y + \ldots + y}_{n \text{ times}} = \frac{x}{m} + l$. Because there is only one choice for i, there is only one choice for l.

Thus $H \models Pr$. Thus $T = Pr_\forall$.

Moreover, suppose that $H' \supseteq G$ and $H' \models Pr$. Let $x \in G$ such that $G \models P_m(x)$. There is $y \in H'$ such that $my = x$. Thus, there is an embedding of H into H' fixing G.

Quantifier elimination will follow from the next lemma.

Lemma 3.1.20 *If $G, H \models \mathrm{Pr}$ and $G \subseteq H$, then $G \prec_s H$.*

Proof Let $\bar{a} \in G$, let $\phi(v, \bar{w})$ be a quantifier-free formula, and let $b \in H$ such that $H \models \phi(b, \bar{a})$.

We claim that we may assume that $\phi(v, \bar{a})$ is of the form

$$\bigwedge m_i v = g_i \wedge \bigwedge P_{n_i}(s_i v + h_i) \wedge \bigwedge c_i < v < d_i,$$

where $m_i, n_i, s_i \in \mathbb{Z}$ and $c_i, d_i, g_i, h_i \in G$.

First note that $\neg P_n(x) \leftrightarrow \bigvee_{i=1}^{n-1} P_n(x+i)$. If we first assume that ϕ is in conjunctive normal form, we can then replace all negative occurrences of P_n by a disjunction of positive occurrences. We then use the distributive law to get an equivalent formula in conjunctive normal form with no negative occurrences of P_n. As usual, we need only consider a single disjunct that is satisfied by b. We can also replace $mv > g_i$ by $v > h$, where $h \in G$ and $mh \leq g_i < m(h+1)$.

Thus, without loss of generality, we may assume that $\phi(v, \bar{a})$ is of the form above.

If there is any conjunct of the form $m_i v = g_i$, then $b = \frac{g_i}{m_i} \in G$. Thus, we may assume that there are no conjuncts of the first type.

We can find c and $d \in G$ such that $c_i \leq c < b < d \leq d_i$. If $d - c$ is finite, then $b \in G$. Thus we may assume that $d - c$ is infinite.

For each i, there is a j_i such that $0 \le j_i < n_i$ and $P_{n_i}(h_i - j_i)$ for all i. Thus, $P_{n_i}(s_i v + h_i)$ if and only if $P_{n_i}(s_i v + j_i)$. Then, b is a solution to the system of congruences

$$
\begin{aligned}
s_1 v + j_1 &\equiv 0 \ (\mathrm{mod}\ n_1) \\
s_2 v + j_2 &\equiv 0 \ (\mathrm{mod}\ n_2) \\
&\cdots \\
s_m v + j_m &\equiv 0 \ (\mathrm{mod}\ n_m).
\end{aligned}
$$

Let $N = \prod n_i$ and let $l \in \omega$ such that $P_N(b - l)$. Then l is a solution to this system of congruences. Because $d - c$ is infinite, there is a $g \in G$ such that $c < g < d$ and $P_N(g - l)$. Then g is a solution to the system of congruences above. Thus $G \models \phi(g, \bar{a})$.

Corollary 3.1.21 *Presburger arithmetic is a complete decidable theory with quantifier elimination in the language $\mathcal{L}^*$.*

Proof Because $\mathbb{Z}$ can be embedded in any model of Pr, Pr is complete by Proposition3.1.15. Because we have given a recursive set of axioms Pr is decidable by Lemma 2.2.8.

Corollary 3.1.21 provides an interesting counterpoint to Gödel's Incompleteness Theorem as $\mathrm{Th}(\mathbb{Z}, +, <)$ is decidable whereas $\mathrm{Th}(\mathbb{Z}, +, \cdot)$ is not.

We have provided a number of proofs of quantifier elimination without explicitly explaining how to take an arbitrary formula and produce a quantifier free one. In all of these cases, one can give explicit effective procedures. After the fact, the following lemma tells us that there is an algorithm to eliminate quantifiers.

Proposition 3.1.22 *Suppose that T is a decidable theory with quantifier elimination. Then, there is an algorithm which when given a formula ϕ as input will output a quantifier-free formula ψ such that $T \models \phi \leftrightarrow \psi$.*

Proof Given input $\phi(\bar{v})$ we search for a quantifier-free formula $\psi(\bar{v})$ such that $T \models \forall \bar{v} \ (\phi(\bar{v}) \leftrightarrow \psi(\bar{v}))$. Because T is decidable this is an effective search. Because T has quantifier elimination, we will eventually find ψ.

3.2 Algebraically Closed Fields

We now return to the theory of algebraically closed fields. In Proposition 2.2.5, we proved that the theory of algebraically closed fields of a fixed characteristic is complete. We begin this section by showing that algebraically closed fields have quantifier elimination. The first step is to identify $\mathrm{ACF}_\forall$, the universal consequences of the theory of ACF. We recall that the theory of ACF is formulated in $\mathcal{L}_r$, the language of rings $\{+, -, \cdot, 0, 1\}$.

Lemma 3.2.1 $\mathrm{ACF}_\forall$ *is the theory of integral domains.*

Proof The axioms for integral domains are universal consequences of ACF. If D is an integral domain, then the algebraic closure of the fraction field of D is a model of ACF. Because every integral domain is a subring of an algebraically closed field, $\mathrm{ACF}_\forall$ is the theory of integral domains.

Theorem 3.2.2 ACF *has quantifier elimination.*

Proof We will apply Corollary 3.1.12. If D is an integral domain, then the algebraic closure of the fraction field of D embeds into any algebraically closed field containing D. Thus, ACF has algebraically prime models.

To prove quantifier elimination, we need only show that if K and F are algebraically closed fields, $F \subseteq K$, $\phi(x, \overline{y})$ is quantifier-free, $\overline{a} \in F$, and $K \models \phi(b, \overline{a})$ for some $b \in K$, then $F \models \exists v\, \phi(v, \overline{a})$.

As in Lemma 3.1.7, we may assume that $\phi(x, \overline{y})$ is a conjunction of atomic and negated atomic formulas. In the language of rings, atomic formulas $\theta(v_1, \ldots, v_n)$ are of the form $p(\overline{v}) = 0$, where $p \in \mathbb{Z}[X_1, \ldots, X_n]$. If $p(X, \overline{Y}) \in \mathbb{Z}[X, \overline{Y}]$, we can view $p(X, \overline{a})$ as a polynomial in $F[X]$. Thus, there are polynomials $p_1, \ldots, p_n, q_1, \ldots, q_m \in F[X]$ such that $\phi(v, \overline{a})$ is equivalent to

$$\bigwedge_{i=1}^{n} p_i(v) = 0 \wedge \bigwedge_{i=1}^{m} q_i(v) \neq 0.$$

If any of the polynomials p_i are nonzero, then b is algebraic over F. In this case, because F is algebraically closed, $b \in F$. Thus, we may assume that $\phi(v, \overline{a})$ is equivalent to

$$\bigwedge_{i=1}^{m} q_i(v) \neq 0.$$

But $q_i(X) = 0$ has only finitely many solutions for each $i \leq m$. Thus, there are only finitely many elements of F that do not satisfy F. Because algebraically closed fields are infinite, there is a $c \in F$ such that $F \models \phi(c, \overline{a})$.

Corollary 3.2.3 ACF *is model-complete and* ACF_p *is complete where $p = 0$ or p is prime.*

Proof Model-completeness is an immediate consequence of quantifier elimination.

The completeness of ACF_p was proved in Proposition 2.2.5, but it also follows from quantifier elimination. Suppose that $K, L \models ACF_p$. Let ϕ be any sentence in the language of rings. By quantifier elimination, there is a quantifier-free sentence ψ such that

$$\mathrm{ACF} \models \phi \leftrightarrow \psi.$$

Because quantifier-free sentences are preserved under extension and substructure,

$$K \models \psi \Leftrightarrow \mathbb{F}_p \models \psi \Leftrightarrow L \models \psi,$$

where $\mathbb{F}_p$ is the p-element field if $p > 0$ and the rationals if $p = 0$. Thus,

$$K \models \phi \Leftrightarrow K \models \psi \Leftrightarrow L \models \psi \Leftrightarrow L \models \phi.$$

Thus $K \equiv L$ and ACF_p is complete.

Zariski Closed and Constructible Sets

Quantifier elimination has a geometric interpretation. To state this, we first must review some basic definitions from algebraic geometry. Let K be a field. If $S \subseteq K[X_1, \ldots, X_n]$, let $V(S) = \{a \in K^n : p(a) = 0$ for all $p \in S\}$. If $Y \subseteq K^n$, we let $I(Y) = \{f \in K[X_1, \ldots, X_n] : f(\bar{a}) = 0$ for all $\bar{a} \in Y\}$. We say that $X \subseteq K^n$ is *Zariski closed* if $X = V(S)$ for some $S \subseteq K[X_1, \ldots, X_n]$. We summarize some basic facts about Zariski closed sets. For more details, see [37] or [91].

Lemma 3.2.4 *Let K be a field.*
  i) If $X \subseteq K^n$, then $I(X)$ is a radical ideal.
  ii) If X is Zariski closed, then $X = V(I(X))$.
  iii) If X and Y are Zariski closed and $X \subseteq Y \subseteq K^n$, then $I(Y) \subseteq I(X)$.
  iv) If $X, Y \subseteq K^n$ are Zariski closed, then $X \cup Y = V(I(X) \cap I(Y))$ and $X \cap Y = V(I(X) + I(Y))$.

Proof
  i) Suppose that $p, q \in I(X)$ and $f \in K[X_1, \ldots, X_n]$. If $a \in X$, then $p(a) + q(a) = f(a)p(a) = 0$. Thus, $p + q, fp \in I(X)$ and $I(X)$ is an ideal. If $f^n \in I(X)$ and $a \in X$, then $f^n(a) = 0$ so $f(a) = 0$. Thus, $f \in I(X)$ and $I(X)$ is a radical ideal.

  ii) If $a \in X$ and $p \in I(X)$, then $p(a) = 0$. Thus $X \subseteq V(I(X))$. If $a \in V(I(X)) \setminus X$, then there is $p \in I(X)$ such that $p(a) \neq 0$, a contradiction. Thus $X = V(I(X))$.

  iii) If $p \in I(Y)$ and $a \in X$, then $p(a) = 0$ because $a \in Y$. Thus $I(Y) \subseteq I(X)$. By ii), if $I(X) = I(Y)$, then $X = Y$.

  iv) If $p \in I(X) \cap I(Y)$, then $p(a) = 0$ for $a \in X$ or $a \in Y$. Thus $X \cup Y \subseteq V(I(X) \cap I(Y))$. On the other hand, if $a \notin X \cup Y$, there are $p \in I(X)$ and $q \in I(Y)$ such that $p(a) \neq 0$ and $q(a) \neq 0$. But then $p(a)q(a) \neq 0$. Because $pq \in I(X) \cap I(Y)$, $a \notin V(I(X) \cap I(Y))$.

  If $a \in X \cup Y$, $p \in I(X)$, and $q \in I(Y)$, then $p(a) + q(a) = 0$. Thus $X \cup Y \subseteq V(I(X) + I(Y))$. If $a \notin X$, then there is $p \in I(X) \subseteq I(X) + I(Y)$ such that $p(a) \neq 0$. Thus $a \notin V(I(X) + I(Y))$. Similarly, if $a \notin Y$, then $a \notin V(I(X) + I(Y))$.

By Lemma 3.2.4, the Zariski closed sets are closed under finite unions and intersections. Closure under arbitrary intersections is a consequence of the next important algebraic result (see, for example, [58] VI §2).

Theorem 3.2.5 (Hilbert's Basis Theorem) *If K is a field, then the polynomial ring $K[X_1,\ldots,X_n]$ is a Noetherian ring, (i.e., there are no infinite ascending chains of ideals). In particular, every ideal is finitely generated.*

Corollary 3.2.6 *i) There are no infinite descending sequences of Zariski closed sets.*

ii) If X_i is Zariski closed for $i \in I$, then there is a finite $I_0 \subseteq I$ such that

$$\bigcap_{i \in I} X_i = \bigcap_{i \in I_0} X_i.$$

In particular, an arbitrary intersection of Zariski closed sets is Zariski closed.

Proof

i) If $X_0 \supset X_1 \supset X_2 \supset \ldots$ is a descending sequence of Zariski closed sets, then $I(X_0) \subset I(X_1) \subset I(X_2) \subset \ldots$ is an ascending sequence of prime ideals contradicting Hilbert's Basis Theorem.

ii) Suppose not. Then, we can find $X_1, X_2, \ldots$, Zariski closed such that

$$\bigcap_{i=1}^{n+1} X_i \subset \bigcap_{i=1}^{n} X_i$$

for $n = 1, 2, \ldots$, contradicting i).

Thus, the Zariski closed sets are closed under finite unions and arbitrary intersections. Because $\emptyset = V(1)$ and $K^n = V(0)$, the Zariski closed sets are the closed sets of a topology.

We can now give a geometric description of the definable sets.

Lemma 3.2.7 *Let K be a field. The subsets of K^n defined by atomic formulas are exactly those of the form $V(p)$ for some $p \in K[\overline{X}]$. A subset of K^n is quantifier-free definable if and only if it is a Boolean combination of Zariski closed sets.*

Proof If $\phi(\overline{x}, \overline{y})$ is an atomic $\mathcal{L}_r$-formula, then there is $q(\overline{X}, \overline{Y}) \in \mathbb{Z}[\overline{X}, \overline{Y}]$ such that $\phi(\overline{x}, \overline{y})$ is equivalent to $q(\overline{x}, \overline{y}) = 0$. If $X = \{\overline{x} : \phi(\overline{x}, \overline{a})\}$, then $X = V(q(\overline{X}, \overline{a}))$ and $q(\overline{X}, \overline{a}) \in K[\overline{X}]$. On the other hand, if $p \in K[\overline{X}]$, there is $q \in \mathbb{Z}[\overline{X}, \overline{Y}]$ and $\overline{a} \in K^m$ such that $p(\overline{X}) = q(\overline{X}, \overline{a})$. Then, $V(p)$ is defined by the quantifier-free formula $q(\overline{X}, \overline{a}) = 0$.

If X is Zariski closed, then by Hilbert's Basis Theorem, there are $p_1, \ldots, p_l$ such that

$$X = V(p_1, \ldots, p_n) = V(p_1) \cap \ldots V(p_n).$$

Because the quantifier-free definable sets are exactly finite Boolean combinations of atomic definable sets, the quantifier-free definable sets are exactly the Boolean combinations of Zariski closed sets.

If $X \subseteq K^n$ is a finite Boolean combination of Zariski closed sets we call X *constructible*. If K is algebraically closed, the constructible sets have much stronger closure properties.

Corollary 3.2.8 *Let K be an algebraically closed field.*

i) $X \subseteq K^n$ is constructible if and only if it is definable.

ii) (**Chevalley's Theorem**) *The image of a constructible set under a polynomial map is constructible.*

Proof i) By Lemma 3.2.7, the constructible sets are exactly the quantifier-free definable sets, but by quantifier elimination every definable set is quantifier-free definable.

ii) Let $X \subseteq K^n$ be constructible and $p : K^n \to K^m$ be a polynomial map. Then, the image of $X = \{y \in K^m : \exists x \in K^n \ p(x) = y\}$. This set is definable and hence constructible.

Quantifier elimination has very strong consequences for definable subsets of K.

Corollary 3.2.9 *If K is an algebraically closed field and $X \subseteq K$ is algebraically closed, then either X or $K \setminus X$ is finite. Thus, ACF is strongly minimal.*

Proof By quantifier elimination X is a finite Boolean combination of sets of the form $V(p)$, where $p \in K[X]$. But $V(p)$ is either finite or (if $p = 0$) all of K.

Lemma 3.2.4 shows that the map $X \mapsto I(X)$ is a lattice inverting map from Zariski closed subsets of K^n to radical ideals in $K[X]$. Hilbert's Nullstellensatz says that this is a bijection and for I a radical ideal $I = I(V(I))$. The model-completeness of algebraically closed fields can be used to give a proof of the Nullstellensatz.

We need one fact from commutative algebra ([58] VI §5).

Lemma 3.2.10 (Primary Decomposition) *If $I \subset K[\overline{X}]$ is a radical ideal, then there are prime ideals $P_1, \ldots, P_m$ containing I such that $I = P_1 \cap \ldots \cap P_m$, $I \neq \bigcup_{j \in J}$ for any proper $J \subset \{1, \ldots, m\}$, and if $Q_1, \ldots, Q_n$ is another set of prime ideals with these properties, then $n = m$ and $\{Q_1, \ldots, Q_n\} = \{P_1, \ldots, P_m\}$.*

Theorem 3.2.11 (Hilbert's Nullstellensatz) *Let K be an algebraically closed field. Suppose that I and J are radical ideals in $K[X_1, \ldots, X_n]$ and $I \subset J$. Then $V(J) \subset V(I)$. Thus $X \mapsto I(X)$ is a bijective correspondence between Zariski closed sets and radical ideals.*

Proof Let $p \in J \setminus I$. By Lemma 3.2.10, there is a prime ideal $P \supseteq I$ such that $p \notin P$. We will show that there is $x \in V(P) \subseteq V(I)$ such that $p(x) \neq 0$. Thus $V(I) \neq V(J)$. Because P is prime, $K[\overline{X}]/P$ is a domain and we can take F, the algebraic closure of its fraction field.

Let $q_1, \ldots, q_m \in K[X_1, \ldots, X_n]$ generate J. Let a_i be the element X_i/P in F. Because each $q_i \in P$ and $p \notin P$,

$$F \models \bigwedge_{i=1}^{m} q_i(\overline{a}) = 0 \wedge p(\overline{a}) \neq 0.$$

Thus

$$F \models \exists \overline{w} \bigwedge_{i=1}^{m} q_i(\overline{w}) = 0 \wedge p(\overline{w}) \neq 0$$

and by model-completeness

$$K \models \exists \overline{w} \bigwedge_{i=1}^{m} q_i(\overline{w}) = 0 \wedge p(\overline{w}) \neq 0.$$

Thus there is $\overline{b} \in K^n$ such that $q_1(\overline{b}) = \ldots = q_m(\overline{b}) = 0$ and $p(\overline{b}) \neq 0$. But then $\overline{b} \in V(P) \setminus V(J)$.

Corollary 3.2.12 *If $J \subseteq K[\overline{X}]$ is a radical ideal, then $J = I(V(J))$.*

Proof Clearly $I(V(J)) \supseteq V(J)$. By Lemma 3.2.4, $V(J) = V(I(V(J)))$. Thus, by the Nullstellensatz, $J = I(V(J))$.

Quantifier elimination gives us a powerful tool for analyzing definability in algebraically closed fields. For the moment, we will analyze definable functions and equivalence relations. In Chapter 7.4, we will examine groups definable in algebraically closed fields.

Definition 3.2.13 Let $X \subseteq K^n$. We say that $f : X \to K$ is *quasirational* if either
i) K has characteristic zero and for some rational function $q(\overline{X}) \in K(X_1, \ldots, X_n)$, $f(\overline{x}) = q(\overline{x})$ on X, or
ii) K has characteristic $p > 0$ and for some rational function $q(\overline{X}) \in K(\overline{X})$, $f(\overline{x}) = q(\overline{x})^{\frac{1}{p^k}}$.

Rational functions are easily seen to be definable. In algebraically closed fields of characteristic p, the formula $x = y^p$ defines the function $x \mapsto x^{\frac{1}{p}}$, because every element has a unique p^{th}-root. Thus, every quasirational function is definable.

Proposition 3.2.14 *If $X \subseteq K^n$ is constructible and $f : X \to K$ is definable, then there are constructible sets $X_1, \ldots, X_m$ and quasirational functions $\rho_1, \ldots, \rho_m$ such that $\bigcup X_i = X$ and $f|X_i = \rho_i|X_i$.*

Proof Let $\Gamma(v_1, \ldots, v_n) = \{ f(\overline{x}) \neq \rho(\overline{x}) : \rho$ a quasirational function$\}$ $\cup \{ \overline{v} \in X \} \cup$ ACF $\cup$ Diag(K).

Claim Γ is not satisfiable.

Suppose that Γ is consistent. Let $L \models$ ACF $+$Diag(K) with $b_1, \ldots, b_n \in L$ such that for all $\gamma(\overline{v}) \in \Gamma$, $L \models \gamma(\overline{b})$.

Let K_0 be the subfield of L generated by K and $\overline{b}$. Then, K_0 is the closure of $B = \{ b_1, \ldots, b_n \}$ under the rational functions of K. Let K_1 be the closure of B under all quasirational functions. If K has characteristic 0, then $K_0 = K_1$. If K has characteristic $p > 0$, $K_1 = \bigcup K_0^{\frac{1}{p^n}}$, the perfect closure of K_0.

By model-completeness, $K \prec L$, thus f^L, the interpretation of f in L, is a function from X^L to L, extending f. Because $L \models \Gamma(\overline{b})$, $f(\overline{b})$ is not in K_1. Because K_1 is perfect there is an automorphism α of L fixing K_1 pointwise such that $\alpha(f^L(\overline{b})) \neq f^L(\overline{b})$. But f^L is definable with parameters from K; thus, any automorphism of L which fixes K and fixes $\overline{a}$ must fix $f(\overline{a})$, a contradiction. Thus Γ is unsatisfiable.

Thus, by compactness, there are quasirational functions $\rho_1, \ldots, \rho_m$ such that

$$K \models \forall x \in X \bigwedge f(\overline{x}) = \rho_i(\overline{x}).$$

Let $X_i = \{ \overline{x} \in X : f(\overline{x}) = \rho_i(\overline{x}) \}$. Each X_i is definable.

A similar argument shows the model theoretic notion of "algebraic" introduced in Exercise 1.4.11 agrees with the field-theoretic notion. Recall that if $A \subseteq K$ we say that $b \in$ acl(A) if there is a formula $\phi(x, \overline{y})$ and $\overline{a} \in A$ such that $K \models \phi(b, \overline{a})$ and $\{ x \in K : K \models \phi(x, \overline{a}) \}$ is finite.

Proposition 3.2.15 *Let $K \models$ ACF and $A \subseteq K$. Then, $a \in$ acl(A) if and only if a is algebraic over the subfield of K generated by A.*

Proof Let k be the field generated by A. If a is algebraic over k, there are polynomials $q_0(X_1, \ldots, X_n), \ldots, q_m(X_1, \ldots, X_n) \in \mathbb{Z}[X_1, \ldots, X_n]$ and $b_1, \ldots, b_n \in A$ such that $p(Y) = \sum q_i(b_1, \ldots, b_n) Y^i$ is a nonzero polynomial such that $p(a) = 0$. Let $\phi(x, \overline{y})$ be the $\mathcal{L}_r$-formula

$$\sum q_i(\overline{y}) x^i = 0.$$

Then $\phi(a, \overline{b})$ and $\{ x \in K : K \models \phi(x, \overline{b}) \}$ is finite. Thus $a \in$ acl(A).

On the other hand, suppose that $\overline{b} \in A$, $\{ x \in K : \phi(x, \overline{b}) \}$ is finite, and $K \models \phi(a, \overline{b})$, but a is transcendental over k. Let c be any other element of a that is transcendental over k. Then, there is an automorphism σ of K such that σ is the identity on k but $\sigma(a) = c$. By Theorem 1.1.10, $K \models \phi(c, \overline{b})$. Because there are infinitely many choices for c, we have a contradiction.

Elimination of Imaginaries in Algebraically Closed Fields

We conclude by studying definable equivalence relations in algebraically closed fields. In Section 1.3, we remarked that quotient structures are frequently important in mathematics. For example, if G is a group and H is a subgroup, we study the coset space G/H. If G is a definable group and H is a definable subgroup, then the equivalence relation xEy if and only if $y \in Hx$ is a definable equivalence relation. Another example that arises in algebraic geometry is the construction of projective space. If K is a field, define the equivalence relation $\sim$ on $K^{n+1} \setminus \{0\}$, where $(x_0, \ldots, x_n) \sim (y_0, \ldots, y_n)$ if and only if there is a nonzero $\lambda \in K$ such that $y_i = \lambda x_i$ for $i = 0, \ldots, n$. The equivalence relation $\sim$ is definable and the quotient is $\mathbb{P}_K^n = K^{n+1} \setminus \{0\} \bmod \sim$, projective n-space over K.

Our goal is to show that quotients are constructible. In particular, we will show that if E is a definable equivalence relation on K^n, where K is algebraically closed, then there is a definable function $f : K^n \to K^m$ for some m such that $\overline{x}E\overline{y}$ if and only if $f(\overline{x}) = f(\overline{y})$. Thus, we can identify the quotient K^n/E with the image of f. In other words it allows us to view the quotient of a constructible set by a constructible equivalence relation as a constructible set.

We begin with a very important special case. Let k be a field (in this special case we will not need to assume that k is algebraically closed). We can think of $\overline{x} \in k^{nm}$ as a sequence $(\overline{x}_1, \ldots, \overline{x}_m)$, where $\overline{x}_i \in k^n$. Let E be the equivalence relation $\overline{x}E\overline{y}$ if and only if $(\overline{x}_1, \ldots, \overline{x}_n)$ is a permutation of $(\overline{y}_1, \ldots, \overline{y}_n)$. Clearly, E is a definable equivalence relation.

Lemma 3.2.16 *There is a definable function $f : k^{nm} \to k^l$ for some $l \in \omega$ such that $\overline{c}E\overline{d}$ if and only if $f(\overline{c}) = f(\overline{d})$.*

Proof Suppose that $\overline{c} = (\overline{c}_1, \ldots, \overline{c}_m)$ where $\overline{c}_i = (c_{i,1}, \ldots, c_{i,n})$. Let $q_i^{\overline{c}}$ be the polynomial

$$Y - \sum_{j=1}^n c_{i,j} X_j$$

in $k[X_1, \ldots, X_n, Y]$ and let $p^{\overline{c}} = \prod q_i^{\overline{c}}$. Let $f(\overline{c})$ be the sequence of coefficients of $p^{\overline{c}}$. Because $k[X_1, \ldots, X_n, Y]$ is a unique factorization domain, $p^{\overline{c}} = p^{\overline{d}}$ if and only if $(q_1^{\overline{c}}, \ldots, q_m^{\overline{c}})$ is a permutation of $(q_1^{\overline{d}}, \ldots, q_m^{\overline{d}})$. Thus, $\overline{c}E\overline{d}$ if and only if $f(\overline{c}) = f(\overline{d})$.

We need to do some preparatory work before examining the general case.

Definition 3.2.17 Let $\mathcal{M}$ be an $\mathcal{L}$-structure and E be a definable equivalence relation on M^n. For $\overline{a} \in M^n$, let $\overline{a}/E$ denote the E-equivalence class of $\overline{a}$. For $b_1, \ldots, b_m \in M$ and $c \in M$, we say that c is algebraic over $\overline{a}/E, b_1, \ldots, b_m$ if and only if there is a formula $\phi(x, y_1, \ldots, y_n, z_1, \ldots, z_m)$ such that

i) $\mathcal{M} \models \phi(c, \overline{a}, \overline{b})$,

ii) if $\bar{a}E\bar{a}'$, then $\mathcal{M} \models \phi(x,\bar{a},\bar{b}) \leftrightarrow \phi(x,\bar{a}',\bar{b})$, and

iii) $\{x \in M : \mathcal{M} \models \phi(x,\bar{a},\bar{b})\}$ is finite.

We say that $\bar{c} = (c_1,\ldots,c_l)$ is algebraic over $\bar{a}/E, b_1,\ldots,b_m$ if and only if each c_i is.

This definition agrees with the natural notion of algebraic in $\mathcal{M}^{\mathrm{eq}}$ (see Lemma 1.3.10). Algebraic closure in $\mathcal{M}^{\mathrm{eq}}$ has all of the properties of algebraic closure in $\mathcal{M}$ developed in Exercise 1.4.11.

Lemma 3.2.18 *Suppose that $\bar{c}$ is algebraic over $\bar{a}/E, \bar{d}, \bar{b}$ and $\bar{b}$ is algebraic over $\bar{a}/E, \bar{d}$, then $\bar{c}$ is algebraic over $\bar{a}/E, \bar{d}$.*

Proof Exercise 3.4.16.

Lemma 3.2.19 *Suppose that K is an algebraically closed field and E is a definable equivalence relation on K^n. Let $\psi(\bar{x},\bar{y},\bar{d})$ define E. If $\bar{a} \in K^n$, then there is $\bar{c} \in K^n$ algebraic over $\bar{a}/E, \bar{d}$ such that $\bar{c}E\bar{a}$.*

Proof Let $\psi(\bar{x},\bar{y},\bar{d})$ define E. Let $0 \le m \le n$ be maximal such that there are $c_1,\ldots,c_m$ algebraic over $\bar{a}/E, \bar{d}$ such that

$$K \models \exists v_{m+1} \ldots \exists v_n \; \psi(\bar{c},\bar{v},\bar{a},\bar{d}).$$

Suppose that $m < n$. Consider

$$X = \{x \in K : K \models \exists w_{m+2} \ldots \exists w_n \; \psi(\bar{c},x,\bar{w},\bar{a},\bar{d})\}.$$

If X is finite we can choose $c_{m+1} \in X$ algebraic over $\bar{a}/E, \bar{d}, c_1,\ldots,c_m$ (indeed any element of X would work). By Lemma 3.2.18, c_{m+1} is algebraic over $\bar{a}/E, \bar{d}$, contradicting the maximality of m.

If X is infinite, then by strong minimality $K \setminus X$ is finite. Because the algebraic closure of the prime field is infinite, we can find $c_{m+1} \in X$, which is algebraic over $\emptyset$. This contradicts the maximality of m. Thus, $m = n$ and $\bar{c} = (c_1,\ldots,c_n)$ is the desired element of K^n.

In Lemma 8.2.9 we will examine a generalization of 3.2.19. We can now prove the main theorem.

Theorem 3.2.20 *Suppose that K is an algebraically closed field, $A \subseteq K$, and E is an A-definable equivalence relation on K^n. Then for some l there is an A-definable function $f : K^n \to K^l$ such that $\bar{x}E\bar{y}$ if and only if $f(\bar{x}) = f(\bar{y})$.*

Proof For notational simplicity, we will assume that E is defined over $\emptyset$. For each formula $\phi(\bar{x},\bar{y})$ and $k > 0$, let $\Theta_{\phi,k}(\bar{y})$ be the conjunction of

i) $\forall\bar{x}(\phi(\bar{x},\bar{y}) \to \bar{x}E\bar{y})$;

ii) $\forall\bar{x}\forall\bar{z}(\bar{y}E\bar{z} \to (\phi(\bar{x},\bar{y}) \leftrightarrow \phi(\bar{x},\bar{z})))$;

iii) $|\{\bar{x} : \phi(\bar{x},\bar{y})\}| = k$.

By Lemma 3.2.19, for all $\bar{a} \in K^n$, there are ϕ and k such that $\Theta_{\phi,k}(\bar{a})$. By ii), if $\Theta_{\phi,k}(\bar{a})$ and $\bar{b}E\bar{a}$, then $\Theta_{\phi,k}(\bar{b})$.

Let $X = \{\bar{a} : \Theta_{\phi,k}(\bar{a})\}$. If $\bar{a} \in X$, let $Y_{\bar{a}} = \{\bar{b} : \phi(\bar{b}, \bar{a})\}$. For $\bar{a}, \bar{b} \in X$, $\bar{a}E\bar{b}$ if and only if $Y_{\bar{a}} = Y_{\bar{b}}$. By Lemma 3.2.16, there is a $\emptyset$-definable function $f : X \to K^l$ for some l such that $Y_{\bar{a}} = Y_{\bar{b}}$ if and only if $f(\bar{a}) = f(\bar{b})$.

By compactness, we can find $\phi_1, \ldots, \phi_m$ and $k_1, \ldots, k_m$ such that some $\Theta_{\phi_i,k_i}(\bar{y})$ holds for each element of K^n. Let $X_i = \{\bar{y} : \Theta_{\phi_i,k_i}(\bar{x})\}$. There is $f_i : X_i \to l_i$ such that $\bar{a}E\bar{b}$ if and only if $f_i(\bar{a}) = f_i(\bar{b})$ for $\bar{a}, \bar{b} \in X_i$. Extend f_i to K^n by making $f_i(\bar{x}) = 0$ for $\bar{x} \notin X_i$. Let $f : K^n \to K^{\sum l_i}$ by $f(\bar{x}) = (f_1(\bar{x}), \ldots, f_m(\bar{x}))$. Then, $\bar{a}E\bar{b}$ if and only if $f(\bar{a}) = f(\bar{b})$, as desired.

Exercise 3.4.19 explains why Theorem 3.2.20 is referred to as the "elimination of imaginaries."

3.3   Real Closed Fields

In this section, we will concentrate on the field of real numbers. Unlike algebraically closed fields, the theory of the real numbers does not have quantifier elimination in $\mathcal{L}_r$, the language of rings. The proof of Corollary 3.2.9 shows that any field with quantifier elimination is strongly minimal, whereas in $\mathbb{R}$, if $\phi(x)$ is the formula $\exists z \; z^2 = x$, then ϕ defines an infinite coinfinite definable set (see also Exercise 3.4.24). In 7.2 we will see that algebraically closed fields are the only infinite fields with quantifier elimination in the language of rings.

In fact, the ordering is the only obstruction to quantifier elimination. We will eventually analyze the real numbers in the language $\mathcal{L}_{or}$ and show that we have quantifier elimination in this language. Because the ordering $x < y$ is definable in the real field by the formula

$$\exists z \; (z \neq 0 \wedge x + z^2 = y),$$

any subset of $\mathbb{R}^n$ definable using an $\mathcal{L}_{or}$-formula is already definable using an $\mathcal{L}_r$-formula (see Exercise 1.4.15). We will see that quantifier elimination in $\mathcal{L}_{or}$ leads us to a good geometric understanding of the definable sets.

We begin by reviewing some of the necessary algebraic background on ordered fields. All of the algebraic results stated in this chapter are due to Artin and Schreier and proved in Appendix B.

Definition 3.3.1 We say that a field F is *orderable* if there is a linear order $<$ of F making $(F, <)$ an ordered field.

Although there are unique orderings of the fields $\mathbb{R}$ and $\mathbb{Q}$, orderable fields may have many possible orderings. The field of rational functions $\mathbb{Q}(X)$ has $2^{\aleph_0}$ distinct orderings. To see this, let x be any real number transcendental over $\mathbb{Q}$. The evaluation map $f(X) \mapsto f(x)$ is a field isomorphism between $\mathbb{Q}(X)$ and $\mathbb{Q}(x)$, the subfield of $\mathbb{R}$ generated by x. We can lift the ordering of the reals to an ordering $\mathbb{Q}(X)$ by $f(X) < g(X)$ if and only if $f(x) < g(x)$. Because $X < q$ if and only if $x < q$, choosing a different transcendental real would yield a different ordering. These are not the only orderings. We will see in Exercise 3.4.23 that we can also order $\mathbb{Q}(X)$ by making X infinite or infinitesimally close to a rational.

There is a purely algebraic characterization of the orderable fields.

Definition 3.3.2 We say that F is *formally real* if -1 is not a sum of squares.

In any ordered field all squares are nonnegative. Thus, every orderable field is formally real. The following result shows that the converse is also true.

Theorem 3.3.3 *If F is a formally real field, then F is orderable. Indeed, if $a \in F$ and $-a$ is not a sum of squares of elements of F, then there is an ordering of F where a is positive.*

Because the field of complex numbers is the only proper algebraic extension of the real field, the real numbers have no proper formally real algebraic extensions. Fields with this property will play a key role.

Definition 3.3.4 A field F is *real closed* if it is formally real with no proper formally real algebraic extensions.

Although it is not obvious at first that real closed fields form an elementary class, the next theorem allows us to axiomatize the real closed fields.

Theorem 3.3.5 *Let F be a formally real field. The following are equivalent.*
i) F is real closed.
ii) $F(i)$ is algebraically closed (where $i^2 = -1$).
iii) For any $a \in F$, either a or $-a$ is a square and every polynomial of odd degree has a root.

Corollary 3.3.6 *The class of real closed fields is an elementary class of $\mathcal{L}_r$-structures.*

Proof We can axiomatize real closed fields by:
i) axioms for fields
ii) for each $n \geq 1$, the axiom

$$\forall x_1 \ldots \forall x_n \; x_1^2 + \ldots + x_n^2 + 1 \neq 0$$

iii) $\forall x \exists y \; (y^2 = x \vee y^2 + x = 0)$

iv) for each $n \geq 0$, the axiom

$$\forall x_0 \ldots \forall x_{2n} \exists y \; y^{2n+1} + \sum_{i=0}^{2n} x_i y^i = 0.$$

Although we can axiomatize real closed fields in the language of rings, we already noticed that we do not have quantifier elimination in this language. Instead, we will study real closed fields in $\mathcal{L}_{or}$, the language of ordered rings. If F is a real closed field and $0 \neq a \in F$, then exactly one of a and $-a$ is a square. This allows us to order F by

$$x < y \text{ if and only if } y - x \text{ is a nonzero square.}$$

It is easy to check that this is an ordering and it is the only possible ordering of F.

Definition 3.3.7 We let RCF be the $\mathcal{L}_{or}$-theory axiomatized by the axioms above for real closed fields and the axioms for ordered fields.

The models of RCF are exactly real closed fields with their canonical ordering. Because the ordering is defined by the $\mathcal{L}_r$-formula

$$\exists z \; (z \neq 0 \wedge x + z^2 = y),$$

the next result tells us that using the ordering does not change the definable sets.

Proposition 3.3.8 If F is a real closed field and $X \subseteq F^n$ is definable by an $\mathcal{L}_{or}$-formula, then X is definable by an $\mathcal{L}_r$-formula.

Proof Replace all instances of $t_i < t_j$ by $\exists v \; (v \neq 0 \wedge v^2 + t_i = t_j)$, where t_i and t_j are terms occurring in the definition of X (see Exercise 1.4.15). ∎

The next result suggests another possible axiomatization of RCF.

Theorem 3.3.9 An ordered field F is real closed if and only if whenever $p(X) \in F[X]$, $a, b \in X$, $a < b$, and $p(a)p(b) < 0$, there is $c \in F$ such that $a < c < b$ and $p(c) = 0$.

We will prove quantifier elimination using the test given in Corollary 3.1.12. We first identify $\text{RCF}_\forall$.

Definition 3.3.10 If F is a formally real field, a *real closure* of F is a real closed algebraic extension of F.

By Zorn's Lemma, every formally real field F has a maximal formally real algebraic extension. This maximal extension is a real closure of F.

The real closure of a formally real field may not be unique. Let $F = \mathbb{Q}(X)$, $F_0 = F(\sqrt{X})$, and $F_1 = F(\sqrt{-X})$. By Theorem 3.3.3, F_0 and F_1

are formally real. Let R_i be a real closure of F_i. There is no isomorphism between R_0 and R_1 fixing F because X is a square in R_0 but not in R_1. Thus, some work needs to be done to show that any ordered field $(F, <)$ has a real closure where the canonical order extends the ordering of F.

Lemma 3.3.11 *If* $(F, <)$ *is an ordered field,* $0 < x \in F$, *and* x *is not a square in* F, *then we can extend the ordering of* F *to* $F(\sqrt{x})$.

Proof We can extend the ordering to $F(\sqrt{x})$ by $0 < a + b\sqrt{x}$ if and only if

  i) $b = 0$ and $a > 0$, or
  ii) $b > 0$ and ($a > 0$ or $x > \frac{a^2}{b^2}$), or
  iii) $b < 0$ and ($a < 0$ and $x < \frac{a^2}{b^2}$).

Corollary 3.3.12 *i) If* $(F, <)$ *is an ordered field, there is a real closure* R *of* F *such that the canonical ordering of* R *extends the ordering on* F.
  ii) $\mathrm{RCF}_\forall$ *is the theory of ordered integral domains.*

Proof
i) By successive applications of Lemma 3.3.11, we can find an ordered field $(L, <)$ extending $(F, <)$ such that every positive element of F has a square root in L. We now apply Zorn's Lemma to find a maximal formally real algebraic extension R of L. Because every positive element of F is a square in R, the canonical ordering of R extends the ordering of F.

ii) Clearly, any substructure of a real closed field is an ordered integral domain. If $(D, <)$ is an ordered integral domain and F is the fraction field of F, then we can order F by

$$\frac{a}{b} > 0 \Leftrightarrow a, b > 0 \text{ or } a, b < 0.$$

By i), we can find $(R, <) \models \mathrm{RCF}$ such that $(F, <) \subseteq (R, <)$.

Although a formally real field may have nonisomorphic real closures, if $(F, <)$ is an ordered field there will be a unique real closure compatible with the ordering of F.

Theorem 3.3.13 *If* $(F, <)$ *is an ordered field, and* R_1 *and* R_2 *are real closures of* F *where the canonical ordering extends the ordering of* F, *then there is a unique field isomorphism* $\phi : R_1 \to R_2$ *that is the identity on* F.

Note that because the ordering of a real closed field is definable in $\mathcal{L}_r$, ϕ also preserves the ordering. We often say that any ordered field $(F, <)$ has a unique real closure. By this we mean that there is a unique real closure that extends the given ordering.

Corollary 3.3.14 RCF *has algebraically prime models.*

Proof Let $(D, <)$ be an ordered domain, and let $(R, <)$ be the real closure of the fraction field compatible with the ordering of D. Let $(F, <)$ be any real closed field extension of $(D, <)$. Let $K = \{\alpha \in F : \alpha$ is algebraic over the fraction field of $D\}$. By Theorem 3.3.5, it is easy to see that K is real closed. Because the ordering of K extends $(D, <)$, by Theorem 3.3.13 there is an isomorphism $\phi : F \to K$ fixing D.

We are now ready to prove quantifier elimination.

Theorem 3.3.15 *The theory* RCF *admits elimination of quantifiers in* $\mathcal{L}_{or}$.

Proof Because RCF has algebraically prime models, by Corollary 3.1.12, we need only show that $F \prec_s K$ when $F, K \models$ RCF and $F \subseteq K$. Let $\phi(v, \overline{w})$ be a quantifier-free formula and let $\overline{a} \in F$, $b \in K$ be such that $K \models \phi(b, \overline{a})$. We must find $b' \in F$ such that $F \models \phi(b', \overline{a})$.

Note that

$$p(X) \neq 0 \leftrightarrow (p(\overline{X}) > 0 \vee -p(\overline{X}) > 0)$$

and

$$p(\overline{X}) \not> 0 \leftrightarrow (p(\overline{X}) = 0 \vee -p(\overline{X}) > 0).$$

With this in mind, we may assume that ϕ is a disjunction of conjunctions of formulas of the form $p(v, \overline{w}) = 0$ or $p(v, \overline{w}) > 0$. As in Theorem 3.2.2, we may assume that there are polynomials $p_1, \ldots, p_n$ and $q_1, \ldots, q_m \in F[X]$ such that

$$\phi(v, \overline{a}) \leftrightarrow \bigwedge_{i=1}^{n} p_i(v) = 0 \wedge \bigwedge_{i=1}^{m} q_i(v) > 0.$$

If any of the polynomials $p_i(X)$ is nonzero, then b is algebraic over F. Because F has no proper formally real algebraic extensions, in this case $b \in F$. Thus, we may assume that

$$\phi(v, \overline{a}) \leftrightarrow \bigwedge_{i=1}^{m} q_i(v) > 0.$$

The polynomial $q_i(X)$ can only change signs at zeros of q_i and if all zeros of q_i are in F. Thus, we can find $c_i, d_i \in F$ such that $c_i < b < d_i$ and $q_i(x) > 0$ for all $x \in (c_i, d_i)$. Let $c = \max(c_1, \ldots, c_m)$ and $d = \min(d_1, \ldots, d_m)$. Then, $c < d$ and $\bigwedge_{i=1}^{m} q_i(x) > 0$ whenever $c < x < d$. Thus, we can find $b' \in F$ such that $F \models \phi(b', \overline{a})$.

Corollary 3.3.16 RCF *is complete, model complete, and decidable. Thus* RCF *is the theory of* $(\mathbb{R}, +, \cdot, <)$ *and* RCF *is decidable.*

Proof By quantifier elimination, RCF is model complete.

Every real closed field has characteristic zero; thus, the rational numbers are embedded in every real closed field. Therefore, $\mathbb{R}_{alg}$, the field of real algebraic numbers (i.e., the real closure of the rational numbers) is a subfield of any real closed field. Thus, for any real closed field R, $\mathbb{R}_{alg} \prec R$, so $R \equiv \mathbb{R}_{alg}$.

In particular, $R \equiv \mathbb{R}_{alg} \equiv \mathbb{R}$.

Because RCF is complete and recursively axiomatized, it is decidable.

Semialgebraic Sets

Quantifier elimination for real closed fields has a geometric interpretation.

Definition 3.3.17 Let F be an ordered field. We say that $X \subseteq F^n$ is *semialgebraic* if it is a Boolean combination of sets of the form $\{\overline{x} : p(\overline{x}) > 0\}$, where $p(\overline{X}) \in F[X_1, \ldots, X_n]$.

By quantifier elimination, the semialgebraic sets are exactly the definable sets. The next corollary is a geometric restatement of quantifier elimination. It is analogous to Chevalley's Theorem (3.2.8) for algebraically closed fields.

Corollary 3.3.18 (Tarski–Seidenberg Theorem) *The semialgebraic sets are closed under projection.*

The next corollary is a typical application of quantifier elimination.

Corollary 3.3.19 *If $F \models RCF$ and $A \subseteq F^n$ is semialgebraic, then the closure (in the Euclidean topology) of F is semialgebraic.*

Proof We repeat the main idea of Lemma 1.3.3. Let d be the definable function

$$d(x_1, \ldots, x_n, y_1, \ldots, y_n) = z \text{ if and only if } z \geq 0 \wedge z^2 = \sum_{i=1}^{n} (x_i - y_i)^2.$$

The closure of A is

$$\{\overline{x} : \forall \epsilon > 0 \; \exists \overline{y} \in A \; d(\overline{x}, \overline{y}) < \epsilon\}.$$

Because this set is definable, it is semialgebraic.

We say that a function is semialgebraic if its graph is semialgebraic. The next result shows how we can use the completeness of RCF to transfer results from $\mathbb{R}$ to other real closed fields.

Corollary 3.3.20 *Let F be a real closed field. If $X \subseteq F^n$ is closed and bounded, and f is a continuous semialgebraic function, then $f(X)$ is closed and bounded.*

Proof If $F = \mathbb{R}$, then X is closed and bounded if and only if X is compact. Because the continuous image of a compact set is compact, the continuous image of a closed and bounded set is closed and bounded.

In general, there are $\bar{a}, \bar{b} \in F$ and formulas ϕ and ψ such that $\phi(\bar{x}, \bar{a})$ defines X and $\psi(\bar{x}, y, \bar{b})$ defines $f(\bar{x}) = y$. There is a sentence Φ asserting:

> $\forall \bar{u}, \bar{w}$ [if $\psi(\bar{x}, y, \bar{w})$ defines a continuous function with domain $\phi(\bar{x}, \bar{u})$ and $\phi(\bar{x}, \bar{u})$ is a closed and bounded set, then the range of the function is closed and bounded].

By the remarks above, $\mathbb{R} \models \Phi$. Therefore, by the completeness of RCF, $F \models \Phi$ and the range of f is closed and bounded.

Model-completeness has several important applications. A typical application is Abraham Robinson's simple proof of Artin's positive solution to Hilbert's 17th problem.

Definition 3.3.21 Let F be a real closed field and $f(\overline{X}) \in F(X_1, \dots, X_n)$ be a rational function. We say that f is *positive semidefinite* if $f(\bar{a}) \geq 0$ for all $\bar{a} \in F^n$.

Theorem 3.3.22 (Hilbert's 17th Problem) *If f is a positive semidefinite rational function over a real closed field F, then f is a sum of squares of rational functions.*

Proof Suppose that $f(X_1, \dots, X_n)$ is a positive semidefinite rational function over F that is not a sum of squares. By Theorem 3.3.3, there is an ordering of $F(\overline{X})$ so that f is negative. Let R be the real closure of $F(\overline{X})$ extending this order. Then

$$R \models \exists \bar{v} \; f(\bar{v}) < 0$$

because $f(\overline{X}) < 0$ in R. By model-completeness

$$F \models \exists \bar{v} \; f(\bar{v}) < 0,$$

contradicting the fact that f is positive semidefinite.

In Exercise 3.4.28, we will show how model-completeness can be used to give a real version of the Nullstellensatz.

We will show that quantifier elimination gives us a powerful tool for understanding the definable subsets of a real closed field.

Corollary 3.3.23 RCF *is an o-minimal theory.*

Proof Let $R \models$ RCF. We need to show that every definable subset of R is a finite union of points and intervals with endpoints in $R \cup \{\pm\infty\}$. By quantifier elimination, very definable subset of R is a finite Boolean combination of sets of the form

$$\{x \in R : p(x) = 0\}$$

and
$$\{x \in R : q(x) > 0\}.$$

Solution sets to nontrivial equations are finite, whereas sets of the second form are finite unions of intervals. Thus, any definable set is a finite union of points and intervals.

Next we will show that definable functions in one variable are piecewise continuous. The first step is to prove a lemma about $\mathbb{R}$ that we will transfer to all real closed fields.

Lemma 3.3.24 *If $f : \mathbb{R} \to \mathbb{R}$ is semialgebraic, then for any open interval $U \subseteq \mathbb{R}$ there is a point $x \in U$ such that f is continuous at x.*

Proof
case 1: There is an open set $V \subseteq U$ such that f has finite range on V.

Pick an element b in the range of f such that $\{x \in V : f(x) = b\}$ is infinite. By o-minimality, there is an open set $V_0 \subseteq V$ such that f is constantly b on V.

case 2: Otherwise.

We build a chain $U = V_0 \supset V_1 \supset V_2 \ldots$ of open subsets of U such that the closure $\overline{V}_{n+1}$ of V_{n+1} is contained in V_n. Given V_n, let X be the range of f on V_n. Because X is infinite, by o-minimality, X contains an interval (a, b) of length at most $\frac{1}{n}$. The set $Y = \{x \in V_n : f(x) \in (a, b)\}$ contains a suitable open interval V_{n+1}. Because $\mathbb{R}$ is locally compact,

$$\bigcap_{i=1}^{\infty} V_i = \bigcap_{i=1}^{\infty} \overline{V}_i \neq \emptyset.$$

If $x \in \bigcap_{i=1}^{n} V_i$, then f is continuous at x.

The proof above makes essential use of the completeness of the ordering of the reals. However, because the statement is first order, it is true for all real closed fields, by the completeness of RCF.

Corollary 3.3.25 *Let F be a real closed field and $f : F \to F$ is a semialgebraic function. Then, we can partition F into $I_1 \cup \ldots \cup I_m \cup X$, where X is finite and the I_j are pairwise disjoint open intervals with endpoints in $F \cup \{\pm\infty\}$ such that f is continuous on each I_j.*

Proof Let

$$D = \{x : F \models \exists \epsilon > 0 \ \forall \delta > 0 \ \exists y \ |x - y| < \delta \wedge |f(x) - f(y)| > \epsilon\}$$

be the set of points where f is discontinuous. Because D is definable, by o-minimality D is either finite or has a nonempty interior. By Corollary

3.3.23, D must be finite. Thus, $F \setminus D$ is a finite union of intervals on which F is continuous.

The next result of van den Dries shows that real closed fields have definable Skolem functions.

Corollary 3.3.26 *Let F be a real closed field and $X \subseteq F^{n+m}$ be semialgebraic. There is a semialgebraic function $f : F^n \to F^m$ such that for all $\overline{x} \in F^n$, if there is a $\overline{y} \in F^m$ such that $(\overline{x}, \overline{y}) \in X$, then $(\overline{x}, f(\overline{x})) \in X$.*
In fact, we can choose F such that if

$$\{\overline{y} \in F^m : (\overline{a}, \overline{y}) \in X\} = \{\overline{y} \in F^m : (\overline{b}, \overline{y})\},$$

then $F(\overline{a}) = F(\overline{b})$. We call such an F an invariant Skolem function.

Proof We proceed by induction on m. For each m we will show that our claim holds for all n and for all definable $X \subseteq F^{n+m}$.

Assume that $m = 1$. For $\overline{a} \in F^n$, let $X_{\overline{a}} = \{y : (\overline{a}, y) \in X\}$. By o-minimality, $X_{\overline{a}}$ is a finite union of points and intervals. If $X_{\overline{a}}$ is empty we let $f(\overline{a}) = 0$. If $X_{\overline{a}}$ is nonempty, we define $f(\overline{a})$ by cases.

<u>case 1</u>: If $X_{\overline{a}} = F$, let $f(\overline{a}) = 0$.

<u>case 2</u>: If $X_{\overline{a}}$ has a least element b, let $f(\overline{a}) = b$.

<u>case 3</u>: If the leftmost interval of $X_{\overline{a}} = (c, d)$, let $f(\overline{a}) = \dfrac{d - c}{2}$.

<u>case 4</u>: If the leftmost interval of $X_{\overline{a}} = (-\infty, c)$, let $f(\overline{a}) = c - 1$.

<u>case 5</u>: If the leftmost interval of $X_{\overline{a}} = (c, +\infty)$, let $f(\overline{a}) = c + 1$.

This exhausts all possibilities. Clearly, f is definable and if $X_{\overline{a}} \neq \emptyset$, then $(\overline{a}, f(\overline{a})) \in X$.

Assume that our claim is true for m, and let $X \subseteq F^{n+m+1}$. By induction, there is $f : F^{n+1} \to F^m$ such that if $a_1, \ldots, a_n, b \in F$ and $\exists \overline{z} \in F^m \ (\overline{a}, b, \overline{z}) \in X$, then $(\overline{a}, b, f(\overline{a}, b)) \in X$. Also by induction, there is $g : F^n \to F$ such that if $\exists y \exists \overline{z}(\overline{x}, y, \overline{z}) \in X$, then $\exists \overline{z}(\overline{x}, f(\overline{x}), \overline{z}) \in X$. Let $h : F^n \to F^{m+1}$ by $h(\overline{x}) = (f(\overline{x}), g(\overline{x}), f(\overline{x}))$. If $\overline{a} \in F^n$ and $\exists y \exists \overline{z}(\overline{a}, y, \overline{z}) \in X$, then $(\overline{a}, h(\overline{a})) \in X$.

We leave it to the reader to show that the chosen element of $X_{\overline{a}}$ depends on $X_{\overline{a}}$ but not $\overline{a}$.

Definable Skolem functions give an easy proof of the next result of Milnor's, which was first proved by geometric techniques.

Corollary 3.3.27 (Curve Selection) *Let F be a real closed field. Let $X \subseteq F^n$ be semialgebraic and $\overline{a}$ be a point in the closure of X. There is a continuous semialgebraic function $f : (0, r) \to F^n$ such that for all $\epsilon \in (0, r), f(x) \in X$ and $\lim_{\epsilon \to 0} f(\epsilon) = \overline{a}$.*

Proof Let

$$X = \left\{ (\epsilon, \overline{y}) : \sum_{i=1}^{n} (y_i - a_i)^2 < \epsilon \right\}.$$

For all $\epsilon > 0$, there is $\bar{y} \in F^n$ with $(\epsilon, \bar{y}) \in X$. By Corollary 3.3.26, there is a definable function $f : (0, +\infty) \to F^n$ such that $(\epsilon, f(\epsilon)) \in X$ for all $\epsilon > 0$. By Corollary 3.3.25 there is $r > 0$ such that f is continuous on $(0, r)$. Clearly, $\lim_{\epsilon \to 0} f(\epsilon) = \bar{a}$.

Definable equivalence relations are very easy to analyze in real closed fields.

Corollary 3.3.28 *Let F be a real closed field. Let $E \subseteq F^n \times F^n$ be a definable equivalence relation. There is a definable $X \subset F^n$ such that for all $a \in F^n$ there is a unique $b \in X$ such that aEb. We call X a definable transversal of E.*

Proof Let $f : F^n \to F^n$ be a definable invariant Skolem function. Then, $aEf(a)$ for all $a \in F^n$ and if aEb, then $f(a) = f(b)$. Let X be the range of f.

If F is real closed, then o-minimality tells us what the definable subsets of F look like. Definable subsets of F^n are also relatively simple.

Definition 3.3.29 We inductively define the collection of *cells* as follows.
- $X \subseteq F^n$ is a 0-cell if it is a single point.
- $X \subseteq F$ is a 1-cell if it is an interval (a, b), where $a \in F \cup \{-\infty\}$, $b \in F \cup \{+\infty\}$, and $a < b$.
- If $X \subseteq F^n$ is an n-cell and $f : X \to F$ is a continuous definable function, then $Y = \{(\bar{x}, f(\bar{x})) : \bar{x} \in X\}$ is an n-cell.
- Let $X \subseteq F^n$ be an n-cell. Suppose that f is either a continuous definable function from X to F or identically $-\infty$ and g is either a continuous definable function from X to F such that $f(\bar{x}) < g(\bar{x})$ for all $\bar{x} \in X$ or g is identically $+\infty$; then

$$Y = \{(\bar{x}, y) : \bar{x} \in X \wedge f(\bar{x}) < y < g(\bar{x})\}$$

is an $n + 1$-cell.

In a real closed field, every nonempty definable set is a finite disjoint union of cells. The proof relies on the following lemma.

Lemma 3.3.30 (Uniform Bounding) *Let $X \subseteq F^{n+1}$ be semialgebraic. There is a natural number N such that if $\bar{a} \in F^n$ and $X_{\bar{a}} = \{y : (\bar{a}, y) \in X\}$ is finite, then $|X_{\bar{a}}| < N$.*

Proof First, note that $X_{\bar{a}}$ is infinite if and only if there is no interval (c, d) such that $(c, d) \subseteq X_{\bar{a}}$. Thus $\{(\bar{a}, b) \in X : X_{\bar{a}} \text{ is finite}\}$ is definable. Without loss of generality, we may assume that for all $\bar{a} \in F^n$, $X_{\bar{a}}$ is finite. In particular, we may assume that

$$F \models \forall \bar{x} \forall c \forall d \neg [c < d \wedge \forall y (c < y < d \to y \in X_{\bar{a}})].$$

Consider the following set of sentences in the language of fields with constants added for each element of F and new constants $c_1, \ldots, c_n$. Let Γ be

$$\text{RCF} + \text{Diag}(F) + \left\{ \exists y_1, \ldots, y_m \left[\bigwedge_{i<j} y_i \neq y_j \wedge \bigwedge_{i=1}^{m} y_i \in X_{\bar{c}} \right] : m \in \omega \right\}$$

Suppose that Γ is satisfiable. Then, there is a real closed field $K \supseteq F$ and elements $\bar{c} \in K^n$ such that $X_{\bar{c}}$ is infinite. By model-completeness, $F \prec K$. Therefore

$$K \models \forall \bar{x} \forall c, d \; \neg [c < d \; \wedge \forall y \; (c < y < d \rightarrow y \in X_{\bar{a}})].$$

This contradicts the o-minimality of K. Thus, Γ is unsatisfiable and there is an N such that

$$\text{RCF} + \text{Diag}(F) \models \forall \bar{x} \; \neg \left(\exists y_1, \ldots, y_N \left[\bigwedge_{i<j} y_i \neq y_j \wedge \bigwedge_{i=1}^{N} y_i \in X_{\bar{x}} \right] \right).$$

In particular, for all $\bar{a} \in F^n$, $|X_{\bar{a}}| < N$.

We now state the Cell Decomposition Theorem and give the proof for subsets of F^2. In the exercises, we will outline the results needed for the general case.

Theorem 3.3.31 (Cell Decomposition) *Let $X \subseteq F^m$ be semialgebraic. There are finitely many pairwise disjoint cells $C_1, \ldots, C_n$ such that $X = C_1 \cup \ldots \cup C_n$.*

Proof (for $m = 2$) For each $a \in F$, let

$$C_a = \{x : \forall \epsilon > 0 \exists y, z \in (x - \epsilon, x + \epsilon) \; [(a, y) \in X \wedge (a, z) \notin X]\}.$$

We call C_a the *critical values* above a. By o-minimality, there are only finitely many critical values above a. By uniform bounding, there is a natural number N such that for all $a \in F$, $|C_a| \leq N$. We partition F into $A_0, A_1, \ldots, A_N$, where $A_n = \{a : |C_a| = n\}$.

For each $n \leq N$, we have a definable function $f_n : A_1 \cup \ldots \cup A_n \rightarrow F$ by $f_n(a) = n$th element of C_a. As above, $X_a = \{y : (a, y) \in X\}$.

For $n \leq N$ and $a \in A_n$, we define $P_a \in 2^{2n+1}$, the *pattern* of X above a, as follows.

If $n = 0$, then $P_a(0) = 1$ if and only if $X_a = F$. Suppose that $n > 0$.

$P_a(0) = 1$ if and only if $x \in X_a$ for all $x < f_1(a)$.

$P_a(2i - 1) = 1$ if and only if $f_i(a) \in X$.

For $i < n$, $P_a(2i) = 1$ if and only if $x \in X_a$ for all $x \in (f_i(a), f_{i+1}(a))$.

$P(2n) = 1$ if and only if $x \in X_a$ for all $x > f_n(a)$.

For each possible pattern $\sigma \in 2^{2n+1}$, let $A_{n,\sigma} = \{a \in A_n : P_a = \sigma\}$. Each $A_{n,\sigma}$ is semialgebraic. For each $A_{n,\sigma}$, we will give a decomposition of $\{(x,y) \in X : x \in A_{n,\sigma}\}$ into disjoint cells. Because the $A_{n,\sigma}$ partition F, this will suffice.

Fix one $A_{n,\sigma}$. By Corollary 3.3.25, we can partition $A_{n,\sigma} = C_1 \cup \ldots \cup C_l$, where each C_j is either an interval or a singleton and f_i is continuous on C_j for $i \leq n, j \leq l$. We can now give a decomposition of $\{(x,y) : x \in A_{n,\sigma}\}$ into cells such that each cell is either contained in X or disjoint from X.

For $j \leq l$, let $D_{j,0} = \{(x,y) : x \in C_j$ and $y < f_1(x)\}$.

For $j \leq l$ and $1 \leq i \leq n$, let $D_{j,2i-1} = \{(x, f_i(x)) : x \in C_j\}$.

For $j \leq l$ and $1 \leq i < n$, let $D_{j,2i} = \{(x,y) : x \in C_j, f_i(x) < y < f_{i+1}(x)\}$.

For $j \leq l$, let $D_{j,2n} = \{(x,y) : x \in C_j, y > f_n(x)\}$.

Clearly, each $D_{j,i}$ is a cell, $\bigcup D_{j,i} = \{(x,y) : x \in A_{n,\sigma}\}$, and each $D_{j,i}$ is either contained in X or disjoint from X. Thus, taking the $D_{j,i}$ that are contained in X, we get a partition of $\{(x,y) \in X : x \in A_{n,\sigma}\}$ into disjoint cells.

3.4   Exercises and Remarks

Exercise 3.4.1 Let $\mathcal{L} = \{E\}$ where E is a binary relation symbol. For each of the following theories either prove that they have quantifier elimination or give an example showing that they do not have quantifier elimination and a natural $\mathcal{L}' \supset \mathcal{L}$ in which they do have quantifier elimination.

a) E has infinitely many classes all of size 2.

b) E has infinitely many classes all of which are infinite.

c) E has infinitely many classes of size 2, infinitely many classes of size 3, and every class has size 2 or 3.

d) E has one class of size n for each $n < \omega$.

Exercise 3.4.2 Show that $\mathrm{DAG_V}$ is the theory of torsion-free Abelian groups.

Exercise 3.4.3 a) Show that the theory of $(\mathbb{Z}, s)$ has quantifier elimination where $s(x) = x + 1$. Show that this theory is strongly minimal and that $\mathrm{acl}(A)$ is the set of elements "reachable" from A.

b) Show that the theory of $(\mathbb{N}, s)$ does not have quantifier elimination.

Exercise 3.4.4 Show that the theory of $(\mathbb{N}, <)$ admits quantifier elimination in the language where we add a function symbol s for the function $s(x) = x + 1$ and a constant symbol for 0. Show that every definable $X \subseteq \mathbb{N}$ is either finite or cofinite but that $(\mathbb{N}, <)$ is not strongly minimal.

Exercise 3.4.5 Show that in models of DAG, algebraic closure and definable closure agree and $\mathrm{acl}(A)$ is the $\mathbb{Q}$-vector space span of A.

Exercise 3.4.6 Consider the theory of $(\mathbb{Z}, +, 0, 1)$ in the language where we add the predicates P_n for the elements divisible by n. Axiomatize this theory and show that it has quantifier elimination. We call this the theory of $\mathbb{Z}$-groups.

Exercise 3.4.7 For G a $\mathbb{Z}$-group, there is a natural homomorphism Ψ : $G \to \prod_{n>0} \mathbb{Z}/n\mathbb{Z}$, given by $\Psi(g)(n) = g \bmod n$.

a) Let $D = \ker G$. Show that $D = \bigcap_{n=1}^{\infty} nG$, the subgroup of divisible elements of G.

b) Let $H \subseteq \prod_{n>0} \mathbb{Z}/n\mathbb{Z}$ be the image of ψ. Show that $G \cong D \oplus H$. [Hint: If A, B, C are Abelian groups, A is divisible, and there is a short exact sequence $0 \to A \to B \to C \to 0$, then $B \cong A \oplus C$.]

c) Show that if $\kappa \geq 2^{\aleph_0}$, then there are exactly $2^{2^{\aleph_0}}$ nonisomorphic $\mathbb{Z}$-groups of cardinality κ.

Exercise 3.4.8 It is useful to use a slightly more subtle homomorphism than Ψ. For p a prime, let $\mathbb{Z}_p$ be the additive group of p-adic integers.

a) Show that for each prime p there is a homomorphism $\phi_p : G \to \mathbb{Z}_p$ such that $g \bmod p^n G = \Phi_p(g) \bmod p^n \mathbb{Z}_p$ for all n.

Suppose that $(G, <, 0, 1)$ is an ordered group with least element 1, and for each prime p, $\Phi_p : G \to \mathbb{Z}_p$ such that p does not divide $\Phi_p(1)$. Let H be the divisible hull of G, and let $G^* = \{\frac{g}{n} \in H : g \in G \text{ and } n | \Phi_p(g) \text{ in } \mathbb{Z}_p$ for all primes $p\}$.

b) Show that G^* is a model of Presburger arithmetic.

c) Show that if G is a model of Presburger arithmetic and $\Phi_2, \Phi_3, \ldots$ are as in a), then $G^* = G$.

Exercise 3.4.9 Show that in $(\mathbb{Z}, +, 0, 1)$ we cannot define the ordering. [Hint: Find $G \models \mathrm{Pr}$ and a bijection $\alpha : G \to G$ such that α is a group isomorphism preserving 0 and 1 but not preserving $<$.]

Exercise 3.4.10 Show that in Presburger arithmetic $\exists y\ 2y = x$ is not equivalent to a quantifier-free $\mathcal{L}_r$-formula. [Hint: Find $G \subset H$ models of Pr, and $a \in G$ such that a is divisible by 2 in H but not in G.]

Exercise 3.4.11 Suppose that $\mathcal{M} \models \mathrm{Pr}$ and $A \subseteq M$. What is $\mathrm{acl}(A)$? What is $\mathrm{dcl}(A)$?

Exercise 3.4.12 In Exercise 2.5.17, we introduced the notion of existentially closed structures.

a) Show that if $\mathcal{M} \subseteq \mathcal{N}$ and $\mathcal{M}$ is existentially closed, then there is $\mathcal{M}_1 \models T$ such that $\mathcal{M} \subseteq \mathcal{N} \subseteq \mathcal{M}_1$ with $\mathcal{M} \prec \mathcal{M}_1$. [Note: See Exercise 2.5.15.]

b) Show that T is model-complete if and only if every model of T is existentially closed. [Hint: ($\Leftarrow$) Suppose that $\mathcal{M}_0 \subseteq \mathcal{N}_0$ are models of T. Use a) to build $\mathcal{M}_0 \subseteq \mathcal{N}_0 \subseteq \mathcal{M}_1 \subseteq \mathcal{N}_1 \subseteq \mathcal{M}_2 \subseteq \ldots$, a chain of models of T such that $M_i \prec M_{i+1}$ and $N_i \prec N_{i+1}$.]

c) Suppose that T is a $\forall\exists$-axiomatizable theory with infinite models that is κ-categorical for some infinite cardinal κ. Show that T is model-complete. [Hint: See Exercise 2.5.17.]

d) Show that T is model-complete if and only if for any formula $\phi(\overline{v})$ there is a quantifier-free formula $\psi(\overline{v}, \overline{w})$ such that $T \models \phi(\overline{v}) \leftrightarrow \exists\overline{w}\, \psi(\overline{v}, \overline{w})$. [Hint: Use Exercise 2.5.12.]

e) Show that any model-complete theory has a $\forall\exists$ axiomatization. [Hint: Use Exercise 2.5.15.]

Exercise 3.4.13 Suppose that T and T' are $\mathcal{L}$-theories. We say that T' is a *model companion* of T if

i) T' is model-complete,

ii) every model of T has an extension that is a model of T', and

iii) every model of T' has an extension that is a model of T.

a) Show that any theory has at most one model companion.

b) Show that DLO is the model companion of the theory of discrete linear orders.

c) Suppose that T is $\forall\exists$-axiomatizable. Show that if T' is a model companion of T, then T' is the theory of existentially closed models of T.

Exercise 3.4.14 If T' is a model companion of T and $T' \cup \mathrm{Diag}(\mathcal{M})$ is complete for any $\mathcal{M} \models T$, then T' is a *model completion* of T.

We say that T has the *amalgamation property* if and only if whenever $\mathcal{M}_0, \mathcal{M}_1$ and $\mathcal{M}_2$ are models of T and $f_i : \mathcal{M}_0 \to \mathcal{M}_i$ are embeddings there is $\mathcal{N} \models T$ and $g_i : \mathcal{M}_i \to \mathcal{N}$ such that $f_1 \circ g_1 = f_2 \circ g_2$.

a) Suppose that T' is a model companion of T. Show that T' is a model completion of T if and only if T has the amalgamation property.

b) Suppose that T has a universal axiomatization and T' is a model completion of T. Show that T' has quantifier elimination.

Exercise 3.4.15 Let $\mathcal{M}$ be an $\mathcal{L}$-structure. We say that a definable $X \subseteq M^n$ is *strongly minimal* if X is defined by an $\mathcal{L}_M$-formula $\phi(\overline{v})$ and for any $\mathcal{M} \prec \mathcal{N}$, if $X^{\mathcal{M}}$ is the subset of $\mathcal{N}^n$ defined by $\phi(\overline{v})$, then every definable subset of $X^{\mathcal{N}}$ is finite or cofinite.

Suppose that $X \subset M^n$ is strongly minimal. Let $\mathrm{acl}^X(A) = X \cap \mathrm{acl}(A)$ for $A \subset M$.

a) Show that, if $x \in \mathrm{acl}^X(A \cup \{b\})$ and $x \notin \mathrm{acl}^X(A)$, then $b \in \mathrm{acl}^X(A \cup \{a\})$. (This is called the *exchange principle*.)

b) We say that $A \subseteq X$ is *independent* if $a \notin \mathrm{acl}^X(A \setminus \{a\})$ for all $a \in X$. Show that for any $A \subseteq X$, there is an independent $B \subseteq A$ such that $A = \mathrm{acl}^X(B)$. We call B a *basis* for X.

c) Suppose that B_0 and B_1 are bases for A. Show that $|B_0| = |B_1|$.

Exercise 3.4.16 Prove Lemma 3.2.18. [Hint: Use Exercise 1.4.11 in $\mathcal{M}^{\mathrm{eq}}$.]

Exercise 3.4.17 Let K be an algebraically closed field and let $V \subseteq K^n$ be Zariski closed. We say that V is *irreducible* if there are Zariski closed sets $F_0, F_1 \subset V$ such that $V = F_0 \cup F_1$.

a) Show that V is irreducible if and only if $I(V)$ is a prime ideal.

b) Suppose that $V \subseteq K^n$ is an irreducible Zariski closed set and $K \prec F$. Let $\phi(\overline{v})$ be the system of equations that defines V. Let $V(F) \subseteq F^n$ be the solutions to $\phi(\overline{v})$ in F^n. Use model-completeness to show that $V(F)$ is also irreducible.

c) Show that every Zariski closed set is a finite union of irreducible sets. Indeed, if V is Zariski closed, there are irreducible sets $V_1, \ldots, V_n$ such that $V = V_1 \cup \ldots \cup V_n$, $V \ne \bigcup_{i \in I_0}$ for any proper $I_0 \subset \{1, \ldots, n\}$. This decomposition is unique in that if $W_1, \ldots, W_m$ are irreducible closed sets, $V = W_1 \cup \ldots \cup W_m$, and $W \ne \bigcup_{j \in J_0} W_j$ for any proper $J_0 \subset \{1, \ldots, m\}$, then $n = m$ and $\{V_1, \ldots, V_n\} = \{W_1, \ldots, W_m\}$. We call $V_1, \ldots, V_n$ the *irreducible components* of V. [This can be proved using Lemma 3.2.10 but can also be proved directly (see also Exercise 8.4.15).

Exercise 3.4.18 Suppose that K is an algebraically closed field and $P \subset K[X_1, \ldots, X_n]$ is a maximal ideal. Show that P is generated by $X_1 - a_1, \ldots, X_n - a_n$ for some $a_1, \ldots, a_n \in K$.

Exercise 3.4.19 (Elimination of Imaginaries for ACF) Show that if K is an algebraically closed field, then for all $x \in K^{\mathrm{eq}}$ there is $\overline{y} \in K^n$ such that $x \in \mathrm{dcl}^{\mathrm{eq}}(\overline{y})$ and $\overline{y} \in \mathrm{dcl}^{\mathrm{eq}}(x)$.

Exercise 3.4.20 Let $K \subset L$ be algebraically closed fields. Let $V, W \subseteq L^n$ be Zariski closed sets defined over K. Suppose that there is $f : V \to W$ a bijective polynomial map defined over L. Show that there is $g : V \cap K^n \to W \cap K^n$ a bijective polynomial map defined over K.

Exercise 3.4.21 (Positive Quantifier Elimination) We say that an $\mathcal{L}$-formula $\phi(\overline{v})$ is *positive* if it is in the smallest collection of $\mathcal{L}$-formulas containing the atomic formulas and closed under $\wedge, \vee, \exists$, and $\forall$.

We say that $\eta : \mathcal{M} \to \mathcal{N}$ is a *$\mathcal{L}$-homomorphism* if and only if:

i) $\eta(c^{\mathcal{M}}) = \eta(c^{\mathcal{N}})$ for all constants c in $\mathcal{L}$;

ii) $\eta(f^{\mathcal{M}}(\overline{x})) = f^{\mathcal{N}}(\eta(\overline{x}))$ for all $\overline{x} \in M$ and function symbols f;

iii) if R is a relation symbol of $\mathcal{L}$ and $\overline{x} \in R^{\mathcal{M}}$, then $\eta(\overline{x}) \in R^{\mathcal{N}}$.

a) Show that if $f : \mathcal{M} \to \mathcal{N}$ is a surjective $\mathcal{L}$-homomorphism, $\overline{a} \in M$, $\phi(\overline{v})$ is positive, and $\mathcal{M} \models \phi(\overline{a})$, then $\mathcal{N} \models \phi(\overline{a})$.

Let T be a complete $\mathcal{L}$-theory and $\phi(\overline{v})$ be an $\mathcal{L}$-formula such that $T \models \exists \overline{v} \, \phi(\overline{v})$. We will prove that the following are equivalent.

i) There is a positive quantifier-free formula $\psi(\overline{v})$ such that $T \models \forall \overline{v} \, \phi(\overline{v}) \leftrightarrow \psi(\overline{v})$.

ii) For all $\mathcal{M}, \mathcal{N} \models T$ and $\mathcal{A} \subseteq \mathcal{M}$, if $f : \mathcal{A} \to \mathcal{N}$ is an $\mathcal{L}$-homomorphism, $\overline{a} \in \mathcal{A}$, and $\mathcal{M} \models \phi(\overline{a})$, then $\mathcal{N} \models \phi(\overline{a})$.

b) Show that i)$\Rightarrow$ ii).

Assume that ii) holds. Let $\Gamma(\overline{v}) = \{\psi(\overline{v}) : \psi \text{ is a positive quantifier-free}$ formula and $T \models \psi(\overline{v}) \to \phi(\overline{v})\}$. Let $\Sigma = T \cup \{\neg(\psi(\overline{c})) : \psi \in \Gamma\} \cup \{\phi(\overline{v})\}$.

c) Show that Σ is unsatisfiable. [Hint: Let $\mathcal{M} \models T$ with $\bar{c} \in M$ such that $\mathcal{M} \models \phi(\bar{c})$ and $\mathcal{M} \models \neg\psi(\bar{c})$ for $\psi \in \Gamma$. Let $\Sigma' = T \cup \neg\phi(\bar{c}) \cup \{\theta(\bar{c}) : \mathcal{M} \models \theta(\bar{c}), \theta(\bar{v})$ positive quantifier-free$\}$. Show that Σ' is satisfiable. Let $\mathcal{N} \models \Sigma'$. Let $\mathcal{A}$ be the substructure of $\mathcal{M}$ generated by $\bar{c}$ and apply ii) to get a contradiction.]

d) Show that ii)$\Rightarrow$ i).

Exercise 3.4.22 [†] (Completeness of Projective Varieties) Let K be an algebraically closed field. Suppose that $p_1, \ldots, p_k \in \mathbb{Z}[Y_1, \ldots, Y_n, X_0, \ldots, X_m]$ are homogeneous in $X_0, \ldots, X_m$ (i.e., if t is a new variable, then

$$p_i(\overline{Y}, tX_0, \ldots, tX_m) = t^d p_i(\overline{Y}, X_0, \ldots, X_m)$$

for some d). Let $\phi(\overline{y})$ be the formula

$$\exists \overline{x} \left(\bigwedge_{i=1}^{k} p_i(\overline{y}, \overline{x}) = 0 \wedge \bigvee_{i=0}^{m} x_i \neq 0 \right)$$

asserting that the system of equations $p_1(\overline{y}, \overline{x}) = \ldots = p_k(\overline{y}, \overline{x}) = 0$ has a nontrivial solution.

a) Show that $\phi(\overline{y})$ is equivalent to a positive quantifier-free formula. [Hint: Use Exercise 3.4.21. Suppose that A is a subring of K, L is an algebraically closed field and $\sigma : A \to L$ is a homomorphism. We may without loss of generality assume that A is a valuation ring (see [58] IX §3). If $(x_0, \ldots, x_m)$ is a nontrivial solution in A, there is an i such that each $\frac{x_j}{x_i} \in A$. Show that $L \models \phi(\sigma(\frac{x_0}{x_i}), \ldots, \sigma(\frac{x_n}{x_i}))$.

b) Let $\mathbb{P}^l$ denote projective l-space over K, and let $\pi : \mathbb{P}^n \times \mathbb{P}^m \to \mathbb{P}^m$ be the natural projection map. Show that π is a closed map in the Zariski topology.

Exercise 3.4.23 a) Show that for $q \in \mathbb{Q}$ we can order $\mathbb{Q}(t)$ such that $t - q$ is a positive infinitesimal. [Hint: Let $p(t) > 0$ if and only if there is $\epsilon > 0$ such that $p(x) > 0$ for $x \in (q, q + \epsilon)$.]

b) Show that we can order $\mathbb{Q}(t)$ such that t is infinite.

Exercise 3.4.24 Let x and y be algebraically independent over $\mathbb{R}$.

a) Show that $\mathbb{R}(x, y)$ is formally real and that we can find orders $<_1$ and $<_2$ of $\mathbb{R}(x, y)$ such that $x <_1 y$ and $y <_1 x$.

b) Use a) to show that the ordering $<$ is not quantifier-free definable in $\mathbb{R}$ in the language of rings.

Exercise 3.4.25 Show that the $\mathcal{L}_r$-theory of the real field is model-complete.

Exercise 3.4.26 Show that for real closed fields $\mathrm{dcl}(A) = \mathrm{acl}(A)$.

Exercise 3.4.27 Let F be a real closed field. We say that a function $g : F^n \to F$ is *algebraic* if there is a nonzero polynomial $p(\overline{X}, Y) \in F[\overline{X}, Y]$ such that for all $\overline{a} \in F$, $p(\overline{a}, g(\overline{a})) = 0$.

a) Use quantifier elimination to show that every semialgebraic function is algebraic.

b) Show that if $f : \mathbb{R} \to \mathbb{R}$ is semialgebraic, then there are disjoint intervals $I_1, \ldots, I^n$ and a finite set X such that $\mathbb{R} = I_1 \cup I_m \cup X$ and f is analytic on each I_j. (Hint: Use the Implicit Function Theorem for $\mathbb{R}$.)

Exercise 3.4.28 (Real Nullstellensatz) Let F be a real closed field, and let I be an ideal in $F[\overline{X}]$. Then, $V_F(I)$ is nonempty if and only if whenever $p_1, \ldots, p_m \in F[\overline{X}]$ and $\sum p_i^2 \in I$, then all the $p_i \in I$.

Exercise 3.4.29 Let C be a k-cell. Show that there is a semialgebraic homeomorphism $h : (0, 1)^k \to C$.

Exercise 3.4.30 (Dimension) Let $X \subseteq F^n$ be semialgebraic. In particular, let $\phi(\overline{v}, \overline{w})$ be a formula, and let $\overline{a} \in F^m$ be such that $X = \{\overline{x} \in F^n : \phi(\overline{x}, \overline{a})\}$. If $K \supset F$ is a real closed field, we define $\dim_K(X)$, the algebraic dimension of X in K, to be the maximum transcendence degree of $F(c_1, \ldots, c_n)$ over F, where $\overline{c} \in K^n$ and $K \models \phi(\overline{c}, \overline{a})$. We define $\dim(X)$, the *algebraic dimension* of X, to be the maximum value of $\dim_K(X)$ as K ranges over all real closed extensions of F.

a) Show that every k-cell has algebraic dimension k.

b) Show that $\dim(X_1 \cup \ldots \cup X_n) = \max \dim(X_i)$.

c) Show that if $f : F^n \to F^m$ is semialgebraic and $X \subseteq F^n$ is semialgebraic, then $\dim(X) \geq \dim(f(X))$.

d) Show that if $X \subseteq F^{n+m}$ is semialgebraic, then for all $k \leq m$, $\{\overline{a} \in F^n : \dim(X_{\overline{a}}) = k\}$ is semialgebraic.

e) Show that $X \subseteq F^n$ has dimension n if and only if X has a nonempty interior.

f) Show that if $U \subseteq F^n$ is open and semialgebraic, then U cannot be decomposed into a union of finitely many semialgebraic sets with empty interior.

g) Suppose that $U \subseteq F^n$ is open and semialgebraic and there is a semialgebraic $f : U \to F$. Show that there is $\overline{x} \in U$ such that f is continuous at $\overline{x}$.

Exercise 3.4.31 Prove the Cell Decomposition Theorem. [Hint: It is best to do this by proving a) and b) by simultaneous induction.]

a) Every semialgebraic set in $X \subseteq F^n$ can be written as a finite disjoint union of cells.

b) If $X \subseteq F^n$ is semialgebraic and $f : X \to F$ is semialgebraic, X can be partitioned into disjoint cells $C_1, \ldots, C_m$ such that for all i, $f|C_i$ is continuous.]

Exercise 3.4.32 We say that $X \subseteq F$ is *semialgebraically connected* if there are no semialgebraic open sets U_0 and U_1 such that each $U_i \cap X \neq \emptyset$, $U_0 \cap U_1 \cap X = \emptyset$, and $(U_0 \cup U_1) \cap X = X$.

a) Show that every cell is semialgebraically connected.

b) (Whitney's Finiteness Theorem) If X is semialgebraic, then $X = C_1 \cup \ldots \cup C_m$ where the C_i are pairwise disjoint and each C_i is semialgebraically connected and closed in X.

Exercise 3.4.33 Suppose that $\mathcal{M} = (G, +, <, \ldots)$ is o-minimal and G is an ordered group.
   a) Show that if $H \subseteq G$ is a nontrivial subgroup, then H is convex.
   b) Show that G is Abelian. [Hint: For each x, consider $\{g \in G : gx = xg\}$.]
   c) Show that G is divisible. [Hint: Consider the groups nG.]
   d) Show that G has definable Skolem functions.

Exercise 3.4.34 Suppose that $(F, +, \cdot, <)$ is an o-minimal field. Show that F is real closed. [Hint: Show that F has the intermediate value property.]

Exercise 3.4.35 If K is a field, let $K[[t]]$ denote the field of formal power series over K in variable t, and let $K((t))$ denote its fraction field, the field of formal Laurent series over K. Let

$$K\langle\langle t\rangle\rangle = \bigcup_{n=1}^{\infty} K\left(\left(t^{\frac{1}{n}}\right)\right)$$

be the field of formal *Puiseux series* over K. Series in $K\langle\langle t\rangle\rangle$ are of the form $\sum_{i=m}^{\infty} a_i t^{\frac{i}{n}}$ for some $m, n \in \mathbb{Z}$ with $n > 0$. An important theorem is that if K is algebraically closed, then $K\langle\langle t\rangle\rangle$ is also algebraically closed ([102] IV §3). Suppose that R is real closed.
   a) Show that $R\langle\langle t\rangle\rangle$ is real closed, $R \prec R\langle\langle t\rangle\rangle$, and t is a positive infinitesimal element of $R\langle\langle t\rangle\rangle$.
   b) Suppose that $r \in R$ and $f : (0, r) \to R$ is definable. Show that there is $\mu \in R\langle\langle t\rangle\rangle$ such that $R\langle\langle t\rangle\rangle \models f(t) = \mu$. Suppose that $\mu = at^q +$ higher-degree terms. Show that f is asymptotic to ax^q at 0. In other words, show that

$$R \models \forall \epsilon > 0 \, \exists \delta > 0 \left(0 < x < \delta \to \left|\frac{f(x)}{ax^q} - 1\right| < \epsilon\right).$$

Exercise 3.4.36 Suppose that $(D, +, \cdot, <, 0, 1)$ is an ordered integral domain with least element 1. We say that D is a model of *open induction.* if whenever $\phi(v, \overline{w})$ is a quantifier-free formula and $\overline{a} \in A$, then

$$D \models (\exists v > 0 \, \phi(v, \overline{a})) \to \exists v > 0 \, (\phi(v, \overline{a}) \wedge \forall w \, (0 < w < v \to \neg\phi(w, \overline{a}))).$$

In other words, the positive part of D satisfies the induction axioms for quantifier-free formulas.
   a) Let R be the real closure of the fraction field of D. Show that D is a model of open induction if and only if for every $r \in R$ there is $d \in D$ such that $|r - d| < 1$.

Let $D \subset \mathbb{R}\langle\langle t\rangle\rangle$, be the subring of series of the form

$$\sum_{i=m}^{0} a_i t^{\frac{i}{n}},$$

where $m, n \in \mathbb{Z}$, $m \leq 0$, $n > 0$ and $a_0 \in \mathbb{Z}$.
b) Use a) to show that D is a model of open induction.
c) Show that $D \models \exists a \exists b (b \neq 0 \wedge a^2 = 2b^2)$. This shows that the irrationality of $\sqrt{2}$ is independent of open induction.

Remarks

Tarski first showed completeness and decidability for the fields of real and complex numbers. His proof gave an explicit algorithm for eliminating quantifiers. Robinson showed that quantifier-elimination results could be proved by finding the right embedding theorems. These ideas were further extended by Blum. Robinson also introduced the notion of model-completeness and saw how it could be used to prove Hilbert's Nullstellensatz and answer Hilbert's 17th-problem. All of the results on model-complete theories developed in Exercises 3.4.12 and 3.4.13 are also due to Robinson.

The completeness and decidability of Presburger arithmetic is due to Presburger, although the proof of quantifier elimination given here is due to van den Dries. Theorem 3.2.20 is due to Poizat [84], although the proof given here is due to Lascar and Pillay.

The positive quantifier-elimination result of Exercise 3.4.21 is due to van den Dries [28], who showed how it could be used to prove the completeness of projective varieties and show that any closed semialgebraic set can be built up from unions and intersections of semialgebraic sets of the form $\{\overline{x} : f(\overline{x}) \geq 0\}$.

The real Nullstellensatz was originally proved by Krivine. His model-theoretic proof was not noticed by real algebraic geometers, and the result was proved again later by Dubois and Risler. Model-completeness and quantifier elimination have many applications in real algebraic geometry (see for example [12]). Exercises 3.4.33 and 3.4.34 on o-minimal ordered groups and fields are due to Pillay and Steinhorn. [83] Exercise 3.4.36 is due to Shepherdson. There are many algebraic constructions of models of open induction (see, for example [61]).

O-minimality was introduced by van den Dries, Pillay, and Steinhorn. Surprisingly, o-minimal structures have many of the good topological and geometric properties of strongly minimal sets. This material is developed carefully in [29].

Once we know that the theories of the real and complex fields are decidable, it is natural to wonder about the computational complexity of these theories. Our proofs of decidability give no complexity information, but more direct proofs do yield concrete bounds. Collins gave an explicit quantifier-elimination procedure using cylindric decomposition. His proof leads to a doubly exponential upper bound on the complexity. Although there are some good algorithms for attacking specific subproblems, Fischer and Rabin showed that the decision problem is exponentially hard. Indeed, they proved that there are exponential lower bounds even for the theory of nontrivial torsion-free divisible Abelian groups. See [21] for a collection of fundamental papers on these issues.

Tarski asked whether the theory of the structure $\mathbb{R}_{\exp} = (\mathbb{R}, +, \cdot, \exp, <, 0, 1)$ is decidable. In the early 1980s, van den Dries recast this question, asking whether this structure was o-minimal. This was answered positively by Wilkie [103].

Theorem 3.4.37 *The theory of* $\mathbb{R}_{\exp}$ *is model-complete and o-minimal.*

The decidability of the theory of $\mathbb{R}_{\exp}$ is still an open question, but Macintyre and Wilkie [62] have given a partial positive result.

Conjecture 3.4.38 (Schanuel's Conjecture) *If* $\lambda_1, \ldots \lambda_n \in \mathbb{C}$ *are linearly independent over* $\mathbb{R}$, *then the field* $\mathbb{Q}(\lambda_1, \ldots, \lambda_n, e^{\lambda_1}, \ldots, e^{\lambda_n})$ *has transcendence degree at least* n.

Theorem 3.4.39 *If Schanuel's Conjecture is true, then* $\mathrm{Th}(\mathbb{R}_{\exp})$ *is decidable.*

Although we know that $\mathbb{R}_{\exp}$ is model-complete, it does not have quantifier elimination. Quantifier elimination was proved by van den Dries, Macintyre, and Marker [30] if we add log and all restrictions of analytic functions to compact sets.

Because $\mathrm{Th}(\mathbb{R}_{\exp})$ is model complete it must have a $\forall\exists$-axiomatization. At present we have no clue what such an axiomatization might look like, even assuming Schanuel's Conjecture.

Ax, Kochen, and Ersov [3], [4], [5], and [55] investigated the model theory of the p-adic field. If K is a valued field, let $\overline{K}$ be the residue field, and $G(K)$ be the value group.

Theorem 3.4.40 *i) Let K and L be Henselian valued fields of characteristic zero. Then, $K \equiv L$ if and only if $\overline{K} \equiv \overline{L}$ and $G(K) \equiv G(L)$.*

ii) The theory of $\mathbb{Q}_p$ is decidable. It is exactly the theory of Henselian fields of characteristic zero with residue field $\mathbb{F}_p$ and value group a model of Presburger arithmetic.

One important corollary is that if D is a nonprincipal ultrafilter on the set of prime numbers, then $\prod \mathbb{Q}_p / D \cong \prod \mathbb{F}_p((t))$, where $\mathbb{F}_p((t))$ is the field of formal Laurent series over $\mathbb{F}_p$. This corollary was used to settle the following conjecture of Artin.

Corollary 3.4.41 *For any integer d, there is $M > 0$ such that if $p > M$ is prime and $f \in \mathbb{Q}_p[X_1, \ldots, X_n]$ is a homogeneous polynomial of degree $d < \sqrt{n}$, then there are $x_1, \ldots, x_n \in \mathbb{Q}_p$ not all zero such that $f(x_1, \ldots, x_n) = 0$.*

Macintyre [60] showed that $\mathbb{Q}_p$ has quantifier elimination if we add a predicate P_n for the nth-powers for $n = 2, 3, \ldots$. Macintyre's quantifier elimination was shown to have interesting applications by Denef [25] who used it to prove the rationality of certain p-adic zeta functions.

The model theory of modules is another interesting area that we will not discuss. The interested reader should consult [87].

4
Realizing and Omitting Types

4.1 Types

Suppose that $\mathcal{M}$ is an $\mathcal{L}$-structure and $A \subseteq M$. Let $\mathcal{L}_A$ be the language obtained by adding to $\mathcal{L}$ constant symbols for each $a \in A$. We can naturally view $\mathcal{M}$ as an $\mathcal{L}_A$-structure by interpreting the new symbols in the obvious way. Let $\mathrm{Th}_A(\mathcal{M})$ be the set of all $\mathcal{L}_A$-sentences true in $\mathcal{M}$. Note that $\mathrm{Th}_A(\mathcal{M}) \subseteq \mathrm{Diag}_{\mathrm{el}}(\mathcal{M})$.

Definition 4.1.1 Let p be the set of $\mathcal{L}_A$-formulas in free variables $v_1, \ldots, v_n$. We call p an n-*type* if $p \cup \mathrm{Th}_A(\mathcal{M})$ is satisfiable. We say that p is a *complete* n-*type* if $\phi \in p$ or $\neg \phi \in p$ for all $\mathcal{L}_A$-formulas ϕ with free variables from $v_1, \ldots, v_n$. We let $S_n^{\mathcal{M}}(A)$ be the set of all complete n-types.

We sometimes refer to incomplete types as *partial types*. Also, we often write $p(v_1, \ldots, v_n)$ to stress that p is an n-type.

By the Compactness Theorem, we could replace "satisfiable" by "finitely satisfiable" in Definition 4.1.1.

Consider the example $\mathcal{M} = (\mathbb{Q}, <)$ where A is the set of natural numbers. Let $p(v)$ be the set of formulas $\{v > 1, v > 2, v > 3, \ldots\}$. If Δ is a finite subset of $p(v) \cup \mathrm{Th}_A(\mathcal{M})$, then we see that Δ is satisfiable by interpreting v as a sufficiently large element of $\mathbb{Q}$. By the Compactness Theorem, $p(v) \cup \mathrm{Th}_A(\mathcal{M})$ is satisfiable and $p(v)$ is a 1-type.

For the same structure, let $q(v) = \{\phi(v) \in \mathcal{L}_A : \mathcal{M} \models \phi(\frac{1}{2})\}$. For example the formula $v < 3$ is in $q(v)$, whereas $v > 2$ is not. For any $\mathcal{L}_A$-formula $\psi(v)$, either $\mathcal{M} \models \psi(\frac{1}{2})$ or $\mathcal{M} \models \neg\psi(\frac{1}{2})$. Thus, $q(v)$ is a complete 1-type.

The latter example can be generalized to produce complete types in arbitrary structures. If $\mathcal{M}$ is any $\mathcal{L}$-structure, $A \subset M$, and $\bar{a} = (a_1, \ldots, a_n) \in M^n$, let $\mathrm{tp}^{\mathcal{M}}(\bar{a}/A) = \{\phi(v_1, \ldots, v_n) \in \mathcal{L}_A : \mathcal{M} \models \phi(a_1, \ldots, a_n)\}$. Then, $\mathrm{tp}^{\mathcal{M}}(\bar{a}/A)$ is a complete n-type. We write $\mathrm{tp}^{\mathcal{M}}(\bar{a})$ for $\mathrm{tp}^{\mathcal{M}}(\bar{a}/\emptyset)$.

Definition 4.1.2 If p is an n-type over A, we say that $\bar{a} \in M^n$ *realizes* p if $\mathcal{M} \models \phi(\bar{a})$ for all $\phi \in p$. If p is not realized in $\mathcal{M}$ we say that $\mathcal{M}$ *omits* p.

In the examples given above, $p(v)$ is not realized in $\mathcal{M} = (\mathbb{Q}, <)$, whereas clearly $1/2$ realizes $q(v)$. In fact, there are many realizations of $q(v)$ in $\mathcal{M}$. Suppose that r is any rational number with $0 < r < 1$. We can construct an automorphism σ of $\mathcal{M}$ that fixes every natural number but $\sigma(1/2) = r$. Because σ fixes all elements of A, σ is also an $\mathcal{L}_A$-automorphism. By Theorem 1.1.10,

$$\mathcal{M} \models \phi(1/2) \Leftrightarrow \mathcal{M} \models \phi(r).$$

Thus, r also realizes $q(v)$.

In fact, the elements of $\mathbb{Q}$ that realize $q(v)$ are exactly the rational numbers s such that $0 < s < 1$. If $s \leq 0$, then the formula $0 < v$ is in $q(v)$ but $\mathcal{M} \models \neg(0 < s)$. Thus, s does not realize $q(v)$. Similarly, no $s \geq 1$ realizes $q(v)$.

The Compactness Theorem tells us that every type can be realized in an elementary extension.

Proposition 4.1.3 *Let $\mathcal{M}$ be an $\mathcal{L}$-structure, $A \subseteq M$, and p an n-type over A. There is $\mathcal{N}$ an elementary extension of $\mathcal{M}$ such that p is realized in $\mathcal{N}$.*

Proof Let $\Gamma = p \cup \mathrm{Diag}_{\mathrm{el}}(\mathcal{M})$. We claim that Γ is satisfiable.

Suppose that Δ is a finite subset of Γ. Without loss of generality, Δ is the single formula

$$\phi(v_1, \ldots, v_n, a_1, \ldots, a_m) \wedge \psi(a_1, \ldots, a_m, b_1, \ldots, b_l),$$

where $a_1, \ldots, a_m \in A$, $b_1, \ldots, b_l \in M \setminus A$, $\phi(\bar{v}, \bar{a}) \in p$, and $\mathcal{M} \models \psi(\bar{a}, \bar{b})$. Let $\mathcal{N}_0$ be a model of the satisfiable set of sentences $p \cup \mathrm{Th}_A(\mathcal{M})$. Because $\exists \bar{w} \; \psi(\bar{a}, \bar{w}) \in \mathrm{Th}_A(\mathcal{M})$,

$$\mathcal{N}_0 \models \phi(\bar{v}, \bar{a}) \wedge \exists \bar{w} \; \psi(\bar{a}, \bar{w}).$$

By interpreting $b_1, \ldots, b_l$ as witnesses to $\exists \bar{w} \; \psi(a_1, \ldots, a_m, \bar{w})$, we make $\mathcal{N}_0 \models \Delta$. Thus, Δ is satisfiable.

By the Compactness Theorem, Γ is satisfiable. Let $\mathcal{N} \models \Gamma$. Because $\mathcal{N} \models \mathrm{Diag}_{\mathrm{el}}(\mathcal{M})$, the map that sends $m \in M$ to the interpretation of the constant symbol m in $\mathcal{N}$ is an elementary embedding. Let $c_i \in N$ be the interpretations of v_i. Then, $(c_1, \ldots, c_n)$ is a realization of p.

It is worth noting that if $\mathcal{N}$ is an elementary extension of $\mathcal{M}$, then $\mathrm{Th}_A(\mathcal{M}) = \mathrm{Th}_A(\mathcal{N})$. Thus $S_n^{\mathcal{M}}(A) = S_n^{\mathcal{N}}(A)$. This observation and Proposition 4.1.3 yield a characterization of complete types.

Corollary 4.1.4 $p \in S_n^{\mathcal{M}}(A)$ *if and only if there is an elementary extension $\mathcal{N}$ of $\mathcal{M}$ and $\bar{a} \in N^n$ such that $p = \text{tp}^{\mathcal{N}}(\bar{a}/A)$.*

Proof If $\bar{a} \in N^n$, then $\text{tp}^{\mathcal{N}}(\bar{a}/A) \in S_n^{\mathcal{N}}(A) = S_n^{\mathcal{M}}(A)$. On the other hand if $p \in S_n^{\mathcal{M}}(A)$, then, by Proposition 4.1.3, there is an elementary extension $\mathcal{N}$ of $\mathcal{M}$ and $\bar{a} \in \mathcal{M}$ realizing p. Because p is complete, if $\phi(\bar{v}) \in \mathcal{L}_A$, then exactly one of $\phi(\bar{v})$ and $\neg\phi(\bar{v})$ is in p. Thus, $\phi(\bar{v}) \in \text{tp}^{\mathcal{N}}(\bar{a}/A)$ if and only if $\phi(\bar{v}) \in p$ and $p = \text{tp}^{\mathcal{N}}(\bar{a}/A)$.

Complete types tell us what possible first-order properties elements can have in an elementary extension. What does it mean if two elements of a structure $\mathcal{M}$ realize the same complete type over A? Let us return to the example where $\mathcal{M} = (\mathbb{Q}, <)$ and A is the natural numbers. We showed that $a, b \in \mathbb{Q}$ realize the same complete 1-type over A if and only if there is an automorphism σ of $\mathcal{M}$ fixing A such that $\alpha(a) = b$. Although this is not true in general (see, for example, Exercises 4.5.1 and 4.5.9), it is if we allow passage to an elementary extension.

Proposition 4.1.5 *Suppose that $\mathcal{M}$ is an $\mathcal{L}$-structure and $A \subseteq M$. Let $\bar{a}, \bar{b} \in M^n$ such that $\text{tp}^{\mathcal{M}}(\bar{a}/A) = \text{tp}^{\mathcal{M}}(\bar{b}/A)$. Then, there is $\mathcal{N}$ an elementary extension of $\mathcal{M}$ and σ an automorphism of $\mathcal{N}$ fixing all elements of A such that $\sigma(\bar{a}) = \bar{b}$.*

If $\mathcal{M}$ and $\mathcal{N}$ are $\mathcal{L}$-structures and $B \subseteq M$, we say that $f : B \to N$ is a *partial elementary map* if and only if

$$\mathcal{M} \models \phi(\bar{b}) \Leftrightarrow \mathcal{N} \models \phi(f(\bar{b}))$$

for all $\mathcal{L}$-formulas ϕ and all finite sequences $\bar{b}$ from B. We will prove Proposition 4.1.5 by carefully iterating the following lemma and its corollary.

Lemma 4.1.6 *Let $\mathcal{M}, \mathcal{N}, B$ be as above and let $f : B \to N$ be partial elementary. If $b \in M$, there is an elementary extension $\mathcal{N}_1$ of $\mathcal{N}$ and $g : B \cup \{b\} \to \mathcal{N}_1$ a partial elementary map extending f.*

Proof Let $\Gamma = \{\phi(v, f(a_1), \ldots, f(a_n)) : \mathcal{M} \models \phi(b, a_1, \ldots, a_n), a_1, \ldots, a_n \in B\} \cup \text{Diag}_{\text{el}}(\mathcal{N})$.

Suppose that we find a structure $\mathcal{N}_1$ and an element $c \in N_1$ satisfying all of the formulas in Γ. Because $\mathcal{N}_1 \models \text{Diag}_{\text{el}}(\mathcal{N})$, $\mathcal{N}_1$ is an elementary extension of $\mathcal{N}$. It is also easy to see that we can extend f to a partial elementary map by $b \mapsto c$.

Thus, it suffices to show that Γ is satisfiable. By the Compactness Theorem it suffices to show that every finite subset of Γ is satisfiable in $\mathcal{N}$. Taking conjunctions, it is enough to show that if $\mathcal{M} \models \phi(b, a_1, \ldots, a_n)$, then $\mathcal{N} \models \exists v\, \phi(v, f(a_1), \ldots, f(a_n))$. But this is clear because $\mathcal{M} \models \exists v\, \phi(v, a_1, \ldots, a_n)$ and f is partial elementary.

Corollary 4.1.7 *If $\mathcal{M}$ and $\mathcal{N}$ are $\mathcal{L}$-structures, $B \subseteq M$ and $f : B \to N$ is a partial elementary map, then there is $\mathcal{N}'$ an elementary extension of $\mathcal{N}$ and $g : \mathcal{M} \to \mathcal{N}'$ an elementary embedding.*

Proof Let $\kappa = |M|$, and let $\{a_\alpha : \alpha < \kappa\}$ be an enumeration of M. Let $\mathcal{N}_0 = \mathcal{N}$, $B_0 = B$, and $g_0 = f$. Let $B_\alpha = B \cup \{a_\beta : \beta < \alpha\}$. We inductively build an elementary chain $(\mathcal{N}_\alpha : \alpha < \kappa)$ and $g_\alpha : B_\alpha \to N_\alpha$ partial elementary such that $g_\beta \subseteq g_\alpha$ for $\beta < \alpha$.

If $\alpha = \beta + 1$, and $g_\beta : B_\beta \to N_\beta$ is partial elementary, then, by Proposition 4.1.3, we can find $N_\beta \prec N_\alpha$ and $g_\alpha : B_\alpha \to N_\alpha$ extending g_β.

If α is a limit ordinal, let $N_\alpha = \bigcup_{\beta < \alpha} N_\beta$ and $g_\alpha = \bigcup_{\beta < \alpha} g_\beta$. By Lemma 2.3.11, $\mathcal{N}_\alpha$ is an elementary extension of N_β for $\beta < \alpha$ and f_α is a partial elementary map.

Let $\mathcal{N}' = \bigcup_{\alpha < \kappa} \mathcal{N}_\alpha$ and $g = \bigcup_{\alpha < \kappa} g_\alpha$. Again by Lemma 2.3.11, $\mathcal{N} \prec \mathcal{N}'$ and g is partial elementary. But $\mathrm{dom}(g) = M$, so g is an elementary embedding of $\mathcal{M}$ into $\mathcal{N}'$.

Proof of 4.1.5 Let $f : A \cup \{a\} \to A \cup \{b\}$ such that $f|A$ is the identity and $f(a) = b$. Because $\mathrm{tp}^{\mathcal{M}}(a/A) = \mathrm{tp}^{\mathcal{M}}(b/A)$, f is a partial elementary map. By Corollary 4.1.7 there is $\mathcal{N}_0$ an elementary extension of $\mathcal{M}$ and $f_0 : \mathcal{M} \to \mathcal{N}_0$ an elementary embedding extending f. We will build a sequence of elementary extensions

$$\mathcal{M} = \mathcal{M}_0 \prec \mathcal{N}_0 \prec \mathcal{M}_1 \prec \mathcal{N}_1 \prec \mathcal{M}_2 \prec \mathcal{N}_2 \ldots$$

and elementary embeddings $f_i : \mathcal{M}_i \to \mathcal{N}_i$ such that $f_0 \subseteq f_1 \subseteq f_2 \ldots$ and N_i is contained in the image of f_{i+1}. Having done this, let

$$\mathcal{N} = \bigcup_{i < \omega} \mathcal{N}_i = \bigcup_{i < \omega} \mathcal{M}_i$$

and $\sigma = \bigcup f_i$. By Lemma 2.3.11, $\mathcal{N}$ is an elementary extension of $\mathcal{M}$ and $\sigma : \mathcal{N} \to \mathcal{N}$ is an elementary map such that $\sigma|A$ is the identity and $\sigma(a) = b$. By construction σ is surjective. Thus, σ is the desired automorphism.

We now describe the construction. Given $f_i : \mathcal{M}_i \to \mathcal{N}_i$, we can view f_i^{-1} as a partial elementary map from the image of f_i into $\mathcal{M}_i \prec \mathcal{N}_i$. By Corollary 4.1.7, we can find $\mathcal{M}_{i+1}$ an elementary extension of $\mathcal{N}_i$ and extend f_i^{-1} to an elementary embedding $g_i : \mathcal{N}_i \to \mathcal{M}_{i+1}$. We can view g_i^{-1} as a partial elementary map from the image of g into $\mathcal{N}_i \prec \mathcal{M}_{i+1}$. Again by Corollary 4.1.7, we can find $\mathcal{N}_{i+1}$ an elementary extension of $\mathcal{M}_{i+1}$ and an elementary embedding $f_{i+1} : \mathcal{M}_{i+1} \to \mathcal{N}_{i+1}$ extending g_i^{-1}. Because $f_{i+1} \supseteq g_i^{-1}$ and $g_i \supseteq f_i^{-1}$, $f_{i+1} \supseteq f_i$. Because N_i is the domain of g_i, N_i is in the range of f_{i+1}.

Stone Spaces

There is a natural topology on the space of complete n-types $S_n^{\mathcal{M}}(A)$. For ϕ an $\mathcal{L}_A$-formula with free variables from $v_1, \ldots, v_n$, let

$$[\phi] = \{p \in S^{\mathcal{M}}(A) : \phi \in p\}.$$

If p is a complete type and $\phi \vee \psi \in p$, then $\phi \in p$ or $\psi \in p$. Thus $[\phi \vee \psi] = [\phi] \cup [\psi]$. Similarly, $[\phi \wedge \psi] = [\phi] \cap [\psi]$.

The *Stone topology* on $S_n^{\mathcal{M}}(A)$ is the topology generated by taking the sets $[\phi]$ as basic open sets. For complete types p, exactly one of ϕ and $\neg\phi$ is in p. Thus, $[\phi] = S_n^{\mathcal{M}}(A) \setminus [\neg\phi]$ is also closed. We refer to sets that are both closed and open as *clopen*.

The topology of the type spaces will eventually play an important role. The next lemmas summarize some of the basic topological properties.

Lemma 4.1.8 *i)* $S_n^{\mathcal{M}}(A)$ *is compact.*

ii) $S_n^{\mathcal{M}}(A)$ *is totally disconnected, that is if* $p, q \in S_n^{\mathcal{M}}(A)$ *and* $p \neq q$, *then there is a clopen set* X *such that* $p \in X$ *and* $q \notin X$.

Proof

i) It suffices to show that every cover of $S_n^{\mathcal{M}}(A)$ by basic open sets has a finite subcover. Suppose not. Let $C = \{[\phi_i(\overline{v})] : i \in I\}$ be a cover of $S_n^{\mathcal{M}}(A)$ by basic open sets with no finite subcover. Let

$$\Gamma = \{\neg\phi_i(\overline{v}) : i \in I\}.$$

We claim that $\Gamma \cup \mathrm{Th}_A(\mathcal{M})$ is satisfiable. If I_0 is a finite subset of I, then because there is no finite subcover of C, there is a type p such that

$$p \notin \bigcup_{i \in I_0} [\phi_i].$$

Let $\mathcal{N}$ be an elementary extension of $\mathcal{M}$ containing a realization $\overline{a}$ of p. Then

$$\mathcal{N} \models \mathrm{Th}_A(\mathcal{M}) \cup \bigwedge_{i \in I_0} \neg\phi_i(\overline{a}).$$

We have shown that Γ is finitely satisfiable and hence, by the Compactness Theorem, satisfiable.

Let $\mathcal{N}$ be an elementary extension of $\mathcal{M}$, and let $\overline{a} \in \mathcal{N}$ realize Γ. Then

$$\mathrm{tp}^{\mathcal{N}}(\overline{a}/A) \in S_n^{\mathcal{M}}(A) \setminus \bigcup_{i \in I} [\phi_i(\overline{v})],$$

a contradiction.

ii) If $p \neq q$, there is a formula ϕ such that $\phi \in p$ and $\neg\phi \in q$. Thus, $[\phi]$ is a basic clopen set separating p and q.

Natural operations on types often give rise to continuous operations on the type space.

Lemma 4.1.9 *i) If $A \subseteq B \subset M$ and $p \in S_n^{\mathcal{M}}(B)$, let $p|A$ be the set of $\mathcal{L}_A$-formulas in p. Then, $p|A \in S_n^{\mathcal{M}}(A)$ and $p \mapsto p|A$ is a continuous map from $S_n^{\mathcal{M}}(B)$ onto $S_n^{\mathcal{M}}(A)$.*

ii) If $f : \mathcal{M} \to \mathcal{N}$ is an elementary embedding and $p \in S_n^{\mathcal{M}}(A)$, let

$$f(p) = \{\phi(\overline{v}, f(\overline{a})) : \phi(\overline{v}, \overline{a}) \in p\}.$$

Then, $f(p) \in S_n^{\mathcal{N}}(f(A))$ and $p \mapsto f(p)$ is continuous.

iii) If $f : A \to \mathcal{N}$ is partial elementary, then $S_n^{\mathcal{M}}(A)$ is homeomorphic to $S_n^{\mathcal{N}}(f(A))$.

Proof

i) Because $p|A \cup \mathrm{Th}_A(\mathcal{M}) \subseteq p \cup \mathrm{Th}_B(\mathcal{M})$, $p|A \cup \mathrm{Th}_A(\mathcal{M})$ is satisfiable. Because $p|A$ is the set of all $\mathcal{L}_A$-formulas in p, $p|A$ is complete. If ϕ is an $\mathcal{L}_A$-formula, then

$$\{p \in S_n^{\mathcal{M}}(B) : \phi \in p\} = [\phi].$$

Thus, ϕ is continuous.

If $q \in S_n^{\mathcal{M}}(A)$, there is an elementary extension $\mathcal{N}$ of $\mathcal{M}$ and $\overline{a} \in N$ realizing q. Then, $p = \mathrm{tp}^{\mathcal{N}}(\overline{a}/B) \in S_n^{\mathcal{M}}(B)$ and $p|A = q$. Thus, the restriction map is surjective.

ii) Suppose that Δ is a finite subset of $f(p)$. Say

$$\Delta = \{\phi_1(\overline{v}, f(\overline{a})), \ldots, \phi_m(\overline{v}, f(\overline{a}))\}$$

where $\phi_1(\overline{v}, \overline{a}), \ldots, \phi_m(\overline{v}, \overline{a}) \in p$. Because $p \cup \mathrm{Th}_A(\mathcal{M})$ is consistent,

$$\mathcal{M} \models \exists \overline{v} \bigwedge_{i=1}^{m} \phi_i(\overline{v}, \overline{a}).$$

Because f is elementary,

$$\mathcal{N} \models \exists \overline{v} \bigwedge_{i=1}^{m} \phi_i(\overline{v}, f(\overline{a}))$$

and $f(p) \cup \mathrm{Th}_{f(A)}(\mathcal{N})$ is consistent. It is easy to see that $f(p)$ is complete. Because

$$\{p \in S_n^{\mathcal{M}}(A) : \phi(\overline{v}, f(\overline{a})) \in f(p)\} = [\phi(\overline{v}, \overline{a})],$$

$p \mapsto f(p)$ is continuous.

iii) Exercise 4.5.12.

Definition 4.1.10 We say that $p \in S_n^{\mathcal{M}}(A)$ is *isolated* if $\{p\}$ is an open subset of $S_n^{\mathcal{M}}(A)$.

Isolated points will play an important role in Section 4.2.

Proposition 4.1.11 *Let $p \in S_n^{\mathcal{M}}(A)$. The following are equivalent.*
i) p is isolated.
ii) $\{p\} = [\phi(\overline{v})]$ for some $\mathcal{L}_A$-formula $\phi(\overline{v})$. We say that $\phi(\overline{v})$ isolates p.
iii) There is an $\mathcal{L}_A$-formula $\phi(\overline{v}) \in p$ such that for all $\mathcal{L}_A$-formulas $\psi(\overline{v})$,
$\psi(\overline{v}) \in p$ if and only if

$$\mathrm{Th}_A(\mathcal{M}) \models \phi(\overline{v}) \to \psi(\overline{v}).$$

Proof
i) $\Rightarrow$ ii) If X is open, then

$$X = \bigcup_{i \in I} [\phi_i]$$

for some collection of formulas $(\phi_i : i \in I)$. If $\{p\}$ is open, then $\{p\} = [\phi]$ for some formula ϕ.

ii) $\Rightarrow$ iii) Suppose that $\{p\} = [\phi(\overline{v})]$. Suppose that $\psi(\overline{v}) \in p$. We claim that $\mathrm{Th}_A(\mathcal{M}) \models \phi(\overline{v}) \to \psi(\overline{v})$. If not, then there is an elementary extension $\mathcal{N}$ of $\mathcal{M}$ and $\overline{a} \in N$ such that $\mathcal{N} \models \phi(\overline{a}) \wedge \neg\psi(\overline{a})$. Let $q = \mathrm{tp}^{\mathcal{N}}(\overline{a}/A) \in S_n^{\mathcal{M}}(A)$. Because $\phi(\overline{v}) \in q$, $q = p$. But $\neg\psi(\overline{v}) \in q$, a contradiction.
    If, on the other hand, $\psi(\overline{v}) \notin p$, then $\neg\psi(\overline{v}) \in p$ and, by the argument above, $\mathrm{Th}_A(\mathcal{M}) \models \phi(\overline{v}) \to \neg\psi(\overline{v})$. Because $\mathrm{Th}_A(\mathcal{M}) \cup \{\phi(\overline{v})\}$ is satisfiable, $\mathrm{Th}_A(\mathcal{M}) \not\models \phi(\overline{v}) \to \psi(\overline{v})$.

iii) $\Rightarrow$ i) We claim that $[\phi(\overline{v})] = \{p\}$. Clearly, $p \in [\phi(\overline{v})]$. Suppose that $q \in [\phi(\overline{v})]$ and $\psi(\overline{v})$ is an $\mathcal{L}_A$-formula. If $\psi(\overline{v}) \in p$, then $\mathrm{Th}_A(\mathcal{M}) \models \phi(\overline{v}) \to \psi(\overline{v})$ and $\psi(\overline{v}) \in q$. On the other hand, if $\psi(\overline{v}) \notin p$, then $\neg\psi(\overline{v}) \in p$ and, by the argument above, $\psi(\overline{v}) \notin q$. Thus $p = q$.

Examples

We conclude this section by giving concrete descriptions of $S_n^{\mathcal{M}}(A)$ for several important examples.

Example 4.1.12 *Dense Linear Orders*

Let $\mathcal{L} = \{<\}$. Let $\mathcal{M} = (M, <)$ be a dense linear order without endpoints and let $A \subseteq M$. Let $p \in S_1^{\mathcal{M}}(A)$. If $a \in A$, then, because p is a complete type, exactly one of the formulas $v = a$, $v < a$, or $v > a$ is in p.

<u>case 1</u>: p is realized in A.
    In other words, the formula $v = a \in p$ for some $a \in A$. In this case, $p = \{\psi(v) : \mathcal{M} \models \psi(a)\}$ and p is isolated by the formula $v = a$.

<u>case 2</u>: Otherwise.
    Let $L_p = \{a \in A : a < v \in p\}$ and $U_p = \{a \in A : v < a \in p\}$. If $a < v, v < b \in p$, then, because $p \cup \mathrm{Th}_A(\mathcal{M})$ is satisfiable, $a < b$. Thus,

$a < b$ for $a \in L_p$ and $b \in U_p$ and L_p and U_p determine a cut in the ordering $(A, <)$.

Also note that if A is the disjoint union of L and U where $a < b$ for $a \in L$ and $b \in U$, then $\text{Th}_A(\mathcal{M}) \cup \{a < v : a \in L\}$ and $\{v < b : b \in U\}$ is satisfiable. Thus, there is a type p with $L_p = L$ and $U_p = U$.

We claim that the cut completely determines p; that is,

$$\{p\} = \bigcap_{a \in L_p} [a < v] \cap \bigcap_{a \in U_p} [v < b].$$

Suppose that $q \neq p$, $L_p = L_q$ and $U_p = U_q$. Because the only atomic formulas are $u = v$ and $u < v$, p and q determine the same cut in A, and they contain the same atomic formulas. Because quantifier-free formulas are Boolean combinations of atomic formulas, p and q contain the same quantifier-free formulas. Because every formula is equivalent to a quantifier-free formula, $p = q$.

Using the identification between types and cuts, we can give a complete description of all types in $S_1^{\mathbb{Q}}(\mathbb{Q})$.

For $a \in \mathbb{Q}$, let p_a be the unique type containing $v = a$.

Let $p_{+\infty}$ be the unique type p with $L_p = \mathbb{Q}$ and $U_p = \emptyset$, and let $p_{-\infty}$ be the unique type p with $L_p = \emptyset$ and $U_p = \mathbb{Q}$. For $r \in \mathbb{R} \setminus \mathbb{Q}$, let p_r be the unique type p with $L_p = \{a \in \mathbb{Q} : a < r\}$ and $U_p = \{b \in \mathbb{Q} : r < b\}$. Finally, for $c \in \mathbb{Q}$, let p_{c+} be the unique type p with $L_p = \{a \in \mathbb{Q} : a \leq c\}$ and $U_p = \{b \in \mathbb{Q} : c < b\}$, and let p_{c-} be the unique type p with $L_p = \{a \in \mathbb{Q} : a < c\}$ and $U_p = \{b \in \mathbb{Q} : c \leq b\}$. These are all possible types. Note in particular that $|S_1^{\mathbb{Q}}(\mathbb{Q})| = 2^{\aleph_0}$.

We return to the general case where $\mathcal{M} \models \text{DLO}$ and $A \subseteq M$ is nonempty. Aside from the types realized by elements of A, what types in $S_1^{\mathcal{M}}(A)$ are isolated? Suppose that L_p has a largest element a and U_p has a smallest element b. Then $p \in [a < v < b]$. Moreover, $\text{Th}_A(\mathcal{M}) \models a < v < b \rightarrow c < v < d$ for all $c \in L_p$ and $d \in U_p$. Thus, $a < v < b$ isolates p. Similarly, if $U_p = \emptyset$ and L_p has a greatest element a, then $a < v$ isolates p, and if U_p has a smallest element b and $L_p = \emptyset$, then $v < b$ isolates p.

We claim that these are the only possibilities. For example, suppose that $U_p \neq \emptyset$ and has no least element. Suppose that $\phi(v)$ isolates p. Because U_p and L_p determine p,

$$\text{Th}_A(\mathcal{M}) \cup \{a < v : a \in L_p\} \cup \{v < b : v \in U_p\} \models \phi(v).$$

Thus, we can find $a \in L_p \cup \{-\infty\}$ and $b \in U_p$ such that

$$\text{Th}_A(\mathcal{M}) \models \{a < v < b\} \rightarrow \phi(v).$$

There is $c \in U_p$ such that $c < b$. Because $a < c < b$, $\mathcal{M} \models \phi(c)$. But then the type containing $v = c$ is in $[\phi(v)]$ contradicting the fact that $[\phi(v)]$ isolates p. Other cases are similar. We summarize as follows.

Proposition 4.1.13 *Let $\mathcal{M} \models DLO$ and let $A \subseteq M$ be nonempty. Types in $S_1^{\mathcal{M}}(A)$ not realized by elements of A correspond to cuts in the ordering of A. A nonrealized type p is nonisolated if either $U_p \neq \emptyset$ has no least element or $L_p \neq \emptyset$ has no greatest element.*

Example 4.1.14 *Algebraically Closed Fields*

Let $K \models \mathrm{ACF}$, and let $A \subseteq K$. We first argue that, without loss of generality, we may assume that A is a field. Let k be the subfield of K generated by A. If $p \in S_n^K(k)$, then $p|A \in S_n^K(A)$. We claim that the restriction map is a bijection. By Lemma 4.1.9, we know that it is surjective, so we need only show that it is one-to-one. Suppose that $q \in S_n^K(A)$. For $b_1, \ldots, b_l \in k$, there are $a_1, \ldots, a_m \in A$ such that for each i there is $q_i(\overline{X}) \in \mathbb{Z}(X_1, \ldots, X_m)$ such that $b_i = q_i(\overline{a})$. Thus, for any $f(X_1, \ldots, X_l) \in \mathbb{Z}[X_1, \ldots, X_l, \overline{Y}]$ there is $g \in \mathbb{Z}[X_1, \ldots, X_m, \overline{Y}]$ such that $f(b_1, \ldots, b_l, \overline{y}) = 0$ if and only if $g(a_1, \ldots, a_m, \overline{y}) = 0$ for any $\overline{y}$. Thus, by quantifier elimination, for any formula $\phi(\overline{v}, \overline{b})$ with $\overline{b} \in k$, there is a formula $\psi(\overline{v}, \overline{a})$ with $\overline{a} \in A$ such that

$$K \models \phi(\overline{v}, \overline{b}) \leftrightarrow \psi(\overline{v}, \overline{a}).$$

Thus, if $p, q \in S_l^K(k)$ and $p \neq q$, then $p|A \neq q|A$.

Let k be a subfield of K. We will show that n-types over k are determined by prime ideals in $k[X_1, \ldots, X_n]$. For $p \in S_n^K(k)$, let

$$I_p = \{f(\overline{X}) \in k[X_1, \ldots, X_n] : f(\overline{v}) = 0 \in p\}.$$

If $f, g \in I_p$, then $f + g \in I_p$, and if $f \in I_p$ and $g \in k[\overline{X}]$, then $fg \in I_p$. Thus, I_p is an ideal. If $f, g \in k[\overline{X}]$, then

$$K \models \forall \overline{v} \; f(\overline{v})g(\overline{v}) = 0 \rightarrow (f(\overline{v}) = 0 \vee g(\overline{v}) = 0).$$

Thus, if $fg \in I_p$, then either $f \in I_p$ or $g \in I_p$. Hence, I_p is a prime ideal.

On the other hand, suppose that $P \subset k[\overline{X}]$ is a prime ideal. There is a prime ideal $Q \subset K[\overline{X}]$ such that $Q \cap k[\overline{X}] = P$.[1] Let F be the algebraic closure of the fraction field of $K[\overline{X}]/Q$. By model-completeness, F is an elementary extension of K. Let $x_i = X_i/Q$ for $i = 1, \ldots, n$. For $f \in K[\overline{X}]$, $f(\overline{x}) = 0$ if and only if $f \in Q$. Thus, if $p = \mathrm{tp}^F(\overline{x}/k)$, then $I_p = P$. Thus, $p \mapsto I_p$ is a surjective map from $S_n^K(k)$ onto the prime ideals of $k[\overline{X}]$. We claim that $p \mapsto I_p$ is one-to-one. Suppose that $p, q \in S_n^K(k)$ and $p \neq q$.

---

[1] This follows, for example, from [68] 7.5, because $K[\overline{X}]$ is a faithfully flat $k[\overline{X}]$-algebra, but we sketch a more elementary proof. If $K[\overline{X}]P$ is the $K[\overline{X}]$ ideal generated by P, we first claim that $K[\overline{X}]P \cap k[\overline{X}] = P$. Let B be a basis for K as a k-vector space with $1 \in B$. B is also a basis for $K[\overline{X}]$ as a free $k[\overline{X}]$-module. If $f \in K[\overline{X}]$, then $f = \sum_{b \in B} f_b b$, where each $f_b \in k[\overline{X}]$ and all but finitely many $f_b = 0$. If $f \in K[\overline{X}]P$, then each $f_b \in P$. If $f \in K[\overline{X}]P \cap k[\overline{X}]$, then $f = f_1 \in P$. Let S be the multiplicatively closed set $k[\overline{X}] \setminus P$. Let $Q \subset K[\overline{X}]$ be maximal among the ideals containing P and avoiding S. Then, Q is a prime ideal and $Q \cap k[\overline{X}] = P$.

There is a formula $\phi \in p$ such that $\neg\phi \in q$. By quantifier elimination, we may assume that ϕ is

$$\bigvee_{i=1}^{m} \left[\bigwedge_{j=1}^{k} f_{i,j}(\overline{v}) = 0 \wedge \bigwedge_{l=1}^{s} g_{i,l}(\overline{v}) \neq 0 \right],$$

where $f_{i,j}, g_{i,l} \in k[\overline{X}]$. If $I_p = I_q$, then

$$f_{i,j}(\overline{v}) = 0 \in p \Leftrightarrow f_{i,j}(\overline{v}) = 0 \in q$$

and

$$g_{i,l}(\overline{v}) = 0 \in p \Leftrightarrow g_{i,l}(\overline{v}) = 0 \in q.$$

Thus, $\phi \in p$ if and only if $\phi \in q$.

Definition 4.1.15 For A a ring, the *Zariski spectrum* of A is the set of all prime ideals of A. We denote the Zariski Spectrum by $\text{Spec}(A)$ and topologize $\text{Spec}(A)$ by taking basic closed sets $\{P \in \text{Spec}(A) : a_1, \ldots, a_m \in P\}$ for $a_1, \ldots, a_m \in A$. This is called the *Zariski topology* on $\text{Spec}(A)$.

Proposition 4.1.16 *The map* $p \mapsto I_p$ *is a continuous bijection from* $S_n^K(k)$ *to* $\text{Spec}(k[X_1, \ldots, X_n])$.

Proof We have shown that the map is one-to-one so we need only show that it is continuous. Suppose that $f_1, \ldots, f_m \in k[X_1, \ldots, X_n]$. Then, the inverse image of $\{P \in \text{Spec}(k[\overline{X}]) : f_1, \ldots, f_m \in P\}$ is $\{p \in S_n^K(k) : f_1(\overline{v}) = 0 \wedge \ldots \wedge f_m(\overline{v}) = 0 \in p\}$, a clopen set. Thus, $p \mapsto I_p$ is continuous.

Although $p \mapsto I_p$ is continuous, it is not a homeomorphism. In particular, for $f \in k[\overline{X}] \setminus k$, $\{p \in S_n^K(k) : f(\overline{v}) = 0\}$ is clopen in $S_n^K(k)$, whereas the image in $\text{Spec}(A)$ is closed but not open. Although the Stone topology is finer than the Zariski topology, we can use it when studying the Zariski topology.

Corollary 4.1.17 *The Zariski topology on* $\text{Spec}(k[\overline{X}])$ *is compact.*

Proof This is clear because $S_n^K(k)$ is compact and $p \mapsto I_p$ is continuous.

Proposition 4.1.16 also allows us to count types.

Corollary 4.1.18 *Suppose that* $K \models \text{ACF}$ *and* k *is a subfield of* K. *Then* $|S_n^K(k)| = |k| + \aleph_0$.

Proof By Hilbert's Basis Theorem, all ideals in $k[\overline{X}]$ are finitely generated. Thus, there are only $|k| + \aleph_0$ prime ideals.

4.2   Omitting Types and Prime Models

The Compactness Theorem allows us to build models realizing types. It is often also useful to build models that omit certain types. Let $\mathcal{L}$ be a language and T an $\mathcal{L}$-theory. For p an n-type consistent with T, we would like to know whether there is $\mathcal{M} \models T$ omitting p. It is not hard to give a necessary topological condition.

For T an $\mathcal{L}$-theory, we let $S_n(T)$ be the set of all complete n-types p such that $p \cup T$ is satisfiable. If T is complete and $\mathcal{M} \models T$, then $S_n(T) = S_n^{\mathcal{M}}(\emptyset)$. In particular, $S_n(T)$ is a totally disconnected compact topological space with basic open sets

$$[\phi] = \{p : \phi \in p\}.$$

For p a complete type, p is isolated in $S_n(T)$ if and only if $\{p\} = [\phi]$ for some ϕ. We can extend this notion to possibly incomplete types.

Definition 4.2.1   Let $\phi(v_1, \ldots, v_n)$ be an $\mathcal{L}$-formula such that $T \cup \{\phi(\overline{v})\}$ is satisfiable, and let p be an n-type. We say that ϕ *isolates* p if

$$T \models \forall \overline{v}(\phi(\overline{v}) \to \psi(\overline{v}))$$

for all $\psi \in p$.

Note that if p is a complete type and $\phi(\overline{v})$ isolates p, then

$$T \models \phi(\overline{v}) \to \psi(\overline{v}) \quad \Leftrightarrow \quad \psi(\overline{v}) \in p$$

for all formulas $\psi(\overline{v})$. In particular, for all formulas $\psi(\overline{v})$ exactly one of $T + \phi(\overline{v}) \wedge \psi(\overline{v})$ and $T + \phi(\overline{v}) \wedge \neg\psi(\overline{v})$ is satisfiable.

We can only omit an isolated type if we do not witness the isolating formula.

Proposition 4.2.2   *If $\phi(\overline{v})$ isolates p, then p is realized in any model of $T \cup \{\exists \overline{v} \; \phi(\overline{v})\}$. In particular, if T is complete, then every isolated type is realized.*

Proof   If $\mathcal{M} \models T$ and $\mathcal{M} \models \phi(\overline{a})$, then $\overline{a}$ realizes p. If T is complete and $T \cup \{\phi(\overline{v})\}$ is satisfiable, then $T \models \exists \overline{v} \; \phi(\overline{v})$.

For countable languages, this is also a sufficient condition.

Theorem 4.2.3 (Omitting Types Theorem)   *Let $\mathcal{L}$ be a countable language, T an $\mathcal{L}$-theory, and p a (possibly incomplete) nonisolated n-type over $\emptyset$. Then, there is a countable $\mathcal{M} \models T$ omitting p.*

Proof   We will prove this by a modification of the Henkin construction used to prove the Compactness Theorem. Let $C = \{c_0, c_1, \ldots\}$ be countably many new constant symbols, and let $\mathcal{L}^* = \mathcal{L} \cup C$. As in the proof of the Compactness Theorem, we will build $T^* \supseteq T$, a complete $\mathcal{L}^*$-theory with the witness property, and build $\mathcal{M} \models T^*$ as in Lemma 2.1.7. We will arrange the construction such that, for all $d_1, \ldots, d_n \in C$, there is a formula $\phi(\overline{v}) \in p$ such that $T^* \models \neg\phi(d_1, \ldots, d_n)$. This will ensure that $d_1^{\mathcal{M}}, \ldots, d_n^{\mathcal{M}}$ does not realize p. Because every element of M is the interpretation of a constant symbol in C, $\mathcal{M}$ omits p.

We will construct a sequence $\theta_0, \theta_1, \theta_2, \ldots$ of $\mathcal{L}^*$-sentences such that

$$\models \theta_t \to \theta_s$$

for $t > s$ and $T^* = T \cup \{\theta_i : i = 0, 1, \ldots\}$ is a satisfiable extension of T.

Let $\phi_0, \phi_1, \phi_2, \ldots$ list all $\mathcal{L}^*$-sentences. To ensure that T^* is complete, we will either have

$$\models \theta_{3i+1} \to \phi_i$$

or

$$\models \theta_{3i+1} \to \neg\phi_i.$$

If ϕ_i is $\exists v\, \psi(v)$ and $\models \theta_{3i+1} \to \phi_i$, then

$$\models \theta_{3i+2} \to \psi(c)$$

for some $c \in C$. This will ensure that T^* has the witness property. Let $\overline{d}_0, \overline{d}_1, \ldots$ list all n-tuples from C. We will choose θ_{3i+3} to ensure that $\overline{d}_i^{\mathcal{M}}$ does not realize p in the canonical model of T^*.

stage 0: Let θ_0 be $\forall x\ x = x$.

Suppose that we have constructed θ_s such that $T \cup \theta_s$ is satisfiable. There are three cases to consider.

stage $s + 1 = 3i + 1$: (completeness) If $T \cup \{\theta_s, \phi_i\}$ is satisfiable then θ_{s+1} is $\theta_s \wedge \phi_i$; otherwise, θ_{s+1} is $\theta_s \wedge \neg\phi_i$. In either case $T \cup \theta_{s+1}$ is satisfiable.

stage $s + 1 = 3i + 2$: (witness property) Suppose that ϕ_i is $\exists v\, \psi(v)$ for some formula ψ and $T \models \theta_s \to \phi_i$. In this case, we want to find a witness for ψ. Let $c \in C$ be a constant that does not occur in $T \cup \{\theta_s\}$. Because only finitely many constants from C have been used so far, we can always find such a c. Let $\theta_{s+1} = \theta_s \wedge \psi(c)$. If $\mathcal{N} \models T \cup \{\theta_s\}$, then there is $a \in N$ such that $\mathcal{N} \models \psi(a)$. By letting $c^{\mathcal{N}} = a$, we have $\mathcal{N} \models \theta_{s+1}$. Thus, in this case $T \cup \{\theta_{s+1}\}$ is satisfiable.

If ϕ_i is not of the correct form or $T \not\models \theta_s \to \phi_i$, then let θ_{s+1} be θ_s.

stage $s + 1 = 3i + 3$: (omitting p) Let $\overline{d}_i = (e_1, \ldots, e_n)$. Let $\psi(v_1, \ldots, v_n)$ be the $\mathcal{L}$-formula obtained from θ_s by replacing each occurrence of e_i by v_i and then replacing every other constant symbol $c \in C \setminus \{e_0, \ldots, e_n\}$ occurring in θ_s by the variable v_c and putting a $\exists v_c$ quantifier in front. In particular, we get rid of all of the constants in θ_s from C either by replacing them by variables or by quantifying over them. For example, if θ_s is

$$\forall x \exists y\ cx + e_1 e_2 = y^2 + de_2,$$

where c, d, e_1, e_2 are distinct constants in C, then $\psi(v_1, v_2)$ would be the formula

$$\exists v_c \exists v_d \forall x \exists y \ v_c x + v_1 v_2 = y^2 + v_d v_2.$$

Because p is nonisolated, there is a formula $\phi(\overline{v}) \in p$ such that

$$T \not\models \forall \overline{v} \ (\psi(\overline{v}) \to \phi(\overline{v})). \tag{$*$}$$

Let θ_{s+1} be $\theta_s \wedge \neg \phi(\overline{d}_i)$. We must argue that $T \cup \theta_{s+1}$ is satisfiable. By $(*)$ there is $\mathcal{N} \models T$ with $\overline{a} \in N$ such that

$$\mathcal{N} \models \psi(\overline{a}) \wedge \neg \phi(\overline{a}).$$

We can make $\mathcal{N}$ into a model of θ_{s+1} by interpreting the constants $c \in C \setminus \{e_1, \ldots, e_n\}$ as the witnesses to v_c and e_i as a_i.

This completes the construction. Let $T^* = T \cup \{\theta_0, \theta_1, \ldots\}$. Because $T \cup \{\theta_s\}$ is satisfiable for each s, T^* is satisfiable. If ϕ is any $\mathcal{L}$-sentence, then $\phi = \phi_i$ for some i, and at stage $3i + 1$ we ensure that $T^* \models \phi$ or $T^* \models \neg \phi$. Thus, T^* is complete.

If $\psi(v)$ is an $\mathcal{L}$-formula and $T^* \models \exists v \ \psi(v)$, then there is an i such that ϕ_i is $\exists v \ \psi(v)$ and at stage $3i + 2$ we ensure that $T^* \models \psi(c)$ for some $c \in C$. Thus, T^* has the witness property.

If $\mathcal{M}$ is the canonical model of T^* constructed as in Lemma 2.1.7, we claim that $\mathcal{M}$ omits p. Suppose that $\overline{a} \in M^n$. Because every element of M is the interpretation of a constant symbol, there is $\overline{d}_i$ such that $\overline{d}_i^{\mathcal{M}} = \overline{a}$. At stage $3i + 3$, we ensure that $\mathcal{M} \models \neg \phi_i(\overline{d})$ for some $\phi_i \in p$. Thus $\overline{a}$ does not realize p.

The proof of the Omitting Types Theorem can be generalized to omit countably many types at once.

Theorem 4.2.4 *Let $\mathcal{L}$ be a countable language, and let T be an $\mathcal{L}$-theory. Let X be a countable collection of nonisolated types over $\emptyset$. There is a countable $\mathcal{M} \models T$ that omits all of the types $p \in X$.*

Proof (Sketch) Let $p_0, p_1, \ldots$ list X. Let C be as in the proof of Theorem 4.2.3, and let $\overline{d}_0, \overline{d}_1 \ldots$ list all finite sequences from C. Fix $\pi : \mathbb{N} \times \mathbb{N} \to \mathbb{N}$, a bijection.

We do a Henkin-style argument as in the proof of Theorem 4.2.3. If $s = 0$, $3i + 1$, or $3i + 2$, we proceed exactly as above. If $i = \pi(m, n)$, then at stage $s = 3i + 3$ we proceed as above to ensure that $\overline{d}_m$ does not realize p_n.

If $\mathcal{M}$ is the canonical model, we eventually ensure that no finite sequence from M realizes any of the types p_i.

The assumption of countability of $\mathcal{L}$ is necessary in the Omitting Types Theorem. Suppose that $\mathcal{L}$ is the language with two disjoint sets of constant symbols C and D, where C is uncountable and $|D| = \aleph_0$. Let T be the theory $\{a \neq b : a, b \in C, a \neq b\}$ and p be the type $\{v \neq d : d \in D\}$. Because every model of T is uncountable, there is always an element that is not the interpretation of a constant in D. Thus, every model of T realizes p. On the other hand, if $\phi(v)$ is any $\mathcal{L}$-formula, then, because only countably many constants from D occur in $T \cup \{\phi(v)\}$, there is $d \in D$ such that $T \cup \{\phi(d)\}$ is satisfiable. Thus, p is nonisolated.

The necessity of X being countable in Theorem 4.2.4 is more problematic. For example, if $\aleph_0 < \lambda < 2^{\aleph_0}$, we could ask whether for a countable T we can omit a family of λ nonisolated types. This turns out to depend on set theoretic assumptions (see Exercise 4.5.14).

We give one concrete application of the Omitting Types Theorem. Let $\mathcal{L} = \{+, \cdot, <, 0, 1\}$, and let PA be the axioms for Peano arithmetic. Suppose that $\mathcal{M}, \mathcal{N} \models \mathrm{PA}$. We say that $\mathcal{N}$ is an *end extension* of $\mathcal{M}$ if $N \supset M$ and $a < b$ for all $a \in M$ and $b \in N \setminus M$.

Theorem 4.2.5 *If $\mathcal{M}$ is a countable model of PA, then there is $\mathcal{M} \prec \mathcal{N}$ such that $\mathcal{N}$ is a proper end extension of $\mathcal{M}$.*

Proof Consider the language $\mathcal{L}^*$ where we have constant symbols for all elements of M and a new constant symbol c. Let $T = \mathrm{Diag}_{\mathrm{el}}(\mathcal{M}) \cup \{c > m : m \in M\}$, and for $a \in M \setminus \mathbb{N}$ let p_a be the type $\{v < a, v \neq m : m \in M\}$. Any $\mathcal{N} \models T$ is a proper elementary extension of $\mathcal{M}$. If $\mathcal{N}$ omits each p_a, then $\mathcal{N}$ is an end extension of $\mathcal{M}$. By Theorem 4.2.4, it suffices to show that each p_a is nonisolated.

Suppose that $\phi(v)$ is an $\mathcal{L}^*$ formula isolating p_a. Let $\phi(v) = \theta(v, c)$, where θ is an $\mathcal{L}_M$-formula. Then

$$T \cup \theta(v, c) \models v < a.$$

Because $T \cup \{\theta(v, c)\}$ is satisfiable,

$$\mathcal{M} \models \forall x \exists y > x \exists v < a \; \theta(v, y).$$

The Pigeonhole Principle is provable in Peano arithmetic. Thus

$$\mathcal{M} \models [\forall x \exists y > x \exists v < a \; \theta(v, y)] \rightarrow \exists v < a \forall x \exists y > x \; \theta(v, y). \qquad (**)$$

Thus, there is $m < a$ such that

$$\mathcal{M} \models \forall x \exists y > x \; \theta(m, y).$$

We claim that $T \cup \{\theta(m, c)\}$ is satisfiable. If not, there is $n \in M$ such that

$$\mathrm{Diag}_{\mathrm{el}}(\mathcal{M}) + c > n \models \neg\theta(m, c)$$

contradicting $(**)$. Thus, $\phi(v)$ does not isolate p_a, a contradiction.

Prime and Atomic Models

We use the Omitting Types Theorem to study small models of a complete theory. For the remainder of this section, we will assume that $\mathcal{L}$ is a countable language and T is a complete $\mathcal{L}$-theory with infinite models.

Definition 4.2.6 We say that $\mathcal{M} \models T$ is a *prime model* of T if whenever $\mathcal{N} \models T$ there is an elementary embedding of $\mathcal{M}$ into $\mathcal{N}$.

For example, let $T = \mathrm{ACF}_0$. If $K \models \mathrm{ACF}_0$, and F is the algebraic closure of $\mathbb{Q}$, then there is an embedding of F into K. Because ACF_0 is model complete this embedding is elementary. Thus, F is a prime model of ACF_0. Similarly, RCF has a prime model, the real closure of $\mathbb{Q}$.

For a third example, consider $\mathcal{L} = \{+, \cdot, <, 0, 1\}$ and let T be $\mathrm{Th}(\mathbb{N})$, true arithmetic. If $\mathcal{M} \models T$, then we can view $\mathbb{N}$ as an initial segment of $\mathcal{M}$. We claim that this embedding is elementary. We use the Tarski–Vaught test (Proposition 2.3.5). Let $\phi(v, w_1, \ldots, w_m)$ be an $\mathcal{L}$-formula and let $n_1, \ldots, n_m \in \mathbb{N}$ such that $\mathcal{M} \models \exists v\ \phi(v, \bar{n})$. Let ψ be the $\mathcal{L}$-sentence

$$\exists v\ \phi(v, \underbrace{1 + \ldots + 1}_{n_1-\text{times}}, \ldots, \underbrace{1 + \ldots + 1}_{n_m-\text{times}}).$$

Then, $\mathcal{M} \models \psi$ and $\mathbb{N} \models \psi$ because $\mathcal{M} \equiv \mathbb{N}$. But then, for some $s \in \mathbb{N}$,

$$\mathbb{N} \models \phi(s, \underbrace{1 + \ldots + 1}_{n_1-\text{times}}, \ldots, \underbrace{1 + \ldots + 1}_{n_m-\text{times}})$$

and

$$\mathbb{N} \models \phi(\underbrace{1 + \ldots + 1}_{s-\text{times}}, \underbrace{1 + \ldots + 1}_{n_1-\text{times}}, \ldots, \underbrace{1 + \ldots + 1}_{n_m-\text{times}}).$$

Because the latter statement is an $\mathcal{L}$-sentence,

$$\mathcal{M} \models \phi(\underbrace{1 + \ldots + 1}_{s-\text{times}}, \underbrace{1 + \ldots + 1}_{n_1-\text{times}}, \ldots, \underbrace{1 + \ldots + 1}_{n_m-\text{times}})$$

and $\mathcal{M} \models \phi(s, n_1, \ldots, n_m)$. By the Tarski–Vaught test, $\mathbb{N} \prec \mathcal{M}$. Thus, $\mathbb{N}$ is a prime model of T.

Suppose $\mathcal{M}$ is a prime model of T. Suppose that $j : \mathcal{M} \to \mathcal{N}$ is an elementary embedding. If $\bar{a} \in M^n$ realizes $p \in S_n(T)$, then so does $j(\bar{a})$. If $p \in S_n(T)$ is nonisolated, there is $\mathcal{N}$ such that $\mathcal{N}$ omits p. If $\mathcal{M}$ realizes p, then we can not elementarily embed $\mathcal{M}$ into $\mathcal{N}$; thus, $\mathcal{M}$ must also omit p. In particular, if $\bar{a} \in M^n$, then $\mathrm{tp}^{\mathcal{M}}(\bar{a})$ must be isolated. This leads us to the following definition.

Definition 4.2.7 We say that $\mathcal{M} \models T$ is *atomic* if $\mathrm{tp}^{\mathcal{M}}(\bar{a})$ is isolated for all $\bar{a} \in M^n$.

We have just argued that prime models are atomic. For countable models, the converse is also true.

Theorem 4.2.8 *Let $\mathcal{L}$ be a countable language and let T be a complete $\mathcal{L}$-theory with infinite models. Then, $\mathcal{M} \models T$ is prime if and only if it is countable and atomic.*

Proof

($\Rightarrow$) We have argued that prime models are atomic. Because $\mathcal{L}$ is countable, T has a countable model. Thus, the prime model must be countable.

($\Leftarrow$) Let $\mathcal{M}$ be countable and atomic. Let $\mathcal{N} \models T$. We must construct an elementary embedding of $\mathcal{M}$ into $\mathcal{N}$. Let $m_0, m_1, \ldots, m_n, \ldots$ be an enumeration of M. For each i, let $\theta_i(v_0, \ldots, v_i)$ isolate the type of $(m_0, \ldots, m_i)$. We will build $f_0 \subseteq f_1 \subseteq \ldots$ a sequence of partial elementary maps from $\mathcal{M}$ into $\mathcal{N}$ where the domain of f_i is $\{m_0, \ldots, m_{i-1}\}$. Then, $f = \bigcup_{i=0}^{\infty} f_i$ is an elementary embedding of $\mathcal{M}$ into $\mathcal{N}$.

Let $f_0 = \emptyset$. Because $\mathcal{M} \equiv \mathcal{N}$, f_0 is partial elementary.

Given f_s, let $n_i = f(m_i)$ for $i < s$. Because $\theta_s(m_0, \ldots, m_s)$ and f_s is partial elementary,

$$\mathcal{N} \models \exists v \; \theta_s(n_0, \ldots, n_{s-1}, v).$$

Let $n_s \in N$ such that $\mathcal{N} \models \theta_s(n_0, \ldots, n_s)$. Because θ_s isolates $\mathrm{tp}^{\mathcal{M}}(m_0, \ldots, m_s)$,

$$\mathrm{tp}^{\mathcal{M}}(m_0, \ldots, m_s) = \mathrm{tp}^{\mathcal{N}}(n_0, \ldots, n_s).$$

Thus, $f_{s+1} = f_s \cup \{(m_s, n_s)\}$ is a partial elementary map.

Theorem 4.2.8 will lead to a criterion for the existence of prime models. We need one preparatory lemma.

Lemma 4.2.9 *Suppose that $(\bar{a}, \bar{b}) \in M^{m+n}$ realizes an isolated type in $S_{m+n}(T)$. Then $\bar{a}$ realizes an isolated type in $S_m(T)$. Indeed if $A \subseteq M$ and $(\bar{a}, \bar{b}) \in M^{m+n}$ realizes an isolated type in $S_{m+n}^{\mathcal{M}}(A)$, then $\mathrm{tp}^{\mathcal{M}}(\bar{a}/A)$ is isolated.*

Proof Let $\phi(\bar{v}, \bar{w})$ isolate $\mathrm{tp}^{\mathcal{M}}(\bar{a}, \bar{b}/A)$. We claim that $\exists w \; \phi(\bar{v}, \bar{w})$ isolates $\mathrm{tp}^{\mathcal{M}}(\bar{a}/A)$. Let $\psi(\bar{v})$ be any $\mathcal{L}_A$-formula such that $\mathcal{M} \models \psi(\bar{a})$. We must show that

$$\mathrm{Th}_A(\mathcal{M}) \models \exists \bar{w} \; (\phi(\bar{v}, \bar{w}) \to \psi(\bar{v})).$$

Suppose not. Then, there is $\bar{c} \in M^m$ such that

$$\mathcal{M} \models \exists \bar{w} \; (\phi(\bar{c}, \bar{w}) \wedge \neg \psi(\bar{c})).$$

Let $\bar{d} \in M^n$ such that $\mathcal{M} \models \phi(\bar{c}, \bar{d}) \wedge \neg \psi(\bar{c})$. Because $\phi(\bar{v}, \bar{w})$ isolates $\mathrm{tp}^{\mathcal{M}}(\bar{a}, \bar{b}/A)$,

$$\mathrm{Th}_A(\mathcal{M}) \models \phi(\bar{v}, \bar{w}) \to \psi(\bar{v}).$$

This is a contradiction because

$$\psi(\bar{v}) \in \mathrm{tp}^{\mathcal{M}}(\bar{a}/A) \subset \mathrm{tp}^{\mathcal{M}}(\bar{a}, \bar{b}/A).$$

An extension of this lemma is proved in Exercise 4.5.11.

Theorem 4.2.10 *Let $\mathcal{L}$ be a countable language and let T be a complete $\mathcal{L}$-theory with infinite models. Then, the following are equivalent:*
i) T has a prime model;
ii) T has an atomic model $\mathcal{M}$;
iii) the isolated types in $S_n(T)$ are dense for all n.

Proof We have already shown i) $\Leftrightarrow$ ii).

ii) $\Rightarrow$ iii) Let $\phi(\bar{v})$ be an $\mathcal{L}$-formula such that $[\phi(\bar{v})]$ is a nonempty open set in $S_n(T)$. We must show that $[\phi(\bar{v})]$ contains an isolated type. Let $\mathcal{M} \models T$ be atomic. Because T is complete and $T \cup \{\phi(\bar{v})\}$ is satisfiable, $T \models \exists \bar{v}\ \phi(\bar{v})$. Thus, there is $\bar{a} \in M^n$ such that $\mathcal{M} \models \phi(\bar{a})$. Then, $\mathrm{tp}^{\mathcal{M}}(\bar{a}) \in [\phi]$ and, because $\mathcal{M}$ is atomic, $\mathrm{tp}^{\mathcal{M}}(\bar{a})$ is isolated. Therefore, the isolated types are dense.

iii) $\Rightarrow$ ii) Suppose that the isolated types in T are dense. We will build an atomic model of T by a Henkin argument. Let $C = \{c_0, \ldots, c_n, \ldots\}$ be a new set of constant symbols, and let $\mathcal{L}^* = \mathcal{L} \cup C$. Let $\phi_0, \phi_1, \ldots$ list all $\mathcal{L}^*$-sentences. We build $\theta_0, \theta_1, \ldots$ a sequence of $\mathcal{L}^*$-sentences such that $T^* = \{\theta_i : i = 0, 1, \ldots\} \cup T$ is a complete satisfiable theory with the witness property. We do this so that the canonical model of T^* is atomic. We assume inductively that $T \cup \{\theta_s\}$ is satisfiable and $\theta_{s+1} \models \theta_s$.

<u>stage 0</u>: $\theta_0 = \exists x\ x = x$.

<u>stage $s + 1 = 3i + 1$</u>: (completeness) If $T + \theta_s \wedge \phi_i$ is satisfiable, let $\theta_{s+1} = \theta_s \wedge \phi_i$; otherwise, $\theta_{s+1} = \theta_s \wedge \neg\phi_i$.

<u>stage $s + 1 = 3i + 2$</u>: (witness property) If ϕ_i is $\exists v\ \psi(v)$ and $\theta_s \models \phi_i$, let $c \in C$ be a constant symbol not occurring in θ_s, and let $\theta_{s+1} = \theta_s \wedge \psi(c)$. Otherwise, let $\theta_{s+1} = \theta_s$. As in Theorem 4.2.3, $T \cup \{\theta_{s+1}\}$ is satisfiable.

<u>stage $s + 1 = 3i + 3$</u>: Let n be minimal such that all of the constants in C occurring in θ_s are from $\{c_0, \ldots, c_n\}$. Let $\psi(v_0, \ldots, v_n)$ be an $\mathcal{L}$-formula such that $\theta_s = \psi(c_0, \ldots, c_n)$. Clearly, $T \cup \{\psi(v_0, \ldots, v_n)\}$ is satisfiable. Because the isolated types in $S_n(T)$ are dense, there is an isolated type $p \in [\psi(\bar{v})]$. Let $\chi(\bar{v})$ be an $\mathcal{L}$-formula isolating p; in particular, $[\chi(\bar{v})] = \{p\}$ and $T \cup \{\chi(\bar{v})\}$ is satisfiable. Let $\theta_{s+1} = \chi(\bar{c})$. Then, $T \cup \{\theta_{s+1}\}$ is satisfiable. Because $\psi(\bar{v}) \in p$, $\theta_{s+1} \models \theta_s$.

As in Theorem 4.2.3, the theory $T^* = T \cup \{\theta_1, \theta_2, \ldots\}$ is a complete theory with the witness property. Let $\mathcal{M}$ be the canonical model of T^*. We must show $\mathcal{M}$ is atomic. Let $\bar{d} \in \mathcal{M}$. We can find an n and an $s = 3_i + 2$ such that each d_i is in $\{c_0, \ldots, c_n\}$ and n is minimal such that $\{c_0, \ldots, c_n\}$ contains all the constants occuring in θ_s. At stage $s + 1$ we make sure that $(c_0^{\mathcal{M}}, \ldots, c_n^{\mathcal{M}})$ realizes an isolated type. By 4.2.9, $\bar{d}$ realizes an isolated type.

In Exercise 4.5.16 we use Theorem 4.2.10 to give an example of a theory with no prime models. We now give one important case where the isolated types are dense. Note that if $\mathcal{L}$ is countable and A is countable, then $|S_n^{\mathcal{M}}(A)| \leq 2^{\aleph_0}$ because there are only $2^{\aleph_0}$ sets of $\mathcal{L}_A$-formulas. We will show that if there are fewer than the maximal possible number of types, then there are prime models.

Theorem 4.2.11 *Suppose that T is a complete theory in a countable language and $A \subseteq \mathcal{M} \models T$ is countable. If $|S_n^{\mathcal{M}}(A)| < 2^{\aleph_0}$, then*
   i) the isolated types in $S_n^{\mathcal{M}}(A)$ are dense and
   ii) $|S_n^{\mathcal{M}}(A)| \leq \aleph_0$.
In particular, if $|S_n(T)| < 2^{\aleph_0}$, then T has a prime model.

Proof
i) We first prove that the isolated types are dense. Suppose that there is a formula ϕ such that $[\phi]$ contains no isolated types. Because ϕ does not isolate a type, we can find ψ such that $[\phi \wedge \psi] \neq \emptyset$ and $[\phi \wedge \neg \psi] \neq \emptyset$. Because $[\phi]$ does not contain an isolated type, neither does $[\phi \wedge \pm\psi]$.

We build a binary tree of formulas $(\phi_\sigma : \sigma \in 2^{<\omega})$ such that:
i) each $[\phi_\sigma]$ is nonempty but contains no isolated types;
ii) if $\sigma \subset \tau$, then $\phi_\tau \models \phi_\sigma$;
iii) $\phi_{\sigma,i} \models \neg\phi_{\sigma,1-i}$.

Let $\phi_\emptyset = \phi$ for some formula ϕ where $[\phi]$ contains no isolated types. Suppose that $[\phi_\sigma]$ is nonempty but contains no isolated types. As above, we can find ψ such that $[\phi_\sigma \wedge \psi]$ and $[\phi_\sigma \wedge \neg \psi]$ are both nonempty and neither contains an isolated type. Let $\phi_{\sigma,0} = \phi \wedge \psi$ and $\phi_{\sigma,1} = \phi \wedge \neg\psi$.

Let $f : \omega \to 2$. Because

$$[\phi_{f|0}] \supseteq [\phi(f|1)] \supseteq [\phi(f|2)] \supseteq \cdots$$

and $S_n^{\mathcal{M}}(A)$ is compact, there is

$$p_f \in \bigcup_{n=0}^{\infty} [\phi_{f|n}].$$

If $g \neq f$, we can find m such that $f|m = g|m$ but $f(m) \neq g(m)$. By construction, $\phi_{f|m+1} \models \neg\phi_{g|m+1}$; thus $p_f \neq p_g$. Because $f \mapsto p_f$ is a one-to-one function from 2^ω into $S_n^{\mathcal{M}}(A)$, $|S_n^{\mathcal{M}}(A)| = 2^{\aleph_0}$.

ii) Suppose that $|S_n^{\mathcal{M}}(A)| > \aleph_0$. We claim that $|S_n^{\mathcal{M}}(A)| = 2^{\aleph_0}$. Because $|S_n^{\mathcal{M}}(A)| > \aleph_0$ and there are only countably many $\mathcal{L}_A$-formulas, there is a formula ϕ such that $\|[\phi]\| > \aleph_0$.

Claim If $\|[\phi]\| > \aleph_0$, there is an $\mathcal{L}_A$-formula ψ such that $\|[\phi \wedge \psi]\| > \aleph_0$ and $\|[\phi \wedge \neg\psi] > \aleph_0$.

Suppose not. Let $p = \{\psi(\overline{v}) : \|[\phi \wedge \psi]\| > \aleph_0\}$. Clearly, for each ψ either $\psi \in p$ or $\neg\psi \in p$ but not both. We claim that p is satisfiable. Suppose that

$\psi_1, \ldots, \psi_m \in p$. Either $\psi_1 \wedge \ldots \wedge \psi_m \in p$, in which case $\{\psi_1, \ldots, \psi_m\} \cup$ $\mathrm{Th}_A(\mathcal{M})$ is satisfiable, or $\neg\psi_1 \vee \ldots \vee \neg\psi_m \in p$. Because

$$[\neg\psi_1 \vee \ldots \vee \neg\psi_m] = [\neg\psi_1] \cup \ldots \cup [\neg\psi_m],$$

we must have $\|[\neg\psi_i]\| > \aleph_0$ for some $\aleph_0$, a contradiction. Thus $p \in S_n^{\mathcal{M}}(A)$. Moreover, if $\psi \notin p$, then $\|[\phi \wedge \psi]\| \leq \aleph_0$. But

$$[\phi] = \bigcup_{\psi \notin p} [\phi \wedge \psi] \cup \{p\}.$$

Because $[\phi]$ is the union of at most $\aleph_0$ sets each of size at most $\aleph_0$, we have $\|[\phi]\| \leq \aleph_0$, a contradiction.

We build a binary tree of formulas $(\phi_\sigma : \sigma \in 2^{<\omega})$ such that:
i) if $\sigma \subset \tau$ then $\phi_\tau \models \phi_\sigma$;
ii) $\phi_{\sigma,i} \models \neg\phi_{\sigma,1-i}$;
iii) $\|[\phi_\sigma]\| > \aleph_0$.
Let $\phi_\emptyset = \phi$ for some formula ϕ with $\|[\phi]\| > \aleph_0$. Given ϕ_σ where $\|[\phi_\sigma]\| > \aleph_0$, by the claim we can find ψ such that $\|[\phi_\sigma \wedge \psi]\| > \aleph_0$ and $\|[\phi_\sigma \wedge \neg\psi]\| > \aleph_0$. Let $\phi_{\sigma,0} = \phi_\sigma \wedge \psi$ and $\phi_{\sigma,1} = \phi_\sigma \wedge \neg\psi$.

As in i), for each $f \in 2^\omega$ there is a

$$p_f \in \bigcap_{m=0}^{\infty} [\phi_{f|m}],$$

and if $f \neq g$, then $p_f \neq p_g$. Thus $|S_n^{\mathcal{M}}(A)| = 2^{\aleph_0}$.

We note that it is possible for there to be prime models even if $|S_n(T)| = 2^{\aleph_0}$. For example, $\mathrm{Th}(\mathbb{N}, +, \cdot, <, 0, 1)$ and RCF have prime models.

Countable Homogeneous Models

Our next goal is to show that prime models are unique up to isomorphism. This will follow from work on homogeneous models.

Definition 4.2.12 Let κ be an infinite cardinal. We say that $\mathcal{M} \models T$ is κ-*homogeneous* if whenever $A \subset M$ with $|A| < \kappa$, $f : A \to M$ is a partial elementary map, and $a \in M$, there is $f^* \supseteq f$ such that $f^* : A \cup \{a\} \to M$ is partial elementary.

We say that $\mathcal{M}$ is *homogeneous* if it is $|M|$-homogeneous.

In homogeneous models, partial elementary maps are just restrictions of automorphisms.

Proposition 4.2.13 *Suppose that $\mathcal{M}$ is homogeneous, $A \subset M$, $|A| < |M|$, and $f : A \to M$ is a partial elementary map. Then, there is an automorphism σ of $\mathcal{M}$ with $\sigma \supseteq f$.*

In particular, if $\mathcal{M}$ is homogeneous and $\bar{a}, \bar{b} \in M^n$ realize the same n-type, then there is an automorphism σ of $\mathcal{M}$ with $\sigma(\bar{a}) = \bar{b}$.

Proof Let $|M| = \kappa$, and let $(a_\alpha : \alpha < \kappa)$ be an enumeration of M. We build a sequence of partial elementary maps $(f_\alpha : \alpha < \kappa)$ extending f with $f_\alpha \subseteq f_\beta$ for $\alpha < \beta$ such that a_α is in the domain and image of $f_{\alpha+1}$ and $|f_{\alpha+1}| \leq |f_\alpha| + 2 < \kappa$. Then, $\sigma = \bigcup_{\alpha < \kappa} f_\alpha$ is the desired automorphism. Let $f_0 = f$.

If α is a limit ordinal and f_β is partial elementary with

$$|f_\beta| \leq |A| + |\beta| + \aleph_0 < \kappa$$

for all $\beta < \alpha$, let $f_\alpha = \bigcup_{\beta < \alpha} f_\beta$. Then, f_α is partial elementary and

$$|f_\alpha| \leq |\alpha|(|A| + |\alpha| + \aleph_0) \leq |A| + |\alpha| + \aleph_0 < \kappa.$$

Given f_α with $|f_\alpha| < \kappa$, because $\mathcal{M}$ is homogeneous, there is $b \in M$ such that if $g_\alpha = f_\alpha \cup \{(a_\alpha, b)\}$, then g_α is partial elementary. Note that g_α^{-1} is also partial elementary. Thus, because $\mathcal{M}$ is homogeneous there is $c \in M$ such that $g_\alpha^{-1} \cup \{(a_\alpha, c)\}$ is partial elementary. Thus, $f_{\alpha+1} = g_\alpha \cup \{(c, a_\alpha)\}$ is partial elementary, $|f_{\alpha+1}| \leq |f_\alpha| + 2 \leq |A| + |\alpha| + \aleph_0$, and a_α is in the domain and range of $f_{\alpha+1}$.

If $\mathcal{M}$ is homogeneous and $\mathrm{tp}^{\mathcal{M}}(\bar{a}) = \mathrm{tp}^{\mathcal{M}}(\bar{b})$, then $\bar{a} \mapsto \bar{b}$ is a partial elementary map that must extend to an automorphism.

Lemma 4.2.14 *If $\mathcal{M}$ is atomic, then $\mathcal{M}$ is $\aleph_0$-homogeneous. In particular, countable atomic models are homogeneous.*

Proof Suppose that $\bar{a} \mapsto \bar{b}$ is elementary and $c \in M$. Let $\phi(\bar{v}, w)$ isolate $\mathrm{tp}^{\mathcal{M}}(\bar{a}, c)$. Because $\mathcal{M} \models \exists w \ \phi(\bar{a}, w)$ and $\bar{a} \mapsto \bar{b}$ is elementary, $\mathcal{M} \models \exists w \ \phi(\bar{b}, w)$. Suppose that $\mathcal{M} \models \phi(\bar{b}, d)$. Because $\phi(\bar{v}, w)$ isolates a type, $\mathrm{tp}^{\mathcal{M}}(\bar{a}, c) = \mathrm{tp}^{\mathcal{M}}(\bar{b}, d)$. Thus, $\bar{a}, c \mapsto \bar{b}, d$ is elementary.

For countable homogeneous models, there is a simple test for isomorphism. Clearly, if $\mathcal{M} \cong \mathcal{N}$, then $\mathcal{M}$ and $\mathcal{N}$ realize the same types from $S_n(T)$. For countable homogeneous models, this condition is also sufficient.

Theorem 4.2.15 *Let T be a complete theory in a countable language. Suppose that $\mathcal{M}$ and $\mathcal{N}$ are countable homogeneous models of T and $\mathcal{M}$ and $\mathcal{N}$ realize the same types in $S_n(T)$ for $n \geq 1$. Then $\mathcal{M} \cong \mathcal{N}$.*

Proof We build an isomorphism $f : \mathcal{M} \to \mathcal{N}$ by a back-and-forth argument. We will build $f_0 \subset f_1 \subset \ldots$, a sequence of partial elementary maps with finite domain, and let $f = \bigcup_{i=0}^\infty f_i$. Let $a_0, a_1, \ldots$ enumerate M and $b_0, b_1, \ldots$ enumerate N. We will ensure that $a_i \in \mathrm{dom}(f_{2i+1})$ and $b_i \in \mathrm{img}(f_{2i+2})$. Thus, we will have $\mathrm{dom}(f) = M$ and $f : M \to N$, a surjective elementary map, as desired.

stage 0: Let $f_0 = \emptyset$. Because T is complete f_0 is partial elementary.

We inductively assume that f_s is partial elementary. Let $\bar{a}$ be the domain of f_s and $\bar{b} = f_s(\bar{a})$.

stage $s + 1 = 2i + 1$: Let $p = \text{tp}^{\mathcal{M}}(\bar{a}, a_i)$. Because $\mathcal{M}$ and $\mathcal{N}$ realize the same types, we can find $\bar{c}, d \in N$ such that $\text{tp}^{\mathcal{N}}(\bar{c}, d) = p$. Note that $\text{tp}^{\mathcal{N}}(\bar{c}) = \text{tp}^{\mathcal{M}}(\bar{a})$, by choice of $\bar{c}$, and $\text{tp}^{\mathcal{M}}(\bar{a}) = \text{tp}^{\mathcal{N}}(\bar{b})$ because f_s is partial elementary. Thus, $\text{tp}^{\mathcal{N}}(\bar{c}) = \text{tp}^{\mathcal{N}}(\bar{b})$. Because $\mathcal{N}$ is homogeneous, there is $e \in N$ such that $\text{tp}^{\mathcal{N}}(\bar{b}, e) = \text{tp}^{\mathcal{N}}(\bar{c}, d) = p$. Thus, $f_{s+1} = f_s \cup \{(a_i, e)\}$ is partial elementary with a_i in the domain.

stage $s + 1 = 2i + 2$: As in the previous case, we can find $\bar{c}, d \in M$ such that $\text{tp}^{\mathcal{M}}(\bar{c}, d) = \text{tp}^{\mathcal{N}}(\bar{b}, b_i)$. Because $\mathcal{M}$ is homogeneous, there is $e \in M$ such that $\text{tp}^{\mathcal{M}}(\bar{c}, d) = \text{tp}^{\mathcal{M}}(\bar{a}, e)$. Then, $f_{s+1} = f_s \cup \{(e, b_i)\}$ with b_i in the range.

Corollary 4.2.16 *Let T be a complete theory in a countable language. If $\mathcal{M}$ and $\mathcal{N}$ are prime models of T, then $\mathcal{M} \cong \mathcal{N}$.*

Proof By Theorem 4.2.8, $\mathcal{M}$ and $\mathcal{N}$ are atomic. Because the types in $S_n(T)$ realized in an atomic model are exactly the isolated types, $\mathcal{M}$ and $\mathcal{N}$ realize the same types. By Lemma 4.2.14, countable atomic models are homogeneous. Thus, by Theorem 4.2.15, $\mathcal{M} \cong \mathcal{N}$.

Prime Model Extensions of ω-Stable Theories

We conclude this section by looking at a relative notion of prime models. Suppose that $\mathcal{M} \models T$ and $A \subseteq M$. We say that $\mathcal{M}$ is *prime over A* if whenever $\mathcal{N} \models T$ and $f : A \to \mathcal{N}$ is partial elementary, there is an elementary $f^* : \mathcal{M} \to \mathcal{N}$ extending f.

We give three examples. Let L be any linear order. We build $L^* \models \text{DLO}$ prime over L as follows. If L has a least element a, add a copy of $\mathbb{Q}$ below a. If L has a greatest element b, add a copy of $\mathbb{Q}$ above b. If $c, d \in L$ with $c < d$ but there are no elements of L between c and d, add a copy of $\mathbb{Q}$ between c and d. We add no other new elements. It is easy to see that $L^* \models \text{DLO}$ and that if $f : L \to \mathcal{M} \models \text{DLO}$, then f extends to $f^* : L^* \to \mathcal{M}$. Because DLO has quantifier elimination, it is model-complete and f^* is elementary.

For ACF, if R is any integral domain and F is the algebraic closure of the fraction field of R, then F is prime over R and any embedding of R into an algebraically closed field K extends to F. Because ACF is model-complete, this map is elementary. Similarly, if R is an ordered integral domain, then the real closure of the fraction field of R is a model of RCF prime over R. In Exercise 4.5.26, we will give examples of theories without prime model extensions.

There is one very natural class of theories with prime model extensions. This class will play a very important role later in the book.

Definition 4.2.17 Let T be a complete theory in a countable language, and let κ be an infinite cardinal. We say that T is *κ-stable* if whenever $\mathcal{M} \models T$, $A \subseteq M$, and $|A| = \kappa$, then $|S_n^{\mathcal{M}}(A)| = \kappa$.

We say that $\mathcal{M}$ is κ-stable if $\mathrm{Th}(\mathcal{M})$ is κ-stable.

For historical reasons, we will refer to $\aleph_0$-stable theories as being "ω-stable." By Corollary 4.1.18, ACF is ω-stable. On the other hand, $|S_1^{\mathcal{Q}}(\mathbb{Q})| = 2^{\aleph_0}$ so DLO is not ω-stable.

We will show that ω-stable theories have prime model extensions. An important first step is to show that if there are few types over countable sets, then there are few types over arbitrary sets.

Theorem 4.2.18 *Let T be a complete theory in a countable language. If T is ω-stable, then T is κ-stable for all infinite cardinals κ.*

Proof Suppose that $\mathcal{M} \models T$, $A \subseteq M$, $|A| = \kappa$ and $|S_n^{\mathcal{M}}(A)| > \kappa$. Because there are only κ formulas with parameters from A, there is some $\mathcal{L}_A$-formula $\phi_\emptyset(\bar{v})$ such that $\|[\phi_\emptyset]\| > \kappa$. The argument from Theorem 4.2.11 ii) can be extended to show that if $\|[\phi]\| > \kappa$ there is an $\mathcal{L}_A$-formula ψ such that $\|[\phi \wedge \psi]\| > \kappa$ and $\|[\phi \wedge \neg\psi]\| > \kappa$.

As in Theorem 4.2.11 ii), we build a binary tree of formulas $(\phi_\sigma : \sigma \in 2^{<\omega})$ such that:

i) if $\sigma \subset \tau$, then $\phi_\tau \models \phi_\sigma$;

ii) $\phi_{\sigma,i} \models \neg\phi_{\sigma,1-i}$;

iii) $\|[\phi_\sigma]\| > \kappa$.

Let A_0 be the set of all parameters from A occurring in any formula ϕ_σ. Clearly A_0 is a countable set. Arguing as in Theorem 4.2.11 ii), $|S_n^{\mathcal{M}}(A_0)| = 2^{\aleph_0}$, contradicting the ω-stability of T.

Proposition 4.2.19 *Let T be a complete theory in a countable language. If T is ω-stable, then for all $\mathcal{M} \models T$ and $A \subseteq M$, the isolated types in $S_n^{\mathcal{M}}(A)$ are dense.*

Proof Suppose not. We can build a binary tree of formulas as in Theorem 4.2.11 i). As in Theorem 4.2.18, we can find a countable $A_0 \subseteq A$ such that all parameters come from A_0. But then $|S_n^{\mathcal{M}}(A_0)| = 2^{\aleph_0}$, contradicting the ω-stability of T.

Theorem 4.2.20 *Suppose that T is ω-stable. Let $\mathcal{M} \models T$ and $A \subseteq M$. There is $\mathcal{M}_0 \prec \mathcal{M}$, a prime model extension of A. Moreover, we can choose $\mathcal{M}_0$ such that every element of $\mathcal{M}_0$ realizes an isolated type over A.*

Proof We will find an ordinal δ and build a sequence of sets $(A_\alpha : \alpha \le \delta)$ where $A_\alpha \subseteq M$ and

i) $A_0 = A$;

ii) if α is a limit ordinal, then $A_\alpha = \bigcup_{\beta < \alpha} A_\beta$;

iii) if no element of $M \setminus A_\alpha$ realizes an isolated type over A_α, we stop and let $\delta = \alpha$; otherwise, pick a_α realizing an isolated type over A_α, and let $A_{\alpha+1} = A_\alpha \cup \{a_\alpha\}$. Let $\mathcal{M}_0$ be the substructure of $\mathcal{M}$ with universe A_δ.

Claim 1 $\mathcal{M}_0 \prec \mathcal{M}$.

We apply the Tarski–Vaught test. Suppose that $\mathcal{M} \models \phi(v, \bar{a})$, where $\bar{a} \in A_\delta$. By Proposition 4.2.19, the isolated types in $S^{\mathcal{M}}(A_\delta)$ are dense. Thus, there is $b \in M$ such that $\mathcal{M} \models \phi(b, \bar{a})$ and $\text{tp}^{\mathcal{M}}(b/A_\delta)$ is isolated. By choice of δ, $b \in A_\delta$. Thus, by Proposition 2.3.5, $\mathcal{M}_0 \prec \mathcal{M}$.

Claim 2 $\mathcal{M}_0$ is a prime model extension of A.

Suppose that $\mathcal{N} \models T$ and $f : A \to \mathcal{N}$ is partial elementary. We show by induction that there are $f = f_0 \subset \ldots \subset f_\alpha \subset \ldots \subset f_\delta$, where $f_\alpha : A_\alpha \to \mathcal{N}$ is elementary.

If α is a limit ordinal, we let $f_\alpha = \bigcup_{\beta < \alpha} f_\beta$.

Given $f_\alpha : A_\alpha \to \mathcal{N}$ partial elementary, 1 $\phi(v, \bar{a})$ isolate $\text{tp}^{\mathcal{M}_0}(a_\alpha / A_\alpha)$. Because f_α is partial elementary, by Lemma 4.1.9 iii), $\phi(v, f_\alpha(\bar{a}))$ isolates $f_\alpha(\text{tp}^{\mathcal{M}_0}(a_\alpha / A_\alpha))$ in $S_1^{\mathcal{N}}(f_\alpha(A))$. Also, because f_α is partial elementary, there is $b \in N$ with $\mathcal{N} \models \phi(b, f_\alpha(\bar{a}))$. Thus, $f_{\alpha+1} = f_\alpha \cup \{(a_\alpha, \bar{b})\}$ is elementary.

In particular, $f_\delta : \mathcal{M}_0 \to \mathcal{N}$ is elementary. Thus, $\mathcal{M}_0$ is a prime model extension of A.

To see that every element of $\mathcal{M}_0$ realizes an isolated type over A, we must show that $\bar{a}$ realizes an isolated type over A for all $\bar{a} \in A_\alpha$, $\alpha < \delta$. We argue by induction on α. For α a limit ordinal, this is clear. For successor ordinals, it follows from the following lemma.

Lemma 4.2.21 *Suppose that $A \subseteq B \subseteq \mathcal{M} \models T$ and every $\bar{b} \in B^m$ realizes an isolated type in $S_m^{\mathcal{M}}(A)$. Suppose that $\bar{a} \in M^n$ realizes an isolated type in $S_n^{\mathcal{M}}(B)$. Then, $\bar{a}$ realizes an isolated type in $S_n^{\mathcal{M}}(A)$.*

Proof Let $\phi(\bar{v}, \bar{w})$ be an $\mathcal{L}$-formula and $\bar{b} \in B^m$ such that $\phi(\bar{v}, \bar{b})$ isolates $\text{tp}^{\mathcal{M}}(\bar{a}/B)$. Let $\theta(\bar{w})$ be an $\mathcal{L}_A$-formula isolating $\text{tp}^{\mathcal{M}}(\bar{b}/A)$. We first claim that $\phi(\bar{v}, \bar{w}) \wedge \theta(\bar{w})$ isolates $\text{tp}^{\mathcal{M}}(\bar{a}, \bar{b}/A)$.

Suppose that $\mathcal{M} \models \psi(\bar{a}, \bar{b})$. Because $\phi(\bar{v}, \bar{b})$ isolates $\text{tp}^{\mathcal{M}}(\bar{a}/B)$,

$$\text{Th}_A(\mathcal{M}) \models \phi(\bar{v}, \bar{b}) \to \psi(\bar{v}, \bar{b}).$$

Thus, because $\theta(\bar{w})$ isolates $\text{tp}^{\mathcal{M}}(\bar{b}/A)$,

$$\text{Th}_A(\mathcal{M}) \models \theta(\bar{w}) \to (\phi(\bar{v}, \bar{w}) \to \psi(\bar{v}, \bar{w}))$$

and

$$\text{Th}_A(\mathcal{M}) \models (\theta(\bar{w}) \wedge \phi(\bar{v}, \bar{w})) \to \psi(\bar{v}, \bar{w}),$$

as desired.

Because $\text{tp}^{\mathcal{M}}(\bar{a}, \bar{b}/A)$ is isolated, so is $\text{tp}^{\mathcal{M}}(\bar{a}/A)$ by Lemma 4.2.9.

For ω-stable theories (indeed, for theories that are κ-stable for some κ), prime model extensions are unique, although we postpone the proof to Chapter 6.

Theorem 4.2.22 *Let T be ω-stable. Suppose that $\mathcal{M} \models T$ and $\mathcal{N} \models T$ are prime model extensions of A and $\mathrm{Th}_A(\mathcal{M}) = \mathrm{Th}_A(\mathcal{N})$. Then, there is $f : \mathcal{M} \to \mathcal{N}$, an isomorphism fixing A.*

4.3   Saturated and Homogeneous Models

In Section 4.2 we concentrated on models that realize very few types. In this section, we will study models realizing many types. Throughout this section, we will assume that T is a complete theory with infinite models in a countable language $\mathcal{L}$.

Definition 4.3.1 Let κ be an infinite cardinal. We say that $\mathcal{M} \models T$ is *κ-saturated* if, for all $A \subseteq M$, if $|A| < \kappa$ and $p \in S_n^{\mathcal{M}}(A)$, then p is realized in $\mathcal{M}$.

We say that $\mathcal{M}$ is *saturated* if it is $|M|$-saturated.

Proposition 4.3.2 *Let $\kappa \geq \aleph_0$. The following are equivalent:*

i) $\mathcal{M}$ is κ-saturated.

ii) If $A \subseteq M$ with $|A| < \kappa$ and p is a (possibly incomplete) n-type over A, then p is realized in $\mathcal{M}$.

iii) If $A \subseteq M$ with $|A| < \kappa$ and $p \in S_1^{\mathcal{M}}(A)$, then p is realized in $\mathcal{M}$.

Proof

i)$\Rightarrow$ ii) If $\mathcal{M}$ is κ-saturated and p is an incomplete n-type over A where $|A| < \kappa$, then there is a complete type $p^* \in S_n^{\mathcal{M}}(A)$ with $p^* \supseteq p$. Because p^* is realized in $\mathcal{M}$ so is p.

ii) $\Rightarrow$ iii) Clear.

iii) $\Rightarrow$ i) We prove this by induction on n. Let $p \in S_n^{\mathcal{M}}(A)$. Let $q \in S_{n-1}^{\mathcal{M}}$ be the type $\{\phi(v_1, \ldots, v_{n-1}) : \phi \in p\}$. By induction, q is realized by some $\bar{a}$ in $\mathcal{M}$. Let $r \in S_1^{\mathcal{M}}(A \cup \{a_1, \ldots, a_{n-1}\})$ be the type $\{\psi(\bar{a}, w) : \psi(v_1, \ldots, v_n) \in p\}$. By iii), we can realize r by some b in $\mathcal{M}$. Then, $(\bar{a}, b)$ realizes p.

Homogeneity is a weak form of saturation.

Proposition 4.3.3 *If $\mathcal{M}$ is κ-saturated, then $\mathcal{M}$ is κ-homogeneous.*

Proof Suppose that $A \subseteq \mathcal{M}$, $|A| < \kappa$, and $f : A \to M$ is partial elementary. Let $b \in M \setminus A$. Let

$$\Gamma = \{\phi(v, f(\bar{a})) : \bar{a} \in A^m \text{ and } \mathcal{M} \models \phi(b, \bar{a})\}.$$

If $\phi(v, f(\bar{a})) \in \Gamma$, then $\mathcal{M} \models \exists v\, \phi(v, \bar{a})$ and hence, because f is partial elementary, $\mathcal{M} \models \exists v\, \phi(v, f(\bar{a}))$. Thus, because Γ is closed under conjunctions, Γ is satisfiable. Because $\mathcal{M}$ is saturated, there is $c \in M$ realizing Γ. Thus, $f \cup \{(b, c)\}$ is elementary and $\mathcal{M}$ is κ-homogeneous.

Countably Saturated Models

We will begin by examining $\aleph_0$-saturated models. If $\mathcal{M}$ is $\aleph_0$-saturated, then $\mathcal{M}$ realizes every type in $S_n(T)$. We will show that for $\aleph_0$-homogeneous models this condition is also sufficient.

Proposition 4.3.4 *If $\mathcal{M} \models T$, then $\mathcal{M}$ is $\aleph_0$-saturated if and only if $\mathcal{M}$ is $\aleph_0$-homogeneous and $\mathcal{M}$ realizes all types in $S_n(T)$.*

Proof
($\Rightarrow$) Clear.
($\Leftarrow$) Let $\bar{a} \in M^m$ and let $p \in S_n^{\mathcal{M}}(\bar{a})$. Let $q \in S_{n+m}(T)$ be the type $\{\phi(\bar{v}, \bar{w}) : \phi(\bar{v}, \bar{a}) \in p\}$. By assumption, there is $(\bar{b}, \bar{c}) \in M^{n+m}$ realizing q. Because $\mathrm{tp}^{\mathcal{M}}(\bar{c}) = \mathrm{tp}^{\mathcal{M}}(\bar{a})$ and $\mathcal{M}$ is $\aleph_0$-homogeneous, there is $\bar{d} \in \mathcal{M}$ such that $\mathrm{tp}^{\mathcal{M}}(\bar{a}, \bar{d}) = \mathrm{tp}^{\mathcal{M}}(\bar{c}, \bar{b})$. Hence, $\bar{d}$ realizes p and $\mathcal{M}$ is $\aleph_0$-saturated.

Countable saturated models are unique up to isomorphism.

Corollary 4.3.5 *If $\mathcal{M}, \mathcal{N} \models T$ are countable saturated models, then $\mathcal{M} \cong \mathcal{N}$.*

Proof Because $\mathcal{M}$ and $\mathcal{N}$ are $\aleph_0$-homogeneous and both realize all types in $S_n(T)$ for all $n < \omega$, by Theorem 4.2.15, $\mathcal{M} \cong \mathcal{N}$.

The next proposition shows that we can extend models to $\aleph_0$-homogeneous models without increasing the cardinality.

Proposition 4.3.6 *Let $\mathcal{M} \models T$. There is $\mathcal{M} \prec \mathcal{N}$ such that $\mathcal{N}$ is $\aleph_0$-homogeneous and $|N| = |M|$.*

Proof We first argue that we can find $\mathcal{M} \prec \mathcal{N}_1$ such that $|M| = |N_1|$, and if $\bar{a}, \bar{b}, c \in M$ and $\mathrm{tp}^{\mathcal{M}}(\bar{a}) = \mathrm{tp}^{\mathcal{M}}(\bar{b})$, then there is $d \in N_1$ such that $\mathrm{tp}^{\mathcal{N}_1}(\bar{a}, c) = \mathrm{tp}^{\mathcal{N}_1}(\bar{b}, d)$.
Let $((\bar{a}_\alpha, \bar{b}_\alpha, c_\alpha) : \alpha < |M|)$ list all tuples $(\bar{a}, \bar{b}, c)$ where $\bar{a}, \bar{b}, c \in M$ and $\mathrm{tp}^{\mathcal{M}}(\bar{a}) = \mathrm{tp}^{\mathcal{M}}(\bar{b})$. We build an elementary chain $\mathcal{M}_0 \prec \mathcal{M}_1 \ldots \prec \mathcal{M}_\alpha \prec \ldots$ for $\alpha < |M|$.
Let $\mathcal{M}_0 = \mathcal{M}$.
If α is a limit ordinal, let $\mathcal{M}_\alpha = \bigcup_{\beta < \alpha} \mathcal{M}_\beta$.
Given $\mathcal{M}_\alpha$, let $\mathcal{M}_\alpha \prec \mathcal{M}_{\alpha+1}$ with $|M_\alpha| = |M_{\alpha+1}|$ such that there is $d \in \mathcal{M}_\alpha$ with $\mathrm{tp}^{\mathcal{M}_{\alpha+1}}(\bar{b}, d) = \mathrm{tp}^{\mathcal{M}_{\alpha+1}}(\bar{a}, c)$. Let $\mathcal{N}_1 = \bigcup_{\alpha < |M|} \mathcal{M}_\alpha$. Because $\mathcal{N}_1$ is a union of $|M|$ models of size $|M|$, $|N_1| = |M|$.
We now build $\mathcal{N}_0 \prec \mathcal{N}_1 \prec \mathcal{N}_2 \ldots$ such that $|N_i| = |M|$ and if $\bar{a}, \bar{b}, c \in N_i$ and $\mathrm{tp}^{\mathcal{N}_i}(\bar{a}) = \mathrm{tp}^{\mathcal{N}_i}(\bar{b})$, then there is $d \in N_{i+1}$ such that $\mathrm{tp}^{\mathcal{N}_{i+1}}(\bar{a}, c) = \mathrm{tp}^{\mathcal{N}_{i+1}}(\bar{b}, d)$.
Let $\mathcal{N} = \bigcup_{i < \omega} \mathcal{N}_i$. Clearly, $|\mathcal{N}| = |\mathcal{M}|$ and $\mathcal{N}$ is $\aleph_0$-homogeneous.

Propositions 4.3.5 and 4.3.6 allow us to characterize theories with countable saturated models.

Theorem 4.3.7 *T has a countable saturated model if and only if $|S_n(T)| \leq \aleph_0$ for all n.*

Proof We need only show that if $|S_n(T)| \leq \aleph_0$ for all n then T has a countable saturated model. Let $p_0, p_1, \ldots$ list all elements of $\bigcup_{n \in \omega} S_n(T)$. Let $\mathcal{M}_0 \models T$. Iterating Lemma 4.1.3, we build $\mathcal{M}_0 \prec \mathcal{M}_1 \prec \ldots$ such that $\mathcal{M}_i$ is countable and $\mathcal{M}_{i+1}$ realizes p_i. Thus, $\mathcal{M} = \bigcup_{i \in \omega} \mathcal{M}_i$ is countable and contains realizations of all types in $S_n(T)$ for $n < \omega$. By Proposition 4.3.6, there is $\mathcal{M} \prec \mathcal{N}$ such that $\mathcal{N}$ is countable and $\aleph_0$-homogeneous. By Corollary 4.3.5, $\mathcal{N}$ is $\aleph_0$-saturated.

Curiously, theories with large countable models also have small countable models.

Corollary 4.3.8 *i) If T has a countable saturated model, then T has a prime model.*

ii) If T has fewer than $2^{\aleph_0}$ countable models, then T has a countable saturated model and a prime model.

Proof

i) If T has a saturated model, then $|S_n(T)|$ is countable for all n. By Theorem 4.2.11, the isolated types are dense in $S_n(T)$ for all n. Thus, by Theorem 4.2.10, T has a prime model.

ii) It suffices to show that $S_n(T)$ is countable for all $n < \omega$. Suppose not. By Theorem 4.2.11, if $|S_n(T)| > \aleph_0$, then $|S_n(T)| = 2^{\aleph_0}$. Each n-type must be realized in some countable model. Because each countable model realizes only countably many n-types, if there are $2^{\aleph_0}$ n-types, then there must be $2^{\aleph_0}$ nonisomorphic countable models.

We consider several examples.

Example 4.3.9 *Dense Linear Orders*

We will show that $(\mathbb{Q}, <)$ is saturated. Suppose $A \subset \mathbb{Q}$ is finite. Suppose that $A = \{a_1, \ldots, a_m\}$ where $a_1 < \ldots < a_m$. By the analysis of types in DLO given in Section 4.1, there are exactly $2m+1$ types in $S_1(A)$. Each type is isolated by one of the formulas $v = a_i$, $v < a_0$, $a_i < v < a_{i+1}$, or $a_m < v$. Clearly, all of these types are realized in $\mathbb{Q}$. Note that in this case $\mathbb{Q}$ is both saturated and prime! In Section 4.4, we will see that this always happens in $\aleph_0$-categorical theories.

Example 4.3.10 *Algebraically Closed Fields*

Fix p prime or 0. Let k be $\mathbb{F}_p$ if $p > 0$ and $\mathbb{Q}$ if $p = 0$. Because $S_n(\mathrm{ACF}_p)$ is in bijection with $\mathrm{Spec}(k[X_1, \ldots, X_n])$, by Corollary 4.1.18, $|S_n(\mathrm{ACF}_p)| = \aleph_0$. Thus, there is a countable saturated model of ACF_p.

Let q_n be the type corresponding to the 0 ideal in $k[X_1, \ldots, X_n]$. If $a_1, \ldots, a_n$ realizes q_n, then $a_1, \ldots, a_n$ are algebraically independent over k. Thus, any saturated model has infinite transcendence degree. It follows that the countable saturated model of ACF_p is the unique algebraically closed field of characteristic p and transcendence degree $\aleph_0$.

Example 4.3.11 *Real Closed Fields*

Let $r \in \mathbb{R} \setminus \mathbb{Q}$. Let p_r be the set of formulas $\Big\{ \underbrace{v + \ldots + v}_{m-\text{times}} < \underbrace{1 + \ldots + 1}_{n-\text{times}} :$
$m, n \in \mathbb{N}, r < \dfrac{n}{m} \Big\} \cup \Big\{ \underbrace{v + \ldots + v}_{m-\text{times}} > \underbrace{1 + \ldots + 1}_{n-\text{times}} : m, n \in \mathcal{N}, r > \dfrac{n}{m} \Big\}.$
Clearly, p_r is satisfiable. Let $p_r^* \in S_1(\text{RCF})$ with $p_r^* \supseteq p_r$. If $r \neq s$, then $p_r^* \neq p_s^*$. Thus, $|S_1(\text{RCF})| = 2^{\aleph_0}$ and RCF has no saturated model.

Existence of Saturated Models

Next we think about the existence of κ-saturated models for $\kappa > \aleph_0$.

Theorem 4.3.12 *For all $\mathcal{M}$, there is a κ^+-saturated $\mathcal{M} \prec \mathcal{N}$ with $|N| \leq |M|^\kappa$.*

Proof

Claim For any $\mathcal{M}$ there is $\mathcal{M} \prec \mathcal{M}'$ such that $|M'| \leq |M|^\kappa$, and if $A \subseteq M$, $|A| \leq \kappa$ and $p \in S_1^\mathcal{M}(A)$, then p is realized in $\mathcal{M}'$.

We first note that

$$|\{A \subseteq M : |A| \leq \kappa\}| \leq |M|^\kappa$$

because for each such A there is f mapping κ onto A. Also, for each such A, $|S_1^\mathcal{M}(A)| \leq 2^\kappa$. Let $(p_\alpha : \alpha < |M|^\kappa)$ list all types in $S_1^\mathcal{M}(A)$ for $n < \omega$, $A \subseteq M$ with $|A| \leq \kappa$. We build an elementary chain $(\mathcal{M}_\alpha : \alpha < |M|^\kappa)$ as follows:

i) $\mathcal{M}_0 = \mathcal{M}$;

ii) $\mathcal{M}_\alpha = \bigcup_{\beta < \alpha} \mathcal{M}_\beta$ for α a limit ordinal;

iii) $\mathcal{M}_\alpha \prec \mathcal{M}_{\alpha+1}$ with $|M_{\alpha+1}| = |M_\alpha|$, and $\mathcal{M}_{\alpha+1}$ realizes p_α. By induction, we see that $|M_\alpha| \leq |M|^\kappa$ for all α. Let $\mathcal{M}' = \bigcup_{\alpha < |M|^\kappa} \mathcal{M}_\alpha$. Then, $|M'| \leq |M|^\kappa$ and $\mathcal{M}'$ is the desired model. This proves the claim.

We build an elementary chain $(\mathcal{N}_\alpha : \alpha < \kappa^+)$ such that each $|N_\alpha| \leq |M|^\kappa$ and

i) $\mathcal{N}_0 = \mathcal{M}$;

ii) $\mathcal{N}_\alpha = \bigcup_{\beta < \alpha} \mathcal{N}_\beta$ for α a limit ordinal;

iii) $\mathcal{N}_\alpha \prec \mathcal{N}_{\alpha+1}$, $|N_\alpha| \leq |M|^\kappa$, and if $A \subseteq N_\alpha$ with $|A| \leq \kappa$ and $p \in S_n^{\mathcal{N}_\alpha}(A)$, then p is realized in $\mathcal{N}_{\alpha+1}$. This is possible by the claim because, by induction,

$$|N_\alpha|^\kappa \leq (|M|^\kappa)^\kappa = |M|^\kappa.$$

Let $\mathcal{N} = \bigcup_{\alpha < \kappa^+} \mathcal{N}_\alpha$. Because $\kappa^+ \leq |M|^\kappa$, N is the union of at most $|M|^\kappa$ sets of size $|M|^\kappa$ so $|N| \leq |M|^\kappa$. Suppose that $|A| \subseteq N$, $|A| \leq \kappa$, and

$p \in S_n^{\mathcal{N}}(A)$. Because κ^+ is a regular cardinal, there is $\alpha < \kappa^+$ such that $A \subset \mathcal{N}_\alpha$ and p is realized in $\mathcal{N}_{\alpha+1} \prec \mathcal{N}$. Thus, $\mathcal{N}$ is κ^+-saturated.

Theorem 4.3.12 guarantees the existence of saturated models under suitable set-theoretic assumptions.

Corollary 4.3.13 *Suppose that $2^\kappa = \kappa^+$. Then, there is a saturated model of T of size κ^+. In particular, if the Generalized Continuum Hypothesis is true, there are saturated models of size κ^+ for all κ.*

For arbitrary T, some set-theoretic assumption is necessary. For example, if $|S_n(T)| = 2^{\aleph_0}$, then any $\aleph_0$-saturated model has size $2^{\aleph_0}$. If $\aleph_1 < 2^{\aleph_0}$, then there is no saturated model of size $\aleph_1$.

We can extend this a bit further.

Corollary 4.3.14 *Suppose that $\kappa \geq \aleph_1$ is regular and $2^\lambda \leq \kappa$ for $\lambda < \kappa$. Then, there is a saturated model of size κ. In particular, if $\kappa \geq \aleph_1$ is strongly inaccessible, then there is a saturated model of size κ.*

Proof Let $\mathcal{M} \models T$ with $|M| = \kappa$. If $\kappa = \lambda^+$ for $\lambda < \kappa$, then the corollary follows from Corollary 4.3.13. Thus, we may assume that κ is a limit cardinal. We build an elementary chain $(\mathcal{M}_\lambda : \lambda < \kappa, \lambda$ a cardinal). Each $\mathcal{M}_\lambda$ will have cardinality κ. Let $\mathcal{M}_0 = \mathcal{M}$.

Let $\mathcal{M}_\lambda = \bigcup_{\mu < \lambda} \mathcal{M}_\mu$ for λ a limit cardinal. Because $\mathcal{M}_\alpha$ is the union of fewer than κ models of size κ, $|M_\alpha| = \kappa$.

Given $\mathcal{M}_\lambda$, by Theorem 4.3.12 there is $\mathcal{M}_\lambda \prec \mathcal{M}_{\lambda^+}$ such that $\mathcal{M}$ is λ^+-saturated and $|M_{\lambda^+}| \leq \kappa^\lambda = \kappa$ (see Corollary A.17).

Let $\mathcal{N} = \bigcup \mathcal{M}_\lambda$. Because κ is a regular limit cardinal, $\kappa = \aleph_\kappa$ (see Proposition A.13). Thus, because κ is regular, if $A \subset N$ and $|A| < \kappa$, then there is $\lambda < \kappa$ such that $A \subset M_\lambda$. Thus, if $p \in S_n^{\mathcal{N}}(A)$, then p is realized in $\mathcal{M}_{\lambda^+} \prec \mathcal{N}$.

The assumption of regularity is necessary for some T. For example, suppose that $\mathcal{M} \models$ DLO with $|M| = \aleph_\omega$. We claim that $\mathcal{M}$ is not saturated. Let $M = \bigcup_{n < \omega} M_n$ where $|M_n| = \aleph_n$. If $\mathcal{M}$ is saturated, then for each $n < \omega$ we can find $a_n \in M$ such that $a_n > b$ for all $b \in M_n$. One more use of saturation allows us to find $c \in M$ such that $c > a_n$ for $n < \omega$. This is impossible. Similar arguments show that all saturated dense linear orders must have regular cardinality.

If T is κ-stable, then we can eliminate all assumptions about cardinal exponentiation.

Theorem 4.3.15 *Let κ be a regular cardinal. If T is κ-stable, then there is a saturated $\mathcal{M} \models T$ with $|M| = \kappa$. Indeed, if $\mathcal{M}_0 \models T$ with $|M_0| = \kappa$, then there is a saturated elementary extension $\mathcal{M}$ of $\mathcal{M}_0$ with $|M| = \kappa$.*

In particular, if T is ω-stable, then there are saturated models of size κ for all regular cardinals κ.

Proof We build an elementary chain $(\mathcal{M}_\alpha : \alpha < \kappa)$ where $|\mathcal{M}_\alpha| = \kappa$ such that:

i) $\mathcal{M}_0 \models T$ with $|\mathcal{M}_0| = \kappa$;

ii) $\mathcal{M}_\alpha = \bigcup_{\beta < \alpha} \mathcal{M}_\beta$ for α a limit ordinal;

iii) $\mathcal{M}_\alpha \prec \mathcal{M}_{\alpha+1}$ and if $p \in S_1^{\mathcal{M}_\alpha}(M_\alpha)$, then p is realized in $\mathcal{M}_{\alpha+1}$.

Because T is κ-stable, if $|\mathcal{M}_\alpha| = \kappa$, then $|S_1^{\mathcal{M}_\alpha}(M_\alpha)| = \kappa$. Thus, as in Theorem 4.3.12, we can find $\mathcal{M}_\alpha \prec \mathcal{M}_{\alpha+1}$ such that $|M_{\alpha+1}| = \kappa$ and $\mathcal{M}_{\alpha+1}$ realizes all types in $S_1^{\mathcal{M}_\alpha}(M_\alpha)$.

Let $\mathcal{M} = \bigcup \mathcal{M}_\alpha$. Because $\mathcal{M}$ is the union of κ models of size κ, $|M| = \kappa$. We claim that $\mathcal{M}$ is saturated. Let $A \subset M$ with $|A| < \kappa$. Because κ is regular, there is an $\alpha < \kappa$ such that $A \subseteq M_\alpha$. If $p \in S_1^{\mathcal{M}}(A)$, then there is $q \in S_1^{\mathcal{M}}(M_\alpha) = S_1^{\mathcal{M}_\alpha}(M_\alpha)$ with $p \subseteq q$. Because q is realized in $\mathcal{M}_{\alpha+1}$, p is realized in $\mathcal{M}$. Thus, $\mathcal{M}$ is saturated.

Saturated models of singular cardinality exist for ω-stable theories, but the proof is much more subtle. We prove this in Theorem 6.5.4.

Homogeneous and Universal Models

Although prime models elementarily embed into all models of T, saturated models embed all small models.

Definition 4.3.16 We say that $\mathcal{M} \models T$ is κ-*universal* if for all $\mathcal{N} \models T$ with $|N| < \kappa$ there is an elementary embedding of $\mathcal{N}$ into $\mathcal{M}$.
We say that $\mathcal{M}$ is *universal* if it is $|M|^+$-universal.

Lemma 4.3.17 Let $\kappa \geq \aleph_0$. If $\mathcal{M}$ is κ-saturated, then $\mathcal{M}$ is κ^+-universal.

Proof Let $\mathcal{N} \models T$ with $|N| \leq \kappa$. Let $(n_\alpha : \alpha < \kappa)$ enumerate N. Let $A_\alpha = \{n_\beta : \beta < \alpha\}$. We build a sequence of partial elementary maps $f_0 \subset f_1 \subset \ldots \subset f_\alpha \subset \ldots$ for $\alpha < \kappa$ with $f_\alpha : A_\alpha \to \mathcal{M}$.
Let $f_0 = \emptyset$ and, if α is a limit ordinal, let $f_\alpha = \bigcup_{\beta < \alpha} f_\beta$.
Given $f_\alpha : A_\alpha \to \mathcal{M}$ partial elementary, let

$$\Gamma(v) = \{\phi(v, f_\alpha(\bar{a})) : \mathcal{M} \models \phi(n_\alpha, \bar{a})\}.$$

Because f_α is partial elementary and $|A_\alpha| < \kappa$, Γ is satisfiable and, by κ-saturation, realized by some b in $\mathcal{M}$. The $f_{\alpha+1} = f_\alpha \cup \{(n_\alpha, b)\}$ is the desired partial elementary map.
We have constructed $f = \bigcup f_\alpha$, an elementary embedding of $\mathcal{N}$ into $\mathcal{M}$.

Theorem 4.3.18 Let $\kappa \geq \aleph_0$. The following are equivalent.

i) $\mathcal{M}$ is κ-saturated.

ii) $\mathcal{M}$ is κ-homogeneous and κ^+-universal.

If $\kappa \geq \aleph_1$ i) and ii) are also equivalent to:

iii) $\mathcal{M}$ is κ-homogeneous and κ-universal.

Proof By Proposition 4.3.2 and Lemma 4.3.17, i) $\Rightarrow$ ii). Clearly, ii) $\Rightarrow$ iii). We argue that ii) $\Rightarrow$ i) and, if κ is uncountable, iii) $\Rightarrow$ i).

Let $A \subseteq M$ with $|A| < \kappa$, and let $p \in S_1^{\mathcal{M}}(A)$. We can find $\mathcal{N} \models \mathrm{Th}_A(\mathcal{M})$ such that $A \subseteq N$ and there is $a \in N$ realizing p. If $\kappa = \aleph_0$, then we can choose $\mathcal{N}$ with $|N| = \aleph_0$. If $\kappa \geq \aleph_1$, then we can choose $\mathcal{N}$ with $|N| < \kappa$. By assumption, there is an elementary embedding $f : \mathcal{N} \to \mathcal{M}$. Because $f|A$ is partial elementary, by κ-homogeneity, there is $b \in M$ such that

$$\mathrm{tp}^{\mathcal{M}}(b/A) = \mathrm{tp}^{\mathcal{M}}(f((a))/f(A)) = \mathrm{tp}^{\mathcal{N}}(a/A) = p.$$

Thus, $\mathcal{M}$ is κ-saturated.

Corollary 4.3.19 $\mathcal{M}$ *is saturated if and only if it is homogeneous and universal.*

Similar arguments can be used to show that there is at most one saturated model of any particular cardinality.

Theorem 4.3.20 *If $\mathcal{M}$ and $\mathcal{N}$ are saturated models of T of cardinality κ, then $\mathcal{M} \cong \mathcal{N}$.*

Proof By Corollary 4.3.5, we may assume that $\kappa \geq \aleph_1$. Let $(m_\alpha : \alpha < \kappa)$ enumerate $\mathcal{M}$ and $(n_\alpha : \alpha < \kappa)$ enumerate $\mathcal{N}$. We build a sequence of partial embeddings $f_0 \subset \ldots \subset f_\alpha \ldots$ for $\alpha < \kappa$ such that $m_\alpha \in \mathrm{dom}(f_{\alpha+1})$ and $n_\alpha \in \mathrm{img}(f_{\alpha+1})$. Let A_α denote the domain of f_α. We will have $|A_\alpha| \leq |\alpha| + \aleph_0 < \kappa$ for all α.

Let $f_0 = \emptyset$, and let $f_\alpha = \bigcup_{\beta < \alpha} f_\beta$ for β a limit ordinal.

Suppose that f_α is partial elementary. By saturation, we can find $b \in N$ such that

$$\mathcal{N} \models \phi(b, f_\alpha(\bar{a})) \Leftrightarrow \mathcal{M} \models \phi(m_\alpha, \bar{a})$$

for all ϕ and all $\bar{a} \in A_\alpha$. Then $g_\alpha = f_\alpha \cup \{(m_\alpha, b)\}$ is partial elementary. Again by saturation, we can find $a \in M$ such that

$$\mathcal{N} \models \phi(n_\alpha, g(\bar{a})) \Leftrightarrow \mathcal{M} \models \phi(a, \bar{a})$$

for all ϕ and all $\bar{a} \in A_\alpha \cup \{m_\alpha\}$. Then, $f_{\alpha+1} = g_\alpha \cup \{(a, n_\alpha)\}$ is partial elementary and $f = \bigcup f_\alpha$ is an isomorphism from $\mathcal{M}$ to $\mathcal{N}$.

Lemma 4.3.17 and Theorem 4.3.20 are special cases of embedding and uniqueness results on homogeneous models generalizing Theorem 4.2.15.

Lemma 4.3.21 *Suppose that $\mathcal{N} \models T$ is κ-homogeneous where $\kappa \leq |N|$ and $\mathcal{M} \equiv \mathcal{N}$ such that every type in $S_n(T)$ realized in $\mathcal{M}$ is realized in $\mathcal{N}$ for $n < \omega$. If $A \subseteq M$ and $|A| \leq \kappa$, then there is a partial elementary map $f : A \to \mathcal{N}$.*

Proof We prove the claim by induction on $|A|$. Suppose that $|A|$ is finite. Let $A = \{a_1, \ldots, a_n\}$. Because every type realized in $\mathcal{M}$ is realized in

$\mathcal{N}$, there is $\bar{b} \in N^n$ such that $\text{tp}^{\mathcal{M}}(\bar{a}) = \text{tp}^{\mathcal{N}}(\bar{b})$. Then, $\bar{a} \mapsto \bar{b}$ is partial elementary.

Suppose that $|A| = \lambda \le \kappa$ and the claim is true for sets of size $\mu < \lambda$. Let $(a_\alpha : \alpha < \lambda)$ enumerate A. For $\alpha < \lambda$, let $A_\alpha = \{a_\beta : \beta < \alpha\}$. We build a sequence of partial elementary maps $f_0 \subseteq \ldots \subseteq f_\alpha \subseteq \ldots$ where A_α is the domain of f_α for $\alpha < \lambda$.

Let $f_0 = \emptyset$. If α is a limit ordinal, let $f_\alpha = \bigcup_{\beta < \alpha} f_\beta$.

Suppose that we are given f_α. Because $|A_{\alpha+1}| < \lambda$, by the induction assumption, there is a partial elementary $g : A_{\alpha+1} \to \mathcal{N}$. Let B be the image of A_α under f_α and let C be the image of A_α under g. Let $h = f_\alpha \circ g^{-1} : C \to B$. Because f_α and g are partial elementary, $h : C \to B$ is partial elementary. Because $\mathcal{N}$ is homogeneous, we can extend h to a partial elementary $h^* : C \cup \{g(a_\alpha)\} \to \mathcal{N}$. Let $b = h^*(g(a_\alpha))$, and let $f_{\alpha+1} = f_\alpha \cup \{(a_\alpha, b)\}$. Then, $f_{\alpha+1} = h^* \circ g$ is partial elementary.

Clearly, $f = \bigcup_{\alpha < \lambda} f_\alpha : A \to \mathcal{N}$ is partial elementary.

Corollary 4.3.22 *If* $\mathcal{M} \models T$ *is κ-homogeneous and realizes all types in* $S_n(T)$ *for all $n < \omega$, then $\mathcal{M}$ is κ-saturated.*

Proof By Lemma 4.3.21, $\mathcal{M}$ is κ^+-universal. Thus, by Theorem 4.3.18, $\mathcal{M}$ is saturated.

Theorem 4.3.23 *If* $\mathcal{M} \equiv \mathcal{N}$ *are homogeneous models of T of the same cardinality realizing the same types in* $S_n(T)$ *for all $n < \omega$, then $\mathcal{M} \cong \mathcal{N}$.*

Proof If $\mathcal{M}$ and $\mathcal{N}$ are countable, this is Theorem 4.2.15 so we assume that $\kappa = |M| = |N|$ is uncountable. We build an isomorphism $f : \mathcal{M} \to \mathcal{N}$ by a back-and-forth argument. Let $(a_\alpha : \alpha < \kappa)$ enumerate M. Let $(b_\alpha : \alpha < \kappa)$ enumerate N. We build a sequence of partial elementary maps $f_0 \subset \ldots \subset f_\alpha \subset \ldots$ such that the domain of f_α has cardinality at most $|\alpha| + \aleph_0 < \kappa$, a_α is in the domain of $f_{\alpha+1}$, and b_α is in the image of $f_{\alpha+1}$. Then, $f = \bigcup_{\alpha < \kappa} f_\alpha$ is the desired isomorphism.

Let $f_0 = \emptyset$. If α is a limit ordinal, then $f_\alpha = \bigcup_{\beta < \alpha} f_\beta$. Let A be the domain of f_α, and let B be its image. By Lemma 4.3.21, there is a partial elementary $h : A \cup \{a_\alpha\} \to \mathcal{N}$. Let C be the image of A under h, and let $c = h(a_\alpha)$. As in the proof of Lemma 4.3.21, $f_\alpha \circ h^{-1} : C \to B$ is partial elementary and, because $\mathcal{N}$ is homogeneous, we can extend this map to $C \cup \{c\}$. Let b be the image of c under this extension. Then, $g_\alpha = f_\alpha \cup \{(a_\alpha, b)\}$ is partial elementary and a_α is in the domain.

Let D be the image of g_α. Then, $g_\alpha^{-1} : D \to \mathcal{M}$ is partial elementary. By a symmetric argument, we can find $a \in M$ such that $g_\alpha^{-1} \cup \{(b_\alpha, a)\}$ is partial elementary. Let $f_{\alpha+1} = g_\alpha \cup \{(a, b_\alpha)\}$.

Corollary 4.3.24 *i) The number of nonisomorphic homogeneous models of T of size κ is at most $2^{2^{\aleph_0}}$.*

ii) If T has a countable saturated model, then the number of homogeneous models of T of size κ is at most $2^{\aleph_0}$.

Proof Homogeneous models of cardinality κ are determined by the set of types realized. Because $|S_n(T)| \leq 2^{\aleph_0}$, the number of possible sets of types realized in a model is at most $2^{2^{\aleph_0}}$. If T has a saturated model, then $|S_n(T)| \leq \aleph_0$ for all $n < \omega$ and there are at most $2^{\aleph_0}$ possible sets of types.

Applications of Saturated Models

We conclude this section with several applications of saturated and homogeneous models. Saturated models are useful because we can do things in the model that we usually could only do in an elementary extension.

Proposition 4.3.25 *Let $\mathcal{M}$ be saturated. Let $A \subset M$ with $|A| < |M|$. Let $X \subset M^n$ be definable with parameters from M. Then, X is A-definable if and only if every automorphism of $\mathcal{M}$ that fixes A pointwise fixes the X setwise.*

Proof

($\Rightarrow$) If $\bar{a} \in A$, $X = \{\bar{b} \in M^n : \mathcal{M} \models \phi(\bar{b}, \bar{a})\}$ and σ is an automorphism of $\mathcal{M}$, then

$$
\begin{aligned}
\sigma(X) &= \{\bar{c} \in M^n : \mathcal{M} \models \phi(\sigma^{-1}(\bar{c}), \bar{a})\} \\
&= \{\bar{c} \in M^n : \mathcal{M} \models \phi(\bar{c}, \sigma(\bar{a}))\} \quad \text{because } \sigma \text{ is an automorphism} \\
&= \{\bar{c} \in M^n : \mathcal{M} \models \phi(\bar{c}, \bar{a})\} \qquad \text{because } \sigma(\bar{a}) = \bar{a} \\
&= X.
\end{aligned}
$$

($\Leftarrow$) Let $\psi(\bar{v}, \bar{m})$ define X, where $\bar{m} \in M^k$. Consider the type $\Gamma(\bar{v}, \bar{w}) =$

$$\{\psi(\bar{v}, \bar{m}), \neg\psi(\bar{w}, \bar{m})\} \cup \{\phi(\bar{v}) \leftrightarrow \phi(\bar{w}) : \phi \text{ an } \mathcal{L}_A\text{-formula}\}.$$

Suppose that $\Gamma \cup \text{Diag}_{\text{el}}(\mathcal{M})$ is satisfiable. Then, by saturation, we can find $(\bar{a}, \bar{b})$ realizing Γ in $\mathcal{M}$. Let f be the map that is the identity on A and sends $\bar{a}$ to $\bar{b}$. By choice of Γ, f is elementary. Because $\mathcal{M}$ is homogeneous, f extends to an automorphism σ of $\mathcal{M}$. But $\mathcal{M} \models \psi(\bar{a}, \bar{m}) \wedge \neg\psi(\bar{b}, \bar{m})$, thus $\bar{a} \in X$ and $\sigma(\bar{a}) = \bar{b} \notin X$, a contradiction. Thus, $\Gamma \cup \text{Diag}_{\text{el}}(\mathcal{M})$ is not satisfiable.

Therefore, there are $\mathcal{L}_A$-formulas $\phi_1, \ldots \phi_m$ such that

$$\mathcal{M} \models \forall\bar{v}\forall\bar{w} \left(\bigwedge_{i=1}^{n}(\phi_i(\bar{v}) \leftrightarrow \phi_i(\bar{w})) \rightarrow (\psi(\bar{v}, \bar{m}) \leftrightarrow \psi(\bar{w}, \bar{m})) \right). \qquad (*)$$

For $\tau : \{1, \ldots, n\} \rightarrow 2$, let $\theta_\tau(\bar{v})$ be the formula

$$\bigwedge_{\tau(i)=1} \phi_i(\bar{v}) \wedge \bigwedge_{\tau(i)=0} \neg\phi_i(\bar{v}).$$

If $\theta_\tau(\bar{a})$ and $\theta_\tau(\bar{b})$, then, by $(*)$, $\bar{a} \in X$ if and only if $\bar{b} \in X$. Let $S = \{\tau : \{1, \ldots, m\} \to 2 : \mathcal{M} \models \theta_\tau(\bar{a})$ for some $\bar{a}$ in $M^n\}$. Then,

$$\bar{a} \in X \text{ if and only if } \mathcal{M} \models \bigvee_{\tau \in S} \theta_\tau(\bar{v}).$$

Hence, X is definable with parameters from A.

Recall that $b \in M$ is *definable from* A if $\{b\}$ is A-definable. The next corollary is a simple consequence of Proposition 4.3.25.

Corollary 4.3.26 *Let $\mathcal{M}$ be saturated, and let $A \subset M$ with $|A| < |M|$. Then, b is definable from A if and only if b is fixed by all automorphisms of $\mathcal{M}$ that fix A pointwise.*

Proof By Proposition 4.3.25, $\{b\}$ is A-definable if and only if every automorphism that fixes A pointwise fixes the set $\{b\}$.

Recall that b is *algebraic* over A if there is a finite A-definable set X such that $b \in X$.

Proposition 4.3.27 *Let $\mathcal{M}$ be saturated. Let $A \subset M$ with $|A| < |M|$ and $b \in M$. The following are equivalent:*

i) b is algebraic over A;

ii) b has only finitely many images under automorphisms of $\mathcal{M}$ fixing A pointwise;

iii) $\mathrm{tp}^{\mathcal{M}}(b/A)$ has finitely many realizations.

Proof

i)$\Rightarrow$ ii) Let X be a finite A-definable set with $b \in A$. By Proposition 4.3.25, any automorphism of $\mathcal{M}$ that fixes A pointwise permutes the elements of the finite set X.

ii)$\Rightarrow$ iii) If c realizes $\mathrm{tp}^{\mathcal{M}}(b/A)$, then, because $\mathcal{M}$ is homogeneous, there is an automorphism of $\mathcal{M}$ fixing A pointwise and mapping b to c. Thus, if b has only finitely many images under automorphisms fixing A, then $\mathrm{tp}^{\mathcal{M}}(b/A)$ has only finitely many realizations.

iii) $\Rightarrow$ i) Suppose that $p = \mathrm{tp}^{\mathcal{M}}(b/A)$ has exactly n realizations. Let

$$\Gamma = \mathrm{Th}_A(\mathcal{M}) \cup \{\phi(v_i) : \phi \in p, i = 0, \ldots, n\} \cup \{\bigwedge_{0 \leq i < j \leq n} v_i \neq v_j\}.$$

Because p has only n realizations in $\mathcal{M}$ and $\mathcal{M}$ is saturated, Γ is not satisfiable. Thus there are $\phi_1, \ldots, \phi_m \in p$ such that

$$\mathcal{M} \models \left(\bigwedge_{k=1}^{m} \bigwedge_{i=0}^{n} \phi_k(v_i)\right) \to \bigvee_{i \neq j} v_i = v_j.$$

In particular $\{c \in M : \mathcal{M} \models \bigwedge_{j=1}^{m} \phi_j(c)\}$ is an A-definable set of size n containing b, so b is algebraic over A.

Saturated models can be used to give a new test for quantifier elimination.

Proposition 4.3.28 *If $\mathcal{L}$ is a language containing a constant symbol and T is an $\mathcal{L}$-theory, then T has quantifier elimination if and only if whenever $\mathcal{M} \models T$, $A \subseteq M$, $\mathcal{N} \models T$ is $|M|^+$-saturated, and $f : A \to \mathcal{N}$ is a partial embedding, f extends to an embedding of $\mathcal{M}$ into $\mathcal{N}$.*

Proof

($\Rightarrow$) By quantifier elimination f is a partial elementary embedding. As in the proof of Lemma 4.3.17, we can extend f to an elementary embedding of $\mathcal{M}$ into $\mathcal{N}$.

($\Leftarrow$) We use the quantifier elimination criterion from Corollary 3.1.6. Suppose that $\mathcal{M}, \mathcal{N} \models T$, $A \subseteq M \cap N$, and $\mathcal{M} \models \phi(b, \bar{a})$, where ϕ is quantifier-free, $\bar{a} \in A$, and $b \in M$. Let $\mathcal{N} \prec \mathcal{N}'$ be an $|M|^+$-saturated model of T. By assumption the identity map on A extends to an embedding $f : \mathcal{M} \to \mathcal{N}'$. Because f is the identity on A, $\mathcal{N}' \models \phi(f(b), \bar{a})$. Because $\mathcal{N} \prec \mathcal{N}'$, $\mathcal{N} \models \exists v\, \phi(v, \bar{a})$, as desired.

Quantifier Elimination for Differentially Closed Fields

We will show how to apply Proposition 4.3.28 in one very interesting case. A *derivation* on a commutative ring R is a map $\delta : R \to R$ such that

$$\delta(x + y) = \delta(x) + \delta(y)$$

and

$$\delta(xy) = x\delta(y) + y\delta(x).$$

We often write $a', a'', \ldots, a^{(n)}$ for $\delta(a), \delta(\delta(a)), \ldots$.

If (R, δ) is a differential ring, we form the ring of differential polynomials $R\{X\} = R[X, X', X'', \ldots, X^{(n)}, \ldots]$. There is a natural extension of the derivation δ to $R\{X\}$ where $\delta(X^{(n)}) = X^{(n+1)}$. For f in $R\{X\} \setminus R$, the order of f is the least n such that $f \in R[X, \ldots, X^{(n)}]$, whereas if $f \in R$ we say that f has order $-\infty$.

We will consider differential fields, which we always assume have characteristic zero.

Definition 4.3.29 We say that K is a *differentially closed field* if K is a differential field of characteristic zero such that if $f, g \in K\{X\} \setminus \{0\}$ and the order of f is less than the order of g, then there is $x \in K$ such that $f(x) = 0$ and $g(x) \neq 0$.

In particular, if f has order 0, there is $x \in K$ with $f(x) = 0$, so K is algebraically closed. We can give axioms for DCF, the theory of differentially closed fields, in the language $\mathcal{L} = \{+, -, \cdot, \delta, 0, 1\}$, where δ is a unary function symbol for the derivation. Our goal is to show that DCF has quantifier elimination.

Let $k \subseteq K$ be differential fields, we say that $a \in K$ is *differentially algebraic over k* if $f(a) = 0$ for some nonzero $f \in k\{X\}$. Otherwise, we say that a is *differentially transcendental* over k.

The next proposition summarizes some basic algebra of differential fields that we will need. We assume that all of our fields have characteristic zero. If $k \subset K$ are differential fields and $a \in K$, we let $k\langle a \rangle$ be the differential subfield of K generated by a over k.

Proposition 4.3.30 *Let $k \subset K$ be differential fields of characteristic zero.*

i) Suppose that $f(X, X', \ldots, X^{(n)}) \in k\{X\} \setminus 0$ and $a, b \in K$ such that $f(a) = f(b) = 0$, $a, \ldots, a^{(n-1)}$ are algebraically independent over k, $b, \ldots, b^{(n-1)}$ are algebraically independent over k, and $g(a) \neq 0$, $g(b) \neq 0$ for any g of order n of lower degree in $X^{(n)}$. Then, $k\langle a \rangle$ and $k\langle b \rangle$ are isomorphic over k.

ii) If $a \in K$ is differentially algebraic over k, then there is $f \in k\{X\} \setminus \{0\}$ such that $f(a) = 0$ and if $g \in k\{X\} \setminus \{0\}$ has lower order, then $g(a) \neq 0$. Moreover, we can choose f such that if $f(b) = 0$ and $g(b) \neq 0$ for any lower order g, then $k\langle a \rangle$ and $k\langle b \rangle$ are isomorphic over k.

iii) If $f \in k\{X\}$, there is a differential field $F \supset k$ and $a \in F$ such that $f(a) = 0$ and $g(a) \neq 0$ for all $g \in k\{X\} \setminus \{0\}$ where the order of g is less than the order of f.

Proof

i) Certainly, $k(a, \ldots, a^{(n)})$ and $k(b, \ldots, b^{(n)})$ are isomorphic as fields. We need only show that the isomorphism preserves the derivation. For $i < n$ we have $\delta(a^{(i)}) = a^{(i+1)}$ and $\delta(b^{(i)}) = b^{(i+1)}$. Because $f(a, \ldots, a^n) = 0$, we must have $\delta(f(a, \ldots, a^n)) = 0$, but an easy calculation shows that

$$\delta(f(a, \ldots, a^{(n)})) = f^\delta(a, \ldots, a^{(n)}) + \sum_{i=0}^{n} \frac{\partial f}{\partial X^{(i)}}(a, \ldots, a^n)\delta(a^{(i)}),$$

where f^δ is the polynomial obtained by differentiating the coefficients of f. Because $f(a, \ldots, a^{(n-1)}, Y)$ is irreducible,

$$\frac{\partial f}{\partial X^{(n)}}(a, \ldots, a^{(n)}) \neq 0.$$

Thus

$$\delta(a^{(n)}) = \frac{-f^\delta(a, \ldots, a^{(n)}) - \sum_{i=0}^{n-1} \frac{\partial f}{\partial X^{(i)}}(a, \ldots, a^{(n)})\delta(a^{(i)})}{\frac{\partial f}{\partial X^{(n)}}(a, \ldots, a^n)}.$$

Similarly,

$$\delta(b^{(n)}) = \frac{-f^\delta(b, \ldots, b^{(n)}) - \sum_{i=0}^{n-1} \frac{\partial f}{\partial X^{(i)}}(b, \ldots, b^{(n)})\delta(b^{(i)})}{\frac{\partial f}{\partial X^{(n)}}(b, \ldots, b^n)}.$$

Thus, the natural field isomorphism is a differential field isomorphism.

ii) Let n be minimal such that $a, a', \ldots, a^{(n)}$ are algebraically dependent over k and let $f(X, \ldots, X^n) \in k[X, \ldots, X^{(n)}]$ be of minimal degree such that $f(a, a', \ldots, a^{(n)}) = 0$. Clearly, $g(a) \neq 0$ for any $g \in k\{X\} \setminus \{0\}$ of order less than n.

Suppose that $f(b) = 0$ and $g(b) \neq 0$ for any lower order g. Then, $b, \ldots, b^{(n-1)}$ are algebraically independent over k and b_n is a solution to the irreducible polynomial $f(b, \ldots, b_{n-1}, Y)$. Thus, by i), $k\langle a \rangle$ and $k\langle b \rangle$ are isomorphic over k.

iii) Let n be the order of f. By taking an irreducible factor of f of maximal order, we may assume that f is irreducible. Let K_0 be the field obtained from k by first adding elements $a, a', \ldots, a^{(n-1)}$ algebraically independent over k. Let K be the algebraic extension of K_0 obtained by adding a solution $a^{(n)}$ to the irreducible algebraic equation $f(a, a', \ldots, a^{(n-1)}, Y) = 0$. We must extend the derivation δ from k to K. For $i < n$, let $\delta(a^{(i)}) = a^{(i+1)}$. As in i), we let

$$\delta(a^{(n)}) = \frac{-f^{\delta}(a, \ldots, a^{(n)}) - \sum_{i=0}^{n-1} \frac{\partial f}{\partial X^{(i)}}(a, \ldots, a^{(n)}) \delta(a^{(i)})}{\frac{\partial f}{\partial X^{(n)}}(a, \ldots, a^{(n)})}.$$

Because $a, \ldots, a^{(n-1)}$ are algebraically independent over k, a satisfies no differential polynomial over k of order less than n.

Corollary 4.3.31 *If k is a differential field of characteristic zero, then there is $K \supseteq k$ with $K \models \mathrm{DCF}$.*

Proof If $f, g \in k\{X\} \setminus \{0\}$ with g of lower order than f, then by Proposition 4.3.30 iii) we can find $k_1 \supset k$ with $a \in k_1$ where $f(a) = 0$ and $g(a) \neq 0$. Iterating this process, we build $K \supset k$ differentially closed.

We can now prove quantifier elimination.

Theorem 4.3.32 DCF *has quantifier elimination.*

Proof Let K, L be differential closed fields where L is $|K|^+$-saturated. Let R be a differential subring of K, and let $f : R \to L$ be a differential ring embedding. We must show that f extends to an embedding of K into L. Because there is a unique extension of the derivation from R to its fraction field k, we may as well assume that $R = k$ is a field. By induction, it suffices to show that if $f : k \to L$ is a differential field embedding and $a \in K \setminus k$, there is a differential field embedding of $k\langle a \rangle$ into L extending f. Identifying k with $f(k)$, we may assume that $k \subset L$ and f is the identity on k. There are two cases to consider.

case 1: a is differentially algebraic over k.

Let f be as in Proposition 4.3.30 ii). Let n be the order of f. Let p be the type $\{f(v) = 0\} \cup \{g(v) \neq 0 : g \text{ is nonzero of order less than } n\}$. If

$g_1, \ldots, g_m$ are nonzero differential polynomials of order less than n, then there is $x \in L$ such that $f(x) = 0$ and $\prod g_i(x) \neq 0$. Thus p is satisfiable. If $b \in L$ realizes p, then $b, b', \ldots, b_{(n-1)}$ are algebraically independent over k; thus, by i), we can extend the embedding by sending a to b.

<u>case 2</u>: a is differentially transcendental over k.

We claim that there is $b \in L$ differentially transcendental over k. Let p be the type $\{f(v) \neq 0 : f \in k\{X\} \setminus 0\}$. Let $f_1, \ldots, f_n \in k\{X\} \setminus \{0\}$. Let N be greater than the order of f_i for $i = 1, \ldots, N$. Because L is differentially closed, there is $x \in L$ such that $x^{(N)} = 0$ and $\prod f_i(x) \neq 0$. Thus p is consistent and must be realized in L by some element b differentially transcendental over k. Because a and b are differentially transcendental over k, $k\langle a \rangle$ and $k\langle b \rangle$ are isomorphic to the fraction field of the differential polynomial ring $k\{X\}$ over k. In particular, we can extend the embedding by sending a to b.

Vaught's Two-Cardinal Theorem

We conclude this section with an application of homogeneous models that will be useful in Chapter 6. If $\mathcal{M}$ is an $\mathcal{L}$-structure and $\phi(v_1, \ldots, v_n)$ is an $\mathcal{L}$-formula, we let $\phi(\mathcal{M}) = \{\overline{x} \in M^n : \mathcal{M} \models \phi(\overline{x})\}$.

Definition 4.3.33 Let $\kappa > \lambda \geq \aleph_0$. We say that an $\mathcal{L}$-theory T has a (κ, λ)-*model* if there is $\mathcal{M} \models T$ and $\phi(\overline{v})$ an $\mathcal{L}$-formula such that $|M| = \kappa$ and $|\phi(\mathcal{M})| = \lambda$.

(κ, λ)-models are an obstruction to κ-categoricity. If T is a theory in a countable language with infinite models, then an easy compactness argument shows that there is $\mathcal{M} \models T$ of cardinality κ where every $\emptyset$-definable subset of $\mathcal{M}$ has cardinality κ. If T also has a (κ, λ)-model, then T is not κ-categorical. Our main goal is the following theorem of Vaught.

Theorem 4.3.34 *If T has a (κ, λ)-model where $\kappa > \lambda \geq \aleph_0$, then T has an $(\aleph_1, \aleph_0)$-model.*

We will prove Theorem 4.3.34 by first showing that the existence of a (κ, λ)-model has interesting implications for the countable models of T.

Definition 4.3.35 We say that $(\mathcal{N}, \mathcal{M})$ is a *Vaughtian pair* of models of T if $\mathcal{M} \prec \mathcal{N}$, $M \neq N$, and there is an $\mathcal{L}_M$-formula ϕ such that $\phi(\mathcal{M})$ is infinite and if $\phi(\mathcal{M}) = \phi(\mathcal{N})$.

For example, if $\mathcal{M}$ and $\mathcal{N}$ are nonstandard models of Peano arithmetic and $\mathcal{N}$ is a proper elementary end extension of $\mathcal{M}$, then $(\mathcal{N}, \mathcal{M})$ is a Vaughtian pair. If a is any infinite element of $\mathcal{M}$, then the formula $v < a$ defines an infinite set containing no elements of $N \setminus M$.

Lemma 4.3.36 *If T has a (κ, λ)-model where $\kappa > \lambda \geq \aleph_0$, then there is $(\mathcal{N}, \mathcal{M})$ a Vaughtian pair of models of T.*

Proof Let $\mathcal{N}$ be a (κ, λ)-model. Suppose that $X = \phi(\mathcal{N})$ has cardinality λ. By the Löwenheim–Skolem Theorem, there is $\mathcal{M} \prec \mathcal{N}$ such that $X \subseteq M$ and $|M| = \lambda$. Because $X \subseteq M$, $(\mathcal{N}, \mathcal{M})$ is a Vaughtian pair.

We would like to show that if there is a Vaughtian pair, then there is a Vaughtian pair of countable models. In the right context, this is a simple Löwenheim–Skolem argument.

Let $\mathcal{L}^* = \mathcal{L} \cup \{U\}$, where U is a unary predicate symbol. If $\mathcal{M} \subseteq \mathcal{N}$ are $\mathcal{L}$-structures, we consider the pair $(\mathcal{N}, \mathcal{M})$ as an $\mathcal{L}^*$-structure by interpreting U as M.

If $\phi(v_1, \ldots, v_n)$ is an $\mathcal{L}$-formula, we define $\phi^U(\overline{v})$, the restriction of ϕ to U, inductively as follows:

i) if ϕ is atomic, then ϕ^U is $U(v_1) \wedge \ldots \wedge U(v_n) \wedge \phi$;
ii) if ϕ is $\neg\psi$, then ϕ^U is $\neg\psi^U$;
iii) if ϕ is $\psi \wedge \theta$, then ϕ^U is $\psi^U \wedge \theta^U$;
iv) if ϕ is $\exists v \ \psi$, then ϕ^U is $\exists v \ U(v) \wedge \psi^U$.

An easy induction shows that if $\mathcal{M} \subset \mathcal{N}$, $\overline{a} \in M^k$ and we view $(\mathcal{N}, \mathcal{M})$ as an $\mathcal{L}^*$-structure, then $\mathcal{M} \models \phi(\overline{a})$ if and only if $(\mathcal{N}, \mathcal{M}) \models \phi^U(\overline{a})$.

Lemma 4.3.37 *If $(\mathcal{N}, \mathcal{M})$ is a Vaughtian pair for T, then there is a Vaughtian pair $(\mathcal{N}_0, \mathcal{M}_0)$ where $\mathcal{N}_0$ is countable.*

Proof Let ϕ be an $\mathcal{L}_M$-formula such that $\phi(\mathcal{M})$ is infinite and $\phi(\mathcal{M}) = \phi(\mathcal{N})$. Let $\overline{m}_0$ be the parameters from M occurring in ϕ. By the Löwenheim–Skolem Theorem, there is $(\mathcal{N}_0, \mathcal{M}_0)$ a countable $\mathcal{L}^*$-structure such that $\overline{m} \in M_0$ and $(\mathcal{N}_0, \mathcal{M}_0) \prec (\mathcal{N}, \mathcal{M})$. Because $\mathcal{M} \prec \mathcal{N}$, for any formula $\psi(v_1, \ldots, v_k)$

$$(\mathcal{N}, \mathcal{M}) \models \forall \overline{v} \left(\left(\bigwedge_{i=1}^{k} U(v_i) \wedge \psi(\overline{v}) \right) \rightarrow \psi^U(\overline{v}) \right).$$

Because $(\mathcal{N}_0, \mathcal{M}_0) \prec (\mathcal{N}, \mathcal{M})$, these sentences are also true in $(\mathcal{N}_0, \mathcal{M}_0)$, so $\mathcal{N}_0 \prec \mathcal{M}_0$.

Let $\phi(\overline{v})$ be an $\mathcal{L}_M$-formula with infinitely many realizations in $\mathcal{M}$ and none in $\mathcal{N} \setminus \mathcal{M}$, witnessing that $(\mathcal{N}, \mathcal{M})$ is a Vaughtian pair. For each k, the sentences

$$\exists \overline{v}_1 \ldots \exists \overline{v}_k \left(\bigwedge_{i<j} \overline{v}_i \neq \overline{v}_j \wedge \bigwedge_{i=1}^{k} \phi(v_i) \right)$$

hold in $(\mathcal{N}, \mathcal{M})$, as do the sentences $\exists x \ \neg U(x)$ and

$$\forall \overline{v} \ (\phi(\overline{v}) \rightarrow \bigwedge U(v_i)).$$

Because these sentences also hold in $(\mathcal{N}_0, \mathcal{M}_0)$, this structure is also a Vaughtian pair.

We need one more lemma before proving Vaught's Theorem.

Lemma 4.3.38 *Suppose that $\mathcal{M}_0 \prec \mathcal{N}_0$ are countable models of T. We can find $(\mathcal{N}_0, \mathcal{M}_0) \prec (\mathcal{N}, \mathcal{M})$ such that $\mathcal{N}$ and $\mathcal{M}$ are countable, homogeneous, and realize the same types in $S_n(T)$. By Theorem 4.2.15 $\mathcal{M} \cong \mathcal{N}$.*

Proof

Claim 1 If $\bar{a} \in M_0$ and $p \in S_n(\bar{a})$ is realized in $\mathcal{N}_0$, then there is $(\mathcal{N}_0, \mathcal{M}_0) \prec (\mathcal{N}', \mathcal{M}')$ such that p is realized in $\mathcal{M}'$.

Let $\Gamma(\bar{v}) = \{\phi^U(\bar{v}, \bar{a}) : \phi(\bar{v}, \bar{a}) \in p\} \cup \mathrm{Diag}_{\mathrm{el}}(\mathcal{N}_0, \mathcal{M}_0)$. If $\phi_1, \ldots, \phi_m \in p$, then $\mathcal{N}_0 \models \exists \bar{v} \bigwedge \phi_i(\bar{v}, \bar{a})$, thus $\mathcal{M}_0 \models \exists \bar{v} \bigwedge \phi_i(\bar{v}, \bar{a})$ and $(\mathcal{N}_0, \mathcal{M}_0) \models \exists \bar{v} \bigwedge \phi_i^U(\bar{v}, \bar{a})$. Thus, $\Gamma(\bar{v})$ is satisfiable. Let $(\mathcal{N}', \mathcal{M}')$ be a countable elementary extension realizing Γ.

By iterating Claim 1, we can find $(\mathcal{N}_0, \mathcal{M}_0) \prec (\mathcal{N}^*, \mathcal{M}^*)$ countable such that if $\bar{a} \in M_0$ and $p \in S_n(\bar{a})$ is realized in $\mathcal{N}_0$, then p is realized in $\mathcal{M}^*$.

Claim 2 If $\bar{b} \in N_0$ and $p \in S_n(\bar{b})$, then there is $(\mathcal{M}_0, \mathcal{N}_0) \prec (\mathcal{N}', \mathcal{M}')$ such that p is realized in $\mathcal{N}'$.

Let $\Gamma(\bar{v}) = p \cup \mathrm{Diag}_{\mathrm{el}}(\mathcal{N}_0, \mathcal{M}_0)$. If $\phi_1, \ldots, \phi_m \in p$, then $\mathcal{N}_0 \models \exists \bar{v} \bigwedge \phi_i(\bar{v}, \bar{b})$; thus, we can find a countable elementary extension of $(\mathcal{N}_0, \mathcal{M}_0)$ realizing p.

We build an elementary chain of countable models

$$(\mathcal{N}_0, \mathcal{M}_0) \prec (\mathcal{N}_1, \mathcal{M}_1) \prec \ldots$$

such that
 i) if $p \in S_n(T)$ is realized in $\mathcal{N}_{3i}$, then p is realized in $\mathcal{M}_{3i+1}$;
 ii) if $\bar{a}, \bar{b}, c \in \mathcal{M}_{3i+1}$ and $\mathrm{tp}^{\mathcal{M}_{3i+1}}(\bar{a}) = \mathrm{tp}^{\mathcal{M}_{3i+1}}(\bar{b})$, then there is $d \in \mathcal{M}_{3i+2}$ such that $\mathrm{tp}^{\mathcal{M}_{3i+2}}(\bar{a}, c) = \mathrm{tp}^{\mathcal{M}_{3i+2}}(\bar{b}, d)$;
 iii) if $\bar{a}, \bar{b}, c \in \mathcal{N}_{3i+2}$ and $\mathrm{tp}^{\mathcal{N}_{3i+2}}(\bar{a}) = \mathrm{tp}^{\mathcal{N}_{3i+2}}(\bar{b})$, then there is $d \in N_{3i+3}$ such that $\mathrm{tp}^{\mathcal{N}_{3i+3}}(\bar{a}, c) = \mathrm{tp}^{\mathcal{N}_{3i+3}}(\bar{b}, d)$.

 i) and ii) are done by using the first claim to build elementary chains, iii) is done by using the second claim to build an elementary chain.

Let $(\mathcal{N}, \mathcal{M}) = \bigcup_{i < \omega} (\mathcal{N}_i, \mathcal{M}_i)$. Then, $(\mathcal{N}, \mathcal{M})$ is a countable Vaughtian pair. By i), $\mathcal{M}$ and $\mathcal{N}$ realize the same types. By ii) and iii), $\mathcal{M}$ and $\mathcal{N}$ are homogeneous and hence isomorphic by Theorem 4.2.15.

Proof of 4.3.34 Suppose that T has a (κ, λ)-model. By the lemmas above, we can find $(\mathcal{N}, \mathcal{M})$ a countable Vaughtian pair such that $\mathcal{M}$ and $\mathcal{N}$ are homogeneous models realizing the same types. Let $\phi(\bar{v})$ be an $\mathcal{L}_M$-formula with infinitely many realizations in M and none in $N \setminus M$.

We build an elementary chain $(\mathcal{N}_\alpha : \alpha < \omega_1)$, each $\mathcal{N}_\alpha$ is isomorphic to $\mathcal{N}$, and $(\mathcal{N}_{\alpha+1}, \mathcal{N}_\alpha) \cong (\mathcal{N}, \mathcal{M})$. In particular, $N_{\alpha+1} \setminus N_\alpha$ contains no elements satisfying ϕ.

Let $\mathcal{N}_0 = \mathcal{N}$. For α a limit ordinal, let $\mathcal{N}_\alpha = \bigcup_{\beta < \alpha} \mathcal{N}_\beta$. Because $\mathcal{N}_\alpha$ is a union of models isomorphic to $\mathcal{N}$, $\mathcal{N}_\alpha$ is homogeneous and realizes the same types as $\mathcal{N}$ so $\mathcal{N}_\alpha \cong \mathcal{N}$ by Theorem 4.2.15.

Given $\mathcal{N}_\alpha \cong \mathcal{N}$, because $\mathcal{N} \cong \mathcal{M}$ there is $\mathcal{N}_{\alpha+1}$ an elementary extension of $\mathcal{N}_\alpha$ such that $(\mathcal{N}, \mathcal{M}) \cong (\mathcal{N}_{\alpha+1}, \mathcal{N}_\alpha)$. Clearly, $\mathcal{N}_{\alpha+1} \cong \mathcal{N}$.

Let $\mathcal{N}^* = \bigcup_{\alpha < \omega_1} \mathcal{N}_\alpha$. Then, $|N^*| = \aleph_1$ and if $\mathcal{N}^* \models \phi(\bar{a})$, then $\bar{a} \in M$; thus, $\mathcal{N}^*$ is an $(\aleph_1, \aleph_0)$-model.

Corollary 4.3.39 *If T is $\aleph_1$-categorical, then T has no Vaughtian pairs and hence no (κ, λ) models for $\kappa > \lambda \geq \aleph_0$.*

If T is ω-stable, we can prove a partial converse to Vaught's Theorem.

Lemma 4.3.40 *Suppose that T is ω-stable, $\mathcal{M} \models T$, and $|M| \geq \aleph_1$. There is a proper elementary extension $\mathcal{N}$ of $\mathcal{M}$ such that if $\Gamma(\overline{w})$ is a countable type over M realized in $\mathcal{N}$, then $\Gamma(\overline{w})$ is realized in $\mathcal{M}$.*

Proof

Claim There is an $\mathcal{L}_M$-formula $\phi(v)$ such that $|[\phi(v)]| \geq \aleph_1$ and for all $\psi(v) \in \mathcal{L}_M$ either $|[\phi(v) \wedge \psi(v)]| \leq \aleph_0$ or $|[\phi(v) \wedge \neg\psi(v)]| \leq \aleph_0$.

Suppose not. Then for any $\mathcal{L}_M$-formula $\phi(v)$ with $|[\phi(v)]| \geq \aleph_1$, there is a formula $\psi(v)$ such that $[\phi(v) \wedge \psi(v)]$ and $[\phi(v) \wedge \neg\psi(v)]$ are both uncountable. Let $\phi_\emptyset$ be the formula $v = v$. Then $[\phi_\emptyset] = |M| \geq \aleph_1$. We can build an infinite tree of formulas $(\phi_\sigma : \sigma \in 2^{<\omega})$ such that for all $\sigma \in 2^{<\omega}$:

i) $|[\phi_\sigma]| \geq \aleph_1$;

ii) $[\phi_{\sigma,0}] \cap [\phi_{\sigma,1}] = \emptyset$.

As in Theorem 4.2.18 we can find a countable $A \subset M$ such that $|S_1^M(A)| = 2^{\aleph_0}$, contradicting ω-stability.

Let $\phi(v)$ be as above. We construct the type p of formulas that are true for "almost all" elements satisfying $\phi(v)$. Let $p = \{\psi(v) : \psi$ an $\mathcal{L}_M$-formula and $|[\phi(v) \wedge \psi(v)]| \geq \aleph_1\}$. If $\psi_1, \ldots, \psi_m \in p$, then $|[\phi(v) \wedge \bigvee \neg\psi_i(v)]| \leq \aleph_0$. Thus, $\bigwedge_{i=1}^m \psi(v) \in p$ and p is finitely satisfiable. Because $|[\phi(v)]| \geq \aleph_1$, for each $\mathcal{L}_M$-formula $\psi(v)$ exactly one of $\psi(v)$ and $\neg\psi(v)$ is in p. Thus, p is a complete type over M.

Let $\mathcal{M}'$ be an elementary extension of $\mathcal{M}$ containing c, a realization of p. By Theorem 4.2.20, there is $\mathcal{N} \prec \mathcal{M}'$ prime over $M \cup \{c\}$ such that every $\bar{a} \in N$ realizes an isolated type over $M \cup \{c\}$.

Let $\Gamma(\overline{w})$ be a countable type over $\mathcal{M}$ realized by $\bar{b} \in \mathcal{N}$. There is an $\mathcal{L}_M$-formula $\theta(\overline{w}, v)$ such that $\theta(\overline{w}, c)$ isolates $\mathrm{tp}^{\mathcal{N}}(\bar{b}/M \cup \{c\})$. Note that $\exists\overline{w}\, \theta(\overline{w}, v) \in p$ and

$$\forall\overline{w}\, (\theta(\overline{w}, v) \rightarrow \gamma(\overline{w})) \in p$$

for all $\gamma(\overline{w}) \in \Gamma$. Let

$$\Delta = \{\exists\overline{w}\, \theta(\overline{w}, v)\} \cup \{\forall\overline{w}\, (\theta(\overline{w}, v) \rightarrow \gamma(\overline{w})) : \gamma \in \Gamma\}.$$

Then, $\Delta \subset p$ is countable and, if c' realizes Δ, then $\exists\overline{w}\, \theta(\overline{w}, c')$, and if $\theta(\bar{b}', c')$, then $\bar{b}'$ realizes Γ.

Let $\delta_0(v), \delta_1(v), \ldots$ enumerate Δ. By choice of p, $|\{x \in M : \phi(x)\}| \geq \aleph_1$ and $|\{x \in M : \phi(x) \wedge \neg(\delta_0(x) \wedge \ldots \wedge \delta_n(x))\}| \leq \aleph_0$ for all $n < \omega$. Thus $|\{x \in M : \phi(x)$ and x realizes $\Delta\}| \geq \aleph_1$. Let $c' \in M$ realize Δ and choose $\bar{b}'$ such that $\mathcal{M} \models \theta(\bar{b}', c')$. Then, $\bar{b}'$ is a realization of Γ in $\mathcal{M}$.

Theorem 4.3.41 *Suppose that T is ω-stable and there is an $(\aleph_1, \aleph_0)$-model of T. If $\kappa > \aleph_1$, then there is a $(\kappa, \aleph_0)$-model of T.*

Proof Let $\mathcal{M} \models T$ with $|M| \geq \aleph_1$ such that $|\phi(\mathcal{M})| = \aleph_0$ and let $\mathcal{M} \prec \mathcal{N}$ be as in Lemma 4.3.40. The type $\Gamma(v) = \{\phi(v)\} \cup \{v \neq m : m \in M$ and $\mathcal{M} \models \phi(m)\}$ is a countable type omitted in $\mathcal{M}$ and hence in $\mathcal{N}$. Thus $\phi(\mathcal{N}) = \phi(\mathcal{M})$.

Iterating this construction, we build an elementary chain $(\mathcal{M}_\alpha : \alpha < \kappa)$ such that $\mathcal{M}_0 = \mathcal{M}$ and $\mathcal{M}_{\alpha+1} \neq \mathcal{M}_\alpha$ but $\phi(\mathcal{M}_\alpha) = \phi(\mathcal{M}_0)$. If $\mathcal{N} = \bigcup_{\alpha < \kappa} \mathcal{M}_\alpha$, then $\mathcal{N}$ is a $(\kappa, \aleph_0)$-model of T.

Without the assumption of ω-stability, Theorem 4.3.41 is false (see Exercise 5.5.7).

4.4   The Number of Countable Models

Throughout this section, T will be a complete theory in a countable language with infinite models.

For any infinite cardinal κ, we let $I(T, \kappa)$ be the number of nonisomorphic models of T of cardinality κ. In this section, we will look at the possible values of $I(T, \aleph_0)$. We have already considered a number of examples.

- $I(\mathrm{DLO}, \aleph_0) = 1$.
- In Exercise 2.5.28, we gave examples of T_n where $I(T_n, \aleph_0) = n$ for $n = 3, 4, \ldots,$.
- $I(\mathrm{ACF}_p, \aleph_0) = \aleph_0$.
- $I(\mathrm{RCF}, \aleph_0) = I(\mathrm{Th}(\mathbb{N}), \aleph_0) = 2^{\aleph_0}$.

Because there are at most $2^{\aleph_0}$ nonisomorphic countable models of T, there are two natural questions:

Can we have $I(T, \aleph_0) = 2$?

Can we have $\aleph_0 < I(T, \aleph_0) < 2^{\aleph_0}$?

Surprisingly, Vaught answered the first question negatively. If the Continuum Hypothesis is true, then the second question has a trivial negative answer. Vaught conjectured that the answer is negative even when the Continuum Hypothesis fails. This remains one of the deep open questions of model theory. Although Vaught's Conjecture has been proved for some special classes of theories (for example, Shelah [93] proved Vaught's Conjecture for ω-stable theories), the best general result is Morley's theorem that if $I(T, \aleph_0) > \aleph_1$, then $I(T, \aleph_0) = 2^{\aleph_0}$.

$\aleph_0$-categorical Theories

We begin by taking a closer look at $\aleph_0$-categorical theories. In particular, we show how to recognize $\aleph_0$-categoricity by looking at the type space.

Theorem 4.4.1 *The following are equivalent:*
i) T *is* $\aleph_0$*-categorical.*
ii) *Every type in* $S_n(T)$ *is isolated for* $n < \omega$.
iii) $|S_n(T)| < \aleph_0$ *for all* $n < \omega$.
iv) *For each* $n < \omega$, *there is a finite list of formulas*

$$\phi_1(v_1, \ldots, v_n), \ldots, \phi_m(v_1, \ldots, v_n)$$

such that for every formula $\psi(v_1, \ldots, v_n)$

$$T \models \phi_i(\overline{v}) \leftrightarrow \psi(\overline{v})$$

for some $i \leq m$.

Proof

i) $\Rightarrow$ ii) If $p \in S_n(T)$ is nonisolated, then there is a countable $\mathcal{M} \models T$ omitting p. There is also a countable $\mathcal{N} \models T$ realizing p. Clearly, $\mathcal{M} \not\cong \mathcal{N}$ so T is not $\aleph_0$-categorical.

ii) $\Rightarrow$ iii) Suppose that $S_n(T)$ is infinite. For each $p \in S_n(T)$, let ϕ_p isolate p. Because $\bigcup_{p \in S_n(T)}[\phi_p] = S_n(T)$ and $S_n(T)$ is compact, there are $p_1, \ldots, p_m$ such that $[\phi_{p_1}] \cup \ldots \cup [\phi_{p_m}] = S_n(T)$. Because $[\phi_p] = \{p\}$, $S_n(T)$ is finite.

iii) $\Rightarrow$ iv) For each i, we can find a formula θ_i such that $\theta_i \in p_i$ and $\neg \theta_i \in p_j$ for $i \neq j$. Then, θ_i isolates p_i. For any formula $\psi(v_1, \ldots, v_n)$,

$$T \models \psi(\overline{v}) \leftrightarrow \bigvee_{\psi \in p_i} \theta_i.$$

Thus, each ψ with free variables $v_1, \ldots, v_n$ is equivalent to $\bigvee_{i \in S} \theta_i$ for some $S \subseteq \{1, \ldots, m\}$. There are at most 2^m such formulas.

iv) $\Rightarrow$ i) Let $\mathcal{M}$ be a countable model of T. If $\overline{a} \in M^n$, let $S_{\overline{a}} = \{i \leq m : \mathcal{M} \models \phi_i(\overline{a})\}$. Then, $\text{tp}^{\mathcal{M}}(\overline{a})$ is isolated by

$$\bigwedge_{i \in S_{\overline{a}}} \phi_i(\overline{v}) \wedge \bigwedge_{i \notin S_{\overline{a}}} \neg \phi(\overline{v}).$$

Thus, $\mathcal{M}$ is atomic and hence, by Theorem 4.2.8, prime. Because there is a unique prime model, T is $\aleph_0$-categorical.

Theorem 4.4.1 tells us a great deal about definability in $\aleph_0$-categorical theories. Recall that b is algebraic over A if there is a formula $\phi(v, \overline{w})$ and $\overline{a} \in A$ such that $\mathcal{M} \models \phi(b, \overline{a})$ and $\{x \in M : \mathcal{M} \models \phi(x, \overline{a})\}$ is finite. Also, $\text{acl}(A) = \{b \in A : b \text{ is algebraic over } A\}$.

Corollary 4.4.2 *Suppose that* T *is* $\aleph_0$*-categorical. There is a function* $f : \mathbb{N} \to \mathbb{N}$ *such that if* $\mathcal{M} \models T$, $A \subset M$, *and* $|A| \leq n$, *then* $|\text{acl}(A)| \leq f(n)$.

Proof By Theorem 4.4.1, $|S_{n+1}(T)|$ is finite. Let $q_1, \ldots, q_k$ list all $n+1$-types. Let $X = \{i \; : \; q_i$ contains a formula $\phi(v, \overline{w})$ such that $\mathcal{M} \models \forall v_0, \ldots, v_N \bigwedge_{i=0}^{N} \phi(v_i, \overline{w}) \rightarrow \bigvee_{i<j \leq N} v_i = v_j$ for some $N\}$. For $i \in X$, let N_i be the least N such that some formula ϕ,

$$\forall v_0, \ldots, v_N \bigwedge_{i=0}^{N} \phi(v_i, \overline{w}) \rightarrow \bigvee_{i<j} v_i = v_j,$$

is in q_i.

If $a, b_1, \ldots, b_n \in M$ and a is algebraic over $\overline{b}$, then $(a, \overline{b})$ realizes some $q_i \in X$ and $|\{x : (x, \overline{b})$ realizes $q_i\}| \leq N_i$. Thus,

$$|\mathrm{acl}(b_1, \ldots, b_n)| \leq \sum_{i \in X} N_i.$$

Let

$$f(n) = \sum_{i \in X} N_i.$$

Corollary 4.4.2 is very useful in understanding algebraic examples.

Corollary 4.4.3 *If F is an infinite field, then the theory of F is not $\aleph_0$-categorical.*

Proof By compactness, we can find an elementary extension K of F such that K contains a transcendental element t. Because $t, t^2, t^3, \ldots$ are distinct, $\mathrm{acl}(t)$ is infinite. Thus, by Corollary 4.4.2, $\mathrm{Th}(F)$ is not $\aleph_0$-categorical.

For groups, the situation is more interesting. We study groups in the multiplicative language $\mathcal{L} = \{\cdot, 1\}$. We say that a group G is *locally finite* if, for any finite $X \subseteq G$, the subgroup generated by X is finite.

Corollary 4.4.4 *Let G be an infinite group.*
i) If $\mathrm{Th}(G)$ is $\aleph_0$-categorical, then G is locally finite. Moreover, there is a number b such that if $g \in G$, then $g^n = 1$ for some $n \leq b$ (we say that G has bounded exponent*).*
ii) If G is an infinite Abelian group of bounded exponent, then $\mathrm{Th}(G)$ is $\aleph_0$-categorical.

Proof
i) By Corollary 4.4.2, there is a function $f : \mathbb{N} \rightarrow \mathbb{N}$ such that if $|X| \leq n$, the group generated by X has size at most $f(n)$. In particular, if $g \in G$, then $g^n = 1$ for some $n \leq f(1)$.
ii) Suppose that G is a countable abelian group of bounded exponent. Then, there are $q_1, \ldots, q_m$ distinct prime powers such that

$$G \cong (\mathbb{Z}/q_1\mathbb{Z})^{n_1} \oplus \ldots \oplus (\mathbb{Z}/q_k\mathbb{Z})^{n_k} \oplus \bigoplus_{i=1}^{\infty} \mathbb{Z}/q_{k+1}\mathbb{Z} \oplus \ldots \oplus \bigoplus_{i=1}^{\infty} \mathbb{Z}/q_m\mathbb{Z}$$

where $n_i \in \mathbb{N}$ for $i \leq k$. Because G is infinite, we must have $k < m$.

Let $q_i = p_i^{l_i}$, where p_i is a prime. The group $(\mathbb{Z}/q_i\mathbb{Z})^{n_i}$ has $p_i^{n_i l_i} - p_i^{n_i(l_i-1)}$ elements of order exactly q_i. If $g \in (\mathbb{Z}/q_i\mathbb{Z})^{n_i}$ has order less than q_i, then there is $h \in (\mathbb{Z}/q_i\mathbb{Z})^{n_i}$ with $p_i h = g$ (i.e., g is p_i-divisible).

Let T be the theory with the following axioms:

i) the axioms for Abelian groups;

ii) $\forall x \; x^{\prod q_i} = 1$;

iii) there are $p_i^{n_i l_i} - p_i^{n_i(l_i-1)}$ elements of order exactly q_i that are not p_i-divisible for $i \leq k$;

iv) there are infinitely many elements of order exactly q_i that are not p_i-divisible for $i > k$.

By the remarks above $G \models T$. If H is a countable model of T, then $H \cong G$. Thus, T is $\aleph_0$-categorical.

We now move on to Vaught's result that $I(T, \aleph_0) \neq 2$. We will use the next lemma, although we leave the proof for the exercises.

Lemma 4.4.5 *Let $\kappa \geq \aleph_0$. Let $A \subset M$ with $|A| < \kappa$. Let $\mathcal{M}_A$ be the $\mathcal{L}_A$-structure obtained from $\mathcal{M}$ by interpreting the new constant symbols in the natural way. If $\mathcal{M}$ is κ-saturated, then so is $\mathcal{M}_A$.*

Theorem 4.4.6 $I(T, \aleph_0) \neq 2$.

Proof Suppose that $I(T, \aleph_0) = 2$. By Corollary 4.3.8 ii), there is $\mathcal{N}$ a prime model of T and $\mathcal{M}$ a countable saturated model of T. Because T is not $\aleph_0$-categorical, by Theorem 4.4.1, there is a nonisolated type $p \in S_n(T)$ for some n. The type p is realized in $\mathcal{M}$ and omitted in $\mathcal{N}$. Let $\bar{a} \in M$ realize p. Let T^* be the $\mathcal{L}_{\bar{a}}$-theory of $\mathcal{M}_{\bar{a}}$ (in the notation of the previous lemma).

By Theorem 4.4.1, there are infinitely many T-inequivalent formulas in the free variables $v_1, \ldots, v_n$. As they are still T^*-inequivalent, T^* is not $\aleph_0$-categorical. By Lemma 4.4.5, $\mathcal{M}_{\bar{a}}$ is a saturated $\mathcal{L}_{\bar{a}}$-structure. Thus, by Corollary 4.3.8 i), T^* has a countable atomic model $\mathcal{A}$. Let $\mathcal{B}$ denote the $\mathcal{L}$-reduct of $\mathcal{B}$. Because $\mathcal{A} \models T^*$, $\mathcal{B}$ contains a realization of p, thus $\mathcal{B} \not\cong \mathcal{N}$. Because T^* is not $\aleph_0$-categorical, there is a nonisolated $\mathcal{L}_{\bar{a}}$-type. This type is not realized in $\mathcal{A}$. Thus $\mathcal{A}$ is not saturated. If $\mathcal{B}$ were saturated, then, by Lemma 4.4.5, $\mathcal{A}$ would be saturated. Thus, $\mathcal{B} \not\cong \mathcal{M}$ and $I(T, \aleph_0) \geq 3$.

Morley's Analysis of Countable Models

Next we prove Morley's theorem that if $I(T, \aleph_0) > \aleph_1$, then $I(T, \aleph_0) = 2^{\aleph_0}$. As in the proof of Theorem 2.4.15, we will use infinitary logic to analyze countable models.

Definition 4.4.7 A *fragment* of $\mathcal{L}_{\omega_1,\omega}$ is a set of $\mathcal{L}_{\omega_1,\omega}$-formulas containing all first-order formulas and closed under subformulas, finite Boolean combinations, quantification, and change of free variables.

If F is a fragment of $\mathcal{L}_{\omega_1,\omega}$, we say that $\mathcal{M} \equiv_F \mathcal{N}$ if

$$\mathcal{M} \models \phi \text{ if and only if } \mathcal{N} \models \phi$$

for all sentences $\phi \in F$.

If F is a fragment of $\mathcal{L}_{\omega_1,\omega}$, we say that $p \subset F$ is an F-*type* if there is a countable $\mathcal{L}$-structure $\mathcal{M}$ and $a_1,\ldots,a_n \in M$ such that $p = \{\phi(v_1,\ldots,v_n) \in F : \mathcal{M} \models \phi(\bar{a})\}$. Let $S_n(F,T)$ be the set of all F-types realized by some n-tuple in some countable model of T.

We will count models by counting types for various fragments. If $|S_n(F,T)| = 2^{\aleph_0}$ for some countable fragment F, then, because a countable model can realize only countably many types, we must have $I(T,\aleph_0) = 2^{\aleph_0}$.

Next, we look at a case where we have the minimal number of types for all countable fragments.

Definition 4.4.8 We say that an $\mathcal{L}$-theory T is *scattered* if $|S_n(F,T)|$ is countable for all countable fragments F of $\mathcal{L}_{\omega_1,\omega}$ and all $n < \omega$.

In particular, if T is scattered, then for countable fragments F, there are only countably many $\equiv_F$-classes of countable models of T. We will show that if T is scattered, then $I(T,\aleph_0) \leq \aleph_1$.

Suppose that T is scattered. We build a sequence of countable fragments $(L_\alpha : \alpha < \omega_1)$ as follows. Let L_0 be all first-order $\mathcal{L}$-formulas. If α is a limit ordinal, then $L_\alpha = \bigcup_{\beta < \alpha} L_\beta$.

Suppose that L_α is a countable fragment. For $p \in S_n(L_\alpha,T)$, let $\Phi_p(v_1,\ldots,v_n)$ be the $\mathcal{L}_{\omega_1,\omega}$-formula $\bigwedge_{\phi \in p} \phi$. This is an $\mathcal{L}_{\omega_1,\omega}$-formula because L_α is countable. Let $L_{\alpha+1}$ be the smallest fragment containing Φ_p for $p \in S_n(L_\alpha,T)$, $n < \omega$. Because T is scattered, $L_{\alpha+1}$ is a countable fragment.

If $\mathcal{M}$ is a countable model of T and $a_1,\ldots,a_n \in M$, let $\operatorname{tp}_\alpha^{\mathcal{M}}(\bar{a}) \in S_n(L_\alpha,T)$ be the L_α-type realized by $\bar{a}$ in $\mathcal{M}$.

Lemma 4.4.9 *For each countable $\mathcal{M} \models T$, there is an ordinal $\gamma < \omega_1$ such that if $\bar{a},\bar{b} \in M^n$ and $\operatorname{tp}_\gamma^{\mathcal{M}}(\bar{a}) = \operatorname{tp}_\gamma^{\mathcal{M}}(\bar{b})$, then $\operatorname{tp}_\alpha^{\mathcal{M}}(\bar{a}) = \operatorname{tp}_\alpha^{\mathcal{M}}(\bar{b})$ for all $\alpha < \omega_1$.*

We call the least such γ the height *of $\mathcal{M}$.*

Proof Note first that if $\operatorname{tp}_\alpha^{\mathcal{M}}(\bar{a}) \neq \operatorname{tp}_\alpha^{\mathcal{M}}(\bar{b})$, then $\operatorname{tp}_\beta^{\mathcal{M}}(\bar{a}) \neq \operatorname{tp}_\beta^{\mathcal{M}}(\bar{b})$ for all $\beta > \alpha$. For $\bar{a},\bar{b}$ in M^n, let

$$f(\bar{a},\bar{b}) = \begin{cases} -1 & \operatorname{tp}_\alpha^{\mathcal{M}}(\bar{a}) = \operatorname{tp}_\alpha^{\mathcal{M}}(\bar{b}) \text{ for all } \alpha < \omega_1 \\ \alpha & \text{if } \alpha \text{ is least } \operatorname{tp}_\alpha^{\mathcal{M}}(\bar{a}) \neq \operatorname{tp}_\alpha^{\mathcal{M}}(\bar{b}). \end{cases}$$

Because $\mathcal{M}$ is countable, we can find $\gamma < \omega_1$ such that $\gamma > f(\bar{a},\bar{b})$ for all $\bar{a},\bar{b} \in M^n$ and $n < \omega$.

Lemma 4.4.10 *Suppose that $\mathcal{M}$ and $\mathcal{N}$ are countable models of T such that $\mathcal{M}$ has height γ and $\mathcal{M} \equiv_{L_{\gamma+1}} \mathcal{N}$. If $\bar{a}, \bar{b} \in N^n$ and $\operatorname{tp}^{\mathcal{N}}_{\gamma}(\bar{a}) = \operatorname{tp}^{\mathcal{N}}_{\gamma}(\bar{b})$, then $\operatorname{tp}^{\mathcal{N}}_{\gamma+1}(\bar{a}) = \operatorname{tp}^{\mathcal{N}}_{\gamma+1}(\bar{b})$.*

Proof Let $p = \operatorname{tp}^{\mathcal{N}}_{\gamma}(\bar{a}) = \operatorname{tp}^{\mathcal{N}}_{\gamma}(\bar{b})$ and let $\psi(\bar{v})$ be an $\mathcal{L}_{\gamma+1}$-formula. Let Θ be the $\mathcal{L}_{\gamma+1}$-formula

$$\forall \bar{v} \forall \bar{w} \left((\Phi_p(\bar{v}) \wedge \Phi_p(\bar{w})) \to (\psi(\bar{v}) \leftrightarrow \psi(\bar{w})) \right).$$

Because γ is the height of $\mathcal{M}$, $\mathcal{M} \models \Theta$. Because $\mathcal{N} \equiv_{L_{\gamma+1}} \mathcal{M}$, $\mathcal{N} \models \Theta$. Thus $\operatorname{tp}^{\mathcal{N}}_{\gamma+1}(\bar{a}) = \operatorname{tp}^{\mathcal{N}}_{\gamma+1}(\bar{b})$.

Lemma 4.4.11 *If $\mathcal{M}$ and $\mathcal{N}$ are countable models of T such that $\mathcal{M}$ has height γ and $\mathcal{M} \equiv_{L_{\gamma+1}} \mathcal{N}$, then $\mathcal{M} \cong \mathcal{N}$.*

Proof Let $a_0, a_1, \ldots$ list M and let $b_0, b_1, \ldots$ list N. We build a sequence of finite partial embeddings $f_0 \subseteq f_1 \subseteq \ldots$ such that if $\bar{a}$ is the domain of f_n, then $\operatorname{tp}^{\mathcal{M}}_{\gamma}(\bar{a}) = \operatorname{tp}^{\mathcal{N}}_{\gamma}(f_n(\bar{a}))$. We will ensure that a_n is in the domain of f_{n+1} and b_n is in the image of f_{n+1}. Then $f = \bigcup f_n$ is the desired isomorphism.

Let $f_0 = \emptyset$. Suppose that $\bar{a}$ is the domain of f_n and $f_n(\bar{a}) = \bar{b}$. Let $p = \operatorname{tp}^{\mathcal{M}}_{\gamma}(\bar{a}, a_n)$. We must find $e \in N$ such that $\operatorname{tp}^{\mathcal{N}}_{\gamma}(\bar{b}, e) = p$. Because

$$\mathcal{M} \models \exists \bar{v} \exists w \bigwedge_{\phi \in p} \phi(\bar{v}, w)$$

and this is an $L_{\gamma+1}$-sentence,

$$\mathcal{N} \models \exists \bar{v} \exists w \bigwedge_{\phi \in p} \phi(\bar{v}, w).$$

Let $(\bar{c}, d) \in N$ realize p. Because $\bar{a}$ and $\bar{c}$ realize the same L_{γ}-type, $\bar{c}$ and $\bar{b}$ realize the same L_{γ}-type. By Lemma 4.4.10, $\bar{c}$ and $\bar{b}$ realize the same $L_{\gamma+1}$-type. Because

$$\mathcal{N} \models \exists w \bigwedge_{\phi \in p} \phi(\bar{c}, w)$$

and this is an $L_{\gamma+1}$-formula,

$$\mathcal{N} \models \exists w \bigwedge_{\phi \in p} \phi(\bar{b}, w).$$

Thus, there is $e \in N$ such that $p = \operatorname{tp}^{\mathcal{N}}_{\gamma}(\bar{b}, e)$.

By a symmetric argument, we can find $s \in M$ such that $\operatorname{tp}^{\mathcal{M}}_{\gamma}(\bar{a}, a_n, s) = \operatorname{tp}^{\mathcal{N}}_{\gamma}(\bar{b}, e, b_n)$. Let $f_{n+1} = f_n \cup \{(a_n, e), (s, b_n)\}$.

Theorem 4.4.12 *If T is scattered, then $I(T, \aleph_0) \leq \aleph_1$.*

Proof For each countable $\mathcal{M} \models T$, let $i(\mathcal{M}) = (\gamma, \mathrm{tp}_{\gamma+1}^{\mathcal{M}}(\emptyset))$, where γ is the height of $\mathcal{M}$. Note that $\mathcal{M} \equiv_{L_\alpha} \mathcal{N}$ if and only if $\mathrm{tp}_\alpha^{\mathcal{M}}(\emptyset) = \mathrm{tp}_\alpha^{\mathcal{N}}(\emptyset)$. By Lemma 4.4.11, if $\mathcal{M}$ and $\mathcal{N}$ are countable models of T, then $\mathcal{M} \cong \mathcal{N}$ if and only if $i(\mathcal{M}) = i(\mathcal{N})$. There are only $\aleph_1$ possible heights and, for any given α, there are only $\aleph_0$ possibilities for $\mathrm{tp}_\alpha^{\mathcal{M}}(\emptyset)$. Thus $I(T, \aleph_0) \le \aleph_1$.

To finish the proof of Morley's theorem, we will show that if T is not scattered, then $|S_n(F, T)| = 2^{\aleph_0}$ for some countable fragment F. Although this is a generalization of Theorem 4.2.11 i), complications arise because we do not have the Compactness Theorem in $\mathcal{L}_{\omega_1, \omega}$. The proof requires some ideas from descriptive set theory.

Suppose F is a fragment of $\mathcal{L}_{\omega_1, \omega}$. We will consider $\mathcal{L}$-structures where the universe of the models is ω. If $\mathcal{M} = (\omega, \dots)$ is an $\mathcal{L}$-structure, the F-*diagram* of $\mathcal{M}$ is $\{\phi(v_0, \dots, v_n) \in F : \mathcal{M} \models \phi(0, 1, \dots, n)\}$.

We consider $D(F, T)$ the set of all possible F-diagrams of models of T. There is a natural bijection between the power set $\mathcal{P}(F)$ and 2^F, the set of all functions from F to $\{0, 1\}$ (identifying a set with its characteristic function). Because $D(F, T)$ is a set of subsets of F, we can view $D(F, T)$ as a subset of 2^F. If we think of $\{0, 1\}$ as the two-element space with the discrete topology, then we can give 2^F the product topology. The topology on 2^F has a basis of clopen sets of the form $\{f \in 2^F : \forall x \in F_0 \ f(x) = \sigma(x)\}$ where $F_0 \subseteq F$ is finite and $\sigma : F_0 \to 2$. If F is countable, then 2^F is homeomorphic to 2^ω.

Lemma 4.4.13 *If F is a countable fragment of $\mathcal{L}_{\omega_1, \omega}$, then $D(F, T)$ is a Borel subset of 2^F.*[2]

Proof Let

$$
\begin{aligned}
E_0 &= \{f \in 2^F : f(\phi) = 1 \Leftrightarrow f(\neg\phi) = 0 \text{ for all } \phi \in F\} \\
&= \bigcap_{\phi \in F} \{f \in 2^F : (f(\phi) = 0 \wedge f(\neg\phi) = 1) \vee (f(\phi) = 1 \wedge f(\neg\phi) = 0)\}.
\end{aligned}
$$

Because E_0 is an intersection of clopen sets, E_0 is closed.

Let $E_1 = \{f \in 2^F : f(\exists v \phi(v)) = 1$ if and only if $f(\phi(v_i)) = 1$ for some i for all $\phi \in F$ with one free variable$\}$. If

$$
E_{1,\phi}^+ = \{f \in 2^F : f(\exists v \phi(v)) = 1\} \cap \bigcup_{i=0}^{\infty} \{f \in 2^F : f(\phi(v_i)) = 1\}
$$

and

$$
E_{1,\phi}^- = \{f \in 2^F : f(\exists v \phi(v)) = 0\} \cap \bigcap_{i=0}^{\infty} \{f \in 2^F : f(\phi(v_i)) = 0\},
$$

---

[2]Recall that the collection of Borel subsets of 2^F is the smallest collection of sets containing the open sets and closed under complement and countable unions and intersections.

then

$$E_1 = \bigcap_{\phi \in F} (E_{1,\phi}^+ \cup E_{1,\phi}^-)$$

and E_1 is Borel.

If $\psi = \bigwedge_{i \in I} \phi_i$ and $\psi \in F$, let

$$E_{2,\psi} = \{f \in 2^F : f(\psi) = 1 \text{ if and only if } f(\phi_i) = 1 \text{ for all } i \in I\}.$$

Because I is countable, we argue as above that $E_{2,\psi}$ is Borel. Thus

$$E_2 = \bigcap \Big\{ E_{2,\psi} : \psi = \bigwedge_{i \in I} \phi_i \text{ and } \psi \in F \Big\}$$

is Borel. Similarly the following sets are Borel:

$E_3 = \{f \in 2^F : f(v_i = v_j) = 0 \text{ for all } i \neq j\}$,
$E_4 = \{f \in 2^F : f(v_i = v_i) = 1 \text{ for all } i\}$,
$E_5 = \{f \in 2^F : f(v_i = v_j \to v_j = v_i) = 1 \text{ for all } i, j\}$,
$E_6 = \{f \in 2^F : f((v_i = v_j \wedge v_j = v_k) \to v_i = v_k) = 1 \text{ for all } i, j, k\}$, and
$E_7 = \{f \in 2^F : f(\phi) = 1 \text{ for all } \phi \in T\}$.

Let $D = E_0 \cap \ldots \cap E_7$. Clearly, D is Borel. We claim that $D = D(F, T)$. It is easy to see that if $\mathcal{M} \models T$ with universe ω, then the F-diagram of $\mathcal{M}$ is in D.

Suppose that $f \in D$. We build an $\mathcal{L}$-structure $\mathcal{M}_f$ with universe ω. If R is an n-ary relation symbol of $\mathcal{L}$, then $(i_1, \ldots, i_n) \in R^{\mathcal{M}_f}$ if and only if $f(R(v_{i_1}, \ldots, v_{i_n})) = 1$. Let g be an n-ary function symbol of $\mathcal{L}$. Because $f \in E_7$, $f(\exists v g(v_{i_1}, \ldots, v_{i_n}) = v) = 1$. Because $f \in E_1$, $f(g(v_{i_1}, \ldots, i_{i_n}) = v_j) = 1$ for some j. Let $g^{\mathcal{M}_f}(i_1, \ldots, i_n) = j$. Because $f \in D$, $f(g(v_{i_1}, \ldots, i_{i_n}) = v_k) = 0$ for $j \neq k$ and $g^{\mathcal{M}_f}$ is well-defined. Now, using the fact that $f \in D$, we can do an induction on formulas to show that

$$\mathcal{M}_f \models \phi(i_1, \ldots, i_n) \Leftrightarrow f(\phi(v_{i_1}, \ldots, v_{i_n})) = 1$$

for all $\phi \in F$. Thus, f is in $D(F, T)$.

We may also view $S_n(F, T)$ as a subset of 2^F. Although this set may not be Borel, it is not much more complicated.

We construct a continuous map Ψ such that $S_n(F, T)$ is the image of $D(F, T)$ under this map. For $f \in 2^F$, let $\Psi(f) \in 2^F$, where

$$\Psi(f)(\phi) = \begin{cases} 1 & \phi \text{ has free variable } v_0, \ldots, v_{n-1} \text{ and } f(\phi) = 1 \\ 0 & \text{otherwise.} \end{cases}$$

Because $\Psi(f)(\phi) = \Psi(g)(\phi)$ if $f(\phi) = g(\phi)$, Ψ is continuous. If $p \in S_n(F, T)$, then there is $\mathcal{M} \models T$ with universe ω such that $(0, 1, \ldots, n-1)$ realizes p in $\mathcal{M}$. Thus, the space of F-types $S_n(F, T)$ is the image of $D(F, T)$ under Ψ.

We now need a classical result from descriptive set theory.

Definition 4.4.14 If $|X| = \aleph_0$, we say that $Y \subseteq 2^X$ is *analytic* if there is a continuous map $\tau : 2^X \to 2^X$ and a Borel set $B \subseteq 2^X$ such that Y is the image of B under τ.

By the remarks above $S_n(F, T)$ is an analytic subset of 2^F for any countable fragment F.

Theorem 4.4.15 *Suppose that X is countable and $Y \subseteq 2^X$ is analytic. If $|Y| > \aleph_0$, then $|Y| = 2^{\aleph_0}$.*

Proof See [52] 14.13.

Theorem 4.4.16 *Let T be a complete theory in a countable language. If $I(T, \aleph_0) > \aleph_1$, then $I(T, \aleph_0) = 2^{\aleph_0}$.*

Proof For any countable fragment F, $S_n(F, T)$ is analytic. Thus, by Theorem 4.4.15, we either have $|S_n(F, T)| \leq \aleph_0$ or $|S_n(F, T)| = 2^{\aleph_0}$. If there is any countable fragment F, where $|S_n(F, T)| = 2^{\aleph_0}$, then $I(T, \aleph_0) = 2^{\aleph_0}$. If not, then T is scattered and, by Theorem 4.4.12, $I(T, \aleph_0) \leq \aleph_1$.

4.5   Exercises and Remarks

We assume throughout that $\mathcal{L}$ is a countable language and that T is an $\mathcal{L}$-theory with only infinite models.

Exercise 4.5.1 a) Let $\mathcal{M} = (X, <)$ be a dense linear order, let $A \subset M$ and $\bar{b}, \bar{c} \in M^n$ with $b_1 < \ldots < b_n$ and $c_1 < \ldots < c_n$. Show that $\operatorname{tp}^{\mathcal{M}}(\bar{a}/A) = \operatorname{tp}^{\mathcal{M}}(\bar{b}/A)$ if and only if $b_i < a \Leftrightarrow c_i < a$ and $b_i > a \Leftrightarrow c_i < a$ for all $i = 1, \ldots, n$ and $a \in A$. In particular, show that any two elements of X realize the same 1-type over $\emptyset$.

b) If $a, b \in \mathbb{Q}$, then $\operatorname{tp}^{\mathbb{Q}}(a/\mathbb{N}) = \operatorname{tp}^{\mathbb{Q}}(b/\mathbb{N})$ if and only if there is an automorphism σ of $\mathbb{Q}$ fixing $\mathbb{N}$ pointwise with $\sigma(a) = b$.

c) Let $A = \{1 - \frac{1}{n} : n = 1, 2, \ldots\} \cup \{2 + \frac{1}{n} : n = 1, 2, \ldots\}$. Show that 1 and 2 realize the same type over A, but there is no automorphism of $\mathbb{Q}$ fixing A pointwise sending 1 to 2.

Exercise 4.5.2 Let T be the theory of $(\mathbb{Z}, s)$ where $s(x) = x + 1$. Determine the types in $S_n(T)$ for each n. Which types are isolated? Do the same for $(\mathbb{Z}, <, s)$.

Exercise 4.5.3 Recall that for $A \subset M$, $\operatorname{dcl}(A)$ denotes the definable closure of A (see Exercise 1.4.10). Show that if $\bar{a}, \bar{b} \in M^n$ and $\operatorname{tp}^{\mathcal{M}}(\bar{a}/A) = \operatorname{tp}^{\mathcal{M}}(\bar{b}/A)$, then $\operatorname{tp}^{\mathcal{M}}(\bar{a}/\operatorname{dcl}(A)) = \operatorname{tp}^{\mathcal{M}}(\bar{b}/\operatorname{dcl}(A))$.

Exercise 4.5.4 Suppose that $\mathcal{M}$ is an $\mathcal{L}$-structure, $A \subseteq M$, $b \in M$, and b is algebraic over A (see Exercise 1.4.11). Show that $\operatorname{tp}^{\mathcal{M}}(b/A)$ is isolated.

Exercise 4.5.5 Let K be an algebraically closed field and k be a subfield of K. What are the isolated types in $S_1^K(k)$?

Exercise 4.5.6 Let R be a real closed field. Show that 1-types over R correspond to cuts in the ordering $(R, <)$.

Exercise 4.5.7 Let T be a complete extension of Peano arithmetic. Show that $|S_1(T)| = 2^{\aleph_0}$. [Hint: Let p_n be the nth prime number. For $X \subseteq \mathbb{N}$, let $\Gamma_X(v) = \{p_n \text{ divides } v : n \in X\} \cup \{p_n \text{ does not divide } v : n \notin X\}$.]

Exercise 4.5.8 [†] If A is a commutative ring, then an ideal $P \subset A$ is *real* if whenever $a_1^2 + \ldots + a_n^2 \in P$, then $a_1, \ldots, a_n \in P$.

a) Show that a prime ideal P is real if and only if A/P is orderable.

Let $\mathrm{Spec}_r(A) = \{(P, <) : P \subset A \text{ is a real prime ideal and } < \text{ is an ordering of } A/P\}$. We call $\mathrm{Spec}_r(A)$ the *real spectrum* of A. If $a \in A$, let $X_a = \{(P, <) \in \mathrm{Spec}_r(A) : a/P > 0 \text{ in } A/P\}$. We topologize $\mathrm{Spec}_r(A)$ by taking the weakest topology in which the sets X_a are open.

If R is a real closed field, k is a subfield of R, and $p \in S_n^R(k)$, let $P_p = \{f \in k[X_1, \ldots, X_n] : f(v_1, \ldots, v_n) = 0 \in p\}$.

b) Show that P_p is a real prime ideal.

c) Show that we can order $k[\overline{X}]/P_p$ by $f(\overline{X})/P_p <_p g(\overline{X})/P_p$ if and only if $f(\overline{v}) < g(\overline{v}) \in p$. Thus $(P_p, <_p) \in \mathrm{Spec}_r(k[\overline{X}])$.

d) Show that $p \mapsto (P_p, <_p)$ is a continuous bijection between $S_n^R(k)$ and $\mathrm{Spec}_r(k[\overline{X}])$.

e) Show that $\mathrm{Spec}_r(k[\overline{X}])$ is compact.

f) What are the isolated types in $S_1^R(k)$?

Exercise 4.5.9 Let x and y be algebraically independent over $\mathbb{R}$. Order $\mathbb{R}(x, y)$ such that $x > r$ for all $r \in \mathbb{R}$ and $y > x^n$ for all $n > 0$. Let F be the real closure of $\mathbb{R}(x, y)$. Show that $\mathrm{tp}^F(x) = \mathrm{tp}^F(y)$, but there is no automorphism of F sending x to y.

Exercise 4.5.10 Suppose that $A \subseteq B$, $\theta(\overline{v})$ is a formula with parameters from A, and θ isolates $\mathrm{tp}^{\mathcal{M}}(\overline{a}/B)$. Then, θ isolates $\mathrm{tp}^{\mathcal{M}}(\overline{a}/A)$.

Exercise 4.5.11 Suppose that $A \subset M$, $\overline{a}, \overline{b} \in M$ such that $\mathrm{tp}^{\mathcal{M}}(\overline{a}, \overline{b}/A)$ is isolated. Show that $\mathrm{tp}^{\mathcal{M}}(\overline{a}/A, \overline{b})$ is isolated.

Combining this with Lemma s4.2.9 and 4.2.21, we have shown that $\mathrm{tp}^{\mathcal{M}}(\overline{a}, \overline{b}/A)$ is isolated if and only if $\mathrm{tp}^{\mathcal{M}}(\overline{a}/A, \overline{b})$ is isolated and $\mathrm{tp}^{\mathcal{M}}(\overline{b}/A)$ is isolated.

Exercise 4.5.12 Prove Lemma 4.1.9 iii).

Exercise 4.5.13 Let Δ be a set of $\mathcal{L}$-formulas closed under $\wedge, \vee, \neg$ and let $\mathcal{M}$ be an $\mathcal{L}$-structure. Let $S_n^\Delta(T) = \{\Sigma \subset \Delta : \Sigma \cup T \text{ is satisfiable and } \phi \in \Sigma \text{ or } \neg\phi \in \Sigma \text{ for all } \phi \in \Delta\}$.

a) Show that for all $p \in S_n^\Delta(T)$ there is $q \in S_n(T)$ with $p \subseteq q$.

b) Suppose that for each n and each $p \in S_n^\Delta(T)$ there is a unique $q \in S_n(T)$ with $p \subseteq q$. Show that for every $\mathcal{L}$-formula $\phi(\overline{v})$ there is $\psi(\overline{v}) \in \Delta$

such that $T \models \phi(\bar{v}) \leftrightarrow \psi(\bar{v})$. In particular, if every quantifier-free type has a unique extension to a complete type, then T has quantifier elimination.

Exercise 4.5.14 [†] We continue with the notation from Exercise 2.5.24. Suppose that p is a non-isolated n-type over $\emptyset$ and $c_1, \ldots, c_n$ are constants in $\mathcal{L}^*$. Let $D'_{p,\bar{c}} = \{\Sigma \in P : \neg\phi(\bar{c}) \in \Sigma$ for some $\phi(v_1, \ldots, v_n) \in p\}$.

a) Show that $D'_{p,\bar{c}}$ is dense.

b) Use a) and Exercise 2.5.24 to give another proof of Theorem 4.2.4.

c) Assume that Martin's Axiom is true (see Appendix A). Suppose that $\mathcal{L}$ is a countable language, T is an $\mathcal{L}$-theory, and X is a collection of nonisolated types over $\emptyset$ with $|X| < 2^{\aleph_0}$. Show that there is a countable $\mathcal{M} \models T$ that omits all of the types $p \in X$.

Exercise 4.5.15 We say that a linear order $(X, <)$ is $\aleph_1$-*like* if $|X| = \aleph_1$ but $|\{y : y < x\}| \leq \aleph_0$ for all $x \in X$.

Show that there is an $\aleph_1$-like model of Peano arithmetic.

Exercise 4.5.16 Let $\mathcal{L}_n = \{U_0, U_1, \ldots, U_n\}$, where $U_0, U_1, \ldots, U_n$ are unary predicates. Let T_n be the $\mathcal{L}_n$-theory that asserts that for each $X \subseteq \{0, \ldots, n\}$ there are infinitely many x such that $U_i(x)$ for $i \in X$ and $\neg U_i(x)$ for $i \notin X$.

a) For which κ is T_n κ-categorical?

b) Show that T_n is complete.

c) Show that T_n has quantifier elimination. [Remark: It is probably easiest to do this explicitly.]

Let $\mathcal{L} = \bigcup \mathcal{L}_n$. For X and Y finite subsets of $\mathbb{N}$, let $\Phi_{X,Y}$ be the sentence

$$\exists x \bigwedge_{i \in X} U_i(x) \wedge \bigwedge_{i \in Y} \neg U_i(x).$$

Let T be the $\mathcal{L}$-theory $\{\Phi_{X,Y} : X, Y$ disjoint finite subsets of $\mathbb{N}\}$.

d) Suppose that $\mathcal{M} \models T$. Show that $\mathcal{M} \models T_n$ for all n.

e) Show that T is complete and has quantifier elimination.

f) For $X \subset \mathbb{N}$, let $\Gamma_X = \{U_i(v) : i \in X\} \cup \{\neg U_i(v) : i \notin X\}$. Show that there is a unique 1-type p_X over $\emptyset$ with $p_X \supset \Gamma_X$.

g) Show that $X \mapsto p_X$ is a bijection between 2^ω and $S_1^{\mathcal{M}}(\emptyset)$.

h) Show that $S_1^{\mathcal{M}}(\emptyset)$ has no isolated points and hence has no prime model.

i) If 2^ω is given the product topology, then $X \mapsto p_X$ is a homeomorphism between 2^ω and $S_1^{\mathcal{M}}(\emptyset)$.

j) Describe all 2-types over $\emptyset$.

k) Show that T is κ-stable for all $\kappa \geq 2^{\aleph_0}$.

Exercise 4.5.17 Show that every algebraically closed field is homogeneous. Show that any uncountable algebraically closed field is saturated.

Exercise 4.5.18 Suppose that $\mathcal{M} = (R, +, \cdot, <, 0, 1)$ is a real closed field. Show that $\mathcal{M}$ is κ-saturated if and only if the ordering $(R, <)$ is κ-saturated.

Exercise 4.5.19 a) Show that the theory of $\mathbb{Z}$-groups is κ-stable for all $\kappa \geq 2^{\aleph_0}$.

b) Does the theory of $\mathbb{Z}$-groups have prime models over sets?

Exercise 4.5.20 Let $\mathcal{L} = \{E\}$ be the language with a single binary relation symbol. Let T be the theory of an equivalence relation where for each $n \in \omega$ there is a unique equivalence class of size n.

a) Show that T is ω-stable but not $\aleph_1$-categorical.

b) Exhibit a Vaughtian pair of models of T.

Exercise 4.5.21 Show that DLO is not κ-stable for any infinite κ.

Exercise 4.5.22 Let $\mathcal{L} = \{E_1, E_2, E_3, \ldots\}$, and let T be the theory asserting that:

i) each E_n is an equivalence relation where every equivalence class is infinite;

ii) if $xE_{i+1}y$, then xE_iy.

We say that $E_1, E_2, \ldots$ is a family of refining equivalence relations.

Let $T^2 \supset T$ be the theory that asserts that E_1 has two classes and each E_i class is the union of two infinite E_{i+1} classes.

Let $T^\infty \supset T$ be the theory that asserts that E_1 has infinitely many classes and each E_i class is the union of infinitely many infinite E_{i+1} classes.

For example, if E_n is the equivalence relation $f|n = g|n$ on ω^ω, then $(\omega^\omega, E_1, E_2, \ldots) \models T^\infty$ and $(2^\omega, E_1, E_2, \ldots) \models T^2$. Both T^2 and T^∞ are complete theories with quantifier elimination.

a) Show that T^2 is κ-stable for all $\kappa \geq 2^{\aleph_0}$.

b) Show that T^∞ is κ-stable if and only if κ such that $\kappa^{\aleph_0} = \kappa$.

Exercise 4.5.23 Suppose that $\mathcal{M}$ is interpretable in $\mathcal{N}$ and $\kappa \geq \aleph_0$.

a) Show that if $\mathcal{M}$ is κ-stable, then $\mathcal{N}$ is κ-stable.

b) Show that if $\mathcal{M}$ is κ-stable, then $\mathcal{M}^{\mathrm{eq}}$ is κ-stable.

c) Show that if $\mathcal{M}$ is κ-saturated, then $\mathcal{N}$ is κ-saturated.

Exercise 4.5.24 We say that $\mathcal{M} \models T$ is minimal if $\mathcal{M}$ has no proper elementary submodels.

a) Show that the field of algebraic numbers is a minimal model of ACF and that the field of real algebraic numbers is a minimal model of RCF.

b) Give an example of a theory with a prime model that is not minimal.

Exercise 4.5.25 Suppose that T is a theory in a countable language with a prime model $\mathcal{M}$ that is not minimal. We will show that T has an atomic model of size $\aleph_1$.

a) Show that there is an elementary embedding $j : \mathcal{M} \to \mathcal{M}$ such that $j(\mathcal{M}) \neq \mathcal{M}$.

b) Use a) to show that there is $\mathcal{M} \prec \mathcal{N}$, $\mathcal{M} \cong \mathcal{N}$ and $\mathcal{M} \neq \mathcal{N}$.

c) Show that if $\mathcal{M}_0 \prec \mathcal{M}_1 \prec \mathcal{M}_2 \ldots$ and each $\mathcal{M}_i \cong \mathcal{M}$, then $\bigcup \mathcal{M}_i \cong \mathcal{M}$. [Hint: Use the uniqueness of atomic models.]

d) Use b) and c) to construct an elementary chain $(\mathcal{M}_\alpha : \alpha < \omega_1)$ such that each $\mathcal{M}_\alpha \cong \mathcal{M}$ and $\mathcal{M}_\alpha \neq \mathcal{M}_{\alpha+1}$. Let $\mathcal{M}' = \bigcup_{\alpha < \omega_1} \mathcal{M}_\alpha$. Show that $\mathcal{M}'$ is atomic and $|M'| = \aleph_1$.

e) Show that if T is not $\aleph_0$-categorical, then T has a nonatomic model of size $\aleph_1$. Conclude that if T is $\aleph_1$-categorical, but not $\aleph_0$-categorical, then any prime model is minimal. (We will prove in Corollary 5.2.10 that $\aleph_1$-categorical theories are ω-stable, thus there always is a prime model.)

f) Give an example of a theory that is $\aleph_1$-categorical and $\aleph_0$-categorical but has a prime model that is not minimal.

Exercise 4.5.26 Let $\mathcal{L} = \{U, <\}$, where U is a unary predicate and $<$ is a binary relation symbol. Let T be the $\mathcal{L}$-theory extending DLO where U picks out a subset that is dense and has a dense complement. Let $\mathcal{M} \models T$ and let $A = U^{\mathcal{M}}$. Show that there is no prime model over A.

Exercise 4.5.27 Suppose that $A \subset \mathcal{M}$, $|A| \leq \aleph_0$, $\mathcal{M}_0$, and $\mathcal{M}_1$ are elementary submodels of $\mathcal{M}$ with $A \subseteq M_0 \cap M_1$, and $\mathcal{M}_0$ and $\mathcal{M}_1$ are prime model extensions of A. Then, $\mathcal{M}_0$ and $\mathcal{M}_1$ are isomorphic over A (i.e., there is an isomorphism $f : \mathcal{M}_0 \to \mathcal{M}_1$ that fixes A pointwise).

Exercise 4.5.28 Suppose that T is an o-minimal theory, $\mathcal{M} \models T$, and $A \subseteq M$. Show that the isolated types in $A \subseteq M$ are dense. Conclude that o-minimal theories have prime models over sets.

Exercise 4.5.29 Show that the union of an elementary chain of $\aleph_0$-homogeneous structures is $\aleph_0$-homogeneous.

Exercise 4.5.30 Show that if T is $\aleph_0$-categorical, then any homogeneous model is saturated. In particular, a dense linear order is saturated if and only if it is homogeneous.

Exercise 4.5.31 Show that if $\mathcal{M}$ is κ-saturated, then every infinite definable subset of M^k has cardinality at least κ.

Exercise 4.5.32 Prove Lemma 4.4.5.

Exercise 4.5.33 Suppose that $\mathcal{M}$ is κ-saturated, $A \subset M$, and $|A| < \kappa$. If $p \in S_n^{\mathcal{M}}(A)$ has only finitely many realizations in $\mathcal{M}$ and $\bar{a}$ realizes p, then $\bar{a} \in \text{acl}(p)$.

Exercise 4.5.34 Suppose that $\mathcal{M}$ is κ-saturated, and $(\phi_i(\bar{v}) : i \in I)$ and $(\theta_j(\bar{v}) : j \in J)$ are sequences of $\mathcal{L}_M$-formulas such that $|I|, |J| < \kappa$ and

$$\mathcal{M} \models \bigvee_{i \in I} \phi_i(\bar{v}) \leftrightarrow \neg \left(\bigvee_{j \in J} \theta_j(\bar{v}) \right).$$

Show that there are finite sets $I_0 \subseteq I$ and $J_0 \subseteq J$ such that

$$\mathcal{M} \models \bigvee_{i \in I} \phi_i(\bar{v}) \leftrightarrow \bigvee_{i \in I_0} \phi_i(\bar{v}).$$

Exercise 4.5.35 (Expandability of Saturated Models) Suppose that $\kappa \geq \aleph_0$ and $\mathcal{M}$ is a saturated $\mathcal{L}$-structure of cardinality κ. Let $\mathcal{L}^* \supset \mathcal{L}$ with $|\mathcal{L}^*| \leq \kappa$. Suppose that T is an $\mathcal{L}^*$-theory consistent with $\text{Th}(\mathcal{M})$. We show that we can interpret the symbols in $\mathcal{L}^* \setminus \mathcal{L}$ to obtain an expansion $\mathcal{M}^*$ of $\mathcal{M}$ with $\mathcal{M}^* \models T$.

Let $\mathcal{L}_M^*$ be the language obtained by adding to $\mathcal{L}^*$ constants for every element of $\mathcal{M}$. Let $(\phi_\alpha : \alpha < \kappa)$ enumerate all $\mathcal{L}_M^*$-sentences. We build an increasing sequence of $\mathcal{L}^*$-theories $(T_\alpha : \alpha < \kappa)$ such that $T_\alpha \cup T \cup \text{Diag}_{\text{el}}(\mathcal{M})$ is satisfiable and $|T_\alpha| < \kappa$ for all $\alpha < \kappa$ (indeed $|T_{\alpha+1}| \leq |T_\alpha| + 2$).

Let $T_0 = \emptyset$. For α a limit ordinal, let $T_\alpha = \bigcup_{\beta < \alpha} T_\beta$. Suppose that we have T_α such that $|T_\alpha| < \kappa$ and $T_\alpha \cup T \cup \text{Diag}_{\text{el}}(\mathcal{M})$ is satisfiable.

a) Show that either $T_\alpha \cup \{\phi_\alpha\} \cup T \cup \text{Diag}_{\text{el}}(\mathcal{M})$ is satisfiable or $T_\alpha \cup \{\neg\phi_\alpha\} \cup T \cup \text{Diag}_{\text{el}}(\mathcal{M})$ is satisfiable.

b) Show that if ϕ_α is $\exists v\, \psi(v)$ and $T_\alpha \cup \{\phi_\alpha\} \cup T \cup \text{Diag}_{\text{el}}(\mathcal{M})$ is satisfiable, then for some $a \in M$, $T_\alpha \cup \{\phi_\alpha, \psi(a)\} \cup T \cup \text{Diag}_{\text{el}}(\mathcal{M})$ is satisfiable. [Hint: Let $A \subset M$ be all parameters from $\mathcal{M}$ occurring in formulas in $T_\alpha \cup \{\phi_\alpha\}$. Let $\Gamma(v)$ be all of the $\mathcal{L}_A$-consequences of $T_\alpha \cup \{\phi_\alpha, \psi(v)\} \cup T \cup \text{Diag}_{\text{el}}(\mathcal{M})$. Show that $\Gamma(v)$ is satisfiable and hence, by saturation, must be realized by some a in $\mathcal{M}$. Show that $T_\alpha \cup \{\phi_\alpha, \psi(a)\} \cup T \cup \text{Diag}_{\text{el}}(\mathcal{M})$ is satisfiable.]

c) Show that we can always choose $T_{\alpha+1}$ such that

   i) $T_{\alpha+1} \cup T \cup \text{Diag}_{\text{el}}(\mathcal{M})$ is satisfiable;

   ii) either $\phi_\alpha \in T_{\alpha+1}$ or $\neg\phi_\alpha \in T_{\alpha+1}$;

   iii) if $\phi_\alpha \in T_{\alpha+1}$ and ϕ_α is $\exists v\, \psi(v)$, then $\psi(a) \in T_{\alpha+1}$ for some $a \in M$;

   iv) $|T_{\alpha+1}| \leq |T_\alpha| + 2 < \kappa$.

Let $T^* = \bigcup_{\alpha < \kappa} T_\alpha$.

d) Show that T^* is a complete $\mathcal{L}_M^*$-theory with the witness property and $T^* \supset T \cup \text{Diag}_{\text{el}}(\mathcal{M})$. Let $\mathcal{N}$ be the canonical model of T^*. Show that as an $\mathcal{L}$-structure $\mathcal{N}$ is exactly $\mathcal{M}$. Thus, $\mathcal{N}$ is the desired expansion of $\mathcal{M}$ to a model of T.

Exercise 4.5.36 Let $\mathcal{L} = \{U_0, U_1, \ldots\} \cup \{s_0, s_1, \ldots\}$. We describe an $\mathcal{L}$-structure $\mathcal{M}$ with universe $\mathbb{N} \times \mathbb{Z}$. Let $U_i^{\mathcal{M}} = \{i\} \times \mathbb{Z}$ and

$$s_i((j, x)) = \begin{cases} (j, x) & \text{if } i \neq j \\ (j, x + 1) & \text{if } i = j. \end{cases}$$

Let T be the full theory of $\mathcal{M}$. Basically, T is the theory of countably many copies of $(\mathbb{Z}, s)$.

a) Show that $|S_n(T)| = \aleph_0$ for all n. (Either show or assume that T has quantifier elimination.)

b) Show that $I(T, \aleph_0) = 2^{\aleph_0}$.

Exercise 4.5.37 ($\aleph_1$-saturation of Ultraproducts) Suppose that U is a non principal ultrafilter on ω. Let $(\mathcal{M}_0, \mathcal{M}_1, \ldots)$ be a sequence of $\mathcal{L}$-structures, and let $\mathcal{M}^* = \prod \mathcal{M}_i / U$. We will show that $\mathcal{M}^*$ is $\aleph_1$-saturated.

Let $A \subset M^*$ be countable. For each $a \in A$, choose $f_a \in \prod M_i$ such that $a = f_a / \sim$. Let $\Gamma(v) = \{\phi_i(v) : i < \omega\}$ be a set of $\mathcal{L}_A$-formulas such that $\Gamma(v) \cup \mathrm{Th}_A(\mathcal{M}^*)$ is satisfiable. By taking conjunctions, we may, without loss of generality, assume that $\phi_{i+1}(v) \to \phi_i(v)$ for $i < \omega$. Let $\phi_i(v)$ be $\theta_i(v, a_{i,1}, \ldots, a_{i,m_i})$, where θ_i is an $\mathcal{L}$-formula.

a) Let $D_i = \{n < \omega : \mathcal{M}_n \models \exists v \theta_i(v, f_{a_{i,1}}(n), \ldots, f_{a_{i,m_i}}(n))\}$. Show that $D_i \in U$.

b) Find $g \in \prod M_i$ such that if $i \leq n$ and $n \in D_i$, then

$$\mathcal{M}_n \models \theta_i(g(n), f_{a_{i,1}}(n), \ldots, f_{a_{i,m_i}}(n)).$$

c) Show that g realizes $\Gamma(v)$. Where do you use the fact that U is non-principal? Conclude that $\mathcal{M}^*$ is $\aleph_1$-saturated. Show that if the Continuum Hypothesis holds, then $\mathcal{M}^*$ is saturated.

Exercise 4.5.38 (Recursively Saturated Models) Let $\mathcal{L}$ be a recursive language. We say that $\mathcal{M}$ is *recursively saturated* if whenever $A \subset M$ is finite and Γ is a recursive (possibly incomplete) n-type over A, then Γ is realized in $\mathcal{M}$. In particular, every $\aleph_0$-saturated structure is recursively saturated.

a) Suppose that $\mathcal{N}$ is a countable model of T. Show that there is a countable recursively saturated $\mathcal{M}$ with $\mathcal{N} \prec \mathcal{M}$.

b) Show that if $\mathcal{M}$ is recursively saturated, then $\mathcal{M}$ is $\aleph_0$-homogeneous. [Hint: If $\mathrm{tp}(\bar{a}) = \mathrm{tp}(\bar{b})$, consider the set of formulas $\{\phi(v, \bar{b}) \leftrightarrow \phi(c, \bar{a}) : \phi$ an $\mathcal{L}$-formula$\}$.]

c) Show that if $\mathcal{M}_0 \prec \mathcal{M}_1 \prec \ldots$ is an elementary chain of recursively saturated models, then $\mathcal{M} = \bigcup_{n \in \omega} \mathcal{M}_n$ is recursively saturated.

d) Suppose $\mathcal{M}, \mathcal{N} \models T$ and such that $(\mathcal{M}, \mathcal{N})$ is a countable recursively saturated model of the theory of pairs of models of T (as in our proof of Vaught's Two-Cardinal Theorem). Show $\mathcal{M} \cong \mathcal{N}$. [Hint: Recall that countable $\aleph_0$-homogeneous models are isomorphic if and only if they realize the same types.] Use this to give a simplified proof of Vaught's Two-Cardinal Theorem.

e) Let $\mathcal{M}$ be a recursively saturated $\mathcal{L}$-structure. Suppose that $\mathcal{L}^* \supset \mathcal{L}$ is recursive and T is a recursive $\mathcal{L}^*$-theory such that $\mathrm{Diag}_{\mathrm{el}}(\mathcal{M}) \cup T$ is satisfiable. Show that there is an expansion of $\mathcal{M}^*$ of $\mathcal{M}$ such that $\mathcal{M}^* \models T$. [Hint: Follow the proof of expandability of saturated models.] Show that we can make $\mathcal{M}^*$ recursively saturated.

Exercise 4.5.39 (Robinson's Consistency Theorem) Let $\mathcal{L}_0$ and $\mathcal{L}_1$ be languages, and let $\mathcal{L} = \mathcal{L}_1 \cap \mathcal{L}_2$. Let T be a complete $\mathcal{L}$-theory and let $T_i \supset T$ be a satisfiable $\mathcal{L}_i$-theory for $i = 1, 2$.

a) Show that there is a recursively saturated structure $(\mathcal{M}_1, \mathcal{M}_2)$ where $\mathcal{M}_i \models T_i$ for $i = 1, 2$.

b) Let $\mathcal{N}_i$ be the $\mathcal{L}$-reduct of $\mathcal{M}_i$. Show that $(\mathcal{N}_1, \mathcal{N}_2)$ is still recursively saturated and that $\mathcal{N}_1 \cong \mathcal{N}_2$.

c) Conclude that we can view $\mathcal{M}_1$ and $\mathcal{M}_2$ as expansions of a single $\mathcal{L}$-structure and that $T_1 \cup T_2$ is satisfiable.

Exercise 4.5.40 [†] Let $\mathcal{M}$ be a nonstandard model of Peano arithmetic.

a) Let A be a finite subset of $\mathcal{M}$, and let Γ be a recursive type over A of bounded quantifier complexity (i.e., there is n such that all formulas in Γ have at most n-quantifiers). Show that Γ is realized in $\mathcal{M}$. [Hint: (see [51] §9). There is a formula $S(v, w)$ that is a truth definition for formulas with at most n quantifiers. In other words if $\lceil \phi \rceil$ is the Gödel code for a formula $\phi(v_1, \ldots, v_n)$ and $\lceil \bar{b} \rceil$ codes a sequence $\bar{b} = (b_1, \ldots, b_n) \in M^n$, then $\mathcal{M} \models S(\lceil \phi \rceil, b) \leftrightarrow \phi(b_1, \ldots, b_n)$. Because Γ is recursive, there is a formula $G(\lceil \phi \rceil)$ if and only if $\phi(v, \bar{a}) \in \Gamma$. Because Γ is satisfiable for all $n < \omega$

$$\mathcal{M} \models \exists b \forall m < n \, G(m) \rightarrow S(m, \lceil (b, \bar{a}) \rceil)].$$

Apply overspill (Exercise 2.5.7).]

b) Let $(G, +, <, 0)$ be the ordered additive group of $\mathcal{M}$. Use a) to show that G is a recursively saturated model of Presburger arithmetic.

Exercise 4.5.41 [†] (Tennenbaum's Theorem) If $\mathcal{M}$ is a nonstandard model of Peano arithmetic and $a \in M$, let $r(a) = \{n \in \mathbb{N} : p_n \text{ divides } a\}$, where p_n is the nth prime number. Let $SS(\mathcal{M}) = \{r(a) : a \in \mathcal{M}\}$. We call $SS(\mathcal{M})$ the *Scott set* of $\mathcal{M}$.

a) Suppose that $X \in SS(\mathcal{M})$ and Y is recursive in X, then $Y \in SS(\mathcal{M})$. [Hint: use Exercise 4.5.40.]

b) We say that $T \subseteq 2^{<\omega}$ is a *tree* if whenever $\sigma \in T$ and $\tau \subset \sigma$, then $\tau \in T$. We say that $f \in 2^\omega$ is an infinite *path* through T if $f|n \in T$ for all $n < \omega$. Show that if $X \in SS(\mathcal{M})$ and T is an infinite tree recursive in X, then there is $Y \in SS(\mathcal{M})$ and f an infinite path through T recursive in Y. [Hint: use Exercise 4.5.40.]

c) Let $\phi_0, \phi_1, \ldots$ be a list of all partial recursive functions. We write $\phi_i(n) \downarrow = j$ if on input n Turing machine i halts with output j. Let $A = \{i : \phi_i(i) \downarrow = 0\}$ and $B = \{i : \phi_i(i) \downarrow = 1\}$. Show that there is no recursive set C such that $A \subseteq C$ and $B \cap C = \emptyset$. We call A and B *recursively inseparable*. [Hint: Suppose that ϕ_i is the characteristic function of C and ask whether $i \in C$.]

d) There is a recursive infinite tree $T \subseteq 2^{<\omega}$ with no recursive infinite paths. [Hint: Let $T = \{\sigma \in 2^{<\omega} : \text{if } i < |\sigma| \text{ and Turing machine } i \text{ on input } i \text{ halts by stage } |\sigma| \text{ with output } j \in \{0, 1\}, \text{ then } \sigma(i) = j\}$. Show that if f is a recursive infinite path through T, then $C = \{i : f(i) = 0\}$ contradicts c).]

e) We can find an isomorphic copy of $\mathcal{M}$ with universe ω. Thus, we may assume that $\mathcal{M} = (\omega, \oplus, \otimes)$. Show that $r(a)$ is recursive in $\oplus$ for all $a \in M$. Conclude that $\oplus$ is not recursive.

Exercise 4.5.42 Show that there is no Vaughtian pair of real closed fields.

Exercise 4.5.43 Let k be a differential field, $K \models DCF$, and $k \subseteq K$. The ring of differential polynomials in $\overline{X} = (X_1, \ldots, X_n)$ is the ring

$$k\{X_1, \ldots, X_n\} = k[X_1, \ldots, X_n, X_1', \ldots, X_n', \ldots, X_1^{(m)}, \ldots, X_n^{(m)}, \ldots].$$

We extend the derivation from k to $k\{\overline{X}\}$ by letting $\delta(X_n^{(m)}) = X_n^{(m+1)}$. An ideal $I \subset k\{\overline{X}\}$ is called a differential ideal if $\delta(f) \in I$ whenever $f \in I$.

a) For $p \in S_n^K(k)$, let $I_p = \{f \in k\{\overline{X}\} : f(v_1, \ldots, v_n) = 0 \in p\}$. Show that I_p is a differential prime ideal.

b) Show that if $I \subset k\{\overline{X}\}$ is a differential prime ideal, then $I = I_p$ for some $p \in S_n^K(k)$. Thus, $p \mapsto I_p$ is a bijection between complete n-types over k and differential prime ideals in $k\{\overline{X}\}$.

c) The Ritt–Raudenbusch Basis Theorem (see [50]) asserts that every differential prime ideal in $k\{\overline{X}\}$ is finitely generated. Use this to show that DCF is ω-stable.

d) If $K \models DCF$, we say that $X \subseteq K^n$ is *Kolchin closed* if X is a finite union of sets of the form $\{\overline{x} \in K^n : f_1(\overline{x}) = \ldots = f_m(\overline{x}) = 0\}$ where $f_1, \ldots, f_m \in K\{\overline{X}\}$. Prove that there are no infinite descending chains of Kolchin closed sets.

e) (Differential Nullstellensatz) Suppose that $K \models DCF$, $P \subseteq K\{X_1, \ldots, X_n\}$ is a differential prime ideal and $g \in K\{\overline{X}\} \setminus P$. Show that there is $\overline{a} \in K^n$ such that $f(\overline{a}) = 0$ for all $f \in P$ but $g(\overline{a}) \neq 0$.

f) (Existence of Differential Closures) Suppose that k is a differentially closed field and $k \subseteq K \models DCF$. We say that K is a *differential closure* of k if whenever $k \subseteq L$ and $L \models DCF$ there is a differential field embedding of K into L fixing k. Show that every field has a differential closure. [Hint: Show that differential closures are prime model extensions.]

Exercise 4.5.44 Suppose that $\mathcal{L}$ is a countable language and T is an $\mathcal{L}$-theory. Let $C = \{T' \supseteq T : T'$ a complete $\mathcal{L}$-theory$\}$. Show that if $|C| \geq \aleph_1$, then $|C| = 2^{\aleph_0}$. Argue that if Vaught's Conjecture is true for complete theories, then it is also true for incomplete theories.

Exercise 4.5.45 Suppose $\mathcal{L}$ is a finite language with no function symbols and T is an $\mathcal{L}$-theory with quantifier elimination. Prove that T is $\aleph_0$-categorical.

Exercise 4.5.46 Describe all $\aleph_0$-categorical linear orders.

Remarks

The Omitting Types Theorem is due to Henkin and Orey , each of whom used it to prove the Completeness Theorem for ω-logic. Theorem 4.2.5 is due to MacDowell and Specker. Their proof uses an ultraproduct construction and works for uncountable models as well. See [51] §8.2 for further results on end extensions of models of arithmetic.

The results on prime models, atomic models, and countable saturated models are due to Vaught and appear in [99], one of the most elegant papers in model theory. Theorem 4.3.23 is due to Keisler, and the other basic results on saturated and homogeneous models are due to Morley and Vaught.

Recursively saturated models were introduced by Barwise and Schlipf. Many results that can be proved using saturated models have elegant proofs using recursively saturated models (see [22] §2.4 or [53]). Friedman showed the weak recursive saturation of nonstandard models of arithmetic and used it to prove the following result (see [51] §12.1).

Theorem 4.5.47 *If M is a countable nonstandard model of Peano arithmetic, then there is $\mathcal{I}$ a proper initial segment of M with $\mathcal{I} \cong M$.*

The $\aleph_1$-saturation of ultraproducts is due to Keisler (see [22] §6.1 for generalizations).

Morley introduced ω-stable theories in his proof of Theorem 6.1.1. He also proved that ω-stable theories have prime model extensions. Shelah showed that there are three possibilities for $\{\kappa \geq \aleph_0 : T$ is κ-stable$\}$.

Theorem 4.5.48 *If T is a complete theory in a countable language, then one of the following holds:*
i) there are no cardinals κ such that T is κ-stable,
ii) T is κ-stable for all $\kappa \geq 2^{\aleph_0}$,
iii) T is κ-stable if and only if $\kappa^{\aleph_0} = \kappa$.

A proof of this theorem can be found in [7], [18] or [76]. If i) holds, we say that T is *unstable*; otherwise, we say that T is *stable*. If ii) holds, we say that T is *superstable*. By Theorem 4.2.18, every ω-stable theory is superstable. In Exercise 4.5.22, we gave an example of a superstable theory that is not ω-stable and a stable theory that is not superstable.

The saturated model test for quantifier elimination is due to Blum, who also axiomatized the theory of DCF, proved that DCF is ω-stable, and deduced from that the existence of differential closures.

In Theorem 5.2.15, we will examine another two-cardinal result. There are many interesting two cardinal questions, but most can not be answered in ZFC. The following Theorem gives several interesting examples.

Theorem 4.5.49 *Let $\mathcal{L}$ be a countable language and T an $\mathcal{L}$-theory.*
i) Assume that $V = L$.[3] If $\kappa > \lambda \geq \aleph_0$ and T has a (κ, λ)-model, then T has a (μ^+, μ)-model for all infinite cardinals μ.
ii) Assume that $V = L$. If T has a (κ^{++}, κ)-model for some infinite cardinal κ, then T has a (λ^{++}, λ)-model for all infinite cardinals λ.

---

[3] $V = L$ is Gödel's Axiom of Constructibility asserting that all sets are constructible (see [57] or [47]).

iii) If ZFC is consistent, then it is consistent with ZFC that there is a countable theory with an $(\aleph_1, \aleph_0)$-model but no $(\aleph_2, \aleph_1)$-model.

The first result was proved by Chang (see [22] 7.2.7) for regular μ under the weaker assumption that the Generalized Continuum Hypothesis holds. The general case is due to Jensen who also proved the second result (see [27] §VIII). The third result is due to Mitchell and Silver (see [47] §29).

The characterization of $\aleph_0$-categorical theories was proved independently by Ryll-Nardzewski, Engler, and Svenonius.

Although Vaught's Conjecture is open for arbitrary theories, we do know that it holds for several interesting classes of theories.

Theorem 4.5.50 *Vaught's Conjecture holds for:*
  i) (Shelah [93]) ω-stable theories;
  ii) (Buechler [19]) superstable theories of finite U-rank;
  iii) (Mayer [69]) o-minimal theories;
  iv) (Miller) theories of linear orders with unary predicates;
  v) (Steel [98]) theories of trees.

See [100] for more on iv) and v).

Theorem 4.4.16 also follows from another powerful theorem in descriptive set theory. Consider the equivalence relation on $D(\mathcal{L}, T)$ given by $f E g$ if and only if $\mathcal{M}_f \cong \mathcal{M}_g$. It is easy to argue that E is an analytic subset of $D(\mathcal{L}, T) \times D(\mathcal{L}, T)$. Burgess (see, for example, [98]) proved that any analytic equivalence relation on a Borel subset of 2^ω with at least $\aleph_2$ classes has $2^{\aleph_0}$ classes.

If ϕ is an $\mathcal{L}_{\omega_1, \omega}$-sentence, we can ask about the number of nonisomorphic countable models of ϕ. Burgess' Theorem shows that if there are at least $\aleph_2$ nonisomorphic models, then there are $2^{\aleph_0}$, but it is unknown whether there can be an $\mathcal{L}_{\omega_1, \omega}$-sentence with exactly $\aleph_1 < 2^{\aleph_0}$ models.

Questions around Vaught's Conjecture can be reformulated in a way that does not involve any model theory. We say that a topological space $\mathbf{X}$ is *Polish* if it is a complete separable metric space. Suppose that G is a Polish topological group and G acts continuously on a Borel subset X of a Polish space $\mathbf{X}$. For example, X could be $D(\mathcal{L}, T)$ and G could be the group of permutations of ω topologized by taking subbasic open sets $N_{n,m} = \{f : f(n) = m\}$. The Topological Vaught Conjecture asserts that if G has uncountably many orbits on X, then G has $2^{\aleph_0}$ orbits. See [9] for more on this topic.

5
Indiscernibles

5.1 Partition Theorems

In this chapter, we will develop a powerful method for analyzing and constructing models. We begin by developing some tools from infinite combinatorics that will play a crucial role.

For X a set and κ, λ (possibly finite) cardinals, we let $[X]^\kappa$ be the collection of all subsets of X of size κ. We call $f : [X]^\kappa \to \lambda$ a *partition* of $[X]^\kappa$. We say that $Y \subseteq X$ is *homogeneous* for the partition f if there is $\alpha < \lambda$ such that $f(A) = \alpha$ for all $A \in [Y]^\kappa$ (i.e., f is constant on $[Y]^\kappa$). Finally, for cardinals κ, η, μ, and λ, we write $\kappa \to (\eta)^\mu_\lambda$ if whenever $|X| \geq \kappa$ and $f : [X]^\mu \to \lambda$, then there is $Y \subseteq X$ such that $|Y| \geq \eta$ and Y is homogeneous for f.

The starting point is Ramsey's Theorem.

Theorem 5.1.1 (Ramsey's Theorem) *If $k, n < \omega$, then $\aleph_0 \to (\aleph_0)^n_k$.*

Before proving Ramsey's Theorem, we give several sample applications to give the flavor of the subject.

One simple application is the standard fact that any sequence of real numbers $(r_0, r_1, \ldots)$ has a monotonic subsequence. Let $f : [\mathbb{N}]^2 \to 3$ by

$$f(\{i, j\}) = \begin{cases} 0 & i < j \text{ and } r_i < r_j \\ 1 & i < j \text{ and } r_i = r_j \\ 2 & i < j \text{ and } r_i > r_j \end{cases}.$$

By Ramsey's Theorem, there is $Y \subseteq \mathbb{N}$ an infinite homogeneous set for f. Let $j_0 < j_1 < \ldots$ list Y. There is $c < 3$ such that $f(\{j_m, j_n\}) = c$ for $m < n$. If $c = 0$, the sequence $r_{j_0}, r_{j_1}, \ldots$ is increasing, if $c = 1$ it is constant, and if $c = 2$ it is decreasing.

For another application, suppose that G is an infinite graph. Let $f : [G]^2 \to 2$ by

$$f(\{a, b\}) = \begin{cases} 1 & (a, b) \text{ is an edge of } G \\ 0 & (a, b) \text{ is not an edge of } G \end{cases}.$$

By Ramsey's Theorem, there is an infinite $H \subseteq G$ homogeneous for f. If f is constantly 1 on $[H]^2$, then H is a complete subgraph, and if f is constantly 0, there are no edges between vertices in H. Thus, every infinite graph either has an infinite complete subgraph or an infinite null subgraph.

Proof We prove Ramsey's Theorem by induction on n. For $n = 1$, Ramsey's Theorem asserts that if X is infinite, $k < \omega$, and $f : X \to k$, then $f^{-1}(i)$ is infinite for some $i < k$. This is just the Pigeonhole Principle that if we put infinitely many items into finitely many boxes, at least one of the boxes will contain infinitely many items.

Suppose that we have proved that if $i < n$, $k < \omega$, X is infinite, and $f : [X]^i \to k$, then there is an infinite $Y \subseteq X$ homogeneous for f.

We could always replace X by a countable subset of X; thus, without loss of generality, we may assume that $X = \mathbb{N}$.

Let $f : [\mathbb{N}]^n \to k$. For $a \in \mathbb{N}$, let $f_a : [\mathbb{N} \setminus \{a\}]^{n-1} \to k$ by $f_a(A) = f(A \cup \{a\})$. We build a sequence $0 = a_0 < a_1 < \ldots$ in $\mathbb{N}$ and $\mathbb{N} = X_0 \supset X_1 \supset \ldots$ a sequence of infinite sets as follows. Given a_i and X_i, let $X_{i+1} \subset X_i \setminus \{0, 1, \ldots, a_i\}$ be homogeneous for f_{a_i}. Let a_{i+1} be the least element of X_{i+1}.

Let $c_i < k$ be such that $f_{a_i}(A) = c_i$ for all $A \in [X_{i+1}]^{n-1}$. By the Pigeonhole Principle, there is $c < k$ such that $\{i : c_i = c\}$ is infinite. Let $X = \{a_i : c_i = c\}$. We claim that X is homogeneous for f. Let $x_1 < \ldots < x_n$ where each $x_i \in X$. There is an i such that $x_1 = a_i$ and $x_2, \ldots, x_n \in X_i$. Thus

$$f(\{x_1, \ldots, x_n\}) = f_{x_1}(\{x_2, \ldots, x_n\}) = c_i = c$$

and X is homogeneous for f, as desired.

From Ramsey's Theorem, we can deduce some results of finite combinatorics.

Theorem 5.1.2 (Finite Ramsey Theorem) *For all $k, n, m < \omega$, there is $l < \omega$ such that $l \to (m)^n_k$.*

Proof Suppose that there is no l such that $l \to (m)^n_k$. For each $l < \omega$, let $T_l = \{f : [\{0, \ldots, l-1\}]^n \to k : \text{there is no } X \subseteq \{0, \ldots, l-1\} \text{ of size at least } m, \text{ homogeneous for } f\}$. Clearly, each T_l is finite and if $f \in T_{l+1}$

there is a unique $g \in T_l$ such that $g \subset f$. Thus, if we order $T = \bigcup T_l$ by inclusion, we get a finite branching tree. Each T_l is not empty, so T is an infinite finite branching tree. By König's Lemma (see Lemma A.21) we can find $f_0 \subset f_1 \subset f_2 \subset \ldots$ with $f_i \in T_i$.

Let $f = \bigcup f_i$. Then $f : [\mathbb{N}]^n \to k$. By Ramsey's Theorem, there is an infinite $X \subseteq \mathbb{N}$ homogeneous for f. Let $x_1, \ldots, x_m$ be the first m elements of X and let $s > x_m$. Then $\{x_1, \ldots, x_m\}$ is homogeneous for f_s, a contradiction.

Because the finite version of Ramsey's Theorem is a statement about the natural numbers, it might be more satisfying to give a direct proof that does not use infinite methods. Such proofs are well-known to finite combinatorists (see [36]). The proof we gave, in addition to being quite slick, also allows us to prove stronger finite versions. In Section 5.4, we will show that a small variant of the argument allows us to prove a result that cannot be proved in Peano arithmetic.

When we begin partitioning sets into infinitely many pieces it becomes harder to find homogeneous sets.

Proposition 5.1.3 $2^{\aleph_0} \not\to (3)^2_{\aleph_0}$.

Proof We define $F : [2^\omega]^2 \to \omega$ by $F(\{f,g\})$ is the least n such that $f(n) \neq g(n)$. Clearly, we cannot find $\{f,g,h\}$ such that $f(n) \neq g(n)$, $g(n) \neq h(n)$, and $f(n) \neq h(n)$.

On the other hand, if $\kappa > 2^{\aleph_0}$, then $\kappa \to (\aleph_1)^2_{\aleph_0}$. This is a special case of an important generalization of Ramsey's Theorem. For κ an infinite cardinal and α an ordinal, we inductively define $\beth_\alpha(\kappa)$ by $\beth_0(\kappa) = \kappa$ and

$$\beth_\alpha(\kappa) = \sup_{\beta < \alpha} 2^{\beth_\beta(\kappa)}.$$

In particular $\beth_1(\kappa) = 2^\kappa$. We let $\beth_\alpha = \beth_\alpha(\aleph_0)$. Under the Generalized Continuum Hypothesis, $\beth_\alpha = \aleph_\alpha$.

Theorem 5.1.4 (Erdős–Rado Theorem) $\beth_n(\kappa)^+ \to (\kappa^+)^{n+1}_\kappa$.

Proof We prove this by induction on n. For $n = 0$, $\kappa^+ \to (\kappa^+)^1_\kappa$ is just the Pigeonhole Principle.

Suppose that we have proved the theorem for $n - 1$. Let $\lambda = \beth_n(\kappa)^+$, and let $f : [\lambda]^{n+1} \to \kappa$. For $\alpha < \lambda$, let $f_\alpha : [\lambda \backslash \{\alpha\}]^n \to \kappa$ by $f_\alpha(A) = f(A \cup \{\alpha\})$.

We build $X_0 \subseteq X_1 \subseteq \ldots \subseteq X_\alpha \subseteq \ldots$ for $\alpha < \beth_{n-1}(\kappa)^+$ such that $X_\alpha \subseteq \beth_n(\kappa)^+$ and each X_α has cardinality at most $\beth_n(\kappa)$. Let $X_0 = \beth_n(\kappa)$. If α is a limit ordinal, then $X_\alpha = \bigcup_{\beta < \alpha} X_\beta$.

Suppose we have X_α with $|X_\alpha| = \beth_n(\kappa)$. Because

$$\beth_n(\kappa)^{\beth_{n-1}(\kappa)} = (2^{\beth_{n-1}(\kappa)})^{\beth_{n-1}(\kappa)} = 2^{\beth_{n-1}(\kappa)} = \beth_n(\kappa),$$

there are $\beth_n(\kappa)$ subsets of X_α of cardinality $\beth_{n-1}(\kappa)$. Also note that if $Y \subset X_\alpha$ and $|Y| = \beth_{n-1}(\kappa)$, then there are $\beth_n(\kappa)$ functions $g : [Y]^n \to \kappa$ because

$$\kappa^{\beth_{n-1}(\kappa)} = 2^{\beth_{n-1}(\kappa)} = \beth_n(\kappa).$$

Thus, we can find $X_{\alpha+1} \supseteq X_\alpha$ such that $|X_{\alpha+1}| = \beth_n(\kappa)$, and if $Y \subset X_\alpha$ with $|Y| = \beth_{n-1}(\kappa)$ and $\beta \in \lambda \setminus Y$, then there is $\gamma \in X_{\alpha+1} \setminus Y$ such that $f_\beta|[Y]^n = f_\gamma|[Y]^n$.

Let $X = \bigcup_{\alpha < \beth_{n-1}(\kappa)^+} X_\alpha$. If $Y \subset X$ with $|Y| \le \beth_{n-1}(\kappa)$, then $Y \subset X_\alpha$ for some $\alpha < \beth_n(\kappa)^+$. If $\beta \in \lambda \setminus Y$, then there is $\gamma \in X \setminus Y$ such that $f_\beta|[Y]^n = f_\gamma|[Y]^n$.

Fix $\delta \in \lambda \setminus X$. Inductively construct $Y = \{y_\alpha : \alpha < \beth_{n-1}^+(\kappa)\} \subseteq X$. Let $y_0 \in X$. Suppose that we have constructed $Y_\alpha = \{y_\beta : \beta < \alpha\}$. Choose $y_\alpha \in X$ such that $f_{y_\alpha}|[Y_\alpha]^n = f_\delta|[Y_\alpha]^n$.

By the induction hypothesis, there is $Z \subseteq Y$ such that $|Z| \ge \kappa^+$ and Z is homogeneous for f_δ. Say $f_\delta(B) = \gamma$ for all $B \in [Z]^n$. We claim that Z is homogeneous for f. Let $A \in [Z]^{n+1}$. There are $\alpha_1 < \ldots < \alpha_{n+1}$ such that $A = \{y_{\alpha_1}, \ldots, y_{\alpha_{n+1}}\}$. Then

$$f(A) = f_{y_{\alpha_{n+1}}}(\{y_{\alpha_1}, \ldots, y_{\alpha_n}\}) = f_\delta(\{y_{\alpha_1}, \ldots, y_{\alpha_n}\}) = \gamma.$$

Thus, Z is homogeneous for f.

We will use the following corollary.

Corollary 5.1.5 $\beth_{\alpha+n}^+ \to (\beth_\alpha^+)_{\beth_\alpha}^{n+1}$.

Proof This follows from Erdös–Rado because $\beth_{\alpha+n} = \beth_n(\beth_\alpha)$.

5.2   Order Indiscernibles

Let $\mathcal{M}$ be an $\mathcal{L}$-structure.

Definition 5.2.1 Let I be an infinite set and suppose that $X = \{x_i : i \in I\}$ is a set of distinct elements of $\mathcal{M}$. We say that X is an *indiscernible set* if whenever $i_1, \ldots, i_m$ and $j_1, \ldots, j_m$ are two sequences of m distinct elements of I, then $\mathcal{M} \models \phi(x_{i_1}, \ldots, x_{i_m}) \leftrightarrow \phi(x_{j_1}, \ldots, x_{j_m})$.

For example, suppose that F is an algebraically closed field of infinite transcendence degree and $x_1, x_2, \ldots$ is an infinite algebraically independent set. For any two sequence $i_1, \ldots, i_m$, $j_1, \ldots, j_m$ as above, there is an automorphism σ of F with $\sigma(x_{i_k}) = x_{j_k}$ for $k = 1, \ldots, m$. It follows that $x_1, x_2, \ldots$ is an infinite set of indiscernibles.

Unfortunately, many structures have no infinite sets of indiscernibles. For example if $(A, <)$ is an infinite linear order, then because we cannot have $a < b$ and $b < a$ there is no set of indiscernibles of size 2. Remarkably, this is the only obstruction.

Definition 5.2.2 Let $(I, <)$ be an ordered set, and let $(x_i : i \in I)$ be a sequence of distinct elements of M. We say that $(x_i : i \in I)$ is a sequence of *order indiscernibles* if whenever $i_1 < i_2 < \ldots < i_m$ and $j_1 < \ldots < j_m$ are two increasing sequences from I, then $\mathcal{M} \models \phi(x_{i_1}, \ldots, x_{i_m}) \leftrightarrow \phi(x_{j_1}, \ldots, x_{j_m})$.

For example, in $(\mathbb{Q}, <)$, by quantifier elimination, if $x_1 < \ldots < x_m$ and $y_1 < \ldots < y_m$, then $\mathbb{Q} \models \phi(\overline{x}) \leftrightarrow \phi(\overline{y})$ for all ϕ. Thus $\mathbb{Q}$, itself is a sequence of order indiscernibles.

We can always find models with infinite sequences of order indiscernibles.

Theorem 5.2.3 *Let T be a theory with infinite models. For any infinite linear order $(I, <)$, there is $\mathcal{M} \models T$ containing $(x_i : i \in I)$, a sequence of order indiscernibles.*

Proof Let $\mathcal{L}^* = \mathcal{L} \cup \{c_i : i \in I\}$. Let Γ be the union of
- T;
- $c_i \neq c_j$ for $i, j \in I$ with $i \neq j$;
- $\phi(c_{i_1}, \ldots, c_{i_m}) \rightarrow \phi(c_{j_1}, \ldots, c_{j_m})$ for all $\mathcal{L}$-formulas $\phi(\overline{v})$, where $i_1 < \ldots < i_m$ and $j_1 < \ldots < j_m$ are increasing sequences from I.

If $\mathcal{M} \models \Gamma$, then $(c_i^{\mathcal{M}} : i \in I)$ is an infinite sequence of order indiscernibles. Thus, it suffices to show that Γ is satisfiable. Suppose that $\Delta \subset \Gamma$ is finite. Let I_0 be the finite subset of I such that if c_i occurs in Δ, then $i \in I_0$. Let $\phi_1, \ldots, \phi_m$ be the formulas such that Δ asserts indiscernibility with respect to the formula ϕ_i, $i \leq m$. Let $v_1, \ldots, v_n$ be the free variables from $\phi_1, \ldots, \phi_m$.

Let $\mathcal{M}$ be an infinite model of T. Fix $<$ any linear order of $\mathcal{M}$. We will define a partition $F : [M]^n \rightarrow \mathcal{P}(\{1, \ldots m\})$. If $A = \{a_1, \ldots, a_n\}$ where $a_1 < \ldots < a_n$, then

$$F(A) = \{i : \mathcal{M} \models \phi_i(a_1, \ldots, a_n)\}.$$

Because F partitions $[M]^n$ into at most 2^m sets, we can find an infinite $X \subseteq M$ homogeneous for F. Let $\eta \subseteq \{1, \ldots, m\}$ such that $F(A) = \eta$ for $A \in [X]^n$.

Suppose that I_0 is a finite subset of I. Choose $(x_i : i \in I_0)$ such that each $x_i \in X$ and such that $x_i < x_j$ if $i < j$. If $i_1 < \ldots < i_n$ and $j_1 < \ldots < j_n$, then

$$\mathcal{M} \models \phi_k(x_{i_1}, \ldots, x_{i_n}) \Leftrightarrow k \in \eta \Leftrightarrow \mathcal{M} \models \phi_k(x_{j_1}, \ldots, x_{j_n}).$$

If we interpret c_i as x_i for $i \in I_0$, then we make $\mathcal{M}$ a model of Δ. Because Γ is finitely satisfiable, it is satisfiable.

If $(x_i : i \in I)$ is any sequence of order indiscernibles in M, we can order $X = \{x_i : i \in I\}$ by $x_i < x_j$ if $i < j$. In this way, we frequently identify X and I.

Suppose that $\psi(x, y)$ is a formula in the language such that in some $\mathcal{M} \models T$, ψ linearly orders an infinite set Y. When we did the construction above, we could add the condition that $\psi(c_i, c_j)$ for $i < j$. We would then restrict the partition to $[Y]^m$ and let the ordering $<$ be the ordering determined by ψ. In this way, we would get an infinite sequence of indiscernibles $(x_i : i \in I)$ such that $\psi(x_i, x_j)$ if and only if $i < j$. In this case, ψ is the ordering of the indiscernible sequence.

Ehrenfeucht–Mostowski Models

Suppose that our theory has built-in Skolem functions. Then, when we have a model containing an infinite sequence of order indiscernibles, we can form the elementary submodel generated by the indiscernibles. We can use properties of the indiscernible set to deduce properties of the elementary submodel. For example, automorphisms of the indiscernibles will induce automorphisms of the elementary submodel. If T does not have built in Skolem functions, we will still get interesting information when we study skolemizations of T.

Let T be an $\mathcal{L}$-theory. By Lemma 2.3.6 we can find $\mathcal{L}^* \supseteq \mathcal{L}$, and $T^* \supseteq T$ an $\mathcal{L}^*$-theory with built-in Skolem functions, such that if $\mathcal{M}$ is any model of T, we can interpret the symbols of $\mathcal{L}^*$ such that $\mathcal{M} \models T^*$. Note that if I is a sequence of order indiscernibles for $\mathcal{L}^*$, then I is also a sequence of order indiscernibles for $\mathcal{L}$.

If $\mathcal{M} \models T^*$ and $X \subseteq M$, let $\mathcal{H}(X)$ be the $\mathcal{L}^*$-substructure of $\mathcal{M}$ generated by X. We call $\mathcal{H}(X)$ the *Skolem hull* of X. Because X has built in Skolem functions, $\mathcal{H}(X) \prec \mathcal{M}$. Models built as Skolem hulls of sequences of order indiscernibles are called *Ehrenfeucht–Mostowski models*.

If I is an infinite set of order indiscernibles, then order-preserving permutations of I induce automorphisms of $\mathcal{H}(I)$.

Lemma 5.2.4 *Suppose that T^* is an $\mathcal{L}^*$-theory with built-in Skolem functions. Let $\mathcal{M} \models T^*$. Let $I \subseteq M$ be an infinite sequence of order indiscernibles. Suppose that $\tau : I \to I$ is an order-preserving permutation. Then, there is an automorphism $\sigma : \mathcal{H}(I) \to \mathcal{H}(I)$ extending τ.*

Proof For each element $a \in \mathcal{H}(I)$, there is a Skolem term t and $x_1 < x_2 < \ldots < x_n \in I$ such that $a = t(x_1, \ldots, x_n)$. Let $\sigma(a) = t(\tau(x_1), \ldots, \tau(x_n))$.

We first show that σ is well-defined. Suppose that there is a second Skolem term s such that $a = s(x_1, \ldots, x_n)$. Because

$$\mathcal{M} \models t(x_1, \ldots, x_n) = s(x_1, \ldots, x_n)$$

and τ is order-preserving,

$$\mathcal{M} \models t(\tau(x_1), \ldots, \tau(x_n)) = s(\tau(x_1), \ldots, \tau(x_n)).$$

Thus σ is well defined.

We must show that σ is an automorphism. If $a = t(\overline{x})$ and $b = t(\tau^{-1}(\overline{x}))$, then $\sigma(b) = a$ so σ is surjective.

Let $\phi(v_1, \ldots, v_m)$ be any $\mathcal{L}^*$-formula, and let $a_1, \ldots, a_m \in \mathcal{H}(I)$. There are terms $t_1, \ldots, t_m$ and $\overline{x} \in I$ such that $a_i = t_i(\overline{x})$. By indiscernibility,

$$
\begin{aligned}
\mathcal{M} \models \phi(a_1, \ldots, a_m) \quad &\Leftrightarrow \quad \mathcal{M} \models \phi(t_1(\overline{x}), \ldots, t_m(\overline{x})) \\
&\Leftrightarrow \quad \mathcal{M} \models \phi(t_1(\tau(\overline{x})), \ldots, t_m(\tau(\overline{x}))) \\
&\Leftrightarrow \quad \mathcal{M} \models \phi(\sigma(a_1), \ldots, \sigma(a_m)).
\end{aligned}
$$

Thus, σ is an automorphism.

Lemma 5.2.4 shows that it would be useful to find order indiscernibles where there are many order-preserving permutations. Indeed, once we have an infinite sequence of order indiscernibles, we can find them of any given order type.

Let $X = (x_i : i \in I)$ be a sequence of order indiscernibles in $\mathcal{M}$. Let

$$\mathrm{tp}(I) = \{\phi(v_1, \ldots, v_n) : \mathcal{M} \models \phi(x_{i_1}, \ldots, x_{i_n}), i_1 < \ldots < i_n \in I, n < \omega\}.$$

We call $\mathrm{tp}(X)$ the *type of the indiscernibles.*

Lemma 5.2.5 *Let T^* be an $\mathcal{L}^*$-theory with built-in Skolem functions. Suppose that $X = (x_i : i \in I)$ is an infinite sequence of order indiscernibles in $\mathcal{M} \models T^*$. If $(J, <)$ is any infinite ordered set, we can find $\mathcal{N} \models T$ containing a sequence of order indiscernibles $Y = (y_j : j \in J)$ and $\mathrm{tp}(X) = \mathrm{tp}(Y)$.*

Proof Add to $\mathcal{L}^*$ constant symbols c_j for $j \in J$ and let $\Gamma = T^* \cup \{c_i \neq c_j : i, j \in J, i \neq j\} \cup \{\phi(c_{i_1}, \ldots, c_{i_m}) : i_1 < \ldots < i_m \in J$ and $\phi \in \mathrm{tp}(X)\}$.

If Δ is a finite subset of Γ, then by choosing elements of X we can make $\mathcal{M}$ a model of Δ. Thus, Γ is satisfiable.

If $\mathcal{N} \models \Gamma$, then the interpretation of the $(c_j : j \in J)$ is the desired indiscernible sequence.

Lemma 5.2.6 *Suppose that T^* is an $\mathcal{L}^*$-theory with built-in Skolem functions. If I is a sequence of order indiscernibles in $\mathcal{M} \models T^*$ and J is a sequence of order indiscernibles in $\mathcal{N} \models T^*$ with $\mathrm{tp}(I) = \mathrm{tp}(J)$, then any order-preserving map $\tau : I \to J$ extends to an elementary embedding $\sigma : \mathcal{H}(I) \to \mathcal{H}(J)$.*

Proof If $a = t(x_1, \ldots, x_n)$ for t a term and $x_1, \ldots, x_n \in I$ we let $\sigma(a) = t(\tau(x_1), \ldots, \tau(x_n))$. We then argue as in Lemma 5.2.4 that this map is well-defined and elementary.

We give several applications of this method.

Corollary 5.2.7 *Let T be an $\mathcal{L}$-theory with infinite models. For any $\kappa \geq |\mathcal{L}| + \aleph_0$, there is $\mathcal{N} \models T$ of cardinality κ with 2^κ automorphisms.*

Proof Let $\mathcal{L}^*$ and T^* be as above. We can find $\mathcal{M} \models T^*$ containing an infinite sequence of order indiscernibles I.

Claim There is a linear order $(X, <)$ of size κ with 2^κ order-preserving permutations.

Let $X = \kappa \times \mathbb{Q}$ with the lexicographic ordering $(\alpha, q) < (\beta, r)$ if $\alpha < \beta$ or $\alpha = \beta$ and $q < r$. For each $A \subseteq \kappa$ let σ_A be the order-preserving permutation

$$\sigma_A((\alpha, q)) = \begin{cases} (\alpha, q) & \text{if } \alpha \in A \\ (\alpha, q+1) & \text{if } \alpha \notin A \end{cases}.$$

Clearly, $\sigma_A = \sigma_B$ if and only if $A = B$. Thus, there are 2^κ order-preserving permutations of X.

By Lemma 5.2.5, we can find $\mathcal{N} \models T^*$ containing J a sequence of order indiscernibles of order type $(X, <)$. By Lemma 5.2.4, each order preserving permutation of the indiscernibles induces an automorphism of $\mathcal{H}(J)$. Thus, $\mathcal{H}(J)$ has 2^κ automorphisms and $|\mathcal{H}(J)| = \kappa$. When viewed as an $\mathcal{L}$-structure $\mathcal{N}$ still has 2^κ automorphisms.

Indiscernibles can be used to build large models omitting types.

Corollary 5.2.8 *Suppose that T^* is an $\mathcal{L}^*$-theory with built in Skolem functions, $\mathcal{M} \models T^*$, $\mathcal{M}$ omits p— a type over $\emptyset$, and $\mathcal{M}$ contains an infinite sequence of order indiscernibles I. There are arbitrarily large models of T^* omitting p.*

Proof Let $\kappa \geq \aleph_0$. By Lemma 5.2.5, we can find $\mathcal{N} \models T^*$ containing a sequence of order indiscernibles J with $|J| \geq \kappa$ and $\mathrm{tp}(I) = \mathrm{tp}(J)$. Then $|\mathcal{H}(J)| \geq \kappa$. Suppose that $(a_1, \ldots, a_n) \in \mathcal{H}(J)$ realizes p. Let $a_i = t_i(x_1, \ldots, x_m)$, where t_i is a Skolem term, $x_1 < \ldots < x_m$, and each $x_i \in J$. If $y_1 < \ldots < y_m$ is an increasing sequence in I, then, because $t(I) = t(J)$,

$$\mathcal{M} \models \phi(t_1(\overline{y}), \ldots, t_n(\overline{y})) \Leftrightarrow \mathcal{N} \models \phi(a_1, \ldots, a_n).$$

Thus, $(t_1(\overline{y}), \ldots, t_n(\overline{y}))$ realizes $p \in \mathcal{M}$, a contradiction.

If we are careful about the order type of our sequence of indiscernibles, we can also omit types over sets of parameters.

Theorem 5.2.9 *Let $\mathcal{L}$ be countable and T be an $\mathcal{L}$-theory with infinite models. For all $\kappa \geq \aleph_0$, there is $\mathcal{M} \models T^*$ with $|M| = \kappa$ such that if $A \subseteq M$, then $\mathcal{M}$ realizes at most $|A| + \aleph_0$ types in $S_n^{\mathcal{M}}(A)$.*

Proof For notational simplicity we assume that $n = 1$. This is no loss of generality. Let $\mathcal{L}^*$ and T^* be as above. Let $\mathcal{M} \models T$ be the Skolem hull of a sequence of order indiscernibles I of order type $(\kappa, <)$. Then $|M| = \kappa$.

Let $A \subseteq M$. For each a in A, there is a term t_a and $\overline{x}_a$ a sequence from I such that $a = t_a(\overline{x}_a)$. Let $X = \{x \in I : x \text{ occurs in some } \overline{x}_a\}$. Then $|X| \leq |A| + \aleph_0$.

If $y_1 < \ldots < y_n$ and $z_1 < \ldots < z_n$, we say that $\overline{y} \sim_X \overline{z}$ if $y_i < x$ if and only if, for all $x \in X$, $z_i < x$ and $y_i = x$ if and only if $z_i = x$ for $i = 1, \ldots, n$.

Claim If $\overline{y} \sim_X \overline{z}$ and t is a Skolem term, then $t(\overline{y})$ and $t(\overline{z})$ realize the same type in $S_1^{\mathcal{M}}(A)$.

Let $a_1, \ldots, a_m \in A$. Because $\overline{y}$ and $\overline{z}$ are in the same position in the ordering with respect to X, by indiscernibility,

$$\mathcal{M} \models \phi(t(\overline{y}), a_1, \ldots, a_m) \quad \Leftrightarrow \quad \mathcal{M} \models \phi(t(\overline{y}), t_{a_1}(\overline{x}_{a_1}), \ldots, t_{a_m}(\overline{x}_{a_m}))$$
$$\Leftrightarrow \quad \mathcal{M} \models \phi(t(\overline{z}), t_{a_1}(\overline{x}_{a_1}), \ldots, t_{a_m}(\overline{x}_{a_m}))$$
$$\Leftrightarrow \quad \mathcal{M} \models \phi(t(\overline{z}), a_1, \ldots, a_m).$$

It suffices to show that $|I^n/\sim_X| \leq |A| + \aleph_0$. For $y \in I \setminus X$, let $C_y = \{x \in X : x < y\}$. Then, $\overline{y} \sim_X \overline{z}$ if and only if for each i:

i) if $y_i \in X$, then $y_i = z_i$; and

ii) if $y_i \notin X$, then $z_i \notin X$ and $C_{y_i} = C_{z_i}$.

Because I is well-ordered, $C_y = C_z$ if and only if $C_y = C_z = \emptyset$ or

$$\inf\{i \in I : i > C_y\} = \inf\{i \in I : i > C_z\}.$$

In particular, there are at most $|X| + 1$ possible cuts C_y. It follows that $|I^n/\sim_X| \leq |A| + \aleph_0$ and $\mathcal{M}$ realizes at most $|A| + \aleph_0$ types over A.

From Theorem 5.2.9, we get crucial information about uncountably categorical theories.

Corollary 5.2.10 *Let T be a complete theory in a countable language with infinite models, and let $\kappa \geq \aleph_1$. If T is κ-categorical, then T is ω-stable.*

Proof If T is not ω-stable, then we can find a countable $\mathcal{M} \models T$ with $A \subseteq M$ such that $|S_n^{\mathcal{M}}(A)| > \aleph_0$. By compactness, we can find $\mathcal{M} \prec \mathcal{N}_0$ of cardinality κ realizing uncountably many types in $S_n^{\mathcal{M}}(A)$. By Theorem 5.2.9, we can find $\mathcal{N}_1 \models T$ of cardinality κ such that for all $B \subset M$ if $|B| = \aleph_0$, then $\mathcal{N}_1$ realizes at most $\aleph_0$ types over B. Then, $\mathcal{N}_0 \not\cong \mathcal{N}_1$, contradicting κ-categoricity.

Combining Corollary 5.2.10 with Theorem 4.3.41 allows us to extend Corollary 4.3.39.

Corollary 5.2.11 *Let T be a complete theory in a countable language with infinite models. If $\kappa \geq \aleph_1$ and T is κ-categorical, then T has no Vaughtian pairs and hence no (κ, λ)-models for $\kappa > \lambda \geq \aleph_0$.*

Proof Because T is κ-categorical, T is ω-stable. If there is a Vaughtian pair, then by Theorem 4.3.34 there is an $(\aleph_1, \aleph_0)$-model, and, by Theorem 4.3.41, a $(\kappa, \aleph_0)$-model. Because we can find a model of T of cardinality κ where every infinite definable set has cardinality κ, this is a contradiction.

In Theorem 6.1.18, we will show that Corollaries 5.2.10 and 5.2.11 characterize uncountably categorical theories.

Indiscernibles in Stable Theories

We have seen that, although it is always possible to find infinite sequences of order indiscernibles, for some theories we cannot find infinite indiscernible sets. There are, however, very important classes of theories where every infinite sequence of order indiscernibles is a set of indiscernibles. We need one combinatorial lemma.

Lemma 5.2.12 *For any infinite cardinal κ, there is a dense linear order $(A, <)$ with $B \subset A$ such that B is dense in A and $|B| \leq \kappa < |A|$.*

Proof Let $\lambda \leq \kappa$ be least such that $2^\lambda > \kappa$. Let A be the set of all functions from λ to $\mathbb{Q}$. If we order A by $f < g$ if and only if $f(\alpha) < g(\alpha)$, where α is least such that $f(\alpha) \neq g(\alpha)$, then $(A, <) \models$ DLO.

Let B be the set of sequences in A that are eventually 0. Then

$$|B| = \sup\{\mu < \lambda : 2^\mu\} \leq \kappa$$

and, for all $f, g \in X$, if $f < g$, there is $h \in Y$ such that $f < h < g$.

Theorem 5.2.13 *Suppose that $\mathcal{L}$ is a countable language, κ is an infinite cardinal, and T is a κ-stable $\mathcal{L}$-theory. If $\mathcal{M} \models T$ and $X \subseteq M$ is an infinite sequence of order indiscernibles, then X is a set of indiscernibles.*

Proof Let $\phi(v_1, \ldots, v_n)$ be an $\mathcal{L}$-formula and $x_1, \ldots, x_n$ be an increasing sequence from I such that $\mathcal{M} \models \phi(x_1, \ldots, x_n)$. Let S_n be the group of all permutations of $\{1, \ldots, n\}$. Let $\Gamma_\phi = \{\sigma \in S_n : \mathcal{M} \models \phi(x_{\sigma(1)}, \ldots, x_{\sigma(n)})\}$. To show that X is a set of indiscernibles, we must show that $\Gamma_\phi = S_n$.

Claim $\Gamma_\phi = S_n$.

Suppose not. Because every permutation is a product of transpositions we can find $\sigma \in \Gamma_\phi$ and $\tau \in S_n \setminus \Gamma_\phi$ such that $\tau = \sigma \circ \mu$ for some transposition μ. Say $\mu(y_1, \ldots, y_n) = (y_1, \ldots, y_{m-1}, y_{m+1}, y_m, y_{m+2}, \ldots, y_n)$.

Let $\psi(v_1, \ldots, v_n)$ be the formula $\phi(v_{\sigma(1)}, \ldots, v_{\sigma(n)})$. Then

$$\mathcal{M} \models \psi(x_1, \ldots, x_n)$$

but

$$\mathcal{M} \models \neg\psi(x_1, \ldots, x_{m-1}, x_{m+1}, x_m, x_{m+2}, \ldots, x_n).$$

Let $(A, <)$ and B be as in Lemma 5.2.12. We can find $\mathcal{N} \models T$ containing a sequence of order indiscernibles Y of order type $(A, <)$ with $\text{tp}(Y) = \text{tp}(X)$. Let Y_0 be the subset of Y corresponding to B. If $y_1 < \ldots < y_n$ are in Y, then

$$\mathcal{N} \models \psi(y_1, \ldots, y_n)$$

and

$$\mathcal{N} \models \neg\psi(y_1, \ldots, y_{m-1}, y_{m+1}, y_m, y_{m+2}, \ldots, y_n).$$

If $x, y \in Y$ and $x < y$ we can find $z_1, \ldots, z_{n-1}$ in Y_0 such that $z_1 < \ldots < z_{k-1} < x < z_k < y < z_{k+1} < \ldots z_{n-1}$. Then

$$\mathcal{N} \models \psi(z_1, \ldots, z_{k-1}, x, z_k, \ldots, z_{n-1})$$

but

$$\mathcal{N} \models \neg\psi(z_1, \ldots, z_{k-1}, y, z_k, \ldots, z_{n-1}).$$

Thus, any two elements of Y realize distinct 1-types over Y_0. Because $|Y_0| \leq \kappa < |Y|$, T is not κ-stable, a contradiction.

Applications of Erdös–Rado

So far, we have tried to build interesting models by controlling the order type of the sequence of indiscernibles. There are other constructions where we instead control tp(I). This is more difficult and often requires more complicated combinatorics.

Corollary 5.2.8 showed that if we can omit a type in a model of T^* containing an infinite sequence of order indiscernibles, then we can omit the type in arbitrarily large models, but it is nontrivial to find models with indiscernibles omitting a type. The Erdös–Rado Theorem provides one method for building such models.

Theorem 5.2.14 *Let T be a theory in a countable language. Suppose that for all $\alpha < \omega_1$ there is $\mathcal{M} \models T$ with $|M| > \beth_\alpha$ such that $\mathcal{M}$ omits p, a type over $\emptyset$. Then, there are arbitrarily large models of T omitting p.*

Proof For notational simplicity, we assume, without loss of generality, that p is a 1-type. Let $T^* \supseteq T$ be a skolemization of T in a language $\mathcal{L}^* \supseteq \mathcal{L}$ as in Lemma 2.3.6. By Corollary 5.2.8, it suffices to find $\mathcal{M} \models T^*$ omitting p and containing an infinite set of indiscernibles I. We will do this by building the type of the indiscernibles. Let $C = \{c_i : i < \omega\}$ be a new set of constant symbols. We build an $\mathcal{L}^* \cup C$ theory $\Sigma \supseteq T^*$ such that Σ is satisfiable and:

a) $c_i \neq c_j \in \Sigma$ for all $i \neq j$;

b) for each $\mathcal{L}^*$-formula $\phi(v_1, \ldots, v_n)$ either $\phi(c_{i_1}, \ldots, c_{i_n}) \in \Sigma$ for all $i_1 < \ldots < i_n$ or $\neg\phi(c_{i_1}, \ldots, c_{i_n}) \in \Sigma$ for all $i_1 < \ldots < i_n$;

c) if $t(v_1, \ldots, v_n)$ is a term, then there is $\phi(v) \in p$ such that $\neg\phi(t(c_{i_1}, \ldots, c_{i_n}))$ for all $i_1 < \ldots < i_n$.

Suppose that we have such a theory Σ. If $\mathcal{N} \models \Sigma$, then the interpretation of the c_i are, by a) and b), an infinite order-indiscernible sequence I. Let $\mathcal{M}$ be the Skolem hull of I. If $a = t(c_{i_1}, \ldots, c_{i_n})$, then, by c), a does not realize T, so $\mathcal{M}$ omits p.

The construction of Σ will use the following two claims. Suppose that for $\alpha < \omega_1$ we have $\mathcal{M}_\alpha \models T^*$ omitting p, $X_\alpha \subset M_\alpha$ such that $|X_\alpha| > \beth_\alpha$, and

$<_\alpha$ a linear ordering of each M_α. For notational simplicity, we will drop the subscript and refer to $<_\alpha$ as $<$.

Claim 1 Let $\phi(v_1, \ldots, v_n)$ be an $\mathcal{L}^*$-formula. There is a sequence of models $(\mathcal{M}'_\alpha : \alpha < \omega)$ with $Y_\alpha \subseteq M'_\alpha$, $|Y_\alpha| > \beth_\alpha$ such that for all $\alpha < \omega_1$ there is $\beta \geq \alpha$ such that $\mathcal{M}'_\alpha = \mathcal{M}_\beta$, $Y_\alpha \subseteq X_\beta$, and either

  i) $\mathcal{M}'_\alpha \models \phi(y_1, \ldots, y_n)$ for all $\alpha < \omega_1$ and for all $y_1 < \ldots < y_n$ in Y_α, or

  ii) $\mathcal{M}'_\alpha \models \neg\phi(y_1, \ldots, y_n)$ for all $\alpha < \omega_1$ and for all $y_1 < \ldots < y_n$ in Y_α.

Let $\mathcal{N}_\alpha = \mathcal{M}_{\alpha+n-1}$. We define $F_\alpha : [X_{\alpha+n-1}]^n \to 2$. If $A = \{a_1, \ldots, a_n\}$ where $a_1 < \ldots < a_n$, then $F_\alpha(A) = 0$ if $\mathcal{N}_\alpha \models \phi(a_1, \ldots, a_n)$ and $F_\alpha(A) = 1$ if $\mathcal{N}_\alpha \models \neg\phi(a_1, \ldots, a_n)$. By Corollary 5.1.5, we can find $Z_\alpha \subseteq X_{\alpha+n-1}$ and $i_\alpha \in \{0, 1\}$ such that $|Z_\alpha| > \beth_\alpha$ and $F_\alpha : [Z_\alpha]^n \to \{i_\alpha\}$.

Let $W_i = \{\alpha < \omega_1 : i_\alpha = i\}$ for $i = 0, 1$. If $|W_0| = \aleph_1$, let $\alpha \mapsto \delta_\alpha$ be a nondecreasing map from ω_1 into W_i. Let $\mathcal{M}'_\alpha = \mathcal{N}_{\delta_\alpha}$, and let $Y_\alpha = Z_{\delta_\alpha}$. Then, $Y_\alpha \subseteq X_{\delta_\alpha+n-1}$, $|Y_\alpha| > \beth_{\delta_\alpha} \geq \beth_\alpha$ and i) holds. Similarly, if $|W_1| = \aleph_1$, then we can find $\mathcal{M}'_\alpha$ and Y_α such that ii) holds.

Claim 2 For each Skolem term $t(v_1, \ldots, v_n)$, we can find a sequence of models $(\mathcal{M}'_\alpha : \alpha < \omega_1)$ with $Y_\alpha \subseteq M'_\alpha$, $|Y_\alpha| > \beth_\alpha$ such that for all $\alpha < \omega_1$ there is $\beta \geq \alpha$ such that $\mathcal{M}'_\alpha = \mathcal{M}_\beta$, $Y_\alpha \subseteq X_\beta$, and there is $\phi(v) \in p$ such that $\mathcal{M}'_\alpha \models \neg\phi(t(y_1, \ldots, y_n))$ for all $\alpha < \omega_1$ and all $y_1 < \ldots < y_n \in Y_\alpha$.

Let $\phi_0, \phi_1, \ldots$ list the formulas in p. Let $\mathcal{N}_\alpha = \mathcal{M}_{\alpha+n-1}$. Let $F_\alpha : [X_{\alpha+n-1}]^n \to \omega$ such that if $A = \{a_1, \ldots, a_n\}$ where $a_1 < \ldots < a_n$, then $F_\alpha(A)$ is the least i such that $\mathcal{N}_\alpha \models \neg\phi_i(t(a_1, \ldots, a_n))$. Because each $\mathcal{N}_\alpha$ omits p, this is well-defined. By Corollary 5.1.5, there is $i_\alpha < \omega$ and $Z_\alpha \subseteq X_{\alpha+n-1}$ such that $|Z_\alpha| > \beth_\alpha$ and $F_\alpha : [Z_\alpha]^n \to \{i_\alpha\}$.

As in Claim 1, we can thin this sequence to get $(\mathcal{M}'_\alpha : \alpha < \omega_1)$ and $(Y_\alpha : \alpha < \omega_1)$, as desired.

We construct Σ as the union of a chain $\Sigma_0 \subseteq \Sigma_1 \subseteq \ldots$. At each stage of the construction we will also have $((M_{i,\alpha}, X_{i,\alpha}) : \alpha < \omega_1)$ and $<$ a linear order of $\bigcup M_{i,\alpha}$ such that $X_{i,\alpha} \subseteq M_{i,\alpha}$, $|X_{i,\alpha}| > \beth_\alpha$, and for all $i < j$ and $\alpha < \omega_1$ there is $\beta \geq \alpha$ such that $\mathcal{M}_{j,\alpha} = \mathcal{M}_{i,\beta}$, $X_{i,\alpha} \subseteq X_{j,\beta}$, and if we interpret $c_0, c_1, c_2, \ldots$ as any increasing sequence in $X_{i,\alpha}$, then $\mathcal{M}_{i,\alpha} \models \Sigma_i$.

Let $\phi_0, \phi_1, \ldots$ list all $\mathcal{L}^*$-formulas, and let $t_1, t_2, \ldots$ list all Skolem terms.

<u>stage 0</u>: Let $\Sigma_0 = T^* \cup \{c_i \neq c_j : i < j < \omega\}$. For $\alpha < \omega_1$, let $\mathcal{M}_{0,\alpha} \models T$ omit p with $|M_{0,\alpha}| > \beth_\alpha$. We may interpret the Skolem functions of $\mathcal{L}^*$ in $\mathcal{M}_{0,\alpha}$ so that $\mathcal{M}_{0,\alpha} \models T^*$, let $X_{0,\alpha} = M_{0,\alpha}$.

<u>stage $s + 1 = 2i + 1$</u>: Let $\phi_i = \phi(v_1, \ldots, v_n)$. By the first claim we can find $((M_{s+1,\alpha}, X_{s+1,\alpha}) : \alpha < \omega_1)$ with $|X_{s+1,\alpha}| > \beth_\alpha$ and for each α there is $\beta \geq \alpha$ such that $\mathcal{M}_{s+1,\alpha} = \mathcal{M}_{s,\beta}$ and $X_{s+1,\alpha} \subseteq X_{s,\beta}$ such that either

  i) for all $\alpha < \omega_1$ and for all $x_1 < \ldots < x_n$, an increasing sequence from $X_{s+1,\alpha}$ $\mathcal{M}_{s+1,\alpha} \models \phi(x_1, \ldots, x_n)$, or

  ii) for all $\alpha < \omega_1$ and for all $x_1 < \ldots < x_n$, an increasing sequence from $X_{s+1,\alpha}$ $\mathcal{M}_{s+1,\alpha} \models \neg\phi(x_1, \ldots, x_n)$.

In the first case, let $\Sigma_{s+1} = \Sigma_s \cup \{\phi(c_{i_1}, \ldots, c_{i_n}) : i_1 < \ldots < i_n\}$; otherwise, let $\Sigma_{s+1} = \Sigma_s \cup \{\neg\phi(c_{i_1}, \ldots, c_{i_n}) : i_1 < \ldots < i_n\}$. By construction, for any α, if we interpret the constants c_i as any increasing sequence in $X_{s+1,\alpha}$, then $\mathcal{M}_{s+1,\alpha} \models \Sigma_{s+1}$.

stage $s + 1 = 2i + 2$: Let $t_i = t(v_1, \ldots, v_n)$. By the second claim we can find $\phi(v) \in p$ and $((\mathcal{M}_{s+1,\alpha}, X_{s+1,\alpha}) : \alpha < \omega_1)$ such that $|X_{s+1,\alpha}| > \beth_\alpha$ and for each α there is $\beta \geq \alpha$ such that $\mathcal{M}_{s+1,\alpha} = \mathcal{M}_{s,\beta}$ and $X_{s+1,\alpha} \subseteq X_{s,\beta}$ and $\mathcal{M}_{s+1,\alpha} \models \neg\phi(t(x_1, \ldots, x_n))$ for all $\alpha < \omega_1$ and all increasing sequences $x_1 < \ldots < x_n \in X_{s+1,\alpha}$.

Let $\Sigma_{s+1} = \Sigma_s \cup \{\neg\phi(t(c_{i_1}, \ldots, c_{i_n})) : i_1 < \ldots < i_n\}$. By construction, for any α, if we interpret the constants c_i as any increasing sequence in $X_{s+1,\alpha}$, then $\mathcal{M}_\alpha \models \Sigma_{s+1}$.

Because each Σ_s is satisfiable, so is Σ. Our construction ensures that the desired properties a), b), and c) hold. Thus we can find a model of T^* omitting p and containing an infinite sequence of order indiscernibles.

In Exercise 5.5.8, we show that the bounds in Theorem 5.2.14 are optimal.

We give one further application of Erdös–Rado. Suppose that $\mathcal{L}$ contains a unary predicate U. We say that $\mathcal{M}$ is a (κ, λ)-model if $|M| = \kappa$ and $|U(\mathcal{M})| = \lambda$, where $U(\mathcal{M}) = \{x \in M : \mathcal{M} \models U(x)\}$.

Theorem 5.2.15 *Let T be a theory in a countable language such that T has a $(\beth_n, \aleph_0)$-model for all $n < \omega$. Then, T has a (κ, λ)-model for all $\kappa \geq \lambda \geq \aleph_0$.*

Proof The proof will be similar to the proof of Theorem 5.2.14. Let $\mathcal{L}^*$ and T^* be a skolemization as in Lemma 2.3.6.

Let $\{c_1, c_2, \ldots\}$ be a set of new constant symbols. We will construct a satisfiable theory $\Sigma \supset T^*$ such that:

a) $c_i \neq c_j \in \Sigma$ for $i \neq j$;

b) $\exists v_1 \ldots \exists v_n \left(\bigwedge_{i \neq j} v_i \neq v_j \wedge \bigwedge_{i=1}^n U(v_i) \right) \in \Sigma$ for $n = 1, 2, \ldots$;

c) for each $\mathcal{L}^*$-formula $\phi(v_1, \ldots, v_n)$, either $\phi(c_{i_1}, \ldots, c_{i_n}) \in \Sigma$ for all $i_1 < \ldots < i_n$ or $\neg(c_{i_1}, \ldots, c_{i_n}) \notin \Sigma$ for all $i_1 < \ldots < i_n$;

d) if $t(v_1, \ldots, v_n, w_1, \ldots, w_k)$ is a term, let $\theta_t(v_1, \ldots, v_n, u_1, \ldots, u_n)$ be the formula

$$\forall w_1 \ldots \forall w_k \left(\left(\bigwedge_{i=1}^m U(w_i) \wedge U(t(\overline{v}, \overline{w})) \right) \to t(\overline{v}, \overline{w}) = t(\overline{u}, \overline{w}) \right).$$

Then, $\theta_i(c_{i_1}, \ldots, c_{i_n}, c_{j_1}, \ldots, c_{j_n}) \in \Sigma$ for $i_1 < \ldots < i_n$ and $j_1 < \ldots < j_n$.

We build Σ as a union of chains $\Sigma_0 \subset \Sigma_1 \subset \ldots$. At any stage s, we will have a sequence of models $(\mathcal{M}_{s,n} : n < \omega)$, $X_{s,n} \subseteq M_{s,n}$ and $<$ linear order of each $M_{s,n}$ such that $|X_{s,n}| > \beth_n$, for all n there is $m \geq n$ such that $\mathcal{M}_{s+1,n} = \mathcal{M}_{s,m}$, $X_{s+1,n} \subseteq X_{s,m}$, and if we interpret the c_i as any increasing sequence in $X_{s,n}$, then $\mathcal{M}_{s,n} \models \Sigma_s$.

Let $\phi_0, \phi_1, \ldots$ list all $\mathcal{L}^*$-formulas and $t_0, t_1, \ldots$ list all Skolem terms in variables v_i and w_i, $i < \omega$.

stage 0: Let Σ_0 be T^* union the axioms from a) and b) above. Let $\mathcal{M}_{0,n} \models T^*$ with $|M_{0,n}| > \beth_n$ and $|U(\mathcal{M}_{0,n})| = \aleph_0$.

stage $s + 1 = 2i + 1$: Let ϕ_i be $\phi(v_1, \ldots, v_m)$. As in the odd stages of the previous construction, we can find a sequence $((\mathcal{M}_{s+1,n}, X_{s+1,n}) : n < \omega)$ such that $|X_{s+1,n}| > \beth_n$ and for each n there is $k \geq n$ such that $\mathcal{M}_{s+1,n} = \mathcal{M}_{s,k}$ and $X_{s+1,n} \subseteq X_{s,k}$ such that either

i) for all $n < \omega$ and for all $x_1 < \ldots < x_m$, an increasing sequence from $X_{s+1,n}$, $\mathcal{M}_{s+1,n} \models \phi(x_1, \ldots, x_m)$, or

ii) for all $n < \omega$ and for all $x_1 < \ldots < x_m$, an increasing sequence from $X_{s+1,n}$, $\mathcal{M}_{s+1,n} \models \neg\phi(x_1, \ldots, x_m)$.[1]

In case i), we let $\Sigma_{s+1} = \Sigma_s \cup \{\phi(c_{i_1}, \ldots, c_{i_m}) : i_1 < \ldots < i_m\}$, and in case ii) we let $\Sigma_{s+1} = \Sigma_s \cup \{\neg\phi(c_{i_1}, \ldots, c_{i_m}) : i_1 < \ldots < i_m\}$.

stage $s + 1 = 2i + 2$: Let $t_i = t(v_1, \ldots, v_n, w_1, \ldots, w_k)$. Let $\mathcal{M}_{s+1,i} = \mathcal{M}_{s,i+n}$. Fix $\eta \notin U(\mathcal{M}_{s,i+n})$. For $x_1 < \ldots < x_n$ in $M_{s+1,n}$, let $f_{\bar{x}} : U(\mathcal{M}_{s,i+n})^k \to U(\mathcal{M}_{s,i+n}) \cup \{\eta\}$ by

$$f_{\bar{x}}(\bar{a}) = \begin{cases} t(\bar{x}, \bar{a}) & \text{if } t(\bar{x}, \bar{a}) \in U(\mathcal{M}_{s+1,n}) \\ \eta & \text{otherwise} \end{cases}.$$

Because U is countable, there are at most $2^{\aleph_0}$ such functions. Thus, $\bar{x} \mapsto f_{\bar{x}}$ partitions $[X_{s,i+n}]^k$ into $2^{\aleph_0}$ pieces. By Corollary 5.1.5, there is $X_{s+1,i} \subseteq X_{s,i+n}$ homogeneous of size greater than $\beth_i$.

Let $\Sigma_{s+1} = \Sigma_s \cup \{\theta_t(c_{i_1}, \ldots, c_{i_n}, c_{j_1}, \ldots, c_{j_n}) : i_1 < \ldots < i_n, j_1 < \ldots < j_n\}$. By construction, if we interpret $c_0, c_1, \ldots$ as an increasing sequence in $X_{s+1,i}$, then $\mathcal{M}_{s+1,i} \models \Sigma_{s+1}$.

Thus, the theory Σ that we have constructed is satisfiable. If $\mathcal{M}_0 \models \Sigma$ such that $|U(\mathcal{M}_0)| = \lambda$, the interpretation of the constants $c_0, c_1, \ldots$ in $\mathcal{M}_0$ is an infinite sequence of order indiscernibles J. We can find $\mathcal{M}_0 \prec \mathcal{M}_1$ containing an infinite sequence of order indiscernibles $I \supseteq J$ with $\mathrm{tp}(I) = \mathrm{tp}(J)$ and $|I| = \kappa$. Let $\mathcal{M} = \mathcal{H}(U(\mathcal{M}_0) \cup I)$. Then $|M| = \kappa$. Suppose that $a \in U(M)$. Then, $a = t(x_1, \ldots, x_n, \bar{b})$, where t is a Skolem term, $x_1 < \ldots < x_n \in I$, and $\bar{b} \in U(\mathcal{M}_0)$. By condition d), $a = t(y_1, \ldots, y_n, \bar{b})$ where $y_1 < \ldots < y_n \in J$. Thus $a \in M_0$. Hence, $\mathcal{M}$ is a (κ, λ)-model.

In Exercise 5.5.7 we show that the hypothesis of 5.2.15 is necessary.

---

[1] Note that, in the proof of Theorem 5.2.14, the fact that we have a sequence of models of length ω_1 is only used in the Pigeonhole argument at the end of the second claim.

5.3 A Many-Models Theorem

In this section, we give a taste of one of the main applications of indiscernibles. Let T be a complete theory in a countable language with infinite models. We know that T has at most 2^κ nonisomorphic models of cardinality κ. In a series of results, Shelah forged dividing lines between theories where there is a good structure theory for the models and theories where there is none. One of the main ideas is that theories with no good structure theory have 2^κ nonisomorphic models of size κ for all $\kappa \geq \aleph_1$.

Definition 5.3.1 We say that T is *stable* if it is λ-stable for some $\lambda \geq \aleph_0$; otherwise, we say T is *unstable*.

Stability is the first dividing line for structure vs. nonstructure. Shelah proved that if T is unstable, then T has 2^κ nonisomorphic models of cardinality κ for all $\kappa \geq \aleph_0$. Indeed, he showed that unless T is κ-stable for all $\kappa \geq 2^{\aleph_0}$, then T has the maximal number of nonisomorphic models for each uncountable cardinal κ. We will prove a special case of the first result.

Assumptions For the rest of this section, we will make the following assumptions:
- T is a complete theory in a countable language $\mathcal{L}$ with infinite models;
- there is a binary relation symbol $<$ in the language;
- there is $\mathcal{M} \models T$ containing an infinite set linearly ordered by $<$.

These simplifying assumptions are not particularly strong. In fact, if T is any unstable theory, we can find $\mathcal{M} \models T$ and $I \subset M^k$ an infinite set that is linearly ordered by some $\mathcal{L}$-formula $\phi(\overline{x}, \overline{y})$. The arguments we will give can easily be modified to work in the general unstable context.

Theorem 5.3.2 *If $\kappa \geq \aleph_1$, then there are 2^κ nonisomorphic models of T.*

To simplify the combinatorics of the proof, we will prove this only for regular $\kappa > \aleph_1$. The proof for $\kappa = \aleph_1$ is a bit harder. The proof for κ singular is much more involved. We begin with some results from combinatorial set theory that will be used in the proof.

Definition 5.3.3 Let $\kappa \geq \aleph_1$ be a regular cardinal. We say that $C \subseteq \kappa$ is *closed unbounded* if
  i) for all $\alpha < \kappa$ there is $\beta \in C$ with $\alpha < \beta$, and
  ii) if $X \subset C$ is bounded, then the least upper bound is in C (i.e., C is closed in the order topology).

For example, if $C = \{\alpha < \kappa : \alpha \text{ is a limit ordinal}\}$, then C is closed unbounded.

Lemma 5.3.4 *Let $\kappa \geq \aleph_1$ be a regular cardinal.*
  i) If C_0 and C_1 are closed unbounded subsets of κ, then $C_0 \cup C_1$ and $C_0 \cap C_1$ are closed unbounded.

ii) If $\alpha < \kappa$ and $(C_\beta : \beta < \alpha)$ is a sequence of closed unbounded subsets of κ, then $C = \bigcap_{\beta < \alpha} C_\beta$ is closed unbounded.

Proof

i) Clearly, $C_0 \cup C_1$ is unbounded. If $X \subseteq C_0 \cup C_1$, then $X \cap C_i$ is unbounded in X for some i and the least upper bound of $X \cap C_i$ is the least upper bound of X.

It is easy to see that $C_0 \cap C_1$ is closed. Suppose that $\alpha < \kappa$. We build a sequence $\alpha = \alpha_0 < \alpha_1 < \alpha_2 < \ldots$ such that $\alpha_{2i+1} \in C_0$ and $\alpha_{2i+2} \in C_1$. Because $\beta = \sup \alpha_i \in C_0 \cap C_1$, $C_0 \cap C_1$ is unbounded.

ii) If $X \subseteq C$ is bounded and γ is the least upper bound of X, then $\gamma \in C_\beta$ for all $\beta < \alpha$. Hence, $\gamma \in C$ and we need only show that C is unbounded. We will prove this by induction on α. For $\beta < \alpha$, let $D_\beta = \bigcap_{\gamma < \beta} C_\gamma$. By induction, each D_β is closed unbounded. If $\alpha = \beta + 1$, then $D_\alpha = D_\beta \cap C_\beta$ is closed unbounded by i). Suppose that α is a limit ordinal and $\delta < \kappa$. Define $f(\beta)$ for $\beta < \alpha$ such that $f(0) = \delta$ and $f(\gamma) \in D_{\gamma+1}$ with $f(\gamma) > f(\beta)$ for $\beta < \gamma$. This is possible because $D_{\gamma+1}$ is unbounded and κ is regular. Then

$$\delta = \sup_{\beta < \alpha} f(\beta) \in \bigcap_{\beta < \alpha} D_\beta = \bigcap_{\beta < \alpha} C_\beta.$$

A stronger closure property is true. Suppose that $(X_\alpha : \alpha < \kappa)$ is a sequence of subsets of κ. The *diagonal intersection* of this sequence is the set

$$\triangle X_\alpha = \left\{ \alpha < \kappa : \alpha \in \bigcap_{\beta < \alpha} X_\beta \right\}.$$

Lemma 5.3.5 *If $(C_\alpha : \alpha < \kappa)$ is a sequence of closed unbounded subsets of κ, then $\triangle C_\alpha$ is closed unbounded.*

Proof Suppose that $X \subseteq \triangle C_\alpha$ is bounded. Let $\beta = \sup X$. If $\alpha < \beta$, then $\{\gamma \in X : \alpha < \gamma\}$ is a bounded subset of C_α with supremum β. Thus, $\beta \in C_\alpha$. Because β is in C_α for all $\alpha < \beta$, $\beta \in \triangle C_\alpha$. Thus $\triangle C_\alpha$ is closed.

Let $\alpha < \kappa$. Build a sequence $\alpha = \beta_0 < \beta_1 < \beta_2 < \ldots$ where β_{i+1} is in the closed unbounded set $D_i = \bigcap_{\gamma < \beta_i} C_\gamma$. Let $\beta = \sup \beta_i$. Because

$$\beta \in \bigcap_{i < \omega} D_i = \bigcap_{\gamma < \beta} C_\gamma,$$

$\beta \in \triangle C_\alpha$. Thus, $\triangle C_\alpha$ is unbounded.

Definition 5.3.6 We say that $S \subseteq \kappa$ is *stationary* if $X \cap C \neq \emptyset$ for every closed unbounded $C \subseteq \kappa$.

If C is closed unbounded, then, by Lemma 5.3.4 i), C is stationary. For example, the set $S = \{\alpha < \kappa : \alpha$ has cofinality $\omega\}$ is a stationary set. For any closed unbounded set C, we can find a sequence $\alpha_0 < \alpha_1 < \ldots$ where each $\alpha_i \in C$. Then $\alpha = \sup \alpha_i \in C \cap S$. If $\kappa > \aleph_1$, then S is not closed. Note that if S is stationary and $S' \supseteq S$, then S' is stationary.

Definition 5.3.7 We say that $f : \kappa \to \kappa$ is *regressive* on S if $f(\alpha) < \alpha$ for all $\alpha \in S$.

If f is regressive on a stationary set, then it is constant on a stationary subset.

Lemma 5.3.8 (Fodor's Lemma) *Suppose that* $f : \kappa \to \kappa$ *is regressive on a stationary set S, then, there is $\gamma < \kappa$ such that $S \cap f^{-1}(\gamma)$ is stationary.*

Proof Suppose not. Then, for each $\alpha < \kappa$, we can find a closed unbounded set C_α such that $C_\alpha \cap S \cap f^{-1}(\alpha) = \emptyset$. By Lemma 5.3.5, $\triangle C_\alpha$ is closed unbounded; thus, there is $\beta \in S \cap \triangle C_\alpha$. Let $\gamma = f(\beta) < \beta$. Because $\beta \in \triangle C_\alpha$, $\beta \in C_\gamma$. But then $\beta \in C_\gamma \cap S \cap f^{-1}(\gamma)$, a contradiction.

Corollary 5.3.9 *If S is stationary and $S = \bigcup_{\alpha < \lambda} S_\alpha$ for some $\lambda < \kappa$, then some S_α is stationary.*

Proof Let $f : S \to \lambda$ by $f(\alpha) = $ least $\beta < \lambda$ such that $\alpha \in S_\beta$. Because $\lambda < \kappa$, $f(\alpha) < \alpha$ on the stationary set $\{\alpha \in S : \lambda < \alpha\}$. Thus, by Fodor's Lemma, f is constant on a stationary subset of S.

The following lemma of Ulam's will play an important role in our proof.

Lemma 5.3.10 *Suppose that $\kappa \geq \aleph_1$ is regular. There is a family $(S_\alpha : \alpha < \kappa)$ of disjoint stationary subsets of κ.*

Proof
case 1: $\kappa = \lambda^+$.
For each ordinal $\alpha < \kappa$, let $f_\alpha : \lambda \to \alpha$ be surjective. For $\beta < \lambda$ and $\gamma < \kappa$, let $U_{\beta,\gamma} = \{\alpha < \kappa : f_\alpha(\beta) = \gamma\}$. We think of the sets as the matrix

$$\begin{pmatrix} U_{0,0} & U_{0,1} & \ldots & U_{0,\gamma} & \ldots \\ U_{1,0} & U_{1,1} & \ldots & U_{1,\gamma} & \ldots \\ \vdots & \vdots & \ldots & \vdots & \ldots \\ U_{\beta,0} & U_{\beta,1} & \ldots & U_{\beta,\gamma} & \ldots \\ \vdots & \vdots & \ldots & \vdots & \ldots \end{pmatrix}$$

We first argue that each column contains a stationary set. Fix $\gamma < \kappa$. Let E be the stationary set $\bigcup_{\beta < \lambda} U_{\beta,\gamma} = \{\alpha < \kappa : \alpha > \gamma\}$. Let $g : E \to \lambda$ be $g(\alpha) = \inf\{\beta : \alpha \in U_{\beta,\gamma}\}$. Because $g : \kappa \to \lambda$, g is regressive on the stationary set $\{\alpha \in E : \alpha > \lambda\}$. Thus, by Fodor's Lemma, there is β such that $g^{-1}(\beta) \cap E$ is stationary. Thus, $U_{\beta,\gamma}$ is stationary.

By the Pigeonhole Principle, we can find a single $\widehat{\beta}$ such that $W = \{\gamma : U_{\widehat{\beta},\gamma}$ is stationary$\}$ has cardinality κ. But all of the sets in any fixed row are pairwise disjoint (i.e., if $\gamma \neq \delta$, then $U_{\beta,\gamma} \cap U_{\beta,\delta} = \emptyset$). Thus, $\{U_{\widehat{\beta},\gamma} : \gamma \in W\}$ is a collection of κ pairwise disjoint stationary subsets of S.

case 2: κ is inaccessible.

Recall that $\kappa = \aleph_\kappa$ (see A.17). For each regular infinite cardinal $\lambda < \kappa$, let $S_\lambda = \{\alpha < \kappa : \mathrm{cof}(\alpha) = \lambda\}$. This is a pairwise disjoint family of stationary sets.

We can now begin the proof of Theorem 5.3.2. Let T^* be a skolemization of T. By the remarks after Theorem 5.2.3, we can find a countable $\mathcal{M} \models T^*$ containing an infinite sequence of order-indiscernibles I that is linearly ordered by $<$. We will build 2^κ nonisomorphic models by stretching the indiscernible sequence I. The following lemma will allow us to distinguish models. If I is an ordered set, we say that $I = I_0 + I_1$ if $I = I_0 \cup I_1$ and $I_0 < I_1$ (i.e., $a < b$ for $a \in I_0$ and $b \in I_1$).

Lemma 5.3.11 *Let $\mathcal{N} \models T^*$ contain an infinite sequence of order indiscernibles J linearly ordered by $<$.*

i) Suppose that $J = J_0 + J_1$, where $|J_1| \geq 2$. Then, no element of J_1 is in $\mathcal{H}(J_0)$.

ii) Suppose that $J = J_0 + J_1$, where J_0 has no top element and J_1 has no bottom element. There is no $a \in \mathcal{H}(J)$ with $J_0 < a < J_1$.

Proof

i) Let $a \in J_1$. Suppose that $\bar{c}$ is a sequence from J_0 and t is a Skolem term such that $a = t(\bar{c})$. By indiscernibility, $t(\bar{c}) = b$ for all $b \in J_1$, thus $|J_1| \leq 1$.

ii) Suppose that $J_0 < a < J_1$. Let $a = t(x_1, \ldots, x_n, y_1, \ldots, y_m)$ where $x_1 < \ldots < x_n \in J_0$ and $y_1 < \ldots < y_m \in J_1$. Pick $x \in J_0$ and $y \in J_1$ with $x_n < x < y < y_1$. Because $a < J_1$, $\mathcal{N} \models t(x_1, \ldots, x_n, y_1, \ldots, y_m) < y$. By indiscernibility, $\mathcal{N} \models t(x_1, \ldots, x_n, y_1, \ldots, y_m) < x$, contradicting $J_0 < a$.

We start by building a family of 2^κ linear orders. By Lemma 5.3.10, we can find $(S_\alpha : \alpha < \kappa)$, a sequence of κ disjoint stationary subsets of κ. For $A \subset \kappa$, let $S_A = \bigcup_{\alpha \in A} S_\alpha$. If $A \neq B$, then there is S stationary such that $S \subseteq (S_A \setminus S_B) \cup (S_B \setminus S_A)$.

If X is a linear order, we let X^* be the same set with the ordering reversed. For α an ordinal, let α^* be the order type of a decreasing α-chain. If X_α is a linear order for $\alpha < \kappa$ we let $\sum X_\alpha$ be the order with universe $\{(\alpha, x) : x \in X_\alpha, \alpha < \kappa\}$ with the lexicographic order (in other words, the order obtained when we replace each $\alpha < \kappa$ with a copy of X_α).

For each $A \subseteq \kappa$, we build a linear order L^A as follows. Let

$$L_\alpha^A = \begin{cases} \omega_1^* & \text{if } \alpha \in S_A \\ \omega^* & \text{if } \alpha \notin S_A \end{cases}.$$

and let $L_A = \sum_{\alpha < \kappa} L_\alpha^A$. Clearly, L^A is a linear order of size κ. It will follow from the arguments below that $L_A \not\cong L_B$ if $A \neq B$. The reader is asked to prove this directly in Exercise 5.5.11.

For each $A \subseteq \kappa$, we build a model $\mathcal{M}^A$ that is the Skolem hull of a sequence of order indiscernibles I^A with $t(I^A) = t(I)$ and I^A order-isomorphic to L^A. We will show that $\mathcal{M}^A \not\cong \mathcal{M}^B$ for $A \neq B$. Note that although we build $\mathcal{M}^A$ as an $\mathcal{L}^*$-structure, we must show that $\mathcal{M}^A$ and $\mathcal{M}^B$ are not isomorphic as $\mathcal{L}$-structures.

Suppose that $f : \mathcal{M}^B \to \mathcal{M}^A$ is an $\mathcal{L}$-isomorphism. We let $\mathcal{M}$ be $\mathcal{M}^A$ viewed as an $\mathcal{L}$-structure. Because $|M| = \kappa$ we may assume that the underlying set of $\mathcal{M}$ is κ. Let $I = I^A$ and let $J = f(I^B)$, J is a set of $\mathcal{L}$-indiscernibles of order type L^B. There are two ways to expand $\mathcal{M}$ to make it into an $\mathcal{L}^*$-structure. Let $\mathcal{A}$ be the expansion where we take the $\mathcal{M}^A$-structure and let $\mathcal{B}$ be the expansion where we use f to expand $\mathcal{M}$ so that it is isomorphic to $\mathcal{M}^B$. The structure $\mathcal{A}$ is the Skolem hull of I, and $\mathcal{B}$ is the Skolem hull of J. For t a Skolem term we let t^A denote the interpretation of t in $\mathcal{A}$ and t^B denote the interpretation of t in $\mathcal{B}$.

We have $I = \sum_{\alpha < \kappa} I_\alpha$ and $J = \sum_{\alpha < \kappa} J_\alpha$, where each I_α and J_α has order type ω^* or ω_1^* and there is a stationary set S such that I_α and J_α have different order types for $\alpha \in S$. By Lemma 5.3.9, we may, without loss of generality, assume that $I_\alpha \cong \omega^*$ and $J_\alpha \cong \omega_1^*$ for $\alpha \in S$. We write $I_{<\alpha}$ and $J_{<\alpha}$ for $\sum_{\beta < \alpha} I_\beta$ and $\sum_{\beta < \alpha} J_\beta$, respectively.

For $\alpha < \kappa$ let $\mathcal{A}_\alpha$ be the Skolem hull of $I_{<\alpha}$ in $\mathcal{A}$ and let $\mathcal{B}_\alpha$ be the Skolem hull of $J_{<\alpha}$. If A_α and B_α are the universes of $\mathcal{A}_\alpha$ and $\mathcal{B}_\alpha$ respectively, then

$$\kappa = \bigcup_{\alpha < \kappa} A_\alpha = \bigcup_{\alpha < \kappa} B_\alpha$$

and $|A_\alpha|, |B_\alpha| < \kappa$ for all $\alpha < \kappa$.

Frequently, $A_\alpha = B_\alpha$.

Lemma 5.3.12 *The set $C = \{\alpha < \kappa : A_\alpha = B_\alpha = \alpha\}$ is closed unbounded.*

Proof We show that $C_A = \{\alpha < \kappa : A_\alpha = \alpha\}$ is closed unbounded. Similarly $\{\alpha < \kappa : B_\alpha = \alpha\}$ is closed unbounded, and C is closed unbounded because it is the intersection of these two sets.

Because C_A is clearly closed, we need only show that it is unbounded. Let $\alpha_0 < \kappa$. We build $\alpha_0 < \alpha_1 < \alpha_2 < \ldots$ such that $A_{\alpha_i} \subseteq \alpha_{i+1}$ and $\alpha_i \subseteq A_{\alpha_{i+1}}$. If $\alpha = \sup \alpha_i$, then $A_\alpha = \alpha$.

For each $\alpha < \kappa$, pick $a_\alpha \in I_\alpha$. Note that by Lemma 5.3.11 i), $a_\alpha \notin A_\alpha$. There is a Skolem term $t_\alpha(\bar{v})$ and $\bar{b}_\alpha \in J$ such that $t_\alpha^B(\bar{b}_\alpha) = a_\alpha$. We write $\bar{b}_\alpha$ as $\bar{c}_\alpha, \bar{d}_\alpha$, where $\bar{c}_\alpha \in J_{<\alpha}$ and $\bar{d}_\alpha \in J \setminus J_{<\alpha}$.

Lemma 5.3.13 *The set $D = \{\alpha < \kappa : \bar{d}_\beta \in B_\alpha$ for all $\beta < \alpha\}$ is closed unbounded.*

Proof It is easy to see that D is closed. Let $\alpha_0 < \kappa$. Build a sequence $\alpha_0 < \alpha_1 < \ldots$ such that for all $\beta < \alpha_n$, $\bar{d}_\beta \in B_{\alpha_{n+1}}$. If $\alpha = \sup \alpha_i$, then $\alpha \in D$.

Next, we make several applications of Fodor's Lemma.

Lemma 5.3.14 *i) There is a stationary $S' \subseteq S \cap C \cap D$ and a Skolem term t such that $t_\alpha = t$ for all $\alpha \in S'$.*

ii) There is $\bar{c}$ a sequence from J and a stationary $S'' \subseteq S'$ such that $\bar{c}_\alpha = \bar{c}$ for $\alpha \in S''$.

Proof

i) Because C and D are closed unbounded and S is stationary, $S \cap C \cap D$ is stationary. Because there are only countably many terms, this follows from Lemma 5.3.9.

ii) Suppose that $\bar{c}_\alpha = (c_{\alpha,1}, \ldots, c_{\alpha,m})$. Each

$$c_{\alpha,i} \in J_{<\alpha} \subseteq B_\alpha = \alpha.$$

Thus, the function $\alpha \mapsto c_{\alpha,i}$ is regressive on S'. By repeated applications of Fodor's Lemma, we find $S' \supseteq S'_1 \supseteq \ldots \supseteq S'_m$ and $c_i < \kappa$ such that $c_{\alpha,i} = c_i$ for $\alpha \in S'_i$. Let $S'' = S'_m$ and $\bar{c} = (c_1, \ldots, c_m)$.

Because there are only finitely many possible permutations of each sequence $\bar{d}_\alpha$, by one further application of Corollary 5.3.9 and permuting the variables, we may assume that each $\bar{d}_\alpha = (d_{\alpha,1}, d_{\alpha,2}, \ldots, d_{\alpha,n})$ where $d_{\alpha,1} < \ldots < d_{\alpha,n}$. By replacing S with S'', we may, without loss of generality, assume that $S \subseteq C \cap D$ and there is a Skolem term t and $\bar{c} \in J$ such that $a_\alpha = t^{\mathcal{B}}(\bar{c}, \bar{d}_\alpha)$ for all $\alpha \in S$.

Although S is not closed, it must contain a stationary set of limit points.

Lemma 5.3.15 *The set $S' = \{\alpha \in S : \alpha = \sup(\alpha \cap S)\}$ is stationary.*

Proof The set $X = \{\alpha < \kappa : \alpha = \sup(\alpha \cap S)\}$ is closed unbounded and $S' = X \cap S$.

In particular, $S' \neq \emptyset$. For the remainder of the proof, we fix $\delta \in S'$. In particular, $\delta \in S$ and δ is a limit point of elements of S.

Lemma 5.3.16 *If $\alpha \in S$ and $\alpha < \delta$, then $\bar{d}_\alpha \in J_{<\delta}$.*

Proof Because δ is a limit point of S, there is $\beta \in S$ with $\alpha < \beta < \delta$. Because $\beta \in S$, $\bar{d}_\alpha \in B_\beta$. By Lemma 5.3.11 i), $B_\beta \cap J = J_{<\beta}$. Thus $\bar{d}_\alpha \in J_{<\beta} \subset J_{<\delta}$.

Lemma 5.3.17 *Let $a \in I_\delta$. There is $x \in J_{<\delta}$ and $y \in J_\delta$ such that if $j_1, \ldots, j_n \in J$ with $x < j_1 < \ldots < j_n < y$, then $t^{\mathcal{B}}(\bar{c}, \bar{j}) < a$.*

Proof Because $\delta \in S$, $A_\delta = B_\delta$ and $a \notin B_\delta$. Let $a = s^B(x_1, \ldots, x_k, y_1, \ldots, y_l)$ where s is a Skolem term, $\bar{x} \in J_{<\delta}$, and $\bar{y} \in J \setminus J_{<\delta}$. Note that $l > 0$ because $a \notin B_\delta$. Choose $x \in J_{<\delta}$ and $y \in J_\delta$ such that $x > \sup\{\bar{c}, x_1, \ldots, x_k\}$ and $y < y_i$ for $i = 1, \ldots, l$. By indiscernibility, if $i_1 < \ldots < i_n$ and $j_1 < \ldots < j_n$ are two sequences from J with $x < i_1, j_1$ and $i_n, j_n < y$, then $t^B(\bar{c}, \bar{i}) < a$ if and only if $t^B(\bar{c}, \bar{j}) < a$.

Because δ is a limit point of S, we can find $\alpha < \delta$ with $\alpha \in S$ such that $x < d_{\alpha,1}$ and $d_{\alpha,n} < y$. But then $t^B(\bar{c}, \bar{d}_\alpha) = a_\alpha < a$ and hence $t^B(\bar{c}, \bar{j}) < a$ for all $j_1, \ldots, j_n \in J$ with $x < j_1 < \ldots < j_n < y$.

Finally, we will exploit the fact that because $\delta \in S$, $I_\delta \cong \omega^*$ and $J_\delta \cong \omega_1^*$.

Lemma 5.3.18 *i) If $j_1, \ldots, j_n \in J_\delta$ and $j_1 < \ldots < j_n$, then $t^B(\bar{c}, \bar{j}) > a_\alpha$ for $\alpha \in S$ with $\alpha < \delta$.*

ii) There are $j_1 < \ldots < j_n$ in J_δ such that $t^B(\bar{c}, \bar{j}) < a$ for all $a \in I_\delta$.

Proof

i) Because $\delta \in S$ and $\alpha < \delta$, $\bar{d}_\alpha \in B_\delta$. Because, by Lemma 5.3.11 i), $B_\delta \cap J = J_{<\delta}$, $d_{\delta,1} \notin J_{<\delta}$. Thus $d_{\alpha,n} < d_{\delta,1}$.

Because

$$a_\alpha = t^B(\bar{c}, \bar{d}_\alpha) < t^B(\bar{c}, \bar{d}_\delta),$$

by indiscernibility

$$a_\alpha = t^B(\bar{c}, \bar{d}_\alpha) < t^B(\bar{c}, \bar{j}).$$

ii) Let $z_0 > z_1 > \ldots$ be a cofinal descending sequence in I_δ. For each i, we find $x_i \in J_{<\delta}$ and $y_i \in J_\delta$ such that $t^B(\bar{c}, \bar{j}) < z_i$ for all $j_1, \ldots, j_n \in J$ with $x_i < j_1 < \ldots < j_n < y_n$. Because J_δ has order type ω_1^*, we can find $j_1, \ldots, j_n \in J_\delta$ such that $x_i < j_1 < \ldots < j_n < y_i$ for all $i < n$. Thus, $t^B(\bar{c}, \bar{j}) < a_i$ for $i = 0, 1, 2, \ldots$. Thus, $t^B(\bar{c}, \bar{j}) < a$ for all $a \in I_\delta$.

Thus, there is an element of $\mathcal{M}$ that is above all of the elements of $I_{<\delta}$ but below all of the elements of I_δ. Because $\mathcal{A}$ is the Skolem hull of I, this violates Lemma 5.3.11 ii). Thus, $\mathcal{M}^A$ and $\mathcal{M}^B$ are not isomorphic as $\mathcal{L}$-structures.

In this proof, we needed $\kappa > \aleph_1$ so we could use the ordering ω_1^* and still have $|A_\alpha| < \kappa$. More care is needed to prove the theorem when $\kappa = \aleph_1$.

5.4 An Independence Result in Arithmetic

Gödel's famous Incompleteness Theorem asserts that there are sentences ϕ in the language of arithmetic such that ϕ is true in the natural numbers but unprovable from the Peano Axioms for arithmetic. Indeed, for any consistent recursive extension T of Peano arithmetic, we can find a sentence that is independent from T. The original independent sentences were self-referential sentences that asserted their own unprovability or metamathematical sentences asserting the consistency of the theory. People wondered whether the independent statements could be made more "mathematical." In the late 1970s, Paris and Harrington [73] showed that a slight variant of the finite version of Ramsey's Theorem is true but unprovable in Peano arithmetic. The proof is an interesting application of indiscernibles.

We begin with the combinatorial statement.

Theorem 5.4.1 (Paris–Harrington Principle) *For all natural numbers n, k, m, there is a number l such that if $f : [l]^n \to k$, then there is $Y \subseteq l$ such that Y is homogeneous for f, $|Y| \geq m$, and if y_0 is the least element of Y, then $|Y| \geq y_0$.*

Proof We argue as in the proof of the finite version of Ramsey's Theorem. Suppose that there is no such l. For $l < \omega$, let $T_l = \{f : [\{0, \dots, l-1\}]^n \to k$: there is no Y homogeneous for f with $|Y| \geq m, \min Y\}$. Clearly, each T_l is finite, and if $f \in T_{l+1}$ there is a unique $g \in T_l$ such that $g \subset f$. Thus, if we order $T = \bigcup T_l$ by inclusion, we get a finite branching tree. Because each T_l is nonempty, T is an infinite finite branching tree and by König's Lemma there is $f_0 \subset f_1 \subset f_2 \subset \dots$ with $f_i \in T_i$.

Let $f = \bigcup f_i$. Then $f : [\mathbb{N}]^n \to k$. By Ramsey's Theorem, there is an infinite $X \subseteq \mathbb{N}$ homogeneous for f. Let x_1 be the least element of X, and choose $s \geq x_1, m$. Let $x_1, \dots, x_l$ be the first l-elements of X and let $l > x_l$. Then, $Y = \{x_1, \dots, x_l\}$ is homogeneous for f_s and $|Y| \geq m, \min Y$, a contradiction.

Although the proof above is only a minor variant of the proof of the finite version of Ramsey's Theorem, the use of the infinite version of Ramsey's Theorem is in this case unavoidable. We will show that the Paris–Harrington Principle cannot be proved in Peano arithmetic. The approach we give here is due to Kanamori and McAloon [49].

Definition 5.4.2 Let $X \subseteq \omega$. We say that $f : [X]^n \to \omega$ is *regressive* if $f(A) < \min A$ for all $A \in [X]^n$. We say that $Y \subseteq X$ is *min-homogeneous* for f, if whenever $A, B \in [Y]^n$ and $\min A = \min B$, then $f(A) = f(B)$.

If $a < b$, we let (a, b) and $[a, b]$ denote $\{x : a < x < b\}$ and $\{x : a \leq x \leq b\}$, respectively.

We will consider the combinatorial principle.

$(*)$ For all c, m, n, k, there is d such that if $f_1, \dots, f_k : [d]^n \to d$ are regressive, then there is $Y \subseteq [c, d]$ such that $|Y| \geq m$ and Y is min-homogeneous for each f_i.

We will show that $(*)$ is true but not provable in Peano arithmetic. We begin by giving a finite combinatorial proof that $(*)$ follows from the Paris–Harrington Principle. This proof can be formalized in Peano arithmetic. This tells us that not only is $(*)$ true but also if it is not provable in Peano arithmetic, then neither is the Paris-Harrington Principle.

Lemma 5.4.3 *For all $c, m, n, k < \omega$, there is $d < \omega$ such that if $g : [d]^n \to k$, then there is a homogeneous set $Y \subseteq (c, d)$ with $|Y| \geq m + 2n, \min Y + n + 1$.*

Proof By the Paris–Harrington Principle, there is a d such that for any partition $h : [d]^n \to k + 1$ there is a homogeneous set Z with $|Z| \geq c + m + 2n + 1, \min Z$. Given $g : [l]^n \to k$, we define $h : [l]^n \to k + 1$ by $h(\{a_1, \ldots, a_n\}) = k$ if some $a_i < n + 1$ or some $a_i \leq c$; otherwise, $h(\{a_1, \ldots, a_n\}) = h(\{a_1 + n + 1, \ldots, a_n + n + 1\})$. Let Z be a homogeneous set for h with $|Z| \geq c + m + 2n + 1, \min Z$. Because $|Z| \geq c + m + 2n + 1$, we can find $a_1, \ldots, a_n \in Z$ such that each $a_i \geq c + n + 1$. Because $h(\{a_1, \ldots, a_n\}) \neq k$, we must have $h(A) \neq k$ for all $A \in [Z]^n$. Thus, every element of Z is greater than or equal to $c + n + 1$. Let $Y = \{a - n - 1 : a \in Z\}$. Then $Y \subset (c, d)$, $|Y| = |Z| \geq c + m + 2n + 1, \min Z$. But $\min Z \geq \min Y$.

Lemma 5.4.4 *For all c, m, n, k, there is d such that, if $f_1, \ldots, f_k : [d]^n \to b$ are regressive, then there is $Y \subseteq d$ such that $|Y| \geq m$ and Y is min-homogeneous for each f_i.*

Proof By Lemma 5.4.3, there is a $d < \omega$ such that for all $g : [d]^{n+1} \to 3^k$, there is $Y \subset (c, d)$ homogeneous for g with $|Y| \geq m + n, \min Y + n + 1$.

Suppose that $f_1, \ldots, f_k : [d]^n \to d$ are regressive. For $i \leq l$, define $g_i : [d]^{n+1} \to 3$. Suppose that $A = \{a_0, \ldots, a_n\}$ where $a_0 < a_1 < \ldots < a_n$; then

$$
g_i(A) = \begin{cases} 0 & \text{if } f_i(a_0, a_1, \ldots, a_{n-1}) < f_i(a_0, a_2, \ldots, a_n) \\ 1 & \text{if } f_i(a_0, a_1, \ldots, a_{n-1}) = f_i(a_0, a_2, \ldots, a_n) \\ 2 & \text{if } f_i(a_0, a_1, \ldots, a_{n-1}) > f_i(a_0, a_2, \ldots, a_n). \end{cases}
$$

Let $g : [d]^n \to 3^k$ by $g(A) = (g_1(A), \ldots, g_l(A))$. By Lemma 5.4.3, there is $Y \subseteq (c, d)$ homogeneous for g with $|Y| \geq \min Y + n + 1, m + n$. Clearly, Y is homogeneous for each g_i. Let $y_0 < y_1 < \ldots < y_s$ list Y. For $j = 1, \ldots s - n + 1$, let $\bar{a}_j = (y_t, y_{j+1}, \ldots, y_{j+n-1})$. Because f_i is regressive $f_i(y_0, \bar{a}_j) < y_0$ for each $j < s - n + 1$. But $s + 1 = |Y| \geq y_0 + n + 1$. Thus $s - n + 1 \geq x_0 + 1$. Thus, we must have $f_i(y_0, \bar{a}_j) = f_i(y_0, \bar{a}_l)$ for some $j \neq l$. Because Y is homogeneous, the sequence $f_i(y_0, \bar{a}_1), f_i(y_0, \bar{a}_2), \ldots, f_i(y_0, \bar{a}_{s-n-2})$ is either increasing, decreasing, or constant. Because we know that at least two values are equal, they must all be equal. Thus, g_i is constantly zero on $[Y]^{n+1}$.

Let $z_1 < \ldots < z_{n-1}$ be the largest $n - 1$ elements of Y, and let $X = Y \setminus \{z_1, \ldots, z_n\}$. Because $|Y| \geq m + n$, $|X| > n$. We claim that X is min-homogeneous for each f_i. Suppose that $x_1 < x_2, \ldots < x_n$. Then

$$
\begin{aligned}
f_i(x_1, x_2, \ldots, x_{n-1}, x_n) &= f_i(x_1, x_3, \ldots, x_{n-1}, z_1) \\
&= f_i(x_1, x_4, \ldots, z_1, z_2) \\
&\ \ \vdots \\
&= f_i(x_1, z_1, \ldots, z_{n-1}).
\end{aligned}
$$

But the same argument shows that if $y_2, \ldots, y_{n-1} \in X$ with $x_1 < y_2 < \ldots < y_{n-1}$, then

$$f_i(x_1, y_2, \ldots, y_{n-1}) = f_i(x_1, z_1, \ldots, z_{n-1}) = f_i(x_1, x_2, \ldots, x_{n-1}, x_n).$$

Thus, X is min-homogeneous for each f_i.

Note that aside from appealing to the Paris–Harrington Principle in the proof of Lemma 5.4.3, the two proofs above are straightforward finite combinatorics that could easily be formalized in Peano arithmetic.

The independence proof will use a strong form of indiscernibles. Let Γ be a finite set of formulas in the language of arithmetic and $\mathcal{M}$ be a model of Peano arithmetic. We say that $I \subseteq M$ is a sequence of *diagonal indiscernibles* for Γ if whenever $\phi(u_1, \ldots, u_m, v_1, \ldots, v_n) \in \Gamma$ $x_0, \ldots, x_n, y_1, \ldots, y_n \in I$ with $x_0 < x_1 < \ldots < x_n$ and $x_0 < y_1 < \ldots < y_n$ and $a_1, \ldots, a_m < x_0$, then

$$\mathcal{M} \models \phi(\overline{a}, x_1, \ldots, x_n) \leftrightarrow \phi(\overline{a}, y_1, \ldots, y_n).$$

We first show how the combinatorial principle $(*)$ allows us to find sets of diagonal indiscernibles in the standard model $\mathbb{N}$.

Lemma 5.4.5 *For any l, m, n and formulas $\phi_1(u_1, \ldots, u_k, v_1, \ldots, v_n), \ldots, \phi_l(u_1, \ldots, u_k, v_1, \ldots, v_n)$ in the language of arithmetic, there is a set I of diagonal indiscernibles for $\phi_1, \ldots, \phi_l$ with $|I| \geq m$.*

Proof We may assume that $m > 2n$. By the Finite Ramsey Theorem, we can find w such that $w \to (m + n)_{l+1}^{2n+1}$. By $(*)$, we can find s such that whenever $f_1, \ldots, f_k : [s]^n \to s$ are regressive there is $Y \subseteq s$ with $|Y| \geq w$ and Y is min-homogeneous for each f_j. We define regressive functions $f_j : [s]^{2n+1} \to l$ for $j = 1, \ldots, k$ and a partition $g : [s]^{2n+1} \to l + 1$ as follows. Let $X = \{x_0, \ldots, x_{2n}\}$ where $x_0 < x_1 < \ldots < x_{2n} < l$. If

$$\phi_i(\overline{a}, x_1, \ldots, x_n) \leftrightarrow \phi_i(\overline{a}, x_{n+1}, \ldots, x_{2n})$$

for all $i \leq l$ and $a_1, \ldots, a_m < x_0$, then let $f(X) = 0$ and $g(X) = 0$. Otherwise, let $g(X) = i$ and $(f_1(X), \ldots, f_k(X)) = \overline{a}$ such that

$$\phi_{g(X)}(\overline{a}, x_1, \ldots, x_n) \not\leftrightarrow \phi_{g(X)}(\overline{a}, x_{n+1}, \ldots, x_{2n}).$$

Because each function f_j is regressive, there is $Y \subseteq s$ min-homogeneous for each f_j with $|Y| \geq w$. By choice of w there is $X \subseteq Y$ and $i \leq k$ such that $|X| \geq m + n$ and $g(A) = i$ for $A \in [X]^{2n+1}$.

Suppose that $i > 0$. Because $m > 2n$, $|X| > 3n$. Thus, we can find $x_0 < x_1 < \ldots < x_{3n}$ in X. Because X is min-homogeneous for each f_j, we can find $a_j < x_0$ such that

$$
\begin{aligned}
a_j &= f_j(x_0, x_1, \ldots, x_{2n}) \\
&= f_j(x_0, x_1, \ldots, x_n, x_{2n+1}, \ldots, x_{3n}) \\
&= f_j(x_0, x_{n+1}, \ldots, x_{2n}).
\end{aligned}
$$

Let $\bar{a} = (a_1, \ldots, a_k)$. But then,

$$\phi_i(\bar{a}, x_1, \ldots, x_n) \nleftrightarrow \phi_{\bar{i}}(\bar{a}, x_{n+1}, \ldots, x_{2n}),$$

$$\phi_i(\bar{a}, x_1, \ldots, x_n) \nleftrightarrow \phi_{\bar{i}}(\bar{a}, x_{2n+1}, \ldots, x_{3n})$$

and

$$\phi_{\bar{i}}(\bar{a}, x_{n+1}, \ldots, x_{2n}) \nleftrightarrow \phi_{\bar{i}}(\bar{a}, x_{2n+1}, \ldots, x_{3n}).$$

But this is impossible because at least two of the formulas must have the same truth value. Thus $i = 0$.

Let $z_1 < \ldots < z_n$ be the n-largest elements of X and let $I = X \setminus \{z_1, \ldots, z_n\}$. Then, $|I| \geq m$ and we claim that I is the desired sequence of diagonal indiscernibles. If $x_0 < x_1 < \ldots < x_n$ and $y_1 < \ldots < y_n$ are sequences from I with $x_0 < y_1$ and $a < x_0$, then for any $i \leq k$,

$$\phi_i(\bar{a}, x_1, \ldots, x_n) \leftrightarrow \phi_i(\bar{a}, z_1, \ldots, z_n)$$

and

$$\phi_i(\bar{a}, y_1, \ldots, y_n) \leftrightarrow \phi_i(\bar{a}, z_1, \ldots, z_n).$$

Thus

$$\phi_i(\bar{a}, x_1, \ldots, x_n) \leftrightarrow \phi_i(\bar{a}, y_1, \ldots, y_n)$$

and I is a set of diagonal indiscernibles.

We will look for diagonal indiscernibles for a rather simple class of formulas.

Definition 5.4.6 The set of Δ_0-formulas is the smallest set D of formulas in the language of arithmetic such that:
i) every quantifier-free formula is in D;
ii) if $\phi, \psi \in D$, then $\phi \wedge \psi$, $\phi \vee \psi$, and $\neg\phi$ are in D;
iii) if $\phi \in D$ and t is any term, then $\exists v < t \; \phi$ and $\forall v < t \; \phi$ are in D.

For example, if $\phi(x)$ is $\forall v < x \; \forall w < x \; vw \neq x$ is a Δ_0-formula defining the set of prime numbers. The next lemma is an easy induction on formulas that we leave to exercise 5.5.12.

Lemma 5.4.7 *Suppose that $\mathcal{M}$ is a model of Peano arithmetic and $\mathcal{N} \subseteq \mathcal{M}$ is an initial segment of $\mathcal{N}$ (i.e., if $a \in M$, $b \in N$, and $a < b$, then $a \in N$). If $\phi(\bar{v})$ is a Δ_0-formula and $\bar{a} \in N$, then $\mathcal{M} \models \phi(\bar{a})$ if and only if $\mathcal{N} \models \phi(\bar{a})$.*

Diagonal indiscernibles can be used to find initial segments that are models of Peano arithmetic.

Lemma 5.4.8 *Suppose that $\mathcal{M}$ is a model of Peano arithmetic and $x_0 < x_1 < \ldots$ is a sequence of diagonal indiscernibles for all Δ_0-formulas. Let $N = \{y \in M : y < x_i \text{ for some } i < \omega\}$. Then, N is closed under addition and multiplication, and if $\mathcal{N}$ is the substructure of $\mathcal{M}$ with underlying set N, then $\mathcal{N}$ is a model of Peano arithmetic.*

Proof Suppose that $i < j < k < l$ and $a < x_i$. If $a + x_j \geq x_k$, then we can find $b \leq a$ such that $b + x_j = x_k$. By indiscernibility, $b + x_j = x_l$, so $x_k = x_l$, a contradiction. Thus $a + x_j < x_k$. It follows that N is closed under addition. Indeed $x_i + x_j \leq x_k$.

Suppose that $i < j < k < l$. We claim that $ax_j < x_k$ for all $a < x_i$. If not, then, by induction, we can find $a < x_i$ such that $ax_j < x_k \leq (a+1)x_j$. By indiscernibility, $x_l \leq (a + 1)x_j$. But, adding x_j to the first two terms, we see that $(a+1)x_j < x_k + x_j$. By the remarks above, $x_k + x_j \leq x_l$. Thus, $x_l \leq (a + 1)x_j < x_l$, a contradiction. Thus $ax_j < x_k$. It follows that N is closed under multiplication.

Next, we show that truth of arbitrary formulas in N can be reduced to the truth of Δ_0-formulas in M.

Suppose that $\phi(\overline{w})$ is the formula $\exists v_1 \forall v_2 \exists v_3 \ldots \exists v_n \psi(\overline{w}, v_1, \ldots, v_n)$, where $\psi(\overline{w}, \overline{v})$ is quantifier-free. By adding dummy variables, every formula can be put in this form. Let $\overline{a} < x_i$.

Because the sequence $x_0 < x_1 < \ldots$ is unbounded in I, then $N \models \phi(\overline{a})$ if and only if $\exists i_1 > i \forall i_2 > i_1 \ldots \exists i_n > i_{n-1}$:

$$N \models \exists v_1 < x_{i_1} \forall v_2 < x_{i_2} \ldots \exists v_n < x_{i_n} \ \psi(\overline{a}, v_1, \ldots, v_n).$$

By Lemma 5.4.7, $N \models \phi(\overline{a})$ if and only if $\exists i_1 > i \ \forall i_2 > i_1 \ldots \exists i_n > i_{n-1}$:

$$M \models \exists v_1 < x_{i_1} \forall v_2 < x_{i_2} \ldots \exists v_n < x_{i_n} \ \psi(\overline{a}, v_1, \ldots, v_n).$$

By diagonal indiscernibility, $N \models \phi(\overline{a})$ if and only if

$$M \models \exists v_1 < x_{i+1} \forall v_2 < x_{i+2} \ldots \exists v_n < x_{i+n} \ \psi(\overline{a}, v_1, \ldots, v_n).$$

Next, we show that induction holds in N. Let $\phi(u, \overline{w})$ be a formula in the language of arithmetic. Suppose that $\overline{a}, b \in N$ and $N \models \phi(b, \overline{a})$. Choose i_0 such that $\overline{a}, b < x_{i_0}$. If ϕ is $\exists v_1 \forall v_2 \ldots \exists v_n \ \psi(u, \overline{w}, \overline{v})$ where ψ is Δ_0, then, by the analysis above, if $i < i_1 < \ldots < i_n$, then for $c < x_i$

$$N \models \phi(c, \overline{a}) \Leftrightarrow M \models \exists v_1 < x_{i_1} \forall v_2 < x_{i_2} \ldots \exists v_n < x_{i_n} \ \psi(c, \overline{a}, v_1, \ldots, v_n).$$

Because induction holds in M, there is a least $c < x_{i_0}$ such that $N \models \phi(c, \overline{a})$. Thus, N is a model of Peano Arithmetic.

To prove the independence of the Paris–Harrington Principle from Peano arithmetic, we will assume familiarity with formalizing finite combinatorics and syntactic manipulations in arithmetic via coding. We summarize what we will need and refer the reader to [51] §9 for more complete details.

There are formulas $S(u), l(u, v), e(u, x)$ in the language of arithmetic such that in the standard model $S(u)$ defines the set of codes for finite sequence, $l(u, v)$ if u codes a set of length v, and $e(v, u, i)$ if v is the ith element of the sequence coded by v. All basic properties of finite sets and sequences are provable in Peano arithmetic. Using these predicates, we can formalize the Paris–Harrington Principle and $(*)$ as sentences in the language of arithmetic. We can pick our coding of finite sets such that if $X \subseteq \{0, \ldots, a-1\}$, then the code for X is less than 2^{2^a}.

Next, we use some basic facts about coding syntax in the language of arithmetic. For each formula ϕ, we let $\lceil \phi \rceil$ be the Gödel code for ϕ. There is a formula $Form_0(v)$ that defines the set of Gödel codes for Δ_0-formulas, and there is a formula $Sat_0(u, v, w)$ such that $Sat_0(u, v, w)$ asserts that u is a code for a Δ_0-formula with free variables from $v_1, \ldots, v_w$, v codes a sequence $\bar{a}$ of length w, and the formula with code u holds of the sequence $\bar{a}$. We call Sat_0 a *truth-definition* for Δ_0-formulas. All basic metamathematical properties of formulas and satisfaction for Δ_0-formulas are provable in Peano arithmetic.

Theorem 5.4.9 *The combinatorial principle $(*)$ and the Paris–Harrington Principle are not provable in Peano arithmetic.*

Proof By the remarks after Lemma 5.4.4, it suffices to show that $(*)$ is unprovable. Suppose that $\mathcal{M}$ is a nonstandard model of Peano arithmetic and c is a nonstandard element of $\mathcal{M}$. Suppose that $\mathcal{M} \models (*)$. We will use Lemma 5.4.8 to construct an initial segment of $\mathcal{M}$ where $(*)$ fails.

Because the Finite Ramsey Theorem is provable in Peano arithmetic, there is a least $w \in M$ such that $\mathcal{M} \models w \rightarrow (3c + 1)_c^{2c+1}$. Let $d \in M$ be least such that if $f_1, \ldots, f_c : [d]^{2c+1} \rightarrow d$ are regressive, then there is $Y \subseteq (c, d)$ with $|Y| \geq w$ and Y min-homogeneous for each f_i.

Using the truth predicate for Δ_0-sets, we can follow the proof of Lemma 5.4.5 inside $\mathcal{M}$ and obtain $I \subset (c, d)$ with $|I| \geq c$ such that $\mathcal{M}$ believes I is a set of diagonal indiscernibles for all Δ_0-formulas from $\mathcal{M}$ with Gödel code at most c, free variables from $v_1, \ldots, v_c$, and parameter variables from $w_1, \ldots, w_c$. In particular, I is a set of diagonal indiscernibles for all standard Δ_0-formulas.

Let $x_0 < x_1 < \ldots$ be an initial segment of I, and let $\mathcal{N}$ be the initial segment of $\mathcal{M}$ with universe $N = \{y \in M : y < x_i \text{ for some } i = 1, 2, \ldots\}$. By Lemma 5.4.8, $\mathcal{N}$ is a model of Peano arithmetic. Clearly, $c \in N$ and $d \notin N$. We claim that $w \in N$. Because the finite version of Ramsey's Theorem is provable in Peano arithmetic, there is $w' \in N$ such that $\mathcal{N} \models w' \rightarrow (3c + 1)_c^{2c+1}$. Because all functions from $[w']^{2c+1} \rightarrow c$ and all subsets of w' that are coded in $\mathcal{M}$ are coded in $\mathcal{N}$, $\mathcal{M} \models w' \rightarrow (3c+1)_c^{2c+1}$. Because w was minimal, $w \leq w'$ and $w \in N$. By a similar argument, if $d' \in N$ and $\mathcal{N} \models \forall f_1, \ldots, f_c : [d']^{2c+1} \rightarrow d'$ is regressive, there is $Y \subseteq (c, d')$ min-homogeneous for each f_i with $|Y| \geq w$. Then, this is also true in $\mathcal{M}$; thus, by choice of d, $d \leq d'$. Because $d \notin N$, this is a contradiction. Thus, $(*)$ fails in $\mathcal{N}$ and $(*)$ is not provable from Peano arithmetic.

5.5 Exercises and Remarks

Exercise 5.5.1 Show that $6 \rightarrow (3)^2_2$ (i.e., if there are six people at a party, you can either find three mutual acquaintances or three mutual non-acquaintances).

Exercise 5.5.2 Let $\mathcal{L} = \{E\}$, where E is a binary relation symbol, and let T be the theory of an equivalence relation with infinitely many classes each of which is infinite. Show that in any $\mathcal{M} \models T$ we can find infinite sets of indiscernibles I_0 and I_1 such that $\operatorname{tp}(I_0) \neq \operatorname{tp}(I_1)$, but if J is any other infinite set of indiscernibles, then $\operatorname{tp}(J) = \operatorname{tp}(I_i)$ for $i = 0$ or 1.

Exercise 5.5.3 Let G be the free group on generators X. Show that X is a set of indiscernibles in G.

Exercise 5.5.4 Show that if $\mathcal{M}$ is κ-saturated, then there is $I \subseteq M$, a sequence of order indiscernibles with $|I| = \kappa$.

Exercise 5.5.5 Show that, for any infinite $\mathcal{L}$-structure $\mathcal{M}$, we can find $(\mathcal{N}_n : n < \omega)$, a descending elementary chain (i.e., $\mathcal{N}_{n+1} \prec \mathcal{N}_n$ for each n) of elementary extensions of $\mathcal{M}$, such that $\mathcal{M} = \bigcap_{n<\omega} \mathcal{N}$. [Hint: Let $\mathcal{N}_0$ be the Skolem hull of M and an infinite set of indiscernibles.]

Exercise 5.5.6 We say that a theory T has the *order property* if and only if there is a formula $\phi(v_1, \ldots, v_n, w_1, \ldots, w_n)$ and $\mathcal{M} \models T$ with $\overline{x}_1, \overline{x}_2, \ldots$ in M^n such that $\mathcal{M} \models \phi(\overline{x}_i, \overline{x}_j)$ if and only if $i < j$.

a) Show that if ϕ has the order property in T, then T is not κ-stable for any infinite κ. [Hint: Let $(A, <)$ and B be as in Lemma 5.2.12. Find $\mathcal{N} \models T$ containing $(\overline{x}_a : a \in A)$ such that $\mathcal{N} \models \phi(\overline{x}_a, \overline{x}_b)$ if and only if $a < b$. Argue as in Theorem 5.2.13 that $|S_n(\{\overline{x}_b : b \in B\})| > |B|$.]

b) Show that T has the order property if and only if there is a formula $\psi(\overline{v}, \overline{w})$ and $\mathcal{M} \models T$ with $\overline{a}_1, \overline{b}_1 \overline{a}_2, \overline{b}_n \ldots$ such that $T \models \psi(\overline{a}_i, \overline{b}_j)$ if and only if $i < j$.[Hint: ($\Rightarrow$) Let $\phi(\overline{v}_1, \overline{v}_2, \overline{w}_1, \overline{w}_2)$ be $\psi(\overline{v}_1, \overline{w}_2)$. Let $\overline{c}_i = (\overline{a}_i, \overline{b}_i)$. Show that $\phi(\overline{c}_i, \overline{c}_j)$ if and only if $i < j$. The other direction is even easier.]

Exercise 5.5.7 Let $\mathcal{L} = \{U_0, U_1, \ldots, U_n, E_1, \ldots, E_n\}$, where each U_i is unary and E_i is binary, and let T be the $\mathcal{L}$-theory:

$$\bigvee_{i=1}^{n} U_i(x) \wedge \bigwedge_{i \neq j} \neg(U_i(x) \wedge U_j(x))$$

$$\exists x_1 \ldots \exists x_m \left(\bigwedge_{i \neq j} x_i \neq x_j \wedge \bigwedge_{i=1}^{m} U_0(x_i) \right)$$

$E_i(x, y) \rightarrow (U_{i-1}(x) \wedge U_i(y))$ for $i = 1, 2, \ldots, n$,

$(U_i(x) \wedge U_i(y) \wedge \forall z(E_i(z, x) \leftrightarrow E_i(z, y))) \rightarrow x = y$, for $i = 1, 2, \ldots, n$.

For example, if X is any infinite set, let $U(M_0) = X$, $U(M_{i+1}) = \mathcal{P}(U_i)$, and $E_i^{\mathcal{M}}$ be $\in$ restricted to $U_{i-1} \times U_i$, then $\mathcal{M} \models T$. Show that if $\mathcal{M} \models T$

and $|U_0^{\mathcal{M}}| = \aleph_0$, then $|M| \leq \beth_n$. Thus, T has a $(\beth_n, \aleph_0)$-model but does not have $(\kappa, \aleph_0)$-models for arbitrarily large κ. Thus, Theorem 5.2.15 is optimal.

Exercise 5.5.8 Let $\omega \leq \alpha < \omega_1$. Let $\mathcal{L} = \{O, <, V, E, c_\beta : \beta \leq \alpha\}$, where O is unary, $<$ and V are binary, E is ternary, and c_β is a constant symbol for $\beta < \alpha$. Let T be the following $\mathcal{L}$-theory:

$<$ is a linear order of O;

$O(c_\beta)$ for $\beta \leq \alpha$;

$c_\beta < c_\delta$ for $\delta < \alpha$;

$O(x) \to x \leq c_\alpha$;

$(O(x) \wedge O(y) \wedge x < y \wedge V(z, x)) \to V(z, y)$;

$\forall x V(x, c_\alpha)$;

$\forall x \ O(x) \leftrightarrow V(x, c_0)$;

$E(x, y, z) \to (O(z) \wedge V(y, z) \wedge \exists w < z V(x, w))$;

$(O(x) \wedge \forall y < x \exists z \ y < z < x) \to \forall z (V(z, x) \leftrightarrow \exists y < x \ V(z, y))$;

$(O(x) \wedge \exists y < x \forall z < x \ z \leq y) \to \forall w \forall u \ ((V(u, x) \wedge V(w, x) \wedge \forall v(E(v, u, x) \leftrightarrow E(v, w, x))) \to u = w)$.

There is a natural model of T. Let $X_0 = \{\beta : \beta \leq \alpha\}$. For $\beta \leq \alpha$ a limit ordinal, let $X_\beta = \bigcup_{\delta < \alpha} X_\delta$, and let $X_{\beta+1} = X_\beta \cup \mathcal{P}(X_\beta)$. Let $M = X_\alpha$. Let $O^{\mathcal{M}} = X_0$ with the natural ordering, and interpret c_β as β. Let $V^{\mathcal{M}} = \{(x, \alpha) : x \in X_\alpha\}$ and let $E^{\mathcal{M}}(x, y, \alpha)$ if and only if $y \in X_\alpha$ and $x \in y$. Then, $\mathcal{M} \models T$ and $|M| = \beth_\alpha$.

Let $\Gamma(v)$ be the set of formulas $\{O(v), v \neq c_\beta : \beta \leq \alpha\}$. Show that any model of T omitting $\Gamma(v)$ has cardinality at most $\beth_\alpha$. Thus, the bound on Theorem 5.2.14 is optimal.

Exercise 5.5.9 Let $\mathcal{L}$ be a countable language and $\kappa \geq \aleph_1$ a regular cardinal. Let $\mathcal{M}$ be an $\mathcal{L}$-structure of cardinality κ. Suppose that $\mathcal{M} = \bigcup_{\alpha < \kappa} \mathcal{M}_\alpha$ where $|M_\alpha| < \kappa$. Then, $\{\alpha < \kappa : \mathcal{M}_\alpha \prec \mathcal{M}\}$ is closed unbounded.

Exercise 5.5.10 Recall from Exercise 4.5.15 that a linear order $(X, <)$ is $\aleph_1$-like if and only if $|X| = \aleph_1$ and $|\{y : y < x\}| \leq \aleph_0$ for all $x \in X$. We will show that there are $2^{\aleph_1}$ $\aleph_1$-like dense linear orders.

Let $(S_A : A \subseteq \omega_1)$ be a family of stationary subsets of ω_1 such that $S_A \triangle S_B$ contains a stationary set if $A \neq B$. Fix $(L, <)$ a countable dense linear ordering with a least element but no greatest element. For $A \subseteq \omega_1$, we define a linear order $(X^A, <)$ as follows. For $\alpha < \omega_1$, let

$$X_\alpha^A = \begin{cases} (\mathbb{Q}, <) & \text{if } \alpha \in S_A \\ (L, <) & \text{if } \alpha \notin S_A \end{cases}$$

and let $X^A = \mathbb{Q} + \sum_{\alpha < \omega_1} X_\alpha^A$. We let $X_{<\alpha}^A = \mathbb{Q} + \sum_{\beta < \alpha} X_\beta^A$. We will assume that each X^A has underlying set ω_1.

a) Show that each X^A is an $\aleph_1$-like model of DLO.

b) Show that $\{\alpha < \omega_1 : \text{the underlying set of } X^A_{<\alpha} = \alpha\}$ is closed unbounded.

c) Suppose that $f : \kappa \to \kappa$ is a bijection. Then, $\{\alpha < \kappa : f|\alpha \text{ is a bijection}$ from α onto $\alpha\}$ is closed unbounded.

d) Show that $X^A \not\cong X^B$ for $A \neq B$. [Hint: Suppose that $f : \kappa \to \kappa$ is an isomorphism. Use b) and c) to find $\alpha \in S_A \triangle S_B$ such that $X^A_{<\alpha} = X^B_{<\alpha} = \alpha$ and $f|\alpha$ is an isomorphism between $X^A_{<\alpha}$ and $X^B_{<\alpha}$. Find a contradiction.]

Exercise 5.5.11 Suppose that $\kappa > \aleph_1$ is a regular cardinal, $A, B \subseteq \kappa$ and L_A and L_B are the linear orderings constructed in the proof of 5.3.2. Show that $L_A \not\cong L_B$.

Exercise 5.5.12 Prove Lemma 5.4.7.

Exercise 5.5.13 [†] If $f : \mathbb{N} \to \mathbb{N}$, let $f^{(n)}(x)$ be defined by $f^{(0)}(x) = x$ and $f^{(n+1)}(x) = f(f^{(n)}(x))$. We define a sequence of functions $f_i : \mathbb{N} \to \mathbb{N}$ by $f_0(x) = x + 1$ and $f_{n+1}(x) = f_n^{(x)}(x)$. For example, $f_1(x) = 2x$ and $f_2(x) = 2^x x$. Each f_n is primitive recursive and, if $g : \mathbb{N} \to \mathbb{N}$, then there are $n, m \in \mathbb{N}$ such that $f_n(x) > g(x)$ for all $x > m$ (see [94]).

Consider the function $F : \mathbb{N}^2 \to \mathbb{N}$ given by $F(m, k)$ is the least l such that if $f : [l]^2 \to k$ then there is $X \subseteq l$ homogeneous for f with $|X| > m, \min X$. By the Paris–Harrington Principle, F is a total function. We will show that F grows very fast.

Fix m and k. Let $g : [f_k(m)]^2 \to k$ be defined by $g(\{x, y\}) = 0$ if $x < m$ or $y < m$, and otherwise $g(\{x, y\})$ is the least $i \geq 1$ such that $x, y \in \left[f_i^j(m), f_i^{j+1}(m) \right)$ for some j.

Suppose that X is homogeneous for g with $|X| > m$. Let $i \leq k + 1$ such that $g : [X]^2 \to \{i\}$.

a) Show that $x \geq m$ for all $x \in X$.

b) Show that there is j such that $X \subseteq \left[f_i^{(j)}(m), f_i^{(j+1)}(m) \right)$. Let $p = f_i^{(j)}(m)$.

c) Suppose that $i = 1$. Show that $|X| \leq \min X$.

d) Show that $p = f_{i-1}^{(l)}(m)$ for some l and $f_i(p) = f_{i-1}^{(l+p)}(m)$.

e) Show that if $i > 1$, then

$$X \subseteq [p, f_i(p)) = \bigcup_{j=0}^{p-1} \left[f_{i-1}^{(l+j)}(m), f_{i-1}^{(l+j+1)}(m) \right)$$

and each $\left[f_{i-1}^{(l+j)}(m), f_{i-1}^{(l+j+1)}(m) \right)$ contains at most one element of X. Conclude that $|X| \leq \min X$.

f) Conclude that $F(m, k) > f_k(m)$. Let $g(m) = F(m, m)$. Show that g majorizes all primitive recursive functions.

Remarks

Ehrenfeucht and Mostowski introduced indiscernibles and showed that Skolem hulls of models generated by indiscernibles could be used to build models with many automorphisms. Morley showed that indiscernibles could be used to build large models realizing few types and also showed that the Erdös–Rado Theorem could be used to build models with indiscernibles omitting a type. Theorem 2.11 is due to Vaught.

Shelah has proved much finer many-models theorems than the one sketched in Section 5.3. For example, if T is unstable, then, for any uncountable κ, we can find a family of 2^κ mutually nonelementarily embeddable models of T that are all $\mathcal{L}_{\infty,\omega}$-equivalent. The simplified proof in Section 5.3 is from Poizat's review [85] of Shelah. Proofs of the many-models theorems that we mentioned can be found in Shelah [92]. Hodges [40] contains the proof that unstable theories have 2^κ nonisomorphic models of cardinality κ for all regular uncountable κ (in particular, this includes the case $\kappa = \aleph_1$, which we ignored). Baldwin [7] has the proof that stable unsuperstable theories have the maximum number of models in uncountable cardinals in regular cardinals.

The proof of the independence of the Paris–Harrington Principle given is Section 5.4 is from Kanamori and McAloon [49]. The growth analysis of Exercise 5.5.13 is just the tip of the iceberg because we have only really used Paris–Harrington for partitions of pairs of sets. The sequence $f_0, f_1, \ldots$ can be extended to f_α for $\alpha < \epsilon_0$. For example, $f_\omega(n) = f_n(n)$. Any recursive function that we can prove total in Peano arithmetic is majorized by f_α for some $\alpha < \epsilon_0$. These classic results of proof theory are surveyed in [33]. Let $G(n)$ be the least number m such that whenever $g : [m]^n \to n$ there is $X \subseteq m$ homogeneous for g with $|X| > n + 1, \min X$. By the Paris–Harrington Principle, G is a total recursive function, but G majorizes f_α for all $\alpha < \epsilon_0$. This analysis was done by Ketonen and Solovay and is explained well in [36]. Kanamori and McAloon do a similar analysis for their combinatorial principle.

Indiscernibles also play a central role in the set-theoretic investigation of large cardinals. Rowbottom showed that if there is a measurable cardinal, then there are indiscernibles for Gödel's constructible universe. This idea was developed much further by Silver. These results can be found in [47] §29–30 or [48] §9.

6

ω-Stable Theories

6.1 Uncountably Categorical Theories

Throughout this chapter, T will be a complete theory in a countable language with infinite models.

We say that T is *uncountably categorical* if it is κ-categorical for some uncountable κ. We have already seen several examples of uncountably categorical theories. For example, the theory of algebraically closed fields of a fixed characteristic, the theory of $(\mathbb{Z}, s)$, and the theory of torsion-free divisible Abelian groups are κ-categorical for all uncountable κ but not $\aleph_0$-categorical. On the other hand, the theory of an infinite Abelian group where every element has order 2 is κ-categorical for all infinite cardinals.

Every uncountably categorical theory that we have examined is actually κ-categorical for all uncountable κ. Morley proved that this is true for all countable uncountably categorical theories. Morley's proof was the beginning of modern model theory.

Theorem 6.1.1 (Categoricity Theorem) *If T is κ-categorical for some uncountable κ, then T is κ-categorical for every uncountable κ.*

In Theorems 5.2.10 and 5.2.11, we proved two important facts about uncountably categorical theories.

- If $\kappa \geq \aleph_1$ and T is κ-categorical, then T is ω-stable.
- If $\kappa \geq \aleph_1$ and T is κ-categorical, then T has no Vaughtian pairs.

Baldwin and Lachlan showed that, conversely, if T is ω-stable with no Vaughtian pairs, then T is κ-categorical for all uncountable cardinals κ.

Morley's Theorem follows immediately. We will prove this in Theorem 6.1.18. We begin by considering the simplest uncountably categorical theories.

Strongly Minimal Sets

If $\mathcal{M}$ is an $\mathcal{L}$-structure and $\phi(\bar{v})$ is an $\mathcal{L}_M$-formula, we will let $\phi(\mathcal{M})$ denote the elements of M that satisfy ϕ.

Definition 6.1.2 Let $\mathcal{M}$ be an $\mathcal{L}$-structure and let $D \subseteq M^n$ be an infinite definable set. We say that D is *minimal* in $\mathcal{M}$ if for any definable $Y \subseteq D$ either Y is finite or $D \setminus Y$ is finite. If $\phi(\bar{v}, \bar{a})$ is the formula that defines D, then we also say that $\phi(\bar{v}, \bar{a})$ is minimal.

We say that D and ϕ are *strongly minimal* if ϕ is minimal in any elementary extension $\mathcal{N}$ of $\mathcal{M}$.

We say that a theory T is strongly minimal if the formula $v = v$ is strongly minimal (i.e., if $\mathcal{M} \models T$, then M is strongly minimal).

We have already seen several examples of strongly minimal theories. In particular, we have shown that DAG and ACF$_p$ are strongly minimal theories (see Corollaries 3.1.11 and 3.2.9).

Let $\mathcal{L} = \{E\}$ and consider the $\mathcal{L}$-structure $\mathcal{M}$, where E is an equivalence relation with one class of size n for $n = 1, 2, \ldots$ and no infinite classes. In this structure, $v = v$ is a minimal formula, but suppose that $\mathcal{M} \prec \mathcal{N}$ and $a \in N$ such that the equivalence class of a is infinite. Then, the formula vEa defines an infinite-coinfinite subset of the universe. Thus, the formula $v = v$ is not strongly minimal.

Let $\mathcal{M}$ be an $\mathcal{L}$-structure and $D \subseteq M$ be strongly minimal. We will consider acl$_D$, the algebraic closure relation restricted to D. Recall that b is algebraic over A if there is a formula $\phi(x, \bar{a})$ with $\bar{a} \in A$ such that $\phi(\mathcal{M}, \bar{a})$ is finite and $\phi(b, \bar{a})$. For $A \subseteq D$, we let acl$_D(A) = \{b \in D : b$ is algebraic over $A\}$. We will omit the subscript D when no confusion arises.

We have already examined algebraic closure in several strongly minimal theories. If K is an algebraically closed field and $A \subseteq K$, then acl(A) is the algebraic closure of the subfield generated by A (Proposition 3.2.15). If G is a torsion-free divisible Abelian group, then acl(A) is the $\mathbb{Q}$-vector space span of A (Exercise 3.4.5). If $\mathcal{M} \equiv (\mathbb{Z}, s)$, then acl$(A)$ is the set of points reachable from A (Exercise 3.4.3).

The following properties of algebraic closure are true for any set D (see Exercise 1.4.11).

Lemma 6.1.3 *i)* acl(acl(A)) = acl(A) $\supseteq A$.
ii) If $A \subseteq B$, then acl(A) $\subseteq$ acl(B).
iii) If $a \in$ acl(A), then $a \in$ acl(A_0) for some finite $A_0 \subseteq A$.

A more subtle property is true if D is strongly minimal.

Lemma 6.1.4 (Exchange Principle) *Suppose that $D \subset M$ is strongly minimal, $A \subseteq D$, and $a, b \in D$. If $a \in \text{acl}(A \cup \{b\}) \setminus \text{acl}(A)$, then $b \in \text{acl}(A \cup \{a\})$.*

Proof We write $\text{acl}(A, b)$ for $\text{acl}(A \cup \{b\})$.

Suppose that $a \in \text{acl}(A, b) \setminus \text{acl}(A)$. Suppose that $\mathcal{M} \models \phi(a, b)$, where ϕ is a formula with parameters from A and $|\{x \in D : \phi(x, b)\}| = n$. Let $\psi(w)$ be the formula asserting that $|\{x \in D : \phi(x, w)\}| = n$. If $\psi(w)$ defines a finite subset of D, then $b \in \text{acl}(A)$ and $a \in \text{acl}(A)$, a contradiction. Thus, $\psi(w)$ defines a cofinite subset of D.

If $\{y \in D : \phi(a, y) \wedge \psi(y)\}$ is finite, we are done (because $b \in \text{acl}(A, a)$). Thus, we assume, for purposes of contradiction, that $|D - \{y : \phi(a, y) \wedge \psi(y)\}| = l$ for some l. Let $\chi(x)$ be the formula expressing

$$|D - \{y : \phi(x, y) \wedge \psi(y)\}| = l.$$

If $\chi(x)$ defines a finite subset of D, then $a \in \text{acl}(A)$, a contradiction. Thus, $\chi(x)$ defines a cofinite set.

Choose $a_1, \ldots, a_{n+1}$ such that $\chi(a_i)$. The set $B_i = \{w \in D : \phi(a_i, w) \wedge \psi(w)\}$ is cofinite for $i = 1, \ldots, n + 1$. Choose $\widehat{b} \in \bigcap B_i$. Then, $\phi(a_i, \widehat{b})$ for each i, so $|\{x \in D : \phi(x, \widehat{b})\}| \geq n + 1$, contradicting the fact that $\psi(\widehat{b})$. ∎

In any strongly minimal set, we can define a notion of independence that generalizes linear independence in vector spaces and algebraic independence in algebraically closed fields.

We fix $\mathcal{M} \models T$ and D a strongly minimal set in $\mathcal{M}$.

Definition 6.1.5 We say that $A \subseteq D$ is *independent* if $a \notin \text{acl}(A \setminus \{a\})$ for all $a \in A$. If $C \subset D$, we say that A is *independent over C* if $a \notin \text{acl}(C \cup (A \setminus \{a\}))$ for all $a \in A$.

We will show that infinite independent sets are sets of indiscernibles.

Lemma 6.1.6 *Suppose that $\mathcal{M}, \mathcal{N} \models T$ and $\phi(v)$ is a strongly minimal formula with parameters from A, where either $A = \emptyset$ or $A \subseteq M_0$ where $\mathcal{M}_0 \models T$, $\mathcal{M}_0 \prec \mathcal{M}$, and $\mathcal{M}_0 \prec \mathcal{N}$. If $a_1, \ldots, a_n \in \phi(\mathcal{M})$ are independent over A and $b_1, \ldots, b_n \in \phi(\mathcal{N})$ are independent over A, then $\text{tp}^{\mathcal{M}}(\overline{a}/A) = \text{tp}^{\mathcal{N}}(\overline{b}/A)$.*

Proof We will assume that $\phi(v)$ has parameters from $A \subseteq M_0$ where $\mathcal{M}_0 \prec \mathcal{M}$ and $\mathcal{M}_0 \prec \mathcal{N}$, and leave the case $A = \emptyset$ until Exercise 6.6.4.

We prove this by induction on n. Assume that $n = 1$, $a \in \phi(\mathcal{M}) \setminus \text{acl}(A)$, and $b \in \phi(\mathcal{N}) \setminus \text{acl}(A)$. Let $\psi(v)$ be a formula with parameters from A. Suppose that $\mathcal{M} \models \psi(a)$. Because $a \notin \text{acl}(A)$, $\phi(\mathcal{M}) \cap \psi(\mathcal{M})$ is infinite. Because ϕ is strongly minimal, $\phi(\mathcal{M}) \setminus \psi(\mathcal{M})$ is finite. Thus there is an n such that

$$\mathcal{M} \models |\{x : \phi(x) \wedge \neg\psi(x)\}| = n.$$

Because $\mathcal{M}_0 \prec \mathcal{M}$, $\mathcal{M}_0 \prec \mathcal{N}$, and $b \notin \operatorname{acl}(A)$, $\mathcal{N} \models \psi(b)$. Thus $\operatorname{tp}^{\mathcal{M}}(a/A) = \operatorname{tp}^{\mathcal{N}}(b/A)$.

Suppose that the claim is true for n and $a_1, \ldots, a_{n+1} \in \phi(\mathcal{M})$ and $b_1, \ldots, b_{n+1} \in \phi(\mathcal{N})$ are independent sequences over A. Let $\bar{a} = (a_1, \ldots, a_n)$ and $\bar{b} = (b_1, \ldots, b_n)$. By induction, $\operatorname{tp}^{\mathcal{M}}(\bar{a}/A) = \operatorname{tp}^{\mathcal{N}}(\bar{b}/A)$. Let $\psi(\bar{w}, v)$ be a formula with parameters from A such that $\mathcal{M} \models \psi(\bar{a}, a_{n+1})$. Because $a_{n+1} \notin \operatorname{acl}(A, \bar{a})$, $\phi(\mathcal{M}) \cap \psi(\bar{a}, \mathcal{M})$ is infinite and $\phi(\mathcal{M}) \setminus \psi(\bar{a}, \mathcal{M})$ is finite. There is an n such that

$$\mathcal{M} \models |\{v : \phi(v) \wedge \neg\psi(\bar{a}, v)\}| = n.$$

Because $\mathcal{M}_0 \prec \mathcal{M}$ and $\mathcal{M}_0 \prec \mathcal{N}$ and $\operatorname{tp}^{\mathcal{M}}(\bar{a}/A) = \operatorname{tp}^{\mathcal{N}}(\bar{b}/A)$,

$$\mathcal{N} \models |\{v : \phi(v) \wedge \neg\psi(\bar{b}, v)\}| = n.$$

Because $b_{n+1} \notin \operatorname{acl}(A, \bar{b})$, $\mathcal{N} \models \psi(\bar{b}, b_{n+1})$. Thus $\operatorname{tp}^{\mathcal{M}}(\bar{a}, a_{n+1}/A) = \operatorname{tp}^{\mathcal{N}}(\bar{b}, b_{n+1}/A)$.

Corollary 6.1.7 *If $\mathcal{M}, \mathcal{N} \models T$, A, and $\phi(v)$ are as above, B is an infinite subset of $\phi(\mathcal{M})$ independent over A and C is an infinite subset of $\phi(\mathcal{N})$ independent over A, then B and C are infinite sets of indiscernibles of the same type over A.*

Therefore, cardinality is the only way to distinguish independent subsets of D.

Definition 6.1.8 We say that A is a *basis* for $Y \subseteq D$ if $A \subseteq Y$ is independent and $\operatorname{acl}(A) = \operatorname{acl}(Y)$.

Clearly, any maximal independent subset of Y is a basis for Y. Just as in vector spaces and algebraically closed fields, any two bases have the same cardinality.

Lemma 6.1.9 *Let $A, B \subseteq D$ be independent with $A \subseteq \operatorname{acl}(B)$.*

i) Suppose that $A_0 \subseteq A$, $B_0 \subseteq B$, $A_0 \cup B_0$ is a basis for $\operatorname{acl}(B)$ and $a \in A \setminus A_0$. Then, there is $b \in B_0$ such that $A_0 \cup \{a\} \cup (B_0 \setminus \{b\})$ is a basis for $\operatorname{acl}(B)$.

ii) $|A| \leq |B|$.

iii) If A and B are bases for $Y \subseteq D$, then $|A| = |B|$.

Proof

i) Let $C \subseteq B_0$ be of minimal cardinality such that $a \in \operatorname{acl}(A_0 \cup C)$. Because A is independent, $|C| \geq 1$. Let $b \in C$. By exchange, $b \in \operatorname{acl}(A_0 \cup \{a\} \cup (C \setminus \{b\}))$ and thus $\operatorname{acl}(A_0 \cup \{a\} \cup (B_0 \setminus \{b\})) = \operatorname{acl}(B)$. If $a \in \operatorname{acl}(A_0 \cup (B_0 \setminus \{b\}))$, then $b \in \operatorname{acl}(A_0 \cup (B_0 \setminus \{b\}))$, contradicting the fact that $A_0 \cup B_0$ is a basis. Thus, $A_0 \cup \{a\} \cup (B_0 \setminus \{b\})$ is independent.

ii) Suppose that B is finite. For purposes of contradiction, suppose that $|B| = n$ and $a_1, \ldots, a_{n+1}$ are distinct elements of A. Let $A_0 = \emptyset$ and

$B_0 = B$. Using i) inductively, we can find $b_1, \ldots, b_n \in B$ distinct such that $\{a_1, \ldots, a_i\} \cup (B \setminus \{b_1, \ldots, b_i\})$ is a basis for $\mathrm{acl}(B)$ for $i \leq n$. But then $\mathrm{acl}(a_1, \ldots, a_n) = \mathrm{acl}(B)$. Because $a_{n+1} \in \mathrm{acl}(B)$, this contradicts the independence of A.

If B is infinite, then for any finite $B_0 \subset B$, $A \cap \mathrm{acl}(B_0)$ is finite and

$$A \subseteq \bigcup_{B_0 \subseteq B \ \text{finite}} \mathrm{acl}(B_0).$$

Thus $|A| \leq |B|$.

iii) This is immediate from ii).

Definition 6.1.10 If $Y \subseteq D$, then the *dimension* of Y is the cardinality of a basis for Y. We let $\dim(Y)$ denote the dimension of Y.

Note that if D is uncountable, then $\dim(D) = |D|$ because our language is countable and $\mathrm{acl}(A)$ is countable for any countable $A \subseteq D$.

For strongly minimal theories, every model is determined up to isomorphism by its dimension.

Theorem 6.1.11 *Suppose T is a strongly minimal theory. If $\mathcal{M}, \mathcal{N} \models T$, then $\mathcal{M} \cong \mathcal{N}$ if and only if $\dim(\mathcal{M}) = \dim(\mathcal{N})$.*

More generally, if $\mathcal{M}, \mathcal{N}$ and ϕ are as in Lemma 6.1.6, and $\dim(\phi(\mathcal{M})) = \dim(\phi(\mathcal{N}))$, then there is a bijective partial elementary map $f : \phi(\mathcal{M}) \to \phi(\mathcal{N})$.

Proof Let B be a transcendence basis for $\phi(\mathcal{M})$ and C be a transcendence basis for $\phi(\mathcal{N})$. Because $|B| = |C|$, we can find a bijection $f : B \to C$. By Corollary 6.1.7, f is elementary. Let $I = \{g : B' \to C' : B \subseteq B' \subseteq \phi(\mathcal{M}), C \subseteq C' \subseteq \phi(\mathcal{N}), f \subseteq g \text{ partial elementary}\}$. By Zorn's Lemma, there is a maximal $g : B' \to C'$. Suppose that $b \in \phi(\mathcal{M}) \setminus B'$. Because b is algebraic over B', there is a formula $\psi(v, \overline{d})$ isolating $\mathrm{tp}^{\mathcal{M}}(b/B')$ (see Exercise 4.5.4). Because g is partial elementary, we can find $c \in \phi(\mathcal{N})$ such that $\psi(c, g(\overline{d}))$. Then, $\mathrm{tp}^{\mathcal{M}}(b/B') = \mathrm{tp}^{\mathcal{N}}(c/C')$, and we can extend g by sending b to c. This contradicts the maximality of g. Thus $\phi(\mathcal{M}) = B'$. An analogous argument shows that $C' = \phi(\mathcal{N})$.

Corollary 6.1.12 *If T is a strongly minimal theory, then T is κ-categorical for $\kappa \geq \aleph_1$ and $I(T, \aleph_0) \leq \aleph_0$.*

Proof If $\mathcal{M}$ has cardinality $\kappa \geq \aleph_1$, then any transcendence basis for $\mathcal{M}$ has cardinality κ, whereas if $|\mathcal{M}| = \aleph_0$, then $\dim(\mathcal{M}) \leq \aleph_0$.

Existence of Strongly Minimal Formulas

In ω-stable theories, we can always find minimal formulas.

Lemma 6.1.13 *Let T be ω-stable.*

i) If $\mathcal{M} \models T$, then there is a minimal formula in $\mathcal{M}$.

ii) If $\mathcal{M} \models T$ is $\aleph_0$-saturated and $\phi(\overline{v}, \overline{a})$ is a minimal formula in $\mathcal{M}$, then $\phi(\overline{v}, \overline{a})$ is strongly minimal.

Proof

i) Suppose not. We build a tree of formulas $(\phi_\sigma : \sigma \in 2^{<\omega})$ such that:

if $\sigma \subset \tau$, then $\phi_\tau \models \phi_\sigma$;

$\phi_{\sigma,i} \models \neg \phi_{\sigma,1-i}$;

$\phi_\sigma(\mathcal{M})$ is infinite.

Let $\phi_\emptyset$ be the formula $v = v$. Suppose that we have a formula ϕ_σ such that $\phi_\sigma(\mathcal{M})$ is infinite. Because ϕ_σ is not minimal, we can find a formula ψ such that $(\phi_\sigma \wedge \psi)(\mathcal{M})$ and $(\phi_\sigma \wedge \neg \psi)(\mathcal{M})$ are both infinite. Let $\phi_{\sigma,0} = \phi_\sigma \wedge \psi$, and let $\phi_{\sigma,1}$ be $\phi_\sigma \wedge \neg \psi$. As in Theorem 4.2.18 we can find a countable $A \subseteq M$ such that $|S_1^{\mathcal{M}}(A)| = 2^{\aleph_0}$, contradicting ω-stability.

ii) Suppose not. Let $\mathcal{M} \prec \mathcal{N}$, $\overline{b} \in N$ such that $\psi(\mathcal{N}, \overline{b})$ is an infinite coinfinite subset of $\phi(\mathcal{N}, \overline{a})$. Because $\mathcal{M}$ is $\aleph_0$-saturated, we can find $\overline{b}' \in M$ such that $\text{tp}^{\mathcal{M}}(\overline{a}, \overline{b}') = \text{tp}^{\mathcal{N}}(\overline{a}, \overline{b})$. Then, $\psi(\overline{v}, \overline{b}')$ defines an infinite coinfinite subset of $\phi(\mathcal{M}, \overline{a})$, a contradiction.

We will show that if there are no Vaughtian pairs, then any minimal formula is strongly minimal. The key to the proof is the following lemma showing that we can express "there exist infinitely many" in theories with no Vaughtian pairs.

Lemma 6.1.14 *Suppose that T is an $\mathcal{L}$-theory with no Vaughtian pairs. Let $\mathcal{M} \models T$, and let $\phi(v_1, \ldots, v_k, w_1, \ldots, w_m)$ be a formula with parameters from M. There is a number n such that if $\overline{a} \in M$ and $|\phi(\mathcal{M}, \overline{a})| > n$, then $\phi(\mathcal{M}, \overline{a})$ is infinite.*

Proof Suppose not. Then, for each $n \in \mathbb{N}$, we can find $\overline{a}_n$ in M such that $\phi(\mathcal{M}, \overline{a}_n)$ is a finite set of size at least n. Consider the language $\mathcal{L}^* = \mathcal{L} \cup \{U\}$ for pairs of models of T used in the proof of Lemma 4.3.37, and let $\Gamma(\overline{w}) \supset T$ be the $\mathcal{L}^*$-type asserting:

U defines a proper $\mathcal{L}$-elementary submodel;

$\bigwedge_{i=1}^m U(w_i)$;

there are infinitely many elements $\overline{v}$ such that $\phi(\overline{v}, \overline{w})$;

$\phi(\overline{v}, \overline{w}) \rightarrow \bigwedge_{i=1}^k U(v_i)$.

Let $\mathcal{N}$ be a proper elementary extension of $\mathcal{M}$. Because $\phi(\mathcal{M}, \overline{a}_n)$ is finite and $\mathcal{M} \prec \mathcal{N}$, $\phi(\mathcal{M}, \overline{a}_n) = \phi(\mathcal{N}, \overline{a}_n)$. If $\Delta \subseteq \Gamma(\overline{w})$ is finite, then by choosing n sufficiently large, $\overline{a}_n$ realizes Δ in $(\mathcal{M}, \mathcal{N})$. Thus, by the Compactness Theorem, Γ is satisfiable.

Suppose that $\overline{a}$ realizes $\Gamma(\overline{w})$ in $(\mathcal{N}', \mathcal{M}')$ where $\mathcal{M}' \models T$ and $\mathcal{N}'$ is a proper elementary extension. Then, $\phi(\mathcal{M}', \overline{a})$ is an infinite set such that $\phi(\mathcal{M}', \overline{a}) = \phi(\mathcal{N}', \overline{a})$, contradicting the fact that there are no Vaughtian pairs of models of T.

Note that if n is as in Lemma 6.1.14, then in any elementary extension $\mathcal{N}$ of $\mathcal{M}$, $|\phi(\mathcal{N},\overline{b})|$ is infinite whenever $|\phi(\mathcal{N},\overline{b})| > n$.

Corollary 6.1.15 *If T has no Vaughtian pairs, then any minimal formula is strongly minimal.*

Proof Let $\phi(\overline{v})$ be a minimal formula over $\mathcal{M} \models T$ (possibly with parameters). Suppose, for purposes of contradiction, that there is an elementary extension $\mathcal{N}$ of $\mathcal{M}$, $\overline{b} \in N$, and an $\mathcal{L}$-formula $\psi(\overline{v},\overline{w})$ such that $\psi(\mathcal{N},\overline{b})$ is an infinite coinfinite subset of $\phi(\mathcal{N})$.

By Lemma 6.1.14, there is a number n such that, for any $\mathcal{M} \prec \mathcal{N}'$ and $\overline{a} \in N'$, $\psi(\mathcal{N}',\overline{a})$ is an infinite coinfinite subset of $\phi(\mathcal{N}')$ if and only if $|\psi(\mathcal{N}',\overline{a}) \cap \phi(\mathcal{N}')| > n$ and $|\neg\psi(\mathcal{N}',\overline{a}) \cap \phi(\mathcal{N}')| > n$. But

$$\mathcal{M} \models \forall\overline{w}\left(|\psi(\mathcal{M},\overline{w}) \cap \phi(\mathcal{M})| \leq n \vee |\neg\psi(\mathcal{M},\overline{w}) \cap \phi(\mathcal{M})| \leq n\right).$$

Because this is a first-order statement, it must also be true in $\mathcal{N}$, a contradiction.

Corollary 6.1.16 *If T is ω-stable and has no Vaughtian pairs, then for any $\mathcal{M} \models T$ there is a strongly minimal formula over $\mathcal{M}$. In particular, there is a strongly minimal formula with parameters from $\mathcal{M}_0$, the prime model of T.*

The Categoricity Theorem

Our proof of categoricity will follow the argument in Theorem 6.1.11. We can find a strongly minimal formula $\phi(v)$ over the prime model. If $\mathcal{M}$ and $\mathcal{N}$ are models of T of the same uncountable cardinality, then we can find a partial elementary bijection between $\phi(\mathcal{M})$ and $\phi(\mathcal{N})$. The next lemma allows us to extend this to an isomorphism between $\mathcal{M}$ and $\mathcal{N}$.

Lemma 6.1.17 *If T has no Vaughtian pairs, $\mathcal{M} \models T$, and $X \subseteq M^n$ is infinite and definable, then no proper elementary submodel of $\mathcal{M}$ contains X. If, in addition, T is ω-stable, then $\mathcal{M}$ is prime over X.*

Proof Let $\phi(\overline{v})$ define X. If $\mathcal{N}$ is a proper elementary submodel of $\mathcal{M}$ containing X, then $X = \phi(\mathcal{M}) = \phi(\mathcal{N})$ and $(\mathcal{M},\mathcal{N})$ is a Vaughtian pair.

If T is ω-stable, then by Theorem 4.2.20, there is $\mathcal{N} \prec \mathcal{M}$, a prime model over X. Because T has no Vaughtian pairs, we must have $\mathcal{N} = \mathcal{M}$, so $\mathcal{M}$ is prime over X.

We can now prove the Baldwin–Lachlan characterization of uncountably categorical theories and deduce Morley's Categoricity Theorem.

Theorem 6.1.18 *Let T be a complete theory in a countable language with infinite models, and let κ be an uncountable cardinal. T is κ-categorical if and only if T is ω-stable and has no Vaughtian pairs.*

In particular, if T is κ-categorical for some uncountable cardinal κ, then T is λ-categorical for all uncountable cardinals λ.

Proof

($\Rightarrow$) If T is κ-categorical, then, by Corollaries 5.2.10 and 5.2.11, T is ω-stable and has no Vaughtian pairs.

($\Leftarrow$) Suppose that T is ω-stable and has no Vaughtian pairs. Because T is ω-stable, it has a prime model $\mathcal{M}_0$. By Lemma 6.1.13 and Corollary 6.1.15, there is $\phi(v)$, a strongly minimal formula with parameters from $\mathcal{M}_0$.

Suppose that $\mathcal{M}$ and $\mathcal{N}$ are models of T of cardinality $\kappa \geq \aleph_1$. We can view $\mathcal{M}$ and $\mathcal{N}$ as elementary extensions of $\mathcal{M}_0$. Then $\dim(\phi(\mathcal{M})) = \dim(\phi(\mathcal{N})) = \kappa$. By Theorem 6.1.11, there is $f : \phi(\mathcal{M}) \to \phi(\mathcal{N})$, a partial elementary bijection. By Lemma 6.1.17, $\mathcal{M}$ is prime over $\phi(\mathcal{M})$. Thus, we can extend f to an elementary $f' : \mathcal{M} \to \mathcal{N}$. But, by Lemma 6.1.17, $\mathcal{N}$ has no proper elementary submodels containing $\phi(\mathcal{N})$. Thus f' is surjective and f' is an isomorphism.

Because the Baldwin–Lachlan characterization of κ-categorical theories does not depend on κ, T is κ-categorical for some uncountable cardinal if and only if T is λ-categorical for every uncountable cardinal λ.

The proof shows that if $\mathcal{M}_0$ is the prime model of T, $\phi(v)$ is a strongly minimal formula with parameters from some finite $A \subset \mathcal{M}_0$, $\mathcal{M}, \mathcal{N} \models T$, and $\dim(\phi(\mathcal{M})/A) = \dim(\phi(\mathcal{N})/A)$, then $\mathcal{M} \cong \mathcal{N}$. Because there are only $\aleph_0$ possibilities for $\dim(\mathcal{M}/A)$, we also get an upper bound on the number of countable models.

Proposition 6.1.19 *If T is uncountably categorical, then $I(T, \aleph_0) \leq \aleph_0$.*

Note that in the analysis above we did not assert that if $\mathcal{M} \cong \mathcal{N}$, then $\dim(\phi(\mathcal{M})/A) = \dim(\phi(\mathcal{N})/A)$. This converse is true if $A = \emptyset$. If $\phi(v)$ is a strongly minimal $\mathcal{L}$-formula, then any isomorphism maps $\phi(\mathcal{M})$ onto $\phi(\mathcal{N})$. In this case, $\dim(\phi(\mathcal{M})) = \dim(\phi(\mathcal{N}))$. What values are possible for $\dim(\phi(\mathcal{M}))$?

Lemma 6.1.20 *Suppose that T is an ω-stable $\mathcal{L}$-theory and $\phi(v)$ is a strongly minimal $\mathcal{L}$-formula (with no additional parameters). Suppose that $\mathcal{M} \models T$ and $\dim(\phi(\mathcal{M})) = n < \aleph_0$. Then, for all $m \geq n$ there is $\mathcal{N} \models T$ with $\dim(\phi(\mathcal{N})) = m$.*

Proof Let $\mathcal{M}^* \models T$ be an ω-saturated elementary extension of $\mathcal{M}$. It is easy to see that $\dim(\phi(\mathcal{M}^*)) \geq \aleph_0$. Let $a_1, \ldots, a_m$ be independent elements of $\phi(\mathcal{M}^*)$. By Theorem 4.2.20, there is $\mathcal{N} \prec \mathcal{M}^*$ prime over $\overline{a}$, where every $\overline{b} \in N$ realizes an isolated type over $\overline{a}$.

In $\mathcal{M}$, $\phi(\mathcal{M})$ is infinite and $\phi(\mathcal{M})$ is the algebraic closure of n independent elements. Because $m \geq n$, $\mathrm{acl}(\overline{a})$ is infinite. We claim that $\phi(\mathcal{N}) = \mathrm{acl}(\overline{a})$. Suppose not. Let $b \in \phi(\mathcal{N}) \setminus \mathrm{acl}(\overline{a})$. There is $\theta(v, \overline{a})$ isolating $\mathrm{tp}^{\mathcal{N}}(b/\overline{a})$. Because $b \notin \mathrm{acl}(\overline{a})$, $\phi(\mathcal{N}) \setminus \theta(\mathcal{N}, \overline{a})$ is finite. Because

$\text{acl}(\overline{a})$ is infinite, there is $c \in \text{acl}(\overline{a})$ such that $\phi(c)$ and $\theta(c, \overline{a})$. But then $\text{tp}^{\mathcal{N}}(c/\overline{a}) = \text{tp}^{\mathcal{N}}(b/\overline{a})$, because $\theta(v, \overline{a})$ isolates $\text{tp}^{\mathcal{N}}(b/\overline{a})$. This is impossible because $b \notin \text{acl}(\overline{a})$. Thus, $\phi(\mathcal{N}) = \text{acl}(\overline{a})$ and $\dim(\phi(\mathcal{N})) = m$.

Thus, the set of dimensions for a strongly minimal $\mathcal{L}$-formula in a countable model is either $\{\aleph_0\}$ or $\{n, n+1, \ldots\} \cup \{\aleph_0\}$ for some n.

Corollary 6.1.21 *If T is uncountably categorical and there is a strongly minimal $\mathcal{L}$-formula, then either T is $\aleph_0$-categorical or $I(T, \aleph_0) = \aleph_0$.*

The situation is much more murky if we really need to use parameters from the prime model. Suppose that $\mathcal{M}_0$ is the prime model of T, $\mathcal{M}$ and $\mathcal{N}$ are elementary extensions of $\mathcal{M}_0$, and the strongly minimal formula $\phi(v)$ is $\psi(v, \overline{a})$, where ψ is an $\mathcal{L}$-formula and $\overline{a} \in M_0$. A priori it seems possible that $\dim(\phi(\mathcal{M})) \neq \dim(\phi(\mathcal{N}))$, but there is $\overline{b} \in N$ with $\text{tp}^{\mathcal{M}}(\overline{a}) = \text{tp}^{\mathcal{N}}(\overline{b})$ and $\dim(\psi(\mathcal{M}, \overline{a})/\overline{a}) = \dim(\psi(\mathcal{N}, \overline{b})/\overline{b})$. Then, we could find $f : \mathcal{M} \to \mathcal{N}$, an isomorphism with $f(\overline{a}) = \overline{b}$, and have $\mathcal{M} \cong \mathcal{N}$ despite the fact that $\dim(\phi(\mathcal{M})) \neq \dim(\phi(\mathcal{N}))$. In fact, this does not happen.

Theorem 6.1.22 (Baldwin–Lachlan Theorem) *If T is uncountably categorical but not $\aleph_0$-categorical, then $I(T, \aleph_0) = \aleph_0$.*

A generalization of this was later proved by Lachlan.

Theorem 6.1.23 *If T is superstable but not $\aleph_0$-categorical, then $I(T, \aleph_0) \geq \aleph_0$.*

The proofs of these results use more detailed results from stability theory than we will develop here. The reader can find the proofs in [7] or [18].

6.2 Morley Rank

Throughout this chapter, T is a complete theory with infinite models.

In this section, we will develop Morley rank, one of the most important tools for analyzing ω-stable theories. Morley rank provides a notion of "dimension." We begin with an illustration from linear algebra. Suppose that K is an infinite field and $V \subseteq K^n$ is an m-dimensional vector space. Suppose that f is a linear function that is not constant on V. For $a \in K$, let $V_a = \{x \in V : f(x) = a\}$. Then, $\{V_a : a \in K\}$ is an infinite family of $(m-1)$-dimensional affine subsets of V. Morley rank is an attempt to generalize this property of dimension. The basic idea is that if a definable set X contains infinitely many pairwise disjoint sets of dimension m, then X should have dimension at least $m+1$. To make this precise, we must allow arbitrary ordinals to occur as dimensions.

Definition 6.2.1 Suppose that $\mathcal{M}$ is an $\mathcal{L}$-structure and $\phi(\overline{v})$ is an $\mathcal{L}_M$-formula. We will define $\text{RM}^{\mathcal{M}}(\phi)$ the, *Morley rank* of ϕ in $\mathcal{M}$. First, we inductively define $\text{RM}^{\mathcal{M}}(\phi) \geq \alpha$ for α an ordinal:

i) $\mathrm{RM}^{\mathcal{M}}(\phi) \geq 0$ if and only if $\phi(\mathcal{M})$ is nonempty;

ii) if α is a limit ordinal, then $\mathrm{RM}^{\mathcal{M}}(\phi) \geq \alpha$ if and only if $\mathrm{RM}^{\mathcal{M}}(\phi) \geq \beta$ for all $\beta < \alpha$;

iii) for any ordinal α, $\mathrm{RM}^{\mathcal{M}}(\phi) \geq \alpha + 1$ if and only if there are $\mathcal{L}_M$-formulas $\psi_1(\overline{v}), \psi_2(\overline{v}), \ldots$ such that $\psi_1(\mathcal{M}), \psi_2(\mathcal{M}), \ldots$ is an infinite family of pairwise disjoint subsets of $\phi(\mathcal{M})$ and $\mathrm{RM}^{\mathcal{M}}(\psi_i) \geq \alpha$ for all i.

If $\phi(\mathcal{M}) = \emptyset$, then $\mathrm{RM}^{\mathcal{M}}(\phi) = -1$. If $\mathrm{RM}^{\mathcal{M}}(\phi) \geq \alpha$ but $\mathrm{RM}^{\mathcal{M}}(\phi) \not\geq \alpha + 1$, then $\mathrm{RM}^{\mathcal{M}}(\phi) = \alpha$. If $\mathrm{RM}^{\mathcal{M}}(\phi) \geq \alpha$ for all ordinals α, then $\mathrm{RM}^{\mathcal{M}}(\phi) = \infty$.

We would like to define Morley rank so that it does not depend on which model the parameters come from. The next lemmas show that we can eliminate dependence on the model if we restrict our attention to $\aleph_0$-saturated models. We first show that if $\mathcal{M}$ is $\aleph_0$-saturated, then $\mathrm{RM}^{\mathcal{M}}(\theta(\overline{v}, \overline{a}))$ depends only on $\mathrm{tp}^{\mathcal{M}}(\overline{a})$.

Lemma 6.2.2 *Suppose that $\theta(\overline{v}, \overline{w})$ is an $\mathcal{L}$-formula, $\mathcal{M}$ is $\aleph_0$-saturated, $\overline{a}, \overline{b} \in M$ and $\mathrm{tp}^{\mathcal{M}}(\overline{a}) = \mathrm{tp}^{\mathcal{M}}(\overline{b})$. Then $\mathrm{RM}^{\mathcal{M}}(\theta(\overline{v}, \overline{a})) = \mathrm{RM}^{\mathcal{M}}(\theta(\overline{v}, \overline{b}))$.*

Proof We prove by induction on α that if $\theta(\overline{v}, \overline{w})$ is any $\mathcal{L}$-formula and $\mathrm{tp}^{\mathcal{M}}(\overline{a}) = \mathrm{tp}^{\mathcal{M}}(\overline{b})$, then $\mathrm{RM}^{\mathcal{M}}(\theta(\overline{v}, \overline{a})) \geq \alpha$ if and only if $\mathrm{RM}^{\mathcal{M}}(\theta(\overline{v}, \overline{b})) \geq \alpha$.

Because $\mathrm{tp}^{\mathcal{M}}(\overline{a}) = \mathrm{tp}^{\mathcal{M}}(\overline{b})$, $\theta(\mathcal{M}, \overline{a}) = \emptyset$ if and only if $\theta(\mathcal{M}, \overline{b}) = \emptyset$. Thus, $\mathrm{RM}^{\mathcal{M}}(\theta(\overline{v}, \overline{a})) \geq 0$ if and only if $\mathrm{RM}^{\mathcal{M}}(\theta(\overline{v}, \overline{b})) \geq 0$.

If α is a limit ordinal and the claim is true for all $\beta < \alpha$, then

$$
\begin{aligned}
\mathrm{RM}^{\mathcal{M}}(\theta(\overline{v}, \overline{a})) \geq \alpha \quad &\Leftrightarrow \quad \mathrm{RM}^{\mathcal{M}}(\theta(\overline{v}, \overline{a})) \geq \beta \ \text{ for all } \beta < \alpha \\
&\Leftrightarrow \quad \mathrm{RM}^{\mathcal{M}}(\theta(\overline{v}, \overline{b})) \geq \beta \ \text{ for all } \beta < \alpha, \ \text{ by induction} \\
&\Leftrightarrow \quad \mathrm{RM}^{\mathcal{M}}(\theta(\overline{v}, \overline{b})) \geq \alpha.
\end{aligned}
$$

Suppose that the claim is true for α and $\mathrm{RM}^{\mathcal{M}}(\theta(\overline{v}, \overline{a})) \geq \alpha + 1$. There are $\mathcal{L}_M$-formulas $\psi_1, \psi_2, \ldots$ such that $\psi_1(\mathcal{M}), \psi_2(\mathcal{M}), \ldots$ is an infinite sequence of pairwise disjoint subsets of $\theta(\mathcal{M}, \overline{a})$ and $\mathrm{RM}^{\mathcal{M}}(\psi_i) \geq \alpha$ for all i. For each i, there is an $\mathcal{L}$-formula $\chi_i(\overline{v}, w_1, \ldots, w_{m_i})$ and $\overline{c}_i \in M^{m_i}$ such that $\psi_i(\overline{v})$ is $\chi_i(\overline{v}, \overline{c}_i)$. Because $\mathcal{M}$ is $\aleph_0$-saturated, we can do a back-and-forth argument to find $\overline{d}_1, \overline{d}_2, \ldots$ such that

$$
\mathrm{tp}^{\mathcal{M}}(\overline{a}, \overline{c}_1, \ldots, \overline{c}_m) = \mathrm{tp}^{\mathcal{M}}(\overline{b}, \overline{d}_1, \ldots, \overline{d}_m)
$$

for all $m < \omega$. Then, $\chi_1(\mathcal{M}, \overline{d}_1), \chi_2(\mathcal{M}, \overline{d}_2), \ldots$ is an infinite sequence of pairwise disjoint subsets of $\theta(\mathcal{M}, \overline{b})$ and, by induction, $\mathrm{RM}^{\mathcal{M}}(\chi_i(\overline{v}, \overline{d}_i)) \geq \alpha$. Thus $\mathrm{RM}^{\mathcal{M}}(\theta(\overline{v}, \overline{b})) \geq \alpha + 1$.

A symmetric argument shows that if $\mathrm{RM}^{\mathcal{M}}(\theta(\overline{v}, \overline{b})) \geq \alpha + 1$, then $\mathrm{RM}^{\mathcal{M}}(\theta(\overline{v}, \overline{a})) \geq \alpha + 1$.

Thus, by induction, $\text{RM}^{\mathcal{M}}(\theta(\overline{v}, \overline{a})) \geq \alpha$ if and only if $\text{RM}^{\mathcal{M}}(\theta(\overline{v}, \overline{b})) \geq \alpha$ for all α. Consequently,

$$\text{RM}^{\mathcal{M}}(\theta(\overline{v}, \overline{a})) = \text{RM}^{\mathcal{M}}(\theta(\overline{v}, \overline{b})).$$

Lemma 6.2.3 *Suppose that $\mathcal{M}$ and $\mathcal{N}$ are $\aleph_0$-saturated models of T and $\mathcal{M} \prec \mathcal{N}$. If ϕ is an $\mathcal{L}_M$-formula, then $\text{RM}^{\mathcal{M}}(\phi) = \text{RM}^{\mathcal{N}}(\phi)$.*

Proof We will prove by induction on α that if ϕ is an $\mathcal{L}_M$-formula, then $\text{RM}^{\mathcal{M}}(\phi) \geq \alpha$ if and only if $\text{RM}^{\mathcal{N}}(\phi) \geq \alpha$.
Because $\mathcal{M} \prec \mathcal{N}$, $\phi(\mathcal{M}) = \emptyset$ if and only if $\phi(\mathcal{N}) = \emptyset$. Thus, $\text{RM}^{\mathcal{M}}(\phi) \geq 0$ if and only if $\text{RM}^{\mathcal{N}}(\phi) \geq 0$.
Suppose that α is a limit ordinal. Then

$$\begin{aligned}
\text{RM}^{\mathcal{M}}(\phi(\overline{v})) \geq \alpha \quad &\Leftrightarrow \quad \text{RM}^{\mathcal{M}}(\phi) \geq \beta \text{ for all } \beta < \alpha \\
&\Leftrightarrow \quad \text{RM}^{\mathcal{N}}(\phi) \geq \beta \text{ for all } \beta < \alpha, \text{ by induction} \\
&\Leftrightarrow \quad \text{RM}^{\mathcal{N}}(\phi) \geq \alpha.
\end{aligned}$$

If $\text{RM}^{\mathcal{M}}(\phi) \geq \alpha + 1$, then we can find $\mathcal{L}_M$-formulas $\psi_1, \psi_2, \dots$ such that $\psi_1(\mathcal{M}), \psi_2(\mathcal{M}), \dots$ is an infinite sequence of pairwise disjoint subsets of $\phi(\mathcal{M})$ and $\text{RM}^{\mathcal{M}}(\psi_i) \geq \alpha$ for all i. By induction, $\text{RM}^{\mathcal{N}}(\psi_i) \geq \alpha$. Because $\mathcal{M} \prec \mathcal{N}$, $\psi_1(\mathcal{N}), \psi_2(\mathcal{N}), \dots$ is an infinite sequence of pairwise disjoint subsets of $\phi(\mathcal{N})$. Thus $\text{RM}^{\mathcal{N}}(\phi) \geq \alpha + 1$.
Suppose that $\text{RM}^{\mathcal{N}}(\phi) \geq \alpha + 1$. There are $\mathcal{L}_N$-formulas $\psi_1, \psi_2, \dots$ such that $\psi_1(\mathcal{N}), \psi_2(\mathcal{N}), \dots$ is an infinite family of pairwise disjoint subsets of $\phi(\mathcal{N})$ and $\text{RM}^{\mathcal{N}}(\psi_i) \geq \alpha$ for all i. Let $\overline{a}$ be the parameters from M occurring in the formula ϕ. Let $\psi_i(\overline{v})$ be $\theta_i(\overline{v}, \overline{b}_i)$, where θ_i is an $\mathcal{L}$-formula and $\overline{b}_i \in N$. Because $\mathcal{M}$ is $\aleph_0$-saturated, we can find $\overline{c}_1, \overline{c}_2, \dots$ in M such that

$$\text{tp}^{\mathcal{N}}(\overline{a}, \overline{b}_1, \dots, \overline{b}_m) = \text{tp}^{\mathcal{M}}(\overline{a}, \overline{c}_1, \dots, \overline{c}_m) = \text{tp}^{\mathcal{N}}(\overline{a}, \overline{c}_1, \dots, \overline{c}_m)$$

for $m < \omega$. By Lemma 6.2.2, $\text{RM}^{\mathcal{N}}(\theta_i(\overline{v}, \overline{c}_i)) \geq \alpha$. By induction, $\text{RM}^{\mathcal{M}}(\theta_i(\overline{v}, \overline{c}_i)) \geq \alpha$. Consequently, $\text{RM}^{\mathcal{M}}(\phi) \geq \alpha + 1$.
This concludes the induction. Thus $\text{RM}^{\mathcal{M}}(\phi) = \text{RM}^{\mathcal{N}}(\phi)$.

Corollary 6.2.4 *Suppose that $\mathcal{M}$ is an $\mathcal{L}$-structure, ϕ is an $\mathcal{L}_M$-formula, and $\mathcal{N}_0$ and $\mathcal{N}_1$ are $\aleph_0$-saturated elementary extensions of $\mathcal{M}$. Then $\text{RM}^{\mathcal{N}_0}(\phi) = \text{RM}^{\mathcal{N}_1}(\phi)$.*

Proof By Exercise 2.5.11 there is $\mathcal{N}_2$, a common elementary extension of $\mathcal{N}_0$ and $\mathcal{N}_1$. Let $\mathcal{N}_3$ be an $\aleph_0$-saturated elementary extension of $\mathcal{N}_2$. By Lemma 6.2.3, $\text{RM}^{\mathcal{N}_0}(\phi) = \text{RM}^{\mathcal{N}_3}(\phi) = \text{RM}^{\mathcal{N}_1}(\phi)$.

Corollary 6.2.4 allows us to define the Morley rank of ϕ in a way that does not depend on which model contains the parameters occurring in ϕ.

Definition 6.2.5 If $\mathcal{M}$ is an $\mathcal{L}$-structure and ϕ is any $\mathcal{L}$-formula, we define $\mathrm{RM}(\phi)$, the *Morley rank* of ϕ, to be $\mathrm{RM}^{\mathcal{N}}(\phi)$, where $\mathcal{N}$ is any $\aleph_0$-saturated elementary extension of $\mathcal{M}$.

Morley rank gives us our desired notion of "dimension" for definable sets.

Definition 6.2.6 Suppose that $\mathcal{M} \models T$ and $X \subseteq M^n$ is defined by the $\mathcal{L}_M$-formula $\phi(\overline{v})$. We let $\mathrm{RM}(X)$, the *Morley rank* of X, be $\mathrm{RM}(\phi)$.

In particular, if $\mathcal{M}$ is $\aleph_0$-saturated and $X \subseteq M^n$ is definable, then $\mathrm{RM}(X) \geq \alpha + 1$ if and only if we can find $Y_1, Y_2, \ldots$ pairwise disjoint definable subsets of X of Morley rank at least α.

The next lemma shows that Morley rank has some basic properties that we would want for a good notion of dimension.

Lemma 6.2.7 *Let $\mathcal{M}$ be an $\mathcal{L}$-structure and let X and Y be definable subsets of M^n.*

i) If $X \subseteq Y$, then $\mathrm{RM}(X) \leq \mathrm{RM}(Y)$.
ii) $\mathrm{RM}(X \cup Y)$ is the maximum of $\mathrm{RM}(X)$ and $\mathrm{RM}(Y)$.
iii) If X is nonempty, then $\mathrm{RM}(X) = 0$ if and only if X is finite.

Proof We leave the proofs of i) and ii) as exercises.

iii) Let $X = \phi(\mathcal{M})$. Because X is nonempty, $\mathrm{RM}(\phi) \geq 0$. Because $\phi(\mathcal{M})$ is finite if and only if $\phi(\mathcal{N})$ is finite for any $\mathcal{M} \prec \mathcal{N}$, we may, without loss of generality, assume that $\mathcal{M}$ is $\aleph_0$-saturated. If X is finite, then, because X cannot be partitioned into infinitely many nonempty sets, $\mathrm{RM}(X) \not\geq 1$. Thus $\mathrm{RM}(X) = 0$. If X is infinite, let $a_1, a_2, \ldots$ be distinct elements of X. Then, $\{a_1\}, \{a_2\}, \ldots$ is an infinite sequence of pairwise disjoint definable subsets of X. Thus $\mathrm{RM}(X) \geq 1$.

We will be interested in theories where every formula is ranked.

Definition 6.2.8 A theory T is called *totally transcendental* if, for all $\mathcal{M} \models T$, if ϕ is an $\mathcal{L}_M$-formula, then $\mathrm{RM}(\phi) < \infty$.

The Monster Model

The definition we just gave of Morley rank is rather awkward because even if a formula has parameters from $\mathcal{M} \models T$ we need to work in an $\aleph_0$-saturated elementary extension to calculate the Morley rank. Then, to show that Morley rank is well-defined, we must show that our calculation did not depend on our choice of $\aleph_0$-saturated model. Arguments such as this come up very frequently and tend to be both routine and repetitive. To simplify proofs, we will frequently adopt the expository device of assuming that we are working in a fixed, very large, saturated model of T.

Let $\mathbb{M} \models T$ be saturated of cardinality κ, where κ is "very large." We call $\mathbb{M}$ the *monster model* of T. If $\mathcal{M} \models T$ and $|M| \leq \kappa$, then by Lemma

4.3.17 there is an elementary embedding of $\mathcal{M}$ into $\mathbb{M}$. Moreover, if $\mathcal{M} \prec \mathbb{M}$, $f : \mathcal{M} \to \mathcal{N}$ is elementary, and $|N| < \kappa$, we can find $j : \mathcal{N} \to \mathbb{M}$ elementary such that $j|M$ is the identity. Thus, if we focus attention on models of T of cardinality less than κ, we can view all models as elementary submodels of $\mathbb{M}$.

There are several problems with this approach. First, we really want to prove theorems about all models of T, not just the small ones. But if there are arbitrarily large saturated models of T, then we can prove something about all models of T by proving it for submodels of larger and larger monster models. Second, and more problematic, for general theories T there may not be any saturated models. For our purposes, this is not a serious problem because, for the remainder of this text, we will be focusing on ω-stable theories and, by Theorem 4.3.15, if T is ω-stable, there are saturated models of T of cardinality κ for each regular cardinal κ. If we were considering arbitrary theories, we could get around this by making some extra set-theoretic assumptions. For example, we could assume that for all cardinals λ there is a strongly inaccessible cardinal $\kappa > \lambda$. Then, by Corollary 4.3.14, there are arbitrarily large saturated models.

We will tacitly assume that T has arbitrarily large saturated models, and thus we can prove theorems about all models of T by proving theorems about elementary submodels of saturated models.[1] We will only use this assumption in arguments where, by careful bookkeeping as in the proofs above, we could avoid it.

For the remainder of the chapter, we make the following assumptions:
- $\mathbb{M}$ is a large saturated model of T;
- all $\mathcal{M} \models T$ that we consider are elementary submodels of $\mathbb{M}$ and $|M| < |\mathbb{M}|$;
- all sets A of parameters that we consider are subsets of $\mathbb{M}$ with $|A| < \mathbb{M}$;
- if $\phi(\overline{v}, \overline{a})$ is a formula with parameters, we assume $\overline{a} \in \mathbb{M}$;
- we write $\mathrm{tp}(\overline{a}/A)$ for $\mathrm{tp}^{\mathbb{M}}(\overline{a}/A)$ and $S_n(A)$ for $S_n^{\mathbb{M}}(A)$.

Note that if $\overline{a} \in M$, then, because $\mathcal{M} \prec \mathbb{M}$, $\mathcal{M} \models \phi(\overline{a})$ if and only if $\mathbb{M} \models \phi(\overline{a})$. We will say that $\phi(\overline{a})$ holds if $\mathbb{M} \models \phi(\overline{a})$.

Because $\mathbb{M}$ is saturated, if $A \subset \mathbb{M}$ and $p \in S_n(A)$, then p is realized in $\mathbb{M}$. Moreover, if $f : A \to \mathbb{M}$ is a partial elementary map, then f extends to an automorphism of $\mathbb{M}$.

We could define Morley rank referring only to the monster model. The Morley rank of an $\mathcal{L}_{\mathbb{M}}$-formula is inductively defined as follows:

$\mathrm{RM}(\phi) \geq 0$ if and only if $\phi(\mathbb{M})$ is nonempty;

$\mathrm{RM}(\phi) \geq \alpha + 1$ if and only if there are $\mathcal{L}_{\mathbb{M}}$-formulas $\psi_1, \psi_2, \dots$ such that $\psi_1(\mathbb{M}), \psi_2(\mathbb{M}), \dots$ is an infinite sequence of pairwise disjoint subsets of $\phi(\mathbb{M})$ and $\mathrm{RM}(\psi_i) \geq \alpha$ for each i;

---

[1] There are other approaches to the monster model, which we discuss in the remarks at the end of this chapter.

if α is a limit ordinal, $\mathrm{RM}(\phi) \geq \alpha$ if and only if $\mathrm{RM}(\phi) \geq \beta$ for each $\beta < \alpha$.

Morley Degree

If X is a definable set of Morley rank α, then we cannot partition X into infinitely many pairwise disjoint definable subsets of Morley rank α. Indeed, we will show that there is a number d such that X cannot be partitioned into more than d definable sets of Morley rank α.

Proposition 6.2.9 Let ϕ be an $\mathcal{L}_{\mathbf{M}}$-formula with $\mathrm{RM}(\phi) = \alpha$ for some ordinal α. There is a natural number d such that if $\psi_1, \ldots, \psi_n$ are $\mathcal{L}_{\mathbf{M}}$-formulas such that $\psi_1(\mathbb{M}), \ldots, \psi_n(\mathbb{M})$ are disjoint subsets of $\phi(\mathbb{M})$ such that $\mathrm{RM}(\psi_i) = \alpha$ for all i, then $n \leq d$.
We call d the Morley degree of ϕ and write $\deg_{\mathbf{M}}(\phi) = d$.

Proof We build $S \subseteq 2^{<\omega}$ and $(\phi_\sigma : \sigma \in S)$ with the following properties.
i) If $\sigma \in S$ and $\tau \subseteq \sigma$, then $\tau \in S$.
ii) $\phi_\emptyset = \phi$.
iii) $\mathrm{RM}(\phi_\sigma) = \alpha$ for all $\sigma \in S$.
iv) If $\sigma \in S$, there are two cases to consider. If there is an $\mathcal{L}_{\mathbf{M}}$-formula ψ such that $\mathrm{RM}(\phi_\sigma \wedge \psi) = \mathrm{RM}(\phi_\sigma \wedge \neg\psi) = \alpha$, then $\sigma, 0$ and $\sigma, 1$ are in S, $\phi_{\sigma,0}$ is $\phi_\sigma \wedge \psi$, and $\phi_{\sigma,1}$ is $\phi_\sigma \wedge \neg\psi$. If there is no such ψ, then no $\tau \supset \sigma$ is in S.

The set S is a binary tree. We claim that S is finite. If S is infinite, then, by König's Lemma (see Lemma A.21), there is $f : \omega \to 2$ such that $f|n \in S$ for all n. Let ψ_n be the formula $\phi_{f|n} \wedge \neg\phi_{f|n+1}$ for $n = 1, 2, \ldots$. Then, $\mathrm{RM}(\psi_n) = \alpha$ for all n and $\psi_1(\mathbb{M}), \psi_2(\mathbb{M}), \ldots$ are disjoint subsets of $\phi(\mathbb{M})$. But then $\mathrm{RM}(\phi) \geq \alpha + 1$, a contradiction. Thus, S is finite.

Let $S_0 = \{\sigma \in S : \tau \notin S \text{ for all } \tau \supset \sigma\}$ be the terminal nodes of the tree S. Let $d = |S_0|$, and let $\psi_1, \ldots, \psi_d$ be an enumeration of $\{\phi_\sigma : \sigma \in S_0\}$. Then, $\mathrm{RM}(\psi_i) = \alpha$ for all i, $\phi(\mathcal{M})$ is the disjoint union of $\psi_1(\mathbb{M}), \ldots, \psi_d(\mathbb{M})$, and, for each i, there is no formula χ such that $\mathrm{RM}(\psi_i \wedge \chi) = \mathrm{RM}(\psi_i \wedge \neg\chi) = \alpha$.

Suppose that $\theta_1, \ldots, \theta_n$ is a sequence of $\mathcal{L}_{\mathbf{M}}$-formulas of Morley rank α such that $\theta_1(\mathbb{M}), \ldots, \theta_n(\mathbb{M})$ is a sequence of pairwise disjoint subsets of $\phi(\mathbb{M})$. We claim that $n \leq d$. By our choice of $\psi_1, \ldots, \psi_d$, for each $i \leq d$, there is at most one $j \leq n$ such that $\mathrm{RM}(\psi_i \wedge \theta_j) = \alpha$. If $n > d$, there is $\widehat{j} \leq n$ such that $\mathrm{RM}(\psi_i \wedge \theta_{\widehat{j}}) < \alpha$ for all $i \leq d$. But

$$\mathbb{M} \models \theta_{\widehat{j}} \leftrightarrow \bigvee_{i=1}^{d} \psi_i \wedge \theta_{\widehat{j}}.$$

Thus, by Lemma 6.2.7, $\mathrm{RM}(\theta_{\widehat{j}}) < \alpha$, a contradiction.

Strongly minimal formulas can be described using Morley rank and degree.

Corollary 6.2.10 *A formula ϕ is strongly minimal if and only if* $\mathrm{RM}(\phi) = \deg_M(\phi) = 1$.

Proof If ϕ is strongly minimal, then, because $\phi(M)$ is infinite, $\mathrm{RM}(\phi) \geq 1$. Because $\phi(M)$ cannot be partitioned into two definable infinite sets, $\mathrm{RM}(\phi) = 1$ and $\deg_M(\phi) = 1$.

On the other hand, if $\mathrm{RM}(\phi) = \deg_M(\phi) = 1$, then $\phi(M)$ is infinite and cannot be partitioned into two infinite definable sets. Thus, ϕ is strongly minimal.

Recall from Exercise 5.5.6 that a formula $\phi(\overline{v}, \overline{w})$ has the *order property* if there are $\overline{a}_1, \overline{b}_1, \overline{a}_2, \overline{b}_2, \ldots$ in M such that $M \models \phi(\overline{a}_i, \overline{b}_j)$ if and only if $i < j$. We can use Morley rank and degree to show that in totally transcendental theories no formula has the order property.

Proposition 6.2.11 *If T is totally transcendental, then no formula has the order property.*

Proof Suppose, for purposes of contradiction, that $\phi(\overline{v}, \overline{w})$ has the order property. By compactness and saturation, we can find $(\overline{a}_q, \overline{b}_q : q \in \mathbb{Q})$ such that $M \models \phi(\overline{a}_q, \overline{b}_r)$ if and only if $q < r$. Note that $\{q \in \mathbb{Q} : \phi(\overline{a}_q, \overline{b}_r)\} = (-\infty, r)$ is an infinite convex set. Thus, there is $\psi(\overline{v})$, an $\mathcal{L}_M$-formula of minimal rank and degree such that $C = \{q \in \mathbb{Q} : M \models \psi(\overline{a}_q)\}$ is an infinite convex set. Choose r in the interior of C. Let $\psi_0(\overline{v})$ be $\psi(\overline{v}) \wedge \phi(\overline{v}, \overline{b}_r)$, and let $\psi_1(\overline{v})$ be $\psi(\overline{v}) \wedge \neg \phi(\overline{v}, \overline{b}_r)$. The set $\{q \in \mathbb{Q} : \psi_i(\overline{a}_q)\}$ is infinite and convex for $i = 0, 1$, and if neither ψ_i has lower Morley rank than ψ, then both have lower Morley degree, contradicting the minimality of ψ.

Ranks of Types

We extend the definitions of Morley rank and degree from formulas to types.

Definition 6.2.12 If $p \in S_n(A)$, then $\mathrm{RM}(p) = \inf\{\mathrm{RM}(\phi) : \phi \in p\}$. If $\mathrm{RM}(p)$ is an ordinal, then $\deg_M(p) = \inf\{\deg_M(\phi) : \phi \in p$ and $\mathrm{RM}(\phi) = \mathrm{RM}(p)\}$.

If $\mathrm{RM}(p) < \infty$, then $(\mathrm{RM}(p), \deg_M(p))$ is the minimum element of $\{(\mathrm{RM}(\phi), \deg_M(\phi)) : \phi \in p\}$ in the lexicographic order. For each type p with $\mathrm{RM}(p) < \infty$, we can find a formula $\phi_p \in p$ such that $(\mathrm{RM}(p), \deg_M(p)) = (\mathrm{RM}(\phi_p), \deg_M(\phi_p))$.

Lemma 6.2.13 *If $p, q \in S_n(A)$, $\mathrm{RM}(p), \mathrm{RM}(q) < \infty$, and $p \neq q$, then $\phi_p \neq \phi_q$.*

Proof There is a formula ψ such that $\psi \in p$ and $\psi \notin q$. Because $\phi_p \wedge \psi \in p$, $\mathrm{RM}(\phi_p \wedge \psi) \leq \mathrm{RM}(\phi_p) \leq \mathrm{RM}(p)$. Because $\mathrm{RM}(\phi_p)$ is minimal,

$$\mathrm{RM}(\phi_p \wedge \psi) = \mathrm{RM}(\phi_p) = \mathrm{RM}(p).$$

Similarly,

$$\mathrm{RM}(\phi_q \wedge \neg\psi) = \mathrm{RM}(\phi_q) = \mathrm{RM}(q).$$

If $\phi_p = \phi_q$, then

$$\mathrm{RM}(\phi_p \wedge \psi) = \mathrm{RM}(\phi_p \wedge \neg\psi) = \mathrm{RM}(\phi_p).$$

Thus, $\deg_{\mathrm{M}}(\phi_p \wedge \psi) < \deg_{\mathrm{M}}(\phi_p)$, contradicting our choice of ϕ_p.

Theorem 6.2.14 *If T is ω-stable, then T is totally transcendental. Conversely, if $\mathcal{L}$ is countable and T is totally transcendental, then T is ω-stable.*

Proof

($\Rightarrow$) Suppose, for purposes of contradiction, that $\phi(v_1, \ldots, v_n)$ is an $\mathcal{L}_{\mathrm{M}}$-formula such that $\mathrm{RM}(\phi) = \infty$. Let $\beta = \sup\{\mathrm{RM}(\psi) : \psi$ an $\mathcal{L}_{\mathrm{M}}$-formula and $\mathrm{RM}(\psi) < \infty\}$.

Because $\mathrm{RM}(\phi) = \infty \geq \beta + 2$, we can find an $\mathcal{L}_{\mathrm{M}}$-formula ψ such that $\mathrm{RM}(\phi \wedge \psi) \geq \beta + 1$ and $\mathrm{RM}(\phi \wedge \neg\psi) \geq \beta + 1$. Then $\mathrm{RM}(\phi \wedge \psi) = \mathrm{RM}(\phi \wedge \neg\psi) = \infty$.

Iterating this construction, we can build a binary tree of $\mathcal{L}_{\mathrm{M}}$-formulas $(\phi_\sigma : \sigma \in 2^{<\omega})$ such that:

i) $\phi_\emptyset = \phi$;

ii) $\mathrm{RM}(\phi_\sigma) = \infty$ for all σ;

iii) for each σ there is a formula ψ_σ such that $\phi_{\sigma,0} = \phi_\sigma \wedge \psi_\sigma$ and $\phi_{\sigma,1} = \phi_\sigma \wedge \neg\psi_\sigma$.

There is a countable $A \subseteq M$ such that each ϕ_σ is an $\mathcal{L}_A$-formula. Then, $|S_n(A)| = 2^{\aleph_0}$ so T is not ω-stable.

($\Leftarrow$) Suppose that $|A| \leq \aleph_0$. For each $p \in S_n(A)$, $\mathrm{RM}(p) < \infty$ so there is ϕ_p as above. Because $\phi_p \neq \phi_q$ for $p \neq q$ and there are only countably many choices for ϕ_p, $|S_n(A)| \leq \aleph_0$. Thus, T is ω-stable.

Because we are concentrating on theories in countable languages, we will not mention totally transcendental theories again, but the reader should note that many of the results we state for ω-stable theories only use the existence of ranks and thus follow for totally transcendental theories as well.

Definition 6.2.15 If $A \subset \mathrm{M}$ and $\bar{a} \in \mathrm{M}$, we write $\mathrm{RM}(\bar{a})$ for $\mathrm{RM}(\mathrm{tp}(\bar{a}))$ and $\mathrm{RM}(\bar{a}/A)$ for $\mathrm{RM}(\mathrm{tp}(\bar{a}/A))$.

The following facts are easy to prove and we leave them as exercises.

Lemma 6.2.16 *i) If $X \subseteq \mathrm{M}^n$ is definable, then $\mathrm{RM}(X) = \sup\{\mathrm{RM}(\bar{a}/A) : \bar{a} \in X, A \subset \mathrm{M}, |A| < |\mathrm{M}|, X, A$-definable$\}$.*

ii) *If* $X \subseteq \mathbb{M}^n$ *is definable, and* $\beta < \mathrm{RM}(X)$, *then there is a definable* $Y \subset X$ *with* $\mathrm{RM}(Y) = \beta$. *In particular, if* $\mathrm{RM}(X) = \alpha$ *and* $\beta < \alpha$ *we can find disjoint definable sets* $Y_1, Y_2, \ldots$ *such that each* Y_i *has Morley rank* α.

iii) *For any* $\mathcal{L}_A$-*formula* ϕ, $|\{p \in S_n(\mathbb{M}) : \phi \in p \text{ and } \mathrm{RM}(p) = \mathrm{RM}(\phi)\}| = \deg_{\mathbb{M}}(\phi)$.

Next, we prove an important generalization of Lemma 6.2.7 iii).

Lemma 6.2.17 *Suppose that* $A \subset \mathbb{M}$, $\bar{a}, b \in \mathbb{M}$, *and* b *is algebraic over* $A \cup \{\bar{a}\}$. *Then* $\mathrm{RM}(\bar{a}, b/A) = \mathrm{RM}(\bar{a}/A)$.

Proof Without loss of generality we will assume that $A = \emptyset$. We leave as an exercise the proof that $\mathrm{RM}(\bar{a}, b) \geq \mathrm{RM}(\bar{a})$. We will prove by induction on α that if $\bar{a}, b \in \mathbb{M}$, b is algebraic over $\bar{a}$, and if $\mathrm{RM}(\bar{a}, b) \geq \alpha$, then $\mathrm{RM}(\bar{a}) \geq \alpha$.

By Exercise 6.6.10, $\mathrm{RM}(\bar{a}) \geq 0$. Thus, the claim is true for $\alpha = 0$. If α is a limit ordinal and the claim is true for all $\beta < \alpha$, then it is also true for α.

Suppose that the claim is true for α and $\mathrm{RM}(\bar{a}, b) \geq \alpha + 1$. By induction, $\mathrm{RM}(\bar{a}) \geq \alpha$. Suppose, for purposes of contradiction, that $\mathrm{RM}(\bar{a}) = \alpha$. Let $\phi(\bar{v})$ be an $\mathcal{L}$-formula such that $\mathrm{RM}(\phi(\bar{v})) = \alpha$ and there is no $\mathcal{L}$-formula ϕ_1 such that $\mathrm{RM}(\phi \wedge \phi_1) = \mathrm{RM}(\phi \wedge \neg \phi_1) = \alpha$.

Because b is algebraic over $\bar{a}$, there is a formula $\psi(\bar{v}, w)$ such that $\psi(\bar{a}, b)$ and $|\{x : \psi(\bar{a}, x)\}| = n$. Let $\hat{\phi}(\bar{v}, w)$ be the formula

$$\phi(\bar{v}) \wedge \psi(\bar{v}, w) \wedge |\{x : \psi(\bar{v}, x)\}| = n.$$

Because $\hat{\phi}(\bar{a}, b)$ holds and $\mathrm{RM}(\bar{a}, b) > \alpha$, $\mathrm{RM}(\hat{\phi}) > \alpha$. Suppose that $\theta_1, \theta_2, \ldots$ are $\mathcal{L}_{\mathbb{M}}$-formulas such that $\theta_1(\mathbb{M}), \theta_2(\mathbb{M}), \ldots$ is a sequence of disjoint subsets of $\hat{\phi}(\mathbb{M})$ and $\mathrm{RM}(\theta_i) \geq \alpha$ for all i. Let $\chi_i(\bar{v})$ be the formula $\exists w \; \theta_i(\bar{v}, w)$.

Claim 1 $\mathrm{RM}(\chi_i) \geq \alpha$ for all i.

Because $\mathrm{RM}(\theta_i) \geq \alpha$, there is $\bar{c}, d \in \mathbb{M}$ such that $\theta_i(\bar{c}, d)$ and $\mathrm{RM}(\bar{c}, d) \geq \alpha$. By induction, $\mathrm{RM}(\bar{c}) \geq \alpha$. Because $\chi_i(\bar{c})$, $\mathrm{RM}(\chi_i) \geq \alpha$.

Claim 2 $\mathrm{RM}(\chi_1 \wedge \ldots \wedge \chi_m) \geq \alpha$ for all m.

Suppose that m is least such that this fails. Then

$$\mathrm{RM}(\chi_1 \wedge \ldots \wedge \chi_{m-1}) = \mathrm{RM}\left(\chi_m \wedge \neg \bigwedge_{i=1}^{m-1} \chi_i\right) \geq \alpha,$$

contradicting our assumption that we cannot find an $\mathcal{L}$-formula ϕ_1 such that $\phi \wedge \phi_1$ and $\phi \wedge \neg \phi_1$ both have rank α.

Because $\mathbb{M}$ is saturated, we can find $\bar{c} \in \mathbb{M}$ such that $\chi_i(\bar{c})$ for all i. For each i, we can find d_i such that $\theta_i(\bar{c}, d_i)$. Because the $\theta_i(\mathcal{M})$ are disjoint, $d_1, d_2, \ldots$ are distinct. But $\theta_i(\mathbb{M}) \subseteq \hat{\phi}(\mathbb{M})$. Thus, $\psi(\bar{c}, d_i)$ holds for all i, contradicting the fact that $|\{x : \psi(\bar{c}, x)\}| = n$. Thus, $\mathrm{RM}(\bar{a}) \geq \alpha + 1$ and the lemma follows by induction.

In particular, this shows that Morley rank is preserved by definable bijections or indeed definable finite-to-one functions.

Corollary 6.2.18 *Suppose that T is ω-stable, $\mathcal{M} \models T$, $X \subseteq M^n$, $Y \subseteq M^m$ are definable, and $f : X \to Y$ is a definable finite-to-one function from X onto Y. Then $\mathrm{RM}(X) = \mathrm{RM}(Y)$.*

Proof Let $A \subset M$ such that X, Y and f are definable over A. Suppose that $f(\bar{a}) = \bar{b}$. Then, $\bar{b}$ is definable over $A, \bar{a}$ and, because f is finite-to-one, $\bar{a}$ is algebraic over $A, \bar{b}$. By Lemma 6.2.17

$$\mathrm{RM}(\bar{a}/A) = \mathrm{RM}(\bar{a}, \bar{b}/A) = \mathrm{RM}(\bar{b}/A).$$

If $\bar{a} \in X$ such that $\mathrm{RM}(\bar{a}/A) = \mathrm{RM}(X)$, then

$$\mathrm{RM}(Y) \geq \mathrm{RM}(f(\bar{a})/A) = \mathrm{RM}(\bar{a}/A) = \mathrm{RM}(X).$$

On the other hand, if $\bar{b} \in Y$ such that $\mathrm{RM}(\bar{b}/A) = \mathrm{RM}(Y)$, then, because f is surjective, there is $\bar{a} \in X$ such that $f(\bar{a}) = \bar{b}$ and

$$\mathrm{RM}(X) \geq \mathrm{RM}(\bar{a}/A) = \mathrm{RM}(\bar{b}/A) = \mathrm{RM}(Y).$$

Hence $\mathrm{RM}(X) = \mathrm{RM}(Y)$.

Morley Rank in Strongly Minimal Theories

In strongly minimal theories, Morley rank agrees with the notion of dimension introduced in Definition 6.1.10. Recall that a theory is strongly minimal if the formula $v = v$ is strongly minimal (i.e., if M is a strongly minimal set).

Theorem 6.2.19 *Suppose that T is a strongly minimal theory. If $A \subset M$ and $\bar{a} \in M$, then $\mathrm{RM}(\bar{a}/A) = \dim(\bar{a}/A)$.*

Proof We will first show by induction that if $a_1, \ldots, a_k$ are independent over A, then $\mathrm{RM}(\bar{a}) = k$. We prove this by induction on k.

Suppose that $k = 1$. If $\phi(v) \in \mathrm{tp}(a/A)$, then, because $a \notin \mathrm{acl}(A)$, $\phi(M)$ is infinite and $\mathrm{RM}(\phi) \geq 1$. Because T is strongly minimal $\mathrm{RM}(v = v) = 1$. Thus $\mathrm{RM}(\phi) \leq 1$.

Suppose $k > 1$ and $a_1, \ldots, a_k$ are independent over A. Let $\phi(\bar{v}) \in \mathrm{tp}(\bar{a}/A)$ be a formula of minimal Morley rank. We first argue that $\mathrm{RM}(\bar{a}/A) \geq k$. Let $b_1, b_2, \ldots$ be distinct elements of M that are not in $\mathrm{acl}(A)$. Let $\psi_i(\bar{v})$ be the formula $\phi(\bar{v}) \wedge v_1 = b_i$. Clearly, $\psi_1(M), \psi_2(M), \ldots$ is a family of pairwise disjoint subsets of $\phi(M)$. If $c_2, \ldots, c_k$ are independent over $A \cup \{b_i\}$, then, by Lemma 6.1.6, $\mathrm{tp}(b_i, \bar{c}/A) = \mathrm{tp}(a_1, \ldots, a_k/A)$. In particular, $M \models \phi(b_i, \bar{c})$. Thus $M \models \psi_i(b_i, \bar{c})$. Then

$$\mathrm{RM}(\psi_i) \geq \mathrm{RM}(b_i, \bar{c}/A) \geq \mathrm{RM}(\bar{c}/A)$$

and, by induction, $\mathrm{RM}(\bar{c}/A) \geq k-1$. Thus, $\mathrm{RM}(\psi_i) \geq k-1$ and $\mathrm{RM}(\phi) \geq k$. Hence $\mathrm{RM}(\bar{a}/A) \geq k$.

Next, we show that $\mathrm{RM}(\bar{a}/A) \leq k$. Let $\mathcal{M}$ be an $\aleph_0$-saturated model containing A. If $d_1, \ldots, d_k$ are independent over M, then, by Lemma 6.1.6, $\mathrm{tp}(\bar{a}/A) = \mathrm{tp}(\bar{d}/A)$ so we may without loss of generality assume that $a_1, \ldots, a_k$ are independent over M. Suppose that $\psi(\bar{v})$ is an $\mathcal{L}_M$-formula such that $\psi(\mathbb{M}) \subset \phi(\mathbb{M})$ and $\neg\psi(\bar{a})$. It suffices to show that $\mathrm{RM}(\psi) < k$. If $\psi(\bar{b})$, then, because $\mathrm{tp}(\bar{b}/M) \neq \mathrm{tp}(\bar{a}/M)$, $b_1, \ldots, b_k$ are dependent over M. By permuting variables, we may assume that $b_k \in \mathrm{acl}(M, b_1, \ldots, b_{k-1})$. Let $\theta(\bar{v})$ be an $\mathcal{L}_M$-formula such that $|\{w : \theta(b_1, \ldots, b_{k-1}, w)\}| = s$ for some $s < \omega$. Replacing ψ by

$$\psi(\bar{v}) \wedge |\{w : \theta(v_1, \ldots, v_{k-1}, w)\}| = s$$

we may assume that if $\psi(\bar{c})$, then $c_k \in \mathrm{acl}(M, c_1, \ldots, c_{k-1})$.

Choose $c_1, \ldots, c_k$ such that $\psi(\bar{c})$ and $\mathrm{RM}(\psi) = \mathrm{RM}(\bar{c}/M)$. By permuting variables, we may assume that $c_1, \ldots, c_l$ are independent over M and $\bar{c} \in \mathrm{acl}(M, c_1, \ldots, c_l)$ for some $l \leq k$. But then, by Lemma 6.2.17,

$$\mathrm{RM}(\psi) = \mathrm{RM}(\bar{c}/M) = \mathrm{RM}(c_1, \ldots, c_l/M)$$

and $\mathrm{RM}(c_1, \ldots, c_l/M) = l < k$ by induction.

The equivalence of Morley rank and dimension will allow us to conclude that Morley rank is definable in strongly minimal theories.

Lemma 6.2.20 *Let T be strongly minimal. Suppose that $C \subseteq \mathbb{M}^{m+n}$ is definable. Let $C_{\bar{a}} = \{\bar{x} \in \mathbb{M}^n : (\bar{a}, \bar{x}) \in C\}$ for $\bar{a} \in \mathbb{M}^m$. The set $Y_{n,k} = \{\bar{a} \in \mathbb{M}^m : \mathrm{RM}(C_{\bar{a}}) \geq k\}$ is definable for each $k \leq n$.*

Proof We prove this by induction on n.

Suppose that $n = 1$. We first note that there is a number N such that $|C_{\bar{a}}| < N$ or $|\mathbb{M} \setminus C_{\bar{a}}| < N$ for all $\bar{a} \in \mathbb{M}^m$ because otherwise the type

$$\left\{ \exists v_1, \ldots, v_{2s} \bigwedge_{i \neq j} v_i \neq v_j \wedge \bigwedge_{i=1}^{s} \phi(\bar{w}, v_i) \wedge \bigwedge_{i=s+1}^{2s} \neg\phi(\bar{w}, v_i) : s = 1, 2, \ldots \right\}$$

is satisfiable and a realization violates strong minimality.

Thus, $\mathrm{RM}(C_{\bar{a}}) \geq 1$ if and only if $|C_{\bar{a}}| > N$, so $Y_{1,1}$ is definable. Clearly, $Y_{1,0} = \{\bar{a} : \exists \bar{w} \ \bar{w} \in C_{\bar{a}}\}$.

Suppose that $n = s+1$. We work by induction on k. Clearly, $Y_{n,0} = \mathbb{M}^n$ is definable. For $\bar{a} \in \mathbb{M}^m$, let $B_{\bar{a}} = \{\bar{b} \in \mathbb{M}^s : \exists y \ (\bar{b}, y) \in C_{\bar{a}}\}$. If $\mathrm{RM}(B_{\bar{a}}) \geq k$, then $\mathrm{RM}(C_{\bar{a}}) \geq k$. Suppose that $\mathrm{RM}(B_{\bar{a}}) < k$. If $\bar{b} \in B_{\bar{a}}$ and $(\bar{b}, c) \in C_{\bar{a}}$, then $\dim(\bar{b}, c) = \dim \bar{b} + \dim(c/\bar{b})$. Let $A_{\bar{a}} = \{\bar{b} \in \mathbb{M}^s : \{y : (\bar{b}, y) \in C_{\bar{a}}\}$ is infinite$\}$. As above, there is an N (independent of $\bar{a}$) such that

$$\bar{b} \in A_{\bar{a}} \quad \text{if and only if} \quad |\{y : (\bar{b}, y) \in C_{\bar{a}}\}| > N.$$

Thus, $A_{\overline{a}}$ is definable and $\mathrm{RM}(C_{\overline{a}}) \geq k$ if and only if $\mathrm{RM}(A_{\overline{a}}) \geq k - 1$.

Thus, $\mathrm{RM}(C_{\overline{a}}) \geq k$ if and only if $\mathrm{RM}(B_{\overline{a}}) \geq k$ or $\mathrm{RM}(A_{\overline{a}}) > k - 1$, so, by induction, $Y_{n,k}$ is definable.

Morley Rank in Algebraically Closed Fields

For algebraically closed fields, we will show that Morley rank is equal to the classical notion of dimension from algebraic geometry.

Definition 6.2.21 Let K be an algebraically closed field. Let $V \subseteq K^n$ be an irreducible algebraic variety. Let $I(V)$ be the prime ideal of polynomials in $K[X_1, \ldots, X_n]$ vanishing on V. The *Krull dimension* of V is the largest number m such that there is a chain of prime ideals

$$I(V) = P_0 \subset P_1 \subset P_2 \subset \ldots \subset P_m \subset K[X_1, \ldots, X_n].$$

If V has Krull dimension 0, then $I(V)$ is a maximal ideal. Because K is algebraically closed, $I(V)$ is the ideal generated by $X_1 - a_1, \ldots, X_n - a_n$ for some $a_1, \ldots, a_n \in K$ and V is the single point $\{\overline{a}\}$.

There are many alternative characterizations of dimension. For example, the Krull dimension is the minimum of the vector space dimension of the tangent space to V at points $p \in V$. If $K = \mathbb{C}$ and V is smooth, then the Krull dimension is equal to the dimension of V as a complex manifold. We will use another characterization of dimension. If $V \subseteq K^n$ is an algebraic variety, the *function field* of V is $K(V)$, the fraction field of $K[X_1, \ldots, X_n]/I(V)$.

Theorem 6.2.22 *If K is an algebraically closed field and $V \subseteq K^n$ is an irreducible variety, then the Krull dimension of V is equal to the transcendence degree of the function field $K(V)$ over K.*

Proof See [6] §11.

It follows that in algebraically closed fields, the Morley rank of a variety is equal to the Krull dimension.

Corollary 6.2.23 *If K is an algebraically closed field and $V \subseteq K^n$ is an irreducible variety, then $\mathrm{RM}(V)$ is equal to the Krull dimension of V.*

Proof We prove this by induction on the Krull dimension of V. If V has Krull dimension 0, then V is a point and $\mathrm{RM}(V) = 0$.

Suppose that V has Krull dimension $k > 0$. Let $\phi(\overline{v})$ be the $\mathcal{L}_K$-formula defining V. If $\overline{a} \in \phi(\mathbb{M})$, let $V_{\overline{a}} \subseteq V$ be

$$\{\overline{x} \in K^n : f(\overline{a}) = 0 \to f(\overline{x}) = 0 \text{ for all } f \in K[\overline{X}]\}$$

(i.e., $V_{\bar{a}}$ is $V(I_{\bar{a}})$ where $I_{\bar{a}}$ is the prime ideal of polynomials in $K[\overline{X}]$ vanishing at $\bar{a}$). If $V_{\bar{a}} \subset V$, then V has Krull dimension at most $k - 1$ and, by induction,

$$\mathrm{RM}(\bar{a}/K) \le \mathrm{RM}(V_{\bar{a}}) \le k - 1.$$

Suppose that $V_{\bar{a}} = V$. Then, $I_{\bar{a}} = I(V)$ and the field $K(\bar{a})$ is exactly the function field $K(V)$. By Theorem 6.2.19 $\mathrm{RM}(\bar{a}/K) = \dim(\bar{a}/K)$. By Proposition 3.2.15, the model-theoretic notion of algebraic closure agrees with the classical algebraic notion in algebraically closed fields. Thus, $\mathrm{RM}(\bar{a}/K)$ is the transcendence degree of $K(V)$, which is k. Because $\mathrm{RM}(V) = \max\{\mathrm{RM}(\bar{a}/K) : \bar{a} \in \phi(\mathbb{M})\}$, $\mathrm{RM}(V) = k$.

Suppose that K is an algebraically closed field and $\mathbb{K}$ is a large saturated model with $K \prec \mathbb{K}$. If $V \subseteq K^n$ is an irreducible algebraic variety defined by polynomial equations with coefficients in K, then we can also consider $V(\mathbb{K})$ the subset of $\mathbb{K}^n$ defined by the same system of equations. By model-completeness, $V(\mathbb{K})$ is still irreducible (see Exercise 3.4.17).

Definition 6.2.24 We say that $\bar{a} \in V(\mathbb{K})$ is a *generic point* of V over K if $\bar{a} \notin W(\mathbb{K})$ for any Zariski closed $W \subset V$ defined over K.

If $I_{\bar{a}} = \{f \in K[\overline{X}] : f(\bar{a}) = 0\}$, then $\bar{a}$ is a generic point if and only if $I_{\bar{a}} = I(V)$. The following lemma will motivate an important definition in Chapter 7. We leave the proof as an exercise.

Lemma 6.2.25 *If $V \subseteq K^n$ is an irreducible Zariski closed set and $\bar{a} \in V(\mathbb{K})$, then $\mathrm{RM}(\bar{a}/K) = \mathrm{RM}(V)$ if and only if a is a generic point of V.*

The following lemma will be useful in Chapter 7.

Lemma 6.2.26 *Suppose that $V \subseteq K^n$ is an irreducible closed set, $X \subseteq K^n$ is constructible, and $\mathrm{RM}(X) = \mathrm{RM}(V)$. There is an open $O \subseteq K^n$ such that $O \cap V \subseteq X$ and $O \cap V \neq \emptyset$.*

Proof By quantifier elimination, $X = \bigcup_{i=1}^n F_i \cap O_i$ where $F_i \subseteq F$ is Zariski closed, $O_i \subseteq K^n$ is Zariski open, and $F_i \cap O_i$ is nonempty. Because $\mathrm{RM}(X) = \mathrm{RM}(V)$ and V is irreducible, X is not contained in any proper closed subset of F. Thus, there is an i such that $F_i = F$ and $F \cap O_i \subseteq X$.

6.3 Forking and Independence

For this section we assume that T is a complete ω-stable theory.

Suppose that we have a type $p \in S_n(A)$ and $A \subseteq B$. It will often be important to find $q \in S_n(B)$ with $p \subseteq q$ such that q is as "free" as possible (i.e., q imposes the fewest possible restrictions on its realizations). For example if K is an algebraically closed field, $k \subseteq l$ are subfields and $p \in S_1(k)$

is the type of an element transcendental over k. Then q will be the type of a transcendental over l. Any other $r \in S_1(l)$ will be less "free" as we have asserted that v satisfies algebraic equations that are not imposed by p. The next definition describes the "free" extensions.

Definition 6.3.1 Suppose that $A \subseteq B$, $p \in S_n(A)$, $q \in S_n(B)$, and $p \subseteq q$. If $\mathrm{RM}(q) < \mathrm{RM}(p)$, we say that q is a *forking* extension of p and that q *forks* over A. If $\mathrm{RM}(q) = \mathrm{RM}(p)$, we say that q is a *nonforking* extension of p.

The nonforking extensions will be the "free" ones.[2] Our first goal is to show that nonforking extensions exist.

Theorem 6.3.2 (Existence of nonforking extensions) *Suppose that* $p \in S_n(A)$ *and* $A \subseteq B$.
   i) There is $q \in S_n(B)$ *a nonforking extension of* p.
   ii) There are at most $\deg_M(p)$ *nonforking extensions of* p *in* $S_n(B)$, *and, if* M *is an* $\aleph_0$-*saturated model with* $A \subseteq M$, *there are exactly* $\deg_M(p)$ *nonforking extensions of* p *in* $S_n(M)$.
   iii) There is at most one $q \in S_n(B)$, *a nonforking extension of* p *with* $\deg_M(p) = \deg_M(q)$. *In particular, if* $\deg_M(p) = 1$, *then* p *has a unique nonforking extension in* $S_n(B)$.

Proof Let $\phi(\overline{v}) \in p$ be of minimal Morley rank and degree with $\mathrm{RM}(\phi) = \alpha$.

i) Let M be an $\aleph_0$-saturated model containing B. Let $\psi(\overline{v})$ be an $\mathcal{L}_M$-formula such that $\psi(M) \subseteq \phi(M)$, $\mathrm{RM}(\psi) = \alpha$, and $\deg_M(\psi) = 1$.

Let $q = \{\theta(\overline{v}, \overline{b}) : \theta \text{ is an } \mathcal{L}\text{-formula}, \overline{b} \in B, \text{ and } \mathrm{RM}(\theta(\overline{v}, \overline{b}) \wedge \psi(\overline{v})) = \alpha\}$. Because $\mathrm{RM}(\psi) = \alpha$, for any $\mathcal{L}_B$-formula $\chi(\overline{v})$ either $\chi \in q$ or $\neg\chi \in q$. Because $\deg_M(\psi) = 1$ if $\chi_1, \chi_2 \in q$, then $\chi_1 \wedge \chi_2 \in q$. In particular, for any $\chi(\overline{v})$, exactly one of χ and $\neg\chi$ is in q. Thus $q \in S_n(M)$. Clearly, $p \subseteq q$ and $\mathrm{RM}(q) = \alpha$.

ii) Suppose that $A \subseteq B$ and $q_1, \ldots, q_m$ are distinct nonforking extensions of p. Let $\psi_i(\overline{v})$ be a formula of minimal rank and Morley degree in q_i. Then, $\mathrm{RM}(\psi_i) = \alpha$ while $\mathrm{RM}(\psi_i \wedge \psi_j) < \alpha$ for $i, j \leq d$ with $i \neq j$. Thus $m \leq \deg_M(\phi)$.

If M is an $\aleph_0$-saturated model containing A and $\mathrm{RM}(\phi) = d$, there are $\psi_1, \ldots, \psi_d$ such that $\mathrm{RM}(\psi_i) = \alpha$ while $\mathrm{RM}(\psi_i \wedge \psi_j) < \alpha$ for $i, j \leq d$ with $i \neq j$. As in i), we can find $q_i \in S_n(M)$ extending p such that $\mathrm{RM}(q_i) = \alpha$ and $\psi_i \in q_i$. Because $\psi_j \notin q_i$ for $i \neq j$, the q_i are distinct.

iii) Exercise.

---

[2]It is unfortunately typical of model-theoretic terminology that the desirable property (in this case nonforking) is defined as the negation of the undesirable property (forking).

Definable Types

To analyze nonforking extensions, we need the notion of a definable type.

Definition 6.3.3 We say that a type $p \in S_n(A)$ is *definable* over B if for each $\mathcal{L}$-formula $\phi(\overline{v}, \overline{w})$ there is an $\mathcal{L}_B$-formula $d_p\phi(\overline{w})$ such that

$$\phi(\overline{v}, \overline{a}) \in p \text{ if and only if } d_p\phi(\overline{a})$$

for all $\overline{a} \in A$.

We first show that in an ω-stable theory every type in $S_n(A)$ is definable over A. We need one lemma.

Lemma 6.3.4 *Suppose that $\mathcal{M}$ is $\aleph_0$-saturated, $\phi(\overline{v})$ is an $\mathcal{L}_M$-formula with $\mathrm{RM}(\phi) = \alpha$, and $\psi(\overline{v})$ is an $\mathcal{L}_M$-formula with $\mathrm{RM}(\phi \wedge \psi) = \alpha$. There is $\overline{a} \in M$ such that $\mathbb{M} \models \phi \wedge \psi(\overline{a})$.*

Proof We prove this by induction on α. If $\alpha = 0$, this is clear because $\phi(\mathbb{M})$ is finite and $\phi(\mathcal{M}) = \phi(\mathbb{M})$. Suppose that $\alpha > 0$. If $\deg_M(\phi) = d > 1$, then we can find $\mathcal{L}_M$-formulas $\theta_1(\overline{v}), \ldots, \theta_d(\overline{v})$ of Morley rank α and Morley degree 1 such that $\mathbb{M} \models \phi(\overline{v}) \leftrightarrow \bigvee \theta_i(\overline{v})$. We must have $\mathrm{RM}(\psi \wedge \theta_i) = \alpha$ for some i, and it suffices to find $\overline{a} \in M$ such that $\mathbb{M} \models \psi \wedge \theta_i(\overline{a})$. Thus, without loss of generality, we may assume that $\deg_M(\phi) = 1$.

If $\deg_M(\phi) = 1$, then $\mathrm{RM}(\phi \wedge \neg\psi) = \beta$ for some $\beta < \alpha$. Because $\mathcal{M}$ is $\aleph_0$-saturated, by Lemma 6.2.16 ii) we can find $\mathcal{L}_M$-formulas $\theta_0(\overline{v}), \theta_1(\overline{v}), \ldots$ such that $\mathrm{RM}(\theta_i) = \beta$ and $\theta_0(\mathbb{M}), \theta_1(\mathbb{M}), \ldots$ are pairwise disjoint subsets of $\phi(\mathbb{M})$. Because $\mathrm{RM}(\phi \wedge \neg\psi) = \beta$, $\mathrm{RM}(\neg\psi \wedge \theta_i) < \beta$ for some i (indeed, for all but finitely many i). Thus, $\mathrm{RM}(\psi \wedge \theta_i) = \beta$ and by induction there is $\overline{a} \in M$ such that $\mathbb{M} \models \psi \wedge \theta_i(\overline{a})$ and $\mathbb{M} \models \psi \wedge \phi(\overline{a})$.

Theorem 6.3.5 *Let $\mathcal{M}$ be an $\aleph_0$-saturated model, $\phi(\overline{v})$ be an $\mathcal{L}_M$-formula with $\mathrm{RM}(\phi) = \alpha$, and $\psi(\overline{v}, \overline{w})$ be an $\mathcal{L}$-formula. The set $\{\overline{b} \in \mathbb{M} : \mathrm{RM}(\psi(\overline{v}, \overline{b}) \wedge \phi(\overline{v})) = \alpha\}$ is definable with parameters from $\mathcal{M}$.*

Moreover, if $A \subseteq M$ and ϕ is an $\mathcal{L}_A$-formula, then $\{\overline{b} \in \mathbb{M} : \mathrm{RM}(\psi(\overline{v}, \overline{b}) \wedge \phi(\overline{v})) = \alpha\}$ is definable with parameters from A.

Proof We first argue that we may, without loss of generality, assume $\deg_M(\phi) = 1$. If $\deg_M(\phi) = d > 1$, let $\theta_1, \ldots, \theta_d$ be $\mathcal{L}_M$-formulas of Morley rank α and Morley degree 1 such that $\mathbb{M} \models \phi(\overline{v}) \leftrightarrow \bigvee \theta_i(\overline{v})$. Because

$$\{\overline{b} \in \mathbb{M} : \mathrm{RM}(\psi(\overline{v}, \overline{b}) \wedge \phi(\overline{v})) = \alpha\} = \bigcup_{i=1}^{n} \{\overline{b} \in \mathbb{M} : \mathrm{RM}(\psi(\overline{v}, \overline{b}) \wedge \theta_i(\overline{v})) = \alpha\},$$

it suffices to prove that each set $\{\overline{b} \in \mathbb{M} : \mathrm{RM}(\psi(\overline{v}, \overline{b}) \wedge \theta_i(\overline{v})) = \alpha\}$ is definable.

Claim For each $\overline{c} \in \mathbb{M}$ such that $\mathrm{RM}(\phi(\overline{v}) \wedge \psi(\overline{v}, \overline{c})) = \alpha$, there is a finite set $X_{\overline{c}} \subset \phi(\mathcal{M}) \cap \psi(\mathbb{M}, \overline{c})$ such that for all $\overline{b}$, if $X_{\overline{c}} \subset \psi(\mathbb{M}, \overline{b})$, then $\mathrm{RM}(\phi(\overline{v}) \wedge \psi(\overline{v}, \overline{b})) = \alpha$.

Suppose not. We build a sequence $\bar{a}_0, \bar{b}_0, \bar{a}_1, \bar{b}_1, \ldots$ violating the order property such that, $\bar{a}_i \in \phi(\mathcal{M}) \cap \psi(\mathbb{M}, \bar{c})$ and $\mathrm{RM}(\phi(\bar{v}) \wedge \psi(\bar{v}, \bar{b}_i)) < \alpha$ for all i. Suppose that we have already constructed $\bar{a}_0, \bar{b}_0, \ldots, \bar{a}_n, \bar{b}_n$. Because

$$\mathrm{RM}\left(\phi(\bar{v}) \wedge \psi(\bar{v}, \bar{c}) \wedge \bigwedge_{i=1}^{n} \neg\psi(\bar{v}, \bar{b}_i)\right) = \alpha,$$

by Lemma 6.3.4 there is

$$\bar{a}_{n+1} \in \phi(\mathcal{M}) \cap \psi(\mathbb{M}, \bar{c}) \setminus \bigcup_{i=1}^{n} \psi(\bar{v}, \bar{b}_i).$$

By assumption, we can find $\bar{b}_{n+1}$ such that $\{\bar{a}_0, \ldots, \bar{a}_{n+1}\} \subset \psi(\mathbb{M}, \bar{b}_{n+1})$ such that $\mathrm{RM}(\phi \wedge \psi(\bar{v}, \bar{b}_{n+1})) < \alpha$.

By construction, $\psi(\bar{a}_i, \bar{b}_j)$ if and only if $i < j$. This violates the order property.

Let $Y = \{X \subset \phi(\mathcal{M}) : X \text{ is finite, and if } X \subset \psi(\mathbb{M}, \bar{b}), \text{ then } \mathrm{RM}(\phi(\bar{v}) \wedge \psi(\bar{v}, \bar{b})) = \alpha\}$. For each $X \in Y$, let $\theta_X(\bar{w})$ be the $\mathcal{L}_M$-formula

$$\bigwedge_{\bar{x} \in X} \psi(\bar{x}, \bar{w}).$$

By the claim,

$$\mathrm{RM}(\phi(\bar{v}) \wedge \psi(\bar{v}, \bar{b})) = \alpha \Leftrightarrow \bigvee_{X \in Y} \theta_X(\bar{b}).$$

Thus, $\mathrm{RM}(\phi(\bar{v}) \wedge \psi(\bar{v}, \bar{w})) = \alpha$ is equivalent to an infinite disjunction of $\mathcal{L}_M$-formulas.

A similar argument with $\neg\psi(\bar{v}, \bar{w})$ shows that $\mathrm{RM}(\phi(\bar{v}) \wedge \neg\psi(\bar{v}, \bar{w})) = \alpha$ is equivalent to an infinite disjunction of $\mathcal{L}_M$-formulas.

Because $\mathbb{M}$ is saturated (see Exercise 4.5.34), there is a finite $Y_0 \subseteq Y$ such that

$$\bigvee_{X \in Y_0} \theta_X(\bar{w}) \leftrightarrow \bigvee_{X \in Y} \theta_X(\bar{w}).$$

The formula $\bigvee_{X \in Y_0} \theta_X(\bar{w})$ is an $\mathcal{L}_M$-formula defining $\{\bar{b} : \mathrm{RM}(\phi(\bar{v}) \wedge \psi(\bar{v}, \bar{w})) = \alpha\}$.

Suppose that $A \subseteq M$ and ϕ is an $\mathcal{L}_A$-formula. If σ is any automorphism of $\mathbb{M}$ fixing A pointwise, then, by Lemma 6.2.2,

$$\mathrm{RM}(\phi(\bar{v}) \wedge \psi(\bar{v}, \bar{b})) = \alpha \quad \text{if and only if} \quad \mathrm{RM}(\phi(\bar{v}) \wedge \psi(\bar{v}, \sigma(\bar{b}))) = \alpha$$

for all $\bar{b}$. Because $\{\bar{b} : \mathrm{RM}(\phi(\bar{v}) \wedge \psi(\bar{v}, \bar{b})) = \alpha\}$ is definable and fixed setwise by any automorphism of $\mathbb{M}$ that fixes A pointwise, this set is definable over A by Proposition 4.3.25.

Corollary 6.3.6 *If $p \in S_n(A)$, then p is definable over A_0 for some finite $A_0 \subseteq A$.*

Proof Let $\phi(\overline{v}) \in p$ be of minimal Morley rank and degree. Let $A_0 \subseteq A$ such that ϕ is an $\mathcal{L}_{A_0}$-formula, and let $\mathrm{RM}(\phi) = \alpha$. For any formula $\psi(\overline{v}, \overline{w})$ and $\overline{a} \in M$,

$$\psi(\overline{v}, \overline{a}) \in p \text{ if and only if } \mathrm{RM}(\phi(\overline{v}) \wedge \psi(\overline{v}, \overline{a})) = \alpha.$$

By Theorem 6.3.5, this is definable by an $\mathcal{L}_{A_0}$-formula.

The following corollary shows that if $\mathcal{M}$ is ω-stable and $D \subseteq M$ is $\emptyset$-definable, then any definable subset of D^n can be defined using only parameters from D.

Corollary 6.3.7 *Suppose that $A \subseteq M$ and $X \subseteq M^n$ is A-definable. Then, any $Y \subseteq X^m$ is $A \cup X$-definable.*

Proof Let $\psi(\overline{v}, \overline{b})$ define Y. Then $Y = \{\overline{c} \in X^n : \psi(\overline{c}, \overline{b}) \in \mathrm{tp}(\overline{b}/X)\}$. Because $\mathrm{tp}(\overline{b}/X)$ is definable over X, Y is definable over $A \cup X$.

We can use the definition of a type to understand the nonforking extensions. The situation is most clear for types of Morley degree 1.

Proposition 6.3.8 *Suppose that $p \in S_n(A)$ and $\deg_M(p) = 1$. Then, p is definable over A. If $B \supseteq A$, let*

$$p_B = \{\psi(\overline{v}, \overline{b}) : \mathbb{M} \models d_p\psi(\overline{b}), \overline{b} \in B, \psi \text{ an } \mathcal{L}\text{-formula}\}.$$

Then, q is the unique nonforking extension of p to B and q is definable over A—indeed, we can take $d_q\psi = d_p\psi$ for all $\mathcal{L}$-formulas ψ.

Proof Suppose that $p \in S_n(A)$, $\mathrm{RM}(p) = \alpha$, and $\deg_M(p) = 1$. Let $\phi(\overline{v})$ be an $\mathcal{L}_A$-formula with $\mathrm{RM}(\phi) = \alpha$ and $\deg_M(p) = 1$. If $\psi(\overline{v}, \overline{w})$ is an $\mathcal{L}$-formula, there is an $\mathcal{L}_A$-formula $d_p\psi$ such that $\mathrm{RM}(\phi \wedge \psi(\overline{v}, \overline{v})) = \alpha$ if and only if $\mathbb{M} \models d_p\phi(\overline{a})$. Suppose that $A \subseteq B$. The proof of Theorem 6.3.2 shows that

$$q = \{\psi(\overline{v}, \overline{b}) : \overline{b} \in B \text{ and } \mathrm{RM}(\phi(\overline{v}) \wedge \psi(\overline{v}, \overline{a})) = \alpha, \psi \text{ an } \mathcal{L}\text{-formula}\}$$

is the unique nonforking extension of p, but

$$q = \{\psi(\overline{v}, \overline{b}) : \mathbb{M} \models d_p\psi(\overline{b}), \overline{b} \in B, \psi \text{ an } \mathcal{L}\text{-formula}\}.$$

Thus, q is definable over A. Indeed we can use $d_q\phi = d_p\phi$ to define q.

We could replace A by any $A_0 \subseteq A$ such that there is an $\mathcal{L}_{A_0}$-formula of Morley rank α and Morley degree 1.

If $\deg_M(p) > 1$, the situation is more complicated. We will show that any nonforking extension of $p \in S_n(A)$ is definable over the algebraic closure of A in $\mathbb{M}^{\mathrm{eq}}$.

Suppose that $\mathcal{M}$ is $\aleph_0$-saturated, $A \subseteq M$, $p \in S_n(A)$, $q \in S_n(M)$, $p \subseteq q$, $\mathrm{RM}(p) = \mathrm{RM}(q) = \alpha$, and $\deg_M(q) = 1$. Let $\phi(\overline{v})$ and $\psi(\overline{v}, \overline{w})$ be $\mathcal{L}_A$-formulas and $\overline{b} \in M$ such that $\phi(\overline{v}) \in p$ is a formula of minimal rank and degree and $\psi(\overline{v}, \overline{b}) \in q$ has Morley rank α and Morley degree 1. We may assume that $\psi(\overline{v}, \overline{w})$ implies $\phi(\overline{v})$.

Let $X = \{\overline{c} : \mathrm{RM}(\psi(\overline{v}, \overline{c})) = \alpha$ and $\forall \overline{d}$ if $\mathrm{RM}(\psi(\overline{v}, \overline{d})) = \alpha$, then either $\mathrm{RM}(\psi(\overline{v}, \overline{c}) \wedge \psi(\overline{v}, \overline{d})) < \alpha$ or $\mathrm{RM}(\psi(\overline{v}, \overline{c}) \wedge \neg\psi(\overline{v}, \overline{d})) < \alpha\}$. By Theorem 6.3.5, X is definable over A.

Define $E \subseteq X \times X$ by

$$(\overline{c}, \overline{d}) \in E \Leftrightarrow \mathrm{RM}(\psi(\overline{v}, \overline{c}) \wedge \psi(\overline{v}, \overline{d})) = \alpha.$$

Claim E is an A-definable equivalence relation with at most $\deg_M(\phi)$ equivalence classes.

By Theorem 6.3.5, E is A-definable. It is clear that E is symmetric and reflexive. Suppose $\overline{c}E\overline{d}$ and $\neg(\overline{c}E\overline{e})$. Then, $\mathrm{RM}(\psi(\overline{v}, \overline{d}) \wedge \neg\psi(\overline{v}, \overline{c})) < \alpha$ and $\mathrm{RM}(\psi(\overline{v}, \overline{c}) \wedge \psi(\overline{v}, \overline{e})) < \alpha$. Because $\psi(\overline{v}, \overline{c}) \wedge \psi(\overline{v}, \overline{d})$ implies

$$(\psi(\overline{v}, \overline{d}) \wedge \neg\psi(\overline{v}, \overline{c})) \vee (\psi(\overline{v}, \overline{c}) \wedge \psi(\overline{v}, \overline{e})),$$

$\mathrm{RM}(\psi(\overline{v}, \overline{d}) \wedge \psi(\overline{v}, \overline{e})) < \alpha$. Thus, $\neg(\overline{d}E\overline{e})$ and E is an equivalence relation.

If $\overline{b}_1, \ldots, \overline{b}_m \in X$ are E-inequivalent, then $\psi(\overline{v}, \overline{b}_i)$ defines a Morley rank α subset of $\phi(\mathcal{M})$ and $\mathrm{RM}(\psi(\overline{v}, \overline{b}_i) \wedge \psi(\overline{v}, \overline{b}_j)) < \alpha$ for $i \neq j$. Thus $m \leq \deg_M(\phi)$.

Suppose that $\overline{c}E\overline{b}$. Because $\deg_M(\psi(\overline{v}, \overline{b})) = 1$, $\mathrm{RM}(\psi(\overline{v}, \overline{c})) = \alpha$, and $\mathrm{RM}(\psi(\overline{v}, \overline{c}) \wedge \neg\psi(\overline{v}, \overline{b})) < \alpha$, $\psi(\overline{v}, \overline{c})$ is also a formula in q of Morley rank α and Morley degree 1. Thus, q is definable over $A \cup \overline{c}$. In fact, for any formula $\theta(\overline{v}, \overline{z})$ and $\overline{a} \in M$, $\theta(\overline{v}, \overline{a}) \in q$ if and only if $\mathrm{RM}(\psi(\overline{v}, \overline{c}) \wedge \theta(\overline{v}, \overline{a})) = \alpha$. In particular, there is an $\mathcal{L}_A$-formula $\theta^*(\overline{w}, \overline{z})$ such that $\theta(\overline{v}, \overline{a}) \in q$ if and only if $\theta^*(\overline{c}, \overline{a})$. The definition of q depends on $\overline{b}/E$ rather than the choice of $\overline{c}$.

We think of $\overline{b}/E$ as an element of $\mathcal{M}^{eq}$ (see Lemma 1.3.10). Because E is definable over A and has only finitely many equivalence classes $\overline{b}/E$ is algebraic over A in $\mathcal{M}^{eq}$. We let $\mathrm{acl}^{eq}(A)$ denote the algebraic closure of A in $\mathcal{M}^{eq}$.

We summarize our analysis in the following theorem.

Theorem 6.3.9 *Suppose that $A \subseteq B$, $p \in S_n(A)$, $q \in S_n(B)$, and q does not fork over A. There is $\alpha \in \mathrm{acl}^{eq}(A)$ such that q is definable over $A \cup \{\alpha\}$. In other words, there is an A-definable equivalence relation E with finitely many classes and $\overline{a} \in M$ such that for any $\mathcal{L}$-formula $\phi(\overline{v}, \overline{w})$ there is an $\mathcal{L}_A$-formula $d_p\phi(\overline{w}, \overline{z})$ such that if $\overline{b}E\overline{a}$ then*

$$\phi(\overline{v}, \overline{d}) \in p \text{ if and only if } d_p\phi(\overline{d}, \overline{b})$$

for all $\overline{d} \in B$.

If $\deg_M(p) = 1$, then any nonforking extension of p is definable over A.

Proof We choose $\mathcal{M}$, an $\aleph_0$-saturated model containing B, and $q^* \in S_n(M)$, a nonforking extension of q. The argument above shows that q^*, and hence q, is definable over $\text{acl}^{\text{eq}}(A)$.

Corollary 6.3.10 *If $\mathcal{M} \models T$, $p \in S_n(\mathcal{M})$, $M \subseteq B$ and $q \in S_n(B)$ is a nonforking extension of p, then q is definable over $\mathcal{M}$.*

Proof If E is a definable equivalence relation with finitely many classes, then, because $\mathcal{M} \prec \mathbb{M}$, every $\bar{a} \in \mathbb{M}$ is equivalent to an element of M.

Our analysis can also be used to show that types over models have unique nonforking extensions.

Definition 6.3.11 We say that $p \in S_n(A)$ is *stationary* if, for all $B \supseteq A$, there is a unique nonforking extension of p to B.

Corollary 6.3.12 *Let $\mathcal{M} \models T$, and let $\phi(\bar{v})$ be an $\mathcal{L}_M$-formula with $\text{RM}(\phi) = \alpha$ and $\deg_M(\phi) = d$.*
 i) There is an $\mathcal{L}_M$-formula $\theta(\bar{v})$ such that $\theta(\mathbb{M}) \subseteq \phi(\mathbb{M})$, $\text{RM}(\theta) = \alpha$, and θ has Morley degree 1.
 ii) There are $\mathcal{L}_M$-formulas $\theta_1, \ldots, \theta_d$ such that each θ_i has Morley rank α and Morley degree 1 and $\phi(\mathbb{M})$ is the disjoint union of $\theta_1(\mathbb{M}), \ldots, \theta_d(\mathbb{M})$.
 iii) If $p \in S_n(\mathcal{M})$, then p has Morley degree 1. In particular, p is stationary.

Proof
 i) In our argument above, we had $\psi(\bar{v}, \bar{w})$ and $\bar{b} \in \mathbb{M}$ and an M-definable equivalence relation E with finitely many classes such that if $\bar{a} E \bar{b}$, then $\psi(\mathbb{M}, \bar{a})$ is a Morley rank α, Morley degree 1 subset of $\phi(\mathbb{M})$. Because E has only finitely many classes, there is $\bar{a} \in M$ such that $\bar{a} E \bar{b}$, so $\psi(\bar{v}, \bar{a})$ is the desired formula.

 ii) We inductively define $\theta_1, \ldots, \theta_d$. For $m < d$, if we are given $\theta_1, \ldots, \theta_{m-1}$, use i) to find θ_m defining a Morley rank α, Morley degree 1 subset of

$$\phi(\mathbb{M}) \setminus \bigcup_{i=1}^{m-1} \theta_i(\mathbb{M}).$$

Let

$$\theta_d = \phi \wedge \bigwedge_{i=1}^{d-1} \theta_i.$$

 iii) If $p \in S_n(\mathcal{M})$, choose $\phi(\bar{v}) \in p$ such that $\text{RM}(\phi) = \text{RM}(p)$, and let $\theta_1, \ldots, \theta_d$ be as in ii). Then p must contain one of the formulas θ_i, so p has Morley degree 1. Because the number of nonforking extensions is at most the Morley degree, p is stationary.

Corollary 6.3.12 generalizes from models to sets that are algebraically closed in $\mathbb{M}^{\text{eq}}$. We leave the proof to Exercise 6.6.24.

Corollary 6.3.13 *If $p \in S_n(\mathrm{acl}^{\mathrm{eq}}(A))$, then p is stationary.*

We next prove a partial converse to Theorem 6.3.9.

Lemma 6.3.14 *Suppose that $\mathcal{M} \models T$, $p \in S_n(M)$, $M \subseteq B$, $q \in S_n(B)$, $p \subseteq q$, and q is definable over M. Then, q is a nonforking extension of p.*

Proof There is $r \in S_n(B)$ a nonforking extension of p that is definable over M. Because q and r both extend p,

$$\mathcal{M} \models \forall \overline{w}\ d_r \phi(\overline{w}) \leftrightarrow d_q \phi(\overline{w})$$

for all formulas $\phi(\overline{v}, \overline{w})$. Because $\mathcal{M} \prec M$, $M \models d_r \phi(\overline{b}) \leftrightarrow d_q \phi(\overline{b})$ for all $\overline{b} \in B$. Thus $p = q$.

Independence

Forking can be used to give a notion of independence in ω-stable theories.

Definition 6.3.15 We say that $\overline{a}$ is *independent* from B over A if $\mathrm{tp}(\overline{a}/A)$ does not fork over $A \cup B$. We write $\overline{a} \underset{A}{\downarrow} B$.

This notion of independence has many desirable properties.

Lemma 6.3.16 (Monotonicity) *If $\overline{a} \underset{A}{\downarrow} B$ and $C \subseteq B$, then $\overline{a} \underset{A}{\downarrow} C$.*

Proof Because $\mathrm{RM}(\overline{a}/A) \geq \mathrm{RM}(\overline{a}/A \cup C) \geq \mathrm{RM}(\overline{a}/A \cup B)$, if $\mathrm{RM}(\overline{a}/A) = \mathrm{RM}(\overline{a}/A \cup B)$, then $\mathrm{RM}(\overline{a}/A) = \mathrm{RM}(\overline{a}/A \cup C)$.

Lemma 6.3.17 (Transitivity) *$\overline{a} \underset{A}{\downarrow} \overline{b}, \overline{c}$ if and only if $\overline{a} \underset{A}{\downarrow} \overline{b}$ and $\overline{a} \underset{A,\overline{b}}{\downarrow} \overline{c}$.*

Proof Because $\mathrm{RM}(\overline{a}/A, \overline{b}, \overline{c}) \leq \mathrm{RM}(\overline{a}/A, \overline{b}) \leq \mathrm{RM}(\overline{a}/A)$, $\mathrm{RM}(\overline{a}/A) = \mathrm{RM}(\overline{a}/A, \overline{b}, \overline{c})$ if and only if $\mathrm{RM}(\overline{a}/A) = \mathrm{RM}(\overline{a}/A, \overline{b})$ and $\mathrm{RM}(\overline{a}/A, \overline{b}) = \mathrm{RM}(\overline{a}/A, \overline{b}, \overline{c})$.

Lemma 6.3.18 (Finite Basis) *$\overline{a} \underset{A}{\downarrow} B$ if and only if $\overline{a} \underset{A}{\downarrow} B_0$ for all finite $B_0 \subseteq B$.*

Proof
($\Rightarrow$) This is clear because for any $B_0 \subseteq B$, $\mathrm{RM}(\overline{a}/A) \leq \mathrm{RM}(\overline{a}/A \cup B_0) \leq \mathrm{RM}(\overline{a}/A \cup B)$.

($\Leftarrow$) Suppose that $\overline{a} \underset{A}{\not\downarrow} B$. Then, there is $\phi(\overline{v}) \in \mathrm{tp}(\overline{a}/A \cup B)$ with $\mathrm{RM}(\phi) < \mathrm{RM}(\overline{a}/A)$. Let B_0 be a finite subset of B such that ϕ is an $\mathcal{L}_{A \cup B_0}$-formula. Then $\overline{a} \underset{A}{\not\downarrow} B_0$.

Lemma 6.3.19 (Symmetry) *If $\overline{a} \underset{A}{\downarrow} \overline{b}$, then $\overline{b} \underset{A}{\downarrow} \overline{a}$.*

Proof   Suppose that $\mathrm{RM}(\overline{a}/A\overline{b}) = \mathrm{RM}(\overline{a}/A)$. We must show that $\mathrm{RM}(\overline{b}/A\overline{a}) = \mathrm{RM}(\overline{b}/A)$. Let $\alpha = \mathrm{RM}(\overline{a}/A)$ and $\beta = \mathrm{RM}(\overline{b}/A)$.

We first assume that $A = M$, an $\aleph_0$-saturated model. Let $\phi(\overline{v}) \in \mathrm{tp}(\overline{a}/M)$ such that $\alpha = \mathrm{RM}(\phi)$ and $\deg_M(\phi) = 1$. Let $\psi(\overline{v}) \in \mathrm{tp}(\overline{b}/M)$ such that $\beta = \mathrm{RM}(\psi)$ and $\deg_M(\psi) = 1$. Suppose, for purposes of contradiction, that there is a formula $\theta(\overline{v}, \overline{w})$ such that $\mathbb{M} \models \theta(\overline{a}, \overline{b})$ and $\mathrm{RM}(\theta(\overline{a}, \overline{w})) < \beta$. By Theorem 6.3.5, there is an $\mathcal{L}_M$-formula $\chi(\overline{v})$ defining $\{\overline{x} : \mathrm{RM}(\psi(\overline{w}) \wedge \theta(\overline{x}, \overline{w})) < \beta\}$. Because $\mathrm{RM}(\overline{a}/M, \overline{b}) = \alpha$, the formula

$$\phi(\overline{v}) \wedge \theta(\overline{v}, \overline{b}) \wedge \chi(\overline{v})$$

has rank α. Thus, by Lemma 6.3.4, there is $\overline{a}' \in M$ such that $\theta(\overline{a}', \overline{b})$ and $\mathrm{RM}(\psi(\overline{w}) \wedge \theta(\overline{a}', \overline{w})) < \beta$, contradicting the fact that $\mathrm{RM}(\overline{b}/M) = \beta$.

For the general case, let $\mathcal{M}$ be an $\aleph_0$-saturated model containing A. Let $\overline{b}'$ realize a nonforking extension of $\mathrm{tp}(\overline{b}/A)$ to $\mathcal{M}$. Then $\mathrm{RM}(\overline{b}'/M) = \beta$. Because $\mathbb{M}$ is saturated, there is $\overline{c}$ such that $\mathrm{tp}(\overline{a}, \overline{b}/A) = \mathrm{tp}(\overline{c}, \overline{b}'/A)$. Let $\overline{a}'$ realize a nonforking extension of $\mathrm{tp}(\overline{c}/A, \overline{b}')$ to $M, \overline{b}'$. Then $\mathrm{RM}(\overline{a}'/M, \overline{b}') = \alpha = \mathrm{RM}(\overline{a}'/M)$. By the first part of the proof, $\mathrm{RM}(\overline{b}'/M, \overline{a}') = \mathrm{RM}(\overline{b}'/M) = \beta$. Hence

$$\mathrm{RM}(\overline{b}/A, \overline{a}) = \mathrm{RM}(\overline{b}'/A, \overline{a}') \geq \mathrm{RM}(\overline{b}'/M, \overline{a}') = \beta.$$

Thus $\mathrm{RM}(\overline{b}/A, \overline{a}) = \mathrm{RM}(\overline{b}/A)$.

Corollary 6.3.20 $\overline{a}, \overline{b} \mathrel{\smash{\underset{A}{\not\mkern-1mu\downarrow}}} C$ *if and only if* $\overline{a} \mathrel{\smash{\underset{A}{\not\mkern-1mu\downarrow}}} C$ *and* $\overline{b} \mathrel{\smash{\underset{A,\overline{a}}{\not\mkern-1mu\downarrow}}} C$.

Proof   Because forking occurs over a finite subset, it suffices to assume that C is a finite sequence $\overline{c}$.

$$
\begin{aligned}
\overline{a}, \overline{b} \mathrel{\smash{\underset{A}{\not\mkern-1mu\downarrow}}} \overline{c} \quad &\Leftrightarrow \quad \overline{c} \mathrel{\smash{\underset{A}{\not\mkern-1mu\downarrow}}} \overline{a}, \overline{b} \quad \text{by symmetry} \\
&\Leftrightarrow \quad \overline{c} \mathrel{\smash{\underset{A}{\not\mkern-1mu\downarrow}}} \overline{a} \text{ and } \overline{c} \mathrel{\smash{\underset{A,\overline{a}}{\not\mkern-1mu\downarrow}}} \overline{b} \quad \text{by transitivity} \\
&\Leftrightarrow \quad \overline{a} \mathrel{\smash{\underset{A}{\not\mkern-1mu\downarrow}}} \overline{c} \text{ and } \overline{b} \mathrel{\smash{\underset{A,\overline{a}}{\not\mkern-1mu\downarrow}}} \overline{c} \quad \text{by symmetry.}
\end{aligned}
$$

Symmetry also gives an easy proof that no type forks when it is extended to the algebraic closure.

Corollary 6.3.21   *For any* $\overline{a}$, $\overline{a} \mathrel{\smash{\underset{A}{\not\mkern-1mu\downarrow}}} \mathrm{acl}(A)$.

Proof   Suppose that $\overline{b} \in \mathrm{acl}(A)$. By Lemma 6.2.7 iii), $\mathrm{RM}(\overline{b}/A, \overline{a}) = \mathrm{RM}(\overline{b}/A) = 0$. Thus, $\overline{b} \mathrel{\smash{\underset{A}{\not\mkern-1mu\downarrow}}} \overline{a}$ and, by symmetry, $\overline{a} \mathrel{\smash{\underset{A}{\not\mkern-1mu\downarrow}}} \overline{b}$.

6.4   Uniqueness of Prime Model Extensions

Throughout this section T will be a complete theory in a countable language with infinite models.

In Theorem 4.2.20 we proved the existence of prime model extensions for ω-stable theories. In this section, we will prove that these extensions are unique up to isomorphism.

Constructible Models

Definition 6.4.1 Let $\mathcal{M}$ be an $\mathcal{L}$-structure and let $A \subseteq M$. Let δ be an ordinal and $(a_\alpha : \alpha < \delta)$ be a sequence of elements from M. Let $A_\alpha = A \cup \{a_\beta : \beta < \alpha\}$. We call $(a_\alpha : \alpha < \delta)$ a *construction* over A if $\mathrm{tp}(a_\alpha/A_\alpha)$ is isolated for all $\alpha < \delta$.

We say that $B \subseteq M$ is *constructible* over A if there is a construction $(a_\alpha : \alpha < \delta)$ such that $B = A \cup \{a_\alpha : \alpha < \delta\}$. We say that a model $\mathcal{M}$ is constructible over A if M is constructible over A.

Note that if $(a_\alpha : \alpha < \delta)$ is a construction over A, then it is also a construction over A_α for all $\alpha < \delta$.

Our proof of Theorem 4.2.20 can be broken into the following two assertions about constructible models.

Lemma 6.4.2 *i) If T is ω-stable, $\mathcal{M} \models T$, and $A \subset M$, there is $\mathcal{N} \prec \mathcal{M}$ constructible over A.*

ii) If $\mathcal{M}$ is constructible over A, then $\mathcal{M}$ is prime over A and every type realized in $\mathcal{M}$ is isolated over A.

We will prove the uniqueness of prime model extensions in two steps. We will first prove Ressayre's Theorem that constructible extensions are unique for any theory. We will then prove that in ω-stable theories all prime model extensions are constructible.

Definition 6.4.3 Suppose that $(a_\alpha : \alpha < \delta)$ is a construction of $\mathcal{M}$ over A. For each $\alpha < \delta$, let $\theta_\alpha(v)$ be a formula with parameters from A_α isolating $\mathrm{tp}(a_\alpha/A_\alpha)$. We say that $C \subseteq M$ is *sufficient* if whenever $a_\alpha \in C$, then all parameters from θ_α are in C.

Lemma 6.4.4 *Suppose that $(a_\alpha : \alpha < \delta)$ is a construction of $\mathcal{M}$ over A.*
i) Each A_α is sufficient.
ii) If C_i is sufficient for all $i \in I$, then $\bigcup_{i \in I} C_i$ is sufficient.
iii) If $X \subseteq M$ is finite, then there is a finite sufficient $C \supseteq X$.
iv) If $C \subseteq M$ is sufficient, then M is constructible over $A \cup C$.

Proof
i) and ii) are clear.

iii) We prove by induction on α that if $X \subseteq A_\alpha$ is finite, then there is a finite sufficient $C \supseteq X$. This is clear if $\alpha = 0$ or α is a limit ordinal. Suppose that $\alpha = \beta + 1$ and $X = X_0 \cup \{a_\beta\}$ where $X_0 \subseteq A_\alpha$. Let $B \subseteq A_\alpha$ be the parameters from θ_β. By induction, there is a finite sufficient $C \supseteq X_0 \cup B$. Then, $C \cup \{a_\beta\}$ is a finite sufficient set containing X.

iv) We must show that $\mathrm{tp}(a_\alpha/A_\alpha \cup C)$ is isolated for all $\alpha < \delta$. If $a_\alpha \in C$, this is trivial, so we will assume that $a_\alpha \notin C$. In this case, we claim that θ_α isolates $\mathrm{tp}(a_\alpha/A_\alpha \cup C)$. Suppose not. Then, there is an $\mathcal{L}_A$-formula $\psi(v, \overline{w})$ and $\overline{b} \in C$ such that $\psi(a_\alpha, \overline{b})$, but

$$\mathrm{Th}_{A_\alpha \cup C}(\mathcal{M}) \not\models \theta_\alpha(v) \rightarrow \psi(v, \overline{b}).$$

Thus, θ_α does not isolate $\mathrm{tp}(a_\alpha/A_\alpha, \overline{b})$. By ii) and iii) we may, without loss of generality, assume that $A_\alpha, \overline{b}$ is sufficient. We may assume that $\overline{b} = (a_{\alpha_1}, \ldots, a_{\alpha_n})$ where $\alpha < \alpha_1 < \ldots < \alpha_n$. Note that $A_\alpha \cup \{a_{\alpha_1}, \ldots, a_{\alpha_m}\}$ is sufficient for $m = 1, \ldots, n$.

Claim   For each $i = 1, \ldots, m$, there is an $\mathcal{L}_{A_\alpha}$-formula isolating $\mathrm{tp}(a_{\alpha_1}, \ldots, a_{\alpha_m}/A_{\alpha+1})$.

We prove this by induction on m. Suppose that the claim holds for $l < m$. The formula $\theta_{\alpha_{l+1}}$ isolates $\mathrm{tp}(a_\alpha/A_{\alpha_{l+1}})$. All parameters occurring in $\theta_{\alpha_{l+1}}$ are in $A_\alpha \cup \{a_{\alpha_1}, \ldots, a_{\alpha_l}\}$. Thus, $\theta_{\alpha_{l+1}}$ isolates $\mathrm{tp}(a_{\alpha_{l+1}}/A_{\alpha+1} \cup \{a_{\alpha_1}, \ldots, a_{\alpha_l}\})$.

Let $\theta_{\alpha_{l+1}}$ be $\psi(a_{\alpha_1}, \ldots, a_{\alpha_l}, v_{l+1})$ where $\psi(v_1, \ldots, v_{l+1})$ is an $\mathcal{L}_{A_\alpha}$-formula. By induction, there is an $\mathcal{L}_{A_\alpha}$-formula $\phi(v_1, \ldots, v_l)$ isolating $\mathrm{tp}(a_{\alpha_1}, \ldots, a_{\alpha_l}/A_{\alpha+1})$. As in the proof of Lemma 4.2.21, $\mathrm{tp}(a_{\alpha_1}, \ldots, a_{\alpha_{l+1}}/A_{\alpha+1})$ is isolated by $\phi(v_1, \ldots, v_l) \wedge \psi(v_1, \ldots, v_{l+1})$. This proves the claim.

Let $\psi(\overline{w})$ be an $\mathcal{L}_{A_\alpha}$-formula isolating $\mathrm{tp}(\overline{b}/A_{\alpha+1})$. Then, $\theta_\alpha(v) \wedge \psi(w)$ isolates $\mathrm{tp}(a_\alpha, \overline{b}/A_\alpha)$. Because a_α does not occur as a parameter in ψ, $\theta_\alpha(v)$ isolates $\mathrm{tp}(a_\alpha/A_\alpha, \overline{b})$, as desired.

We can now give Ressayre's proof of the uniqueness of constructible extensions.

Theorem 6.4.5 (Uniqueness of Constructible Models) *Suppose that $A \subseteq \mathbb{M}$, $\mathcal{M} \prec \mathbb{M}$, $\mathcal{N} \prec \mathbb{M}$, and $\mathcal{M}$ and $\mathcal{N}$ are constructible over A. The identity map on A extends to an isomorphism between $\mathcal{M}$ and $\mathcal{N}$.*

Proof   Let $(a_\alpha : \alpha < \delta)$ and $(b_\alpha : \alpha < \gamma)$ be the constructions of $\mathcal{M}$ and $\mathcal{N}$ over A. Let $\kappa = |M|$. Let $I = \{f : X \rightarrow N : f$ is partial elementary, $A \subset X$, X is sufficient in $\mathcal{M}$, $f|A$ is the identity, and $\mathrm{img}(X)$ is sufficient in $\mathcal{N}\}$. Because the identity map on A is in I and the union of a chain of elements of I is an element of I, we may apply Zorn's Lemma to get a maximal $f : X \rightarrow N$ in I. We claim that f is an isomorphism between $\mathcal{M}$ and $\mathcal{N}$. We must show that $\mathrm{dom}(f) = M$ and $\mathrm{img}(f) = N$.

Suppose that $a \in M \setminus X$. Let C_0 be a finite sufficient subset of M with $a \in C_0$. By Lemma 6.4.4 iv) and 6.4.2 ii), $\mathrm{tp}(C_0/X)$ is isolated. Thus, we can extend f to a partial elementary $f_0 : X \cup C_0 \to N$. Let $D_0 \supseteq f_0(C_0)$ be a finite sufficient subset of N. Because $\mathrm{tp}(D_0/f(X))$ is isolated, by Exercise 4.5.11, $\mathrm{tp}(D_0/f(X \cup C_0))$ is isolated. Thus, we can find a finite $C_0' \supseteq C_0$ and a surjective partial elementary $f_0' : X \cup C_0' \to f(X) \cup D_0$. Let C_1 be a finite sufficient set containing C_0'. By the same argument, $\mathrm{tp}(C_1/X \cup C_0')$ is isolated and we may extend f_0' to a partial elementary $f_1 : C_1 \to M$.

Continuing in this manner, we build $C_0 \subseteq C_1 \subseteq C_2 \subseteq \ldots$, a sequence of finite sufficient subsets of M, and $f \subseteq f_0 \subseteq f_1 \subseteq \ldots$ where $f_i : X \cup C_i \to N$ is partial elementary and $f_i(C_i)$ is contained in a sufficient subset of $f_{i+1}(C_{i+1})$. If $g = \bigcup f_i$, then $g \in I$, contradicting the maximality of f. Thus $X = M$.

A symmetric argument shows that $\mathrm{img}(f) = N$. Thus, f is the desired isomorphism.

Prime Models of ω-Stable Theories

It remains to show that if T is ω-stable, then any prime model extension is constructible. We need one lemma relating forking and isolation in ω-stable theories. This lemma is a special case of the Open Mapping Theorem (see Exercise 6.6.30).

Lemma 6.4.6 Let T be ω-stable. If $A \subseteq B$, $p \in S_n(A)$, $p' \in S_n(B)$, p' is a nonforking extension of p, and p' is isolated, then p is isolated.

Proof We work in $\mathbb{M}^{\mathrm{eq}}$. Let $\phi(\overline{v}, \overline{b})$ isolate p'. Let $q = \mathrm{tp}(\overline{b}/\mathrm{acl}^{\mathrm{eq}}(A))$. By Theorem 6.3.9, there is an $\mathcal{L}_{\mathrm{acl}^{\mathrm{eq}}(A)}$-formula $\psi_0(\overline{v})$ such that

$$\psi_0(\overline{a}) \text{ if and only if } \phi(\overline{a}, \overline{w}) \in \mathrm{tp}(\overline{b}/\mathrm{acl}^{\mathrm{eq}}(A))$$

(ψ_0 is just $d_q\phi$, where we interchange the roles of the variables). Let $\psi_0(\overline{v})$ be $\psi_1(\overline{v}, \overline{\alpha})$ where $\psi_1(\overline{v}, \overline{u})$ is an $\mathcal{L}_A$-formula and $\overline{\alpha} \in \mathrm{acl}^{\mathrm{eq}}(A)$. Let $\theta(\overline{u})$ isolate $\mathrm{tp}(\overline{\alpha}/A)$, and let $\psi(\overline{v})$ be $\exists \overline{u} \ (\psi_1(\overline{v}, \overline{u}) \wedge \theta(\overline{u}))$.

We claim that ψ isolates p. Suppose that $r \in S_n(B)$ and $\psi \in r$. Then, there is $\overline{\beta} \in \mathrm{acl}^{\mathrm{eq}}(A)$ such that $\theta(\overline{\beta})$ and $r \cup \{\psi_1(\overline{v}, \overline{\beta})\}$ is satisfiable. Because θ isolates $\mathrm{tp}(\overline{\alpha}/A)$, $r \cup \{\psi_1(\overline{v}, \overline{\alpha})\}$ is satisfiable. Let $r' \in S_n(\mathrm{acl}^{\mathrm{eq}}(A))$ be an extension of r with $\psi_1(\overline{v}, \overline{\alpha}) \in r'$. By Corollary 6.3.21, r' does not fork over A. Let r'' be a nonforking extension of r' to $\mathrm{acl}^{\mathrm{eq}}(A) \cup \{\overline{b}\}$, and let $\overline{a}$ realize r''. Because $\overline{a} \underset{\mathrm{acl}^{\mathrm{eq}}(A)}{\downarrow} \overline{b}$, by symmetry $\overline{b} \underset{\mathrm{acl}^{\mathrm{eq}}(A)}{\downarrow} \overline{a}$. Thus, $\overline{b}$ realizes the unique nonforking extension of q to $\mathrm{acl}^{\mathrm{eq}}(A)$. But this type has the same definition as q. Because $\mathbb{M} \models \psi_1(\overline{a}, \overline{\alpha})$, $\mathbb{M} \models \phi(\overline{a}, \overline{b})$. Because ϕ isolates p', $\overline{a}$ realizes p' and thus $\overline{a}$ realizes $\overline{b}$.

We can now prove the main lemma needed to show that a prime model of an ω-stable theory is constructible.

Lemma 6.4.7 *Suppose that T is ω-stable, $\mathcal{M}$ is constructible over A, and $A \subseteq B \subseteq M$. Then, B is constructible over A.*

Proof Let $(a_\alpha : \alpha < \kappa)$ be an enumeration of M. We start by building $(X_\alpha : \alpha < \kappa)$, a sequence of subsets of M such that:
i) $X_0 = A$, $X_\alpha \subseteq X_\beta$ for $\alpha < \beta < \kappa$, and each X_α is sufficient;
ii) $a_\alpha \in X_{\alpha+1}$ for all $\alpha < \kappa$;
iii) $|X_{\alpha+1} \setminus X_\alpha| \leq \aleph_0$;
iv) if $\bar{d}$ is a sequence from X_α, then $\bar{d} \downharpoonright_{X_\alpha \cap B} B$ for all α.

If α is a limit ordinal, we can take $X_\alpha = \bigcup_{\beta < \alpha} X_\beta$.

Suppose that we have built X_α. Let C_0 be a finite sufficient set containing a_α. Let $\bar{c}_0$ be an enumeration of C_0. We can find $\bar{b} \in B$ such that $\bar{c}_0 \downharpoonright_{X_\alpha, \bar{b}} B$. Let C_1 be a finite sufficient set containing $C_0 \cup \{\bar{b}\}$. Continuing in this way, we build a sequence of finite sufficient sets $C_0 \subseteq C_1 \subseteq C_2 \subseteq \ldots$ such that if $\bar{c}_n$ is an enumeration of C_n, then

$$\bar{c}_n \downharpoonright_{X_\alpha \cup (B \cap C_{n+1})} B. \tag{1}$$

Let $X_{\alpha+1} = X_\alpha \cup \bigcup_{n=0}^{\infty} C_n$. Let $\bar{x} \in X_\alpha$. There is $\bar{y} \in X_\alpha$ such that

$$\bar{c}_n \downharpoonright_{(X_{\alpha+1} \cap B) \cup \{\bar{x}, \bar{y}\}} B. \tag{2}$$

By our inductive assumption,

$$\bar{x}, \bar{y} \downharpoonright_{X_\alpha \cap B} B \text{ and hence } \bar{x}, \bar{y} \downharpoonright_{X_{\alpha+1} \cap B} B. \tag{3}$$

By Corollary 6.3.20, it follows from (2) and (3) that

$$\bar{x}, \bar{y}, \bar{c}_n \downharpoonright_{B \cap X_{\alpha+1}} B.$$

It follows that

$$\bar{d} \downharpoonright_{B \cap X_{\alpha+1}} B)$$

for any sequence $\bar{d} \in X_{\alpha+1}$.

Let $B_\alpha = X_\alpha \cap B$.

Claim If $\bar{b} \in B$, then $\text{tp}(\bar{b}/B_\alpha)$ is isolated for each α.

Because X_α is sufficient, $\mathcal{M}$ is constructible over X_α and $\text{tp}(\bar{b}/X_\alpha)$ is isolated by some formula $\phi(\bar{v}, \bar{d})$ with $\bar{d} \in X_\alpha$. By iv), $\bar{d} \downharpoonright_{B_\alpha} \bar{b}$. By symmetry $\bar{b} \downharpoonright_{B_\alpha} \bar{d}$. Because $\text{tp}(\bar{b}/B_\alpha, \bar{d})$ is isolated $\text{tp}(\bar{b}/B_\alpha)$ is isolated, by Lemma 6.4.6.

If $b_0, b_1, \ldots$ is an enumeration of $B_{\alpha+1} \setminus B_\alpha$, then, by the claim and Exercise 4.5.11, $(b_0, b_1, \ldots)$ is a construction over B_α. Combining these constructions, we see that B is constructible over A.

Theorem 6.4.8 (Uniqueness of Prime Models) *Suppose that T is ω-stable, $A \subseteq M$, $\mathcal{M}_0 \prec M$, $\mathcal{M}_1 \prec M$, and $\mathcal{M}_0$ and $\mathcal{M}_1$ are prime models over A. The identity map on A extends to an isomorphism between $\mathcal{M}_0$ and $\mathcal{M}_1$.*

Proof By Lemma 6.4.2, there is $\mathcal{N} \prec M$, a constructible model over A. Because each $\mathcal{M}_i$ is prime over A, we can find an elementary embedding of $\mathcal{M}_i$ into $\mathcal{N}$. By the previous lemma, each $\mathcal{M}_i$ is constructible. Thus, by the uniqueness of constructible models, there is $f : \mathcal{M}_0 \to \mathcal{M}_1$, an isomorphism fixing A pointwise.

Differential Closures

We give one application of the uniqueness of prime models.

Definition 6.4.9 Let K and k be differential fields with $k \subseteq K$. We say that K is a *differential closure* of k if K is differentially closed and for any differentially closed $L \supseteq k$, there is a differential field embedding $f : K \to L$ fixing k pointwise.

Because the theory of differentially closed fields has quantifier elimination (Theorem 4.3.32), a differential field embedding of differentially closed fields is an elementary embedding. Thus, a differential closure of k is a model of DCF prime over k. In Exercise 4.5.43, we argued that DCF is ω-stable.

Theorem 6.4.10 *Let k be a differential field. There is $K \supseteq k$ a differential closure of k. If K and L are differential closures of k, then K and L are isomorphic over k. If K is a differential closure of k, then $\text{tp}(\bar{a}/k)$ is isolated for all $\bar{a} \in K$.*

Proof Because DCF is ω-stable and a differential closure of k is a prime model of DCF over k, this follows from the existence and uniqueness of prime model extensions for ω-stable theories (Theorems 4.2.20 and 6.4.8).

Because the differential closure is a prime model over k, every $\bar{a} \in K$ realizes an isolated type over k.

6.5   Morley Sequences

In Theorem 4.3.15 we proved that if T is ω-stable and κ is a regular cardinal, then T has a saturated model of cardinality κ. In this section we will extend this result to singular cardinals. The new idea we will need is the notion of a Morley sequence.

Throughout this section T will be a complete ω-stable theory in a countable language and M will be a monster model of T.

Definition 6.5.1 Suppose that $p \in S_1(A)$ and δ is an ordinal. We say that $(a_\alpha : \alpha < \delta)$ is a *Morley sequence* for p over A if for each α the type $\mathrm{tp}(a_\alpha / A \cup \{a_\beta : \beta < \alpha\})$ is an extension of p of the same Morley rank and degree.

In particular each a_α is a realization of p with $a_\alpha \underset{A}{\downarrow} \{a_\beta : \beta < \alpha\}$. Recall from Theorem 6.3.2 that for any $B \subseteq A$ there is at most one nonforking extension of A to B of the same Morley degree. If $\deg_M(p) = 1$, then p is stationary and there is a unique nonforking extension of p to any $B \supseteq A$. We first show that Morley sequences are sets of indiscernibles.

Theorem 6.5.2 *Suppose that* $I = (a_\alpha : \alpha < \delta)$ *is an infinite Morley sequence for* p *over* A. *Then,* I *is an infinite set of indiscernibles.*

Proof By Theorem 5.2.13, it suffices to show that I is a sequence of order indiscernibles. Let $d = \deg_M(p)$. We will show by induction on n that $\mathrm{tp}(a_{\alpha_1}, \ldots, a_{\alpha_n}/A) = \mathrm{tp}(a_{\beta_1}, \ldots, a_{\beta_n}/A)$ if $\alpha_1 < \ldots, \alpha_n < \delta$, $\beta_1 < \ldots <$ $\beta_n < \delta$. For $n = 1$, a_{α_1} and a_{β_1} are both realizations of p so $\mathrm{tp}(a_{\alpha_1}/A) = \mathrm{tp}(a_{\beta_1}/A)$.

Suppose that $\mathrm{tp}(a_{\alpha_1}, \ldots, a_{\alpha_{n-1}}/A) = \mathrm{tp}(a_{\beta_1}, \ldots, a_{\beta_{n-1}}/A)$. Let σ be an automorphism of $\mathbb{M}$ with $\sigma(a_{\alpha_i}) = a_{\beta_i}$ for $i < n$. Because $a_{\alpha_n} \underset{A}{\downarrow} \{a_\gamma : \gamma < \alpha_n\}$, $a_{\alpha_n} \underset{A}{\downarrow} \{a_{\alpha_1}, \ldots, a_{\alpha_{n-1}}\}$ and a_{α_n} realizes the unique nonforking extension p to $A \cup \{a_{\alpha_1}, \ldots, a_{\alpha_{n-1}}\}$ of Morley degree d. Thus, $\sigma(a_{\alpha_n})$ realizes the unique nonforking extension of p to $A \cup \{a_{\beta_1}, \ldots, a_{\beta_{n-1}}\}$ of Morley degree d. Because a_{β_n} also realizes the unique nonforking extension of p to $A \cup \{a_{\beta_1}, \ldots, a_{\beta_{n-1}}\}$ of Morley degree d,

$$\mathrm{tp}(a_{\alpha_1}, \ldots, a_{\alpha_n}/A) = \mathrm{tp}(\sigma(a_{\alpha_1}), \ldots, \sigma(a_{\alpha_n})/A) = \mathrm{tp}(a_{\beta_1}, \ldots, a_{\beta_n}/A),$$

as desired.

Lemma 6.5.3 *Suppose that* I *is an infinite set of indiscernibles over* $A \subset$ $\mathbb{M}$. *For any* $\bar{b} \in \mathbb{M}$, *there is a finite* $J \subset I$ *such that* $I \setminus J$ *is a set of indiscernibles over* $A \cup \{\bar{b}\}$.

Proof Let $p = \mathrm{tp}(\bar{b}/A \cup I)$. There is a finite $J \subset I$ such that $p|A \cup J$ has the same Morley rank and degree as p. Let $x_1, \ldots, x_n, y_1, \ldots, y_n \subset I \setminus J$ with $x_i \neq x_j$ and $y_i \neq y_j$ for $i < j \leq n$. Because I is a set of indiscernibles over A, there is a partial elementary σ with domain $A \cup J \cup \{\bar{x}\}$ fixing $A \cup J$ with $\sigma(x_i) = y_i$ for $i \leq n$. Let $q = \mathrm{tp}(\bar{b}/A \cup J \cup \{\bar{x}\})$ and $r = \mathrm{tp}(\bar{b}/A \cup J \cup \{\bar{y}\})$. By choice of J, $\mathrm{RM}(p) = \mathrm{RM}(q) = \mathrm{RM}(r)$ and $\deg_M(p) = \deg_M(q) = \deg_M(r)$. Because r is the unique extension of p to $A \cup J \cup \{\bar{y}\}$ of the same rank and degree, we must have $\sigma q = r$. Thus, we can extend σ to $A \cup J \cup \{\bar{x}, \bar{b}\}$ with $\sigma(\bar{b}) = \bar{b}$.

In particular,

$$\phi(\bar{x}) \leftrightarrow \phi(\bar{y})$$

for any $\mathcal{L}_{A,\bar{b}}$-formula $\phi(\bar{v})$. Thus, $I \setminus J$ is an infinite set of indiscernibles over $A, \bar{b}$.

We can now prove that there are saturated models of all infinite cardinalities.

Theorem 6.5.4 *Let T be a complete ω-stable theory in a countable language, and let κ be an infinite cardinal. There is a saturated $\mathcal{M} \models T$ with $|M| = \kappa$.*

Proof In Theorem 4.3.15, we proved this when κ is regular. Thus, we may assume that κ is singular. We can use Theorem 4.3.15 to build an elementary chain $(\mathcal{M}_\alpha : \alpha < \kappa)$, and each $\mathcal{M}_\alpha$ is a saturated model of T of cardinality $(\aleph_0 + |\alpha|)^+$. Let $\mathcal{M} = \bigcup_{\alpha < \kappa} \mathcal{M}_\alpha$. Then, $|M| = \kappa$ and we will prove that $\mathcal{M}$ is saturated.

Let $A \subset M$ with $|A| < \kappa$, and let $q \in S_1(A)$. We must show that q is realized in $\mathcal{M}$. There is a finite $A_0 \subseteq A$ such that q does not fork over A_0. Without loss of generality, we may assume that $A_0 \subseteq M_0$. By taking a nonforking extension of q to $A \cup M_0$, we may also, without loss of generality, assume that $M_0 \subseteq A$. Let $p = q|M_0$. Because $\mathcal{M}_0$ is a model, p is stationary by Corollary 6.3.12.

We begin by building a Morley sequence $\{a_\alpha : \alpha < \kappa\}$ for p such that for each $\delta < \kappa$ there is a $\gamma < \kappa$ such that $\{a_\alpha : \alpha < \delta\} \subset M_\gamma$. Given $(a_\alpha : \alpha < \delta) \subset M_\gamma$, choose $\beta \geq \gamma$ such that $\mathcal{M}_\beta$ is $|M_0 \cup \delta|^+$-saturated. Let p_δ be the unique nonforking extension of p to $M_0 \cup \{a_\alpha : \alpha < \delta\}$. Because $\mathcal{M}_\beta$ is saturated, we can find $a_\delta \in \mathcal{M}_\beta$ realizing p_δ.

We now extend this sequence to a Morley sequence in $\mathbb{M}$ of ordinal length $\kappa + \kappa$ as follows. Let p_κ be the unique nonforking extension of p to M. Note that $p_\kappa \supset q$. Let a_κ realize p_κ. For $\kappa < \delta < \kappa + \kappa$, let $a_\delta \in \mathbb{M}$ realize p_δ the unique nonforking extension of p to $M \cup \{a_\alpha : \kappa \leq \alpha < \delta\}$.

Because $I = (a_\alpha : \alpha < \kappa + \kappa)$ is a Morley sequence, by Theorem 6.5.2 it is an infinite set of indiscernibles over M_0. Suppose that $\bar{b}$ is a finite sequence from A. By Lemma 6.5.3, we can find a finite $J_{\bar{b}} \subset I$ such that $I \setminus J_{\bar{b}}$ is a set of indiscernibles over $M_0, \bar{b}$. Let $J = \bigcup_{\bar{b} \in A^{<\omega}} J_{\bar{b}}$. Then, $|J| = |A| < \kappa$ and $I \setminus J$ is a set of indiscernibles over A.

Choose $\alpha < \kappa$ and $\beta > \kappa$ such that $a_\alpha, a_\beta \in I \setminus J$. Because $p_\kappa \supset q$, a_β realizes q. Thus, by indiscernibility, a_α realizes q. But $a_\alpha \in M$, as desired.

We give one further application of Morley sequences to show that all uncountable models of ω-stable theories contain infinite sets of indiscernibles. An interesting feature of this proof is that we do not use any partition theorems.

Theorem 6.5.5 *Suppose that T is ω-stable, $\mathcal{M} \models T$, $A \subset M$, $|A| < |M|$, and $|M| \geq \aleph_1$. There is $I \subset M$ an infinite set of indiscernibles over A.*

Proof If $\kappa = |M|$ is a limit cardinal we could find $\mathcal{N} \prec \mathcal{M}$ such that $A \subset N$, $|A| < |N|$, and $|N|$ is a successor cardinal. Thus, we may, without loss of generality, assume that κ is regular.

If $B \subset M$ and $|B| < \kappa$, then $|S_1(B)| < \kappa$ and, because κ is regular, there is $p \in S_1(B)$ such that p has κ realizations in M. We call such a p *large*. Let (γ, d) be least in the lexicographic order such that there is $M \supseteq B \supseteq A$ with $|B| < \kappa$, $p \in S_1(B)$ large, $\mathrm{RM}(p) = \gamma$, and $\deg_M(p) = d$. Choose A_0 and $p_0 \in S_1(A_0)$ large with $\mathrm{RM}(p_0) = \gamma$ and $\deg_M(p_0) = d$.

Claim i) Suppose that $M \supseteq B \supseteq A_0$ and $|B| < \kappa$. There is a unique $p_B \in S_1(B)$ with $p_B \supseteq p_0$ and p_B large.

ii) $\mathrm{RM}(p_B) = \gamma$ and $\deg_M(p_B) = d$.

iii) If $A_0 \subseteq B \subseteq C \subset M$ and $|C| < \kappa$, then $p_C \supseteq p_B$.

Because p_0 is large and $|S_n(B)| < \kappa$, there is a large $p_B \supseteq p_0$ in $S_n(B)$. Because $p_B \supseteq p_0$, $\mathrm{RM}(p_B) \leq \gamma$. By our choice of A_0, $\mathrm{RM}(p_B) = \gamma$. If $q \in S_n(B)$, $q \neq p_B$, $q \supset p_0$, and q is large, then $\mathrm{RM}(q) = \gamma$ as well. Because q and p_B are both nonforking extensions of p_0 to B, both must have Morley degree less than d, contradicting our choice of A_0. Thus, p_B is unique and $\deg_M(p_B) = d$.

If $A_0 \subseteq B \subseteq C$, then $p_C|B$ is a large extension of p_0. Because p_B is the unique large extension of p_0 to B, $p_B = p_C|B$.

We build a Morley sequence $(a_\alpha : \alpha < \kappa)$ as follows. Let $a_0 \in M$ realize p_0. Given $(a_\alpha : \alpha < \delta)$, let $A_\delta = A_0 \cup \{a_\alpha : \alpha < \delta\}$, and let a_δ realize p_{A_δ}. This is possible because p_{A_δ} is large. By the claim, $(p_{A_\alpha} : \alpha < \kappa)$ is an increasing sequence of types of Morley rank α and Morley degree d. Thus, $(a_\alpha : \alpha < \kappa)$ is a Morley sequence over A_0. By Theorem 6.5.2, $(a_\alpha : \alpha < \kappa)$ is a set of indiscernibles over $A_0 \supseteq A$.

6.6   Exercises and Remarks

Exercise 6.6.1 Show that the theory of the group $(\mathbb{Z}/4\mathbb{Z})^\omega$ is κ-categorical for all uncountable cardinals but not strongly minimal.

Exercise 6.6.2 Give an example of an uncountably categorical theory where there is no strongly minimal formula over $\emptyset$.

Exercise 6.6.3 Let $\mathcal{M} \models \mathrm{Th}(\mathbb{Z}, s)$. We can find a set X such that $\mathcal{M} \cong X \times \mathbb{Z}$ where $s(x, n) = (x, n + 1)$. Show that $\dim(\mathcal{M}) = |X|$.

Exercise 6.6.4 Prove Lemma 6.1.6 when $A = \emptyset$.

Exercise 6.6.5 a) Give examples of uncountably categorical theories where the dimensions of the prime models are $0, 1$, and $\aleph_0$.

b) Let $\mathcal{L} = \{R\}$, where R is a ternary relation symbol. Suppose that V is a $\mathbb{Q}$-vector space. We view V as an $\mathcal{L}$-structure by interpreting R as

$\{(a, b, c) : a + b + c = 0\}$. Show that the $\mathcal{L}$-theory of V is uncountably categorical and the prime model has dimension 2.

Exercise 6.6.6 Prove Lemma 6.2.7 i) and ii).

Exercise 6.6.7 Suppose that T is ω-stable and $\mathcal{M} \models T$ is $\aleph_0$-saturated, and ϕ is an $\mathcal{L}_M$-formula with $\mathrm{RM}(\phi) = \alpha$ an ordinal. Show that there is a maximum d such that there are $\mathcal{L}_M$-formulas $\psi_1, \ldots, \psi_d$ such that $\psi_1(\mathcal{M}), \ldots, \psi_d(\mathcal{M})$ are disjoint subsets of $\phi(\mathcal{M})$ and $\mathrm{RM}(\psi_i) = \alpha$ for $i = 1, \ldots, d$. Show that for any $\mathcal{N} \models T$, if N contains the parameters occurring in ϕ and $\psi_1, \ldots, \psi_n$ are $\mathcal{L}_N$-formulas such that $\psi_1(\mathcal{N}), \ldots, \psi_n(\mathcal{N})$ are disjoint subsets of $\phi(\mathcal{N})$ and $\mathrm{RM}(\psi_i) = \alpha$, then $n \leq d$. Prove these results without using the monster model assumptions. (This is the type of argument one needs to do to completely avoid using monster models.)

Exercise 6.6.8 Suppose that T is ω-stable and ϕ is an $\mathcal{L}_M$-formula with $\mathrm{RM}(\phi) = \alpha$. Show that for all ordinals $\beta < \alpha$ there is an $\mathcal{L}_M$-formula ψ such that $\mathbb{M} \models \psi \rightarrow \phi$ and $\mathrm{RM}(\psi) = \beta$.

Exercise 6.6.9 Suppose that T is a complete ω-stable theory. Show that there is an ordinal $\alpha < \omega_1$ such that for all $A \subset \mathbb{M}$ if $p \in S_n(A)$, then $\mathrm{RM}(p) < \alpha$.

Exercise 6.6.10 Show that $\mathrm{RM}(\overline{a}) \geq 0$ for all $\overline{a} \in \mathbb{M}$.

Exercise 6.6.11 Show that $\mathrm{RM}(\overline{a}, \overline{b}/A) \geq \mathrm{RM}(\overline{a}/A)$.

Exercise 6.6.12 Show that in a strongly minimal theory the notions of independence from Definitions 6.3.15 and 6.1.5 agree.

Exercise 6.6.13 Prove Lemma 6.2.16.

Exercise 6.6.14 Suppose that K is an algebraically closed field, $X \subseteq K^n$ is constructible, and $V \supseteq X$ is the closure of X in the Zariski topology. Show that $\mathrm{RM}(V \setminus X) < \mathrm{RM}(X)$ and hence $\mathrm{RM}(X) = \mathrm{RM}(V)$. [Hint: By quantifier elimination, it suffices to prove this when V is irreducible, O is Zariski open, and $X = V \cap O$.]

Exercise 6.6.15 a) Prove Lemma 6.2.25.

b) Show that if $V \subseteq K^n$ is an irreducible Zariski closed set and $a, b \in V(\mathbb{K})$ are generic points of V, then $\mathrm{tp}(a/K) = \mathrm{tp}(b/K)$.

c) Suppose that $V \subseteq K^n$ is a Zariski closed set, $K \prec \mathbb{K}$, and $a \in V(\mathbb{K})$. Then, $\mathrm{RM}(a/K) = \mathrm{RM}(V)$ if and only if a is the generic point of an irreducible $W \subseteq V$ with $\mathrm{RM}(W) = \mathrm{RM}(V)$.

Exercise 6.6.16 Recall that if F, K and L are fields with $F \subseteq K \cap L$, then K and L are *free* over F if whenever $a_1, \ldots, a_n \in K$ are algebraically dependent over L, they are already algebraically dependent over F.

Let $\mathbb{K}$ be a saturated algebraically closed field, $\overline{a}, \overline{b} \in \mathbb{K}$ and $F \subset \mathbb{K}$ a subfield. Show that $\overline{a} \mathop{\smile\hspace{-0.9em}\vert}_F \overline{b}$ if and only if $F(\overline{a})$ and $F(\overline{b})$ are free over F.

Exercise 6.6.17 We define the Morley rank of a theory to be the Morley rank of the formula $v = v$.

a) Let $\mathcal{L} = \{E\}$, where E is a binary relation symbol. Let T be the theory of an equivalence relation with infinitely many classes, each of which is infinite. Show that $\mathrm{RM}(T) = 2$.

b) Let $\mathcal{L} = \{P_0, P_1, \ldots\}$ where each P_i is a unary predicate. Let T be the theory that asserts $P_0 \supset P_1 \supset \ldots$, $\neg P_0$ is infinite, $P_n \setminus P_{n+1}$ is infinite for each n. Show that $\mathrm{RM}(T) = 2$.

c) For each $n < \omega$, give an example of a theory with $\mathrm{RM}(T) = n$.

d) Let $K \subset F$ be algebraically closed fields of characteristic 0. Let $\mathcal{L} = \{+, \cdot, U, 0, 1\}$, where U is a unary predicate, let $\mathcal{M}$ be the $\mathcal{L}$-structure with universe F where $U(\mathcal{M}) = K$, and let T be the theory of $\mathcal{M}$. Show that K is an ω-stable theory and that the formula $v = v$ has Morley rank ω. [Hint: For example, if $x \in K \setminus F$, $\{y \in K : \exists a, b \in F \ y = ax + b\}$ is in definable bijection with F^2 and has Morley rank at least 2.]

Exercise 6.6.18 † Let K be a differentially closed field. We follow the notation of Exercise 4.5.43. If $p \in S_1(X)$ is a type, let $f(X) \in I_p$ be of minimal order and degree. We let $\mathrm{ord}(p)$ be the order of f, if $I_p \neq \{0\}$. Otherwise, we let $\mathrm{ord}(p) = \omega$. Let $V_p = \{\overline{x} \in K_n : g(\overline{x}) = 0 \text{ for all } g \in I_p\}$. We will need one fact from differential algebra (see [65]). If $V_p \subset V_q$, then $\mathrm{ord}(p) < \mathrm{ord}(q)$.

a) Show that $\mathrm{RM}(p) \leq \mathrm{ord}(p)$ for all $p \in S_1(M)$.

b) Show that the formula $v^{(n)} = 0$ has Morley rank n.

c) Let p be the type of an element differentially transcendental over K. Show that $\mathrm{RM}(p) = \omega$.

d) Conclude that $\mathrm{RM}(\mathrm{DCF}) = \omega$.

Exercise 6.6.19 (Cantor–Bendixson Analysis) Let X be a compact Hausdorff space (for example, a Stone space $S_n(A)$ or the real unit interval). Let $\Gamma(X) = \{x \in X : x \text{ is not isolated in } X\}$. For α an ordinal, we inductively define $\Gamma^\alpha(X)$ as follows:

i) $\Gamma^0(X) = X$;

ii) $\Gamma^{\alpha+1}(X) = \Gamma(\Gamma^{\alpha(X)})$;

iii) $\Gamma^\alpha(X) = \bigcup_{\beta < \alpha} \Gamma^\beta(X)$ if α is a limit ordinal.

We call Γ the *Cantor–Bendixson derivative*.

a) Show that $\Gamma^\alpha(X)$ is a closed subset of X and hence a compact Hausdorff space.

b) Show that there is an ordinal δ such that $\Gamma^\delta(X) = \Gamma^\alpha(X)$ for all $\alpha > \delta$. We let $\Gamma^\infty(X)$ denote $\Gamma^\delta(X)$.

c) Show that $\Gamma^\infty(X)$ is a closed subset of X with no isolated points.

d) Suppose that X is separable (i.e., there is a countable collection of open sets $U_0, U_1, \ldots$ such that for all x and all open V there is an i such that $x \in U_i \subseteq V$) then there is $\alpha < \omega_1$ such that $\Gamma^\alpha(X) = \Gamma^\infty(X)$.

e) If X is separable, then either $\Gamma^\infty(X) = \emptyset$ or $|\Gamma^\infty(X)| = 2^{\aleph_0}$.

f) Show that, for any closed $X \subseteq \mathbb{R}^n$, there is $P \subseteq X$ such that $X \setminus P$ is countable and either $P = \emptyset$ or P is a closed set with no isolated points and $|P| = 2^{\aleph_0}$.

g) Let T be an ω-stable theory, and let $\mathcal{M} \models T$ be $\aleph_0$-saturated. Consider the Cantor–Bendixson derivative Γ on $S_n(M)$. For each type $p \in S_n(M)$, we say that p has *Cantor–Bendixson rank* α if and only if $p \in \Gamma^{\alpha+1}(S_n(M)) \setminus \Gamma^\alpha(S_n(M))$. Show that every type has a Cantor–Bendixson rank and that the Cantor–Bendixson rank is exactly the Morley rank.

Exercise 6.6.20 Prove Theorem 6.3.2 iii).

Exercise 6.6.21 Suppose that T is ω-stable, $\mathcal{M}, \mathcal{N} \models T$, and $\mathcal{M} \prec \mathcal{N}$. If $X \subseteq N^k$ is definable in $\mathcal{N}$, then $X \cap M^k$ is definable in $\mathcal{M}$. [Hint: Let $\phi(\overline{v}, \overline{a})$ define X, and use the definability of $\operatorname{tp}(\overline{a}/M)$.]

Exercise 6.6.22 If K is a differential field, the field of constants of K is the subfield $C_K = \{x \in k : \delta(x) = 0\}$.

a) Show that C_K is algebraically closed in K. (Suppose that $X^n + \sum a_i X^i$ is the minimal polynomial of $\frac{\alpha}{C_K}$. Differentiate $\alpha^n + \sum a_i \alpha^i$.) If K is differentially closed, show that C_K is an algebraically closed field.

b) Suppose that $k \subseteq l$ are differential fields and $c \in C_l$ is algebraic over k; then, c is algebraic over C_l. [Hint: If $X^n + \sum a_i X^i$ is the minimal polynomial of c over k, show that $\sum a_i' c_i = 0$, contradicting minimality unless each $a_i' = 0$.]

c) Suppose K is differentially closed and $X \subseteq C^n$ is definable in K. Show that X is a constructible subset of C^n (i.e., X is already definable in the fields $(C, +, \cdot, 0, 1)$). [Hint: Combine quantifier elimination in DCF with Exercise 6.6.21 in ACF.]

Exercise 6.6.23 a) Suppose that $\mathbb{M}$ is ω-stable, $A, B \subseteq \mathbb{M}^n$ are definable, $\operatorname{RM}(A)$ is finite and $f : A \to B$ is a definable surjective map such that $\operatorname{RM}(f^{-1}(b)) = k$ for all $b \in B$. Show that $\operatorname{RM}(A) \geq \operatorname{RM}(B) + k$. [Hint: Prove by induction on rank that $\operatorname{RM}(f^{-1}(X)) \geq \operatorname{RM}(X) + k$ for all definable $X \subseteq B$.]

b) Suppose that G is an ω-stable group of finite Morley rank and $H \leq G$ is an infinite definable subgroup. Show that $\operatorname{RM}(G) \geq \operatorname{RM}(H) + \operatorname{RM}(G/H)$. In particular, $\operatorname{RM}(G) > \operatorname{RM}(G/H)$.

c) Show that b) is not true for all ω-stable groups. [Hint: Let K be a differentially closed field and consider the derivation $\delta : K \to K$.]

Exercise 6.6.24 Prove Corollary 6.3.13.

Exercise 6.6.25 (U-rank) Suppose that T is ω-stable. Let $A \subset \mathbb{M}$, and let $p \in S_n(A)$; we inductively define $\operatorname{RU}(p)$, the *U-rank* of p, as follows:

$\operatorname{RU}(p) = \sup\{\operatorname{RU}(q) + 1 : \exists B \ A \subset B \subset \mathbb{M} \ q \in S_n(p), p \subset q \text{ and } q \text{ forks over } A\}$.

a) Show that $\mathrm{RU}(p) = 0$ if and only if $\mathrm{RM}(p) = 0$ if and only if p has only finitely many realizations.

b) Show that $\mathrm{RU}(p)$ is well defined for all types and $\mathrm{RU}(p) \leq \mathrm{RM}(p)$.

c) Let T be as in Exercise 6.6.17 b). Let p be the unique 1-type containing $P_i(v)$ for all i. Show that $\mathrm{RM}(p) = 2$ but $\mathrm{RU}(p) = 1$.

Exercise 6.6.26 (Strong Types) The *strong type* of $\bar{a}$ over A is $\mathrm{stp}(\bar{a}/A) = \{\bar{v}E\bar{a} : E$ an A-definable equivalence relation with finitely many classes$\}$.

a) Show that if $\mathrm{stp}(\bar{a}/A) = \mathrm{stp}(\bar{b}/A)$, then $\mathrm{tp}(\bar{a}/A) = \mathrm{tp}(\bar{b}/B)$.

b) Show that $\mathrm{stp}(\bar{a}/A) = \mathrm{stp}(\bar{b}/A)$ if and only if $\mathrm{tp}(\bar{a}/\mathrm{acl}^{\mathrm{eq}}(A)) = \mathrm{tp}(\bar{b}/\mathrm{acl}^{\mathrm{eq}}(A))$.

c) Suppose that T is ω-stable. Show that $\mathrm{stp}(\bar{a}/A) = \mathrm{stp}(\bar{b}/A)$ if and only if there is $\mathcal{M} \models T$ with $A \subseteq M$ such that $\mathrm{tp}(\bar{a}/M) = \mathrm{tp}(\bar{b}/M)$.

Exercise 6.6.27 (Finite Equivalence Relation Theorem) Suppose that T is ω-stable, $A \subseteq B$, $p, q \in S_n(B)$ do not fork over A, and $p \neq q$. Show that there is E an A-definable equivalence relation with finitely many classes such that if $\bar{a}$ realizes p and $\bar{b}$ realizes q, then $\mathrm{stp}(\bar{a}/A) \neq \mathrm{stp}(\bar{b}/A)$.

Exercise 6.6.28 Suppose that $p \in S_n(A)$ and $q_0, q_1 \in S_n(\mathbb{M})$ are non-forking extensions of p. Show that there is an automorphism σ of $\mathbb{M}$ with $\sigma q_0 = q_1$.

Exercise 6.6.29 a) Show that if $p \in S_n(\mathbb{M})$ is definable over A, then p does not fork over A. [Hint: First replace A by a model M.]

b) Show that if $A \subseteq M$ and $p \in S_n(M)$ is definable over $\mathrm{acl}^{\mathrm{eq}}(A)$, then p does not fork over A. [Hint: Use i).]

Exercise 6.6.30 [Open Mapping Theorem] Suppose T that is ω-stable. Let $A \subseteq B$, and let $S_n(B/A)$ be the set of types in $S_n(B)$ that do not fork over A. We give $S_n(B/A)$ the subspace topology. Show that the restriction map $p \mapsto p|A$ is an open map. [Hint: Modify the proof of Lemma 6.4.6.]

Exercise 6.6.31 Prove Lemma 6.4.2.

Exercise 6.6.32 Suppose that k is a differential field and K is the differential closure of k. Show that the constant field of K is the algebraic closure of the constant field of k. [Hint: Let $c \in C_K$. There is a formula $\phi(v)$ that isolates $\mathrm{tp}(c/K)$. Show that c is algebraic over k. Argue that c is algebraic over C_k.]

Exercise 6.6.33 Suppose that $p \in S_1(A)$ is stationary and $\{a_1, \ldots, a_n\}$ is a finite Morley sequence. Show that $\{a_1, \ldots, a_n\}$ extends to an infinite Morley sequence.

Give an example showing that this may not be possible if $\deg_{\mathrm{M}}(p) > 1$.

Exercise 6.6.34 Suppose that T is ω-stable and I is an infinite set of indiscernibles. For any A, we define the *average* of I over A, $\mathrm{Av}(I/A) = \{\phi(v) : \phi$ an $\mathcal{L}_A$-formula such that $\phi(x)$ for all but finitely many $x \in I\}$.

a) Show that $\mathrm{Av}(I/A) \in S_1(A)$.

b) Suppose that $p \in S_1(A_0)$, $A_0 \subseteq A$, and I is a Morley sequence over A_0. Then, $\mathrm{Av}(I/A)$ is a nonforking extension of p.

Exercise 6.6.35 Suppose that $p \in S_1(A)$, $\deg_{\mathrm{M}}(p) = 1$, and $I = (a_\alpha : \alpha < \delta)$ is an infinite Morley sequence for p. Show that if σ is any automorphism of $\mathbb{M}$ fixing I setwise, then $\sigma p = p$.

Exercise 6.6.36 Suppose that $\bar{a}_i \in \mathbb{M}^k$ for $i \in I$. We say that $(\bar{a}_i : i \in I)$ is an *indiscernible set of k-tuples* if

$$\mathbb{M} \models \phi(\bar{a}_{i_1}, \ldots, \bar{a}_{i_n}) \text{ if and only if } \mathbb{M} \models \phi(\bar{a}_{j_1}, \ldots, \bar{a}_{j_n})$$

whenever $i_1, \ldots, i_n \in I$ and $j_1, \ldots, j_n \in I$ are two sequences of distinct elements.

a) Generalize the definition of Morley sequences to stationary k-types.

b) Show that if $I = (\bar{a}_\alpha : \alpha < \kappa)$ is a Morley sequence for a k-type, then I is an indiscernible set of k-tuples.

c) Generalize Lemma 6.5.3 and Exercise 6.6.35 to indiscernible sets of k-tuples.

Remarks

Morley [71] introduced ranks in his proof of the Categoricity Theorem. He originally defined rank using the Cantor–Bendixson derivative as discussed in Exercise 6.6.19.

There are several alternative approaches to the monster model $\mathbb{M}$. Ziegler [104] views the monster model as a proper class—rather than a set—containing all set models of T. Hodges [40] defines a notion of κ-*big models* such that if $\mathcal{M}$ is κ-big, then $\mathcal{M}$ is κ-saturated, and whenever $A, B \subset M$, $|A|, |B| < \kappa$, and $f : A \to B$ is elementary, then f extends to an automorphism of $\mathcal{M}$. He shows that if κ is a regular cardinal greater that $|\mathcal{L}| + \aleph_0$, then there is a κ-big model $\mathcal{M}$ with $|M| \leq \kappa^{<\kappa}$. If we are considering models of cardinality less than κ, we can use a κ-big model as the monster model.

All of Section 6.3 is due to Shelah. Although we have only defined forking and independence for ω-stable theories, these concepts can be defined for arbitrary stable theories so that most of the results of Section 6.3 hold. One notable exception is that, in a stable theory, if $p \in S_n(A)$, then there is a countable $A_0 \subseteq A$ such that p does not fork over A but p might fork over every finite subset of A. There are a number of good references on stable theories—for example, [7], [18] or [75]. Recently, many of these ideas have also been generalized to *simple* unstable theories (see, for example, [54]).

The notion of U-rank from Exercise 6.6.25 was introduced by Lascar. If T is superstable, then $\mathrm{RU}(p) < \infty$ for all $p \in S_n(A)$. Although U-rank is not quite as natural as Morley rank, it has some properties that make it very nice to work with. If α is an ordinal, we can write α as a finite sum

$$\alpha = \sum_{i=1}^{n} \omega^{\alpha_i} m_i,$$

where $\alpha_1 > \alpha_2 > \ldots \alpha_n$ and $m_i \in \mathbb{N}$. If $\alpha = \sum \omega^{\alpha_i} m_i$ and $\beta = \sum \omega^{\alpha_i} n_i$, then $\alpha \oplus \beta$ is defined to be $\sum \omega^{\alpha_i}(m_i + n_i)$. We call $\oplus$ the *symmetric sum* of α and β. Note that $\alpha + \beta \leq \alpha \oplus \beta$ and equality need not hold. For example, $2 + \omega = \omega$ while $2 \oplus \omega = \omega + 2 > \omega$. Lascar proved the following U-rank inequality.

Theorem 6.6.37 *If T is superstable, then*

$$\mathrm{RU}(\bar{a}/A, \bar{b}) + \mathrm{RU}(\bar{b}/A) \leq \mathrm{RU}(\bar{a}, \bar{b}/A) \leq \mathrm{RU}(\bar{a}/A, \bar{b}) \oplus \mathrm{RU}(\bar{b}/A).$$

For a proof, see [18] 6.1.1.

The uniqueness of constructible models is an unpublished result of Ressayre.

In Exercise 4.5.28, we argued that we have prime model extensions in o-minimal theories. Pillay and Steinhorn [83] showed that in o-minimal theories prime model extensions are constructible. Thus, by Ressayre's Theorem, we have unique prime model extensions in o-minimal theories.

The uniqueness of prime models is due to Shelah. The proof of Lemma 6.4.7 can easily be generalized to stable theories, although stable theories need not have prime model extensions.

Blum [11] showed that DCF is ω-stable and stability to show the existence and uniqueness of differential closures. Kolchin [56] later gave an algebraic proof. Kolchin, Rosenlicht, and Shelah gave independent proofs that differential closures need not be minimal. For more on differentially closed fields, see [65].

Morley introduced Morley sequences and proved Theorem 6.5.5. Theorem 6.5.4 is due to Harnik and Shelah.

7

ω-Stable Groups

7.1 The Descending Chain Condition

By an ω-stable group, we mean an ω-stable structure $(G, \cdot, 1, \ldots)$ where $(G, \cdot, 1)$ is a group. We will say that G is a *group of finite Morley rank* if $(G, \cdot, 1, \ldots)$ is ω-stable with $\mathrm{RM}(G) < \omega$.

We have already encountered several simple examples. Of course, all finite groups are ω-stable, but we will focus on infinite groups. If G is a torsion-free divisible Abelian group, then $\mathrm{Th}(G)$ is $\aleph_1$-categorical and hence ω-stable. If p is prime, then an infinite-dimensional vector space over $\mathbb{F}_p$ is categorical in every infinite cardinal and hence ω-stable.

If $\mathcal{M}$ is ω-stable and $(G, \cdot, 1, \ldots)$ is interpretable in $\mathcal{M}$, then G is ω-stable. In particular, any group interpretable in an algebraically closed field is ω-stable. For example, the multiplicative group of an algebraically closed field is ω-stable. We can go much further on these lines.

Algebraic Groups

Let K be an algebraically closed field and let $GL_n(K)$ be the group of invertible $n \times n$ matrices over K. If A is an $n \times n$ matrix over K, we naturally think of A as an element of K^{n^2} and we can identify $GL_n(K)$ with the Zariski closed set $V = \{(A, w) \in K^{n^2+1} : w \det(A) = 1\}$. We can define multiplication on V by $(A, w)(B, v) = (AB, wv)$. This is a polynomial map. By Cramer's rule, the inverse is also a polynomial map.

Definition 7.1.1 A *linear algebraic group* is a Zariski closed subgroup of $GL_n(K)$.

For example, $SL_n(K) = \{(A, w) \in GL_n(K) : w = 1\}$ and

$$\left\{ \begin{pmatrix} a & b \\ 0 & 1 \end{pmatrix} : a, b \in K, a \neq 0 \right\}$$

are linear algebraic groups. Because linear algebraic groups are interpretable in algebraically closed fields, they are ω-stable.

Other groups are interpretable in algebraically closed fields. For example, let $\mathbb{P}^2$ be the projective plane over an algebraically closed field, and let $E \subset \mathbb{P}^2$ be the elliptic curve $\{(X, Y, Z) \in \mathbb{P}^2 : Y^2 Z = X^3 + X Z^2\}$. The curve E has one point at infinity $O = (0, 1, 0)$. We can define a group law on E (see [95]). For example if A, B and C are three collinear points of E, then $A \oplus B \oplus C = 0$. By Theorem 3.2.20, $(E, \oplus, O)$ is an ω-stable group. In Section 7.4, we will give a general definition of algebraic groups that includes both the linear algebraic groups and elliptic curves.

One of our main themes in this chapter will be that ω-stable groups behave very much like algebraic groups. Indeed, the following conjecture is one of the guiding problems in the subject.

Cherlin–Zil'ber Conjecture If G is an infinite simple group of finite Morley rank, then G interprets an algebraically closed field K and G is definably isomorphic to a simple algebraic group defined over K.

Chain Conditions

If G is a group, we write $H \leq G$ if H is a subgroup of G and $H \trianglelefteq G$ if H is a normal subgroup of G.

In algebraically closed fields, there are no infinite descending chains of Zariski closed sets. Thus, in linear algebraic groups there are no infinite descending chains of algebraic subgroups. We will generalize this to arbitrary ω-stable groups.

Suppose that G is an ω-stable group and $H \leq G$ is a definable subgroup. Because $H \subseteq G$, $\mathrm{RM}(H) \leq \mathrm{RM}(G)$. For $a \in G \setminus H$, the coset aH is a subset of G disjoint from H and, because $x \mapsto ax$ is a definable bijection, $\mathrm{RM}(H) = \mathrm{RM}(aH)$. Thus, if $[G : H]$ is infinite, then $\mathrm{RM}(H) < \mathrm{RM}(G)$. If $1 < [G : H] < \aleph_0$, then $\mathrm{RM}(H) = \mathrm{RM}(G)$ but $\deg_M(H) < \deg_M(G)$, indeed, $\deg_M(G) = [G : H]\deg_M(H)$. These easy observations have very important consequences.

Theorem 7.1.2 (Descending Chain Condition) *If G is an ω-stable group, then there is no infinite descending chain of definable subgroups* $G > G_1 > G_2 > \dots$.

Proof Let $\alpha = \mathrm{RM}(G)$. Let $\eta_i = (\mathrm{RM}(G_i), \deg_M(G_i))$. If $G > G_1 > G_2 > \dots$ is a descending chain, then the remarks above show that $\eta_1 >_{\mathrm{lex}}$

$\eta_2 >_{\text{lex}} \ldots$ where $<_{\text{lex}}$ is the lexicographic order on $\alpha \times \omega$. Because this is a well-ordering, there are no infinite descending chains.

A simple corollary shows that in some ways ω-stable groups behave like finite groups.

Corollary 7.1.3 *Suppose that G is an ω-stable group and $\sigma : G \to G$ is a definable injective group homomorphism. Then, σ is surjective.*

Proof If not then, because $\sigma G \cong G$, $G \supset \sigma G \supset \sigma^2 G \supset \ldots$, contradicting the Descending Chain Condition.

Corollary 7.1.4 *If G is an ω-stable group and $\{H_i : i \in I\}$ is a collection of definable subgroups, then there is $I_0 \subseteq I$ finite such that*

$$\bigcap_{i \in I} H_i = \bigcap_{i \in I_0} H_i.$$

Proof If not we can find $i_0, i_1, \ldots$ such that if $G_m = H_{i_0} \cap \ldots \cap H_{i_m}$, then $G_0 > G_1 > G_2 > \ldots$.

We can use the Descending Chain Condition to find some interesting definable subgroups of G.

Suppose that $A \subseteq G$. The *centralizer* of A is $C(A) = \{g \in G : ga = ag$ for all $a \in A\}$. If G is an arbitrary group and A is definable, then $C(A)$ is definable. In ω-stable groups, $C(A)$ is definable even if A is not.

Corollary 7.1.5 *If G is an ω-stable group and $A \subseteq G$, then the centralizer $C(A)$ is definable.*

Proof Because

$$C(A) = \bigcap_{a \in A} C(\{a\}),$$

there are $a_1, \ldots, a_m \in A$ such that $C(A) = \{g \in G : ga_i = a_i g$ for $i = 1, 2, \ldots, m\}$.

Corollary 7.1.6 *If G is an ω-stable group, there is $G^0 \leq G$ the smallest definable finite index subgroup of G. Moreover, G^0 is a normal subgroup of G and definable over $\emptyset$.*

Proof Let $\mathcal{H} = \{H \leq G : H$ definable, $[G : H] < \aleph_0\}$. By Corollary 7.1.4, there are $H_1, \ldots, H_m \in \mathcal{H}$ such that

$$\bigcap_{H \in \mathcal{H}} H = H_1 \cap \ldots \cap H_m.$$

Let $G^0 = H_1 \cap \ldots \cap H_m$. Clearly, G^0 is contained in every finite index subgroup of G. Because G^0 is an intersection of finitely many finite index subgroups of G, $[G : G^0]$ is finite.

We need to show that G^0 is definable over $\emptyset$. Let $[G : G^0] = n$. Suppose that $\phi(v, \overline{w})$ is an $\mathcal{L}$-formula, $\overline{a} \in G$, and $\phi(v, \overline{a})$ defines G^0. Then, $W = \{\overline{b} : \phi(v, \overline{w})$ defines a subgroup of index $n\}$ is defined over $\emptyset$. If $\overline{b} \in W$ and $H = \{g \in G : G \models \phi(g, \overline{b})\}$, then $H \cap G^0$ is a finite index subgroup of G^0. Because G^0 is the smallest definable subgroup of G of finite index, $H = G^0$. Thus

$$G^0 = \{g : \exists \overline{b} \in W \wedge \phi(g, \overline{b})\}$$

is definable over $\emptyset$.

If $h \in G$, then $x \mapsto hxh^{-1}$ is a group automorphism. Thus, hG^0h^{-1} is a definable subgroup with $[G : hG^0h^{-1}] = [G : G^0]$, so $hG^0h^{-1} = G^0$ and G^0 is normal.

Definition 7.1.7 We call G^0 the *connected component* of G. If $G = G^0$, then we say that G is *connected*.

We leave the proof of the following useful lemma as an exercise.

Lemma 7.1.8 *Suppose that G is an ω-stable group and $\sigma : G \to G$ is a definable group automorphism. Then, σ fixes G^0 setwise.*

Stabilizers

We can view the group G as acting on $S_1(G)$ by

$$gp = \{\phi(x) : \phi(gx) \in p\}.$$

If $G \prec G'$, and $a \in G'$ realizes p, then ga realizes gp.

Definition 7.1.9 The *stabilizer* of p is the group $\text{Stab}(p) = \{g \in G : gp = p\}$.

We have considered the "left" action of G on $S_1(G)$. We could also consider the "right" action where $pg = \{\phi(x) : \phi(xg) \in p\}$ and define the right stabilizer of p.

Theorem 7.1.10 *If G is an ω-stable group and $p \in S_1(G)$, then $\text{Stab}(p)$ is a definable subgroup of G.*

Proof For $\phi(v)$ an $\mathcal{L}_G$-formula, let $\text{Stab}^\phi(p) = \{g \in G : \phi(hv) \in p$ if and only if $\phi(hgv) \in p$ for all $h \in G\}$.

Claim 1 $\text{Stab}(p) = \bigcap_{\phi \in p} \text{Stab}^\phi(p)$.

($\subseteq$) Suppose that $g \in \text{Stab}(p)$, $\phi(v) \in p$, and $h \in G$. Let $\psi(v)$ be the formula $\phi(hv)$. Because g stabilizes p,

$$\phi(hv) \in p \Leftrightarrow \psi(v) \in p \Leftrightarrow \psi(gv) \in p \Leftrightarrow \phi(hgv) \in p.$$

Thus $g \in \text{Stab}^\phi(p)$.

($\supseteq$) Suppose that $\phi(v) \in p$ and $g \in \mathrm{Stab}^\phi(p)$; then, (using $h = 1$) $\phi(v) \in p$ if and only if $\phi(gv) \in p$. Thus, if $g \in \mathrm{Stab}^\phi(p)$ for all $\phi(v) \in p$, then $gp = p$.

Claim 2 Each $\mathrm{Stab}^\phi(p)$ is a definable subgroup of G.

If $g_1, g_2 \in \mathrm{Stab}^\phi(p)$ and $h \in G$, then, applying the definition first with h and then with hg_1,

$$\phi(hv) \in p \Leftrightarrow \phi(hg_1 v) \in p \Leftrightarrow \phi(hg_1 g_2 v) \in p.$$

Thus $g_1 g_2 \in \mathrm{Stab}^\phi(p)$. Applying the definition with hg_1^{-1},

$$\phi(hg_1^{-1} v) \in p \Leftrightarrow \phi(hg_1^{-1} g_1 v) \in p \Leftrightarrow \phi(hv) \in p.$$

Thus, $g_1^{-1} \in \mathrm{Stab}^\phi(p)$ and $\mathrm{Stab}^\phi(p)$ is a subgroup of G.

Let $\psi(w, v)$ be the formula $\phi(wv)$. By definability of types (Theorem 6.3.5), there is an $\mathcal{L}_G$-formula $d_p\psi(w)$ such that $\psi(g, v) \in p$ if and only if $G \models d_p\psi(g)$. Thus, $\mathrm{Stab}^\phi(p) = \{g \in G : \forall h \ (d_p\psi(h) \leftrightarrow d_p\psi(hg)))\}$ is a definable subgroup of G.

Thus, $\mathrm{Stab}(p)$ is an intersection of definable subgroups of G. By Corollary 7.1.4, there are $\phi_1, \ldots, \phi_m \in p$ such that $\mathrm{Stab}(p) = \mathrm{Stab}^{\phi_1}(p) \cap \ldots \cap \mathrm{Stab}^{\phi_m}(p)$. Hence, $\mathrm{Stab}(p)$ is a definable subgroup of G.

Suppose that $G \prec G_1$ and p_1 is the unique nonforking extension of p to G_1. The formula defining $\mathrm{Stab}(p)$ in G also defines $\mathrm{Stab}(p_1)$ in G_1 (this is Exercise 7.6.3).

We can bound the rank of the stabilizer by the rank of p.

Lemma 7.1.11 $\mathrm{RM}(\mathrm{Stab}(p)) \leq \mathrm{RM}(p)$.

Proof Let $G \prec G_1$ with $a, b \in G_1$ such that a realizes p, $b \in \mathrm{Stab}(p)$ such that $\mathrm{RM}(b/G) = \mathrm{RM}(\mathrm{Stab}(p))$, and a and b are independent over G. Then $\mathrm{RM}(ba/G, a) = \mathrm{RM}(b/G, a) = \mathrm{RM}(b/G) = \mathrm{RM}(\mathrm{Stab}(p))$. On the other hand, $\mathrm{RM}(ba/G, a) \leq \mathrm{RM}(ba/G) = \mathrm{RM}(p)$.

Lemma 7.1.12 $\mathrm{Stab}(p) \leq G^0$.

Proof Let $a \in \mathrm{Stab}(p)$, and let $\psi(v)$ define G^0. Let $b \in G$ such that $\psi(b^{-1}v) \in p$. Thus $\psi(b^{-1}av) \in p$. Let $G \prec H$ with $c \in H$ realizing p. Then $b^{-1}ac \in H^0$ and $b^{-1}c \in H^0$. Thus

$$(b^{-1}c)^{-1} b^{-1} ac = c^{-1} ac \in H^0.$$

Because H^0 is normal, $a \in H^0$. Thus $a \in G^0$.

7.2 Generic Types

In Lemma 6.2.25 and Exercise 6.6.15 we introduced generic points of algebraic varieties. In this section, we will generalize this notion to arbitrary ω-stable groups. Generics will be a powerful tool for studying ω-stable groups.

Throughout this section, $G = (G, \cdot, \ldots)$ is an infinite ω-stable group. We let $\mathbb{G}$ be a monster model with $G \prec \mathbb{G}$.

Definition 7.2.1 Let $p \in S_1(G)$. We say that p is *generic* if and only if $RM(p) = RM(G)$. We say that $a \in \mathbb{G}$ is *generic* over G if $RM(\text{tp}(a/G)) = RM(G)$.

We begin by proving some of the basic properties that we will use about generic types.

Lemma 7.2.2 *If* $\text{tp}(x/G)$ *is generic and* $a \in G$, *then* $\text{tp}(ax/G)$ *and* $\text{tp}(x^{-1}/G)$ *are generic.*

Proof The maps $x \mapsto ax$ and $x \mapsto x^{-1}$ are definable bijections and hence preserve Morley rank.

Lemma 7.2.3 $p \in S_1(G)$ *is generic if and only if* $[G : \text{Stab}(p)] < \aleph_0$.

Proof

($\Leftarrow$) If $\text{Stab}(p)$ has finite index, $RM(\text{Stab}(p)) = RM(G)$. But $RM(\text{Stab}(p)) \leq RM(p)$. Thus, p is generic.

($\Rightarrow$) Because there are only finitely many types of maximal Morley rank, $\{ap : a \in G\}$ is finite. Choose $b_1, \ldots, b_n$ such that if $a \in G$, then $ap = b_i p$ for some $i \leq n$. If $ap = b_i p$, then $b_i^{-1} a \in \text{Stab}(p)$ and $a \in b_i(\text{Stab}(p))$. Thus $[G : \text{Stab}(p)] \leq n$.

Corollary 7.2.4 $p \in S_1(G)$ *is generic if and only if* $\text{Stab}(p) = G^0$.

Proof

($\Leftarrow$) Clear from Lemma 7.2.3.

($\Rightarrow$) By Lemma 7.2.3, $G^0 \leq \text{Stab}(p)$, and by Lemma 7.1.12, $\text{Stab}(p) \leq G^0$.

We have proved Lemma 7.2.3 and Corollary 7.2.4 for left stabilizers, but symmetric arguments show that they are also true for right stabilizers.

Lemma 7.2.5 *i)* G *has a unique generic type if and only if* G *is connected.*
ii) $\deg_M(G) = [G : G^0]$.

Proof

i) ($\Rightarrow$) Let p be the unique generic type. For all $a \in G$, ap is generic and hence $ap = p$. Thus, $G = \text{Stab}(p)$ and, by Corollary 7.2.4, $G = G^0$.

($\Leftarrow$) Suppose that p and q are distinct generic types. We will get a contradiction by showing that if a and b are independent realizations of p and q, then ba realizes both p and q.

Let G_1 be an elementary extension of G containing b. Let $p_1 \in G_1$ be the unique nonforking extension of p and let a_1 realize p_1. Because a_1 and a both realize the unique nonforking extension of p to $G \cup \{b\}$, $\text{tp}(a, b/G) = \text{tp}(a_1, b/G)$. Because G_1 is connected and p_1 is a generic type of G_1, by Corollary 7.2.4, $\text{Stab}(p_1) = G_1$. Thus, ba_1 realizes p_1. In particular, ba_1 realizes p and hence ba realizes p. A symmetric argument, using right stabilizers, shows that ba realizes q, a contradiction.

ii) Because connected groups have a unique type of maximal Morley rank, the connected component G^0 must have Morley degree 1. Because G is a union of $[G : G^0]$ disjoint translates of G^0, the Morley degree of G is exactly the index $[G : G^0]$.

Next, we show that every element is the product of two generics.

Lemma 7.2.6 *If $g \in G$, there are $a, b \in \mathbb{G}$ generic over G such that $g = ab$.*

Proof Let $a \in \mathbb{G}$ be generic over G. Because $x \mapsto gx^{-1}$ is a definable bijection, $b = ga^{-1}$ is also generic over G and $g = ab$.

Corollary 7.2.7 *Suppose that G is connected and $A \subseteq G$ is a definable subset with $\text{RM}(A) = \text{RM}(G)$. Then $G = A \cdot A = \{ab : a, b \in A\}$.*

Proof Let $\phi(v)$ be an $\mathcal{L}_G$-formula defining A. For any $g \in G$, we can find $a, b \in \mathbb{G}$ generic over G such that $g = ab$. Because there is a unique generic type, $\phi(a)$ and $\phi(b)$. Thus $\mathbb{G} \models \exists x \exists y \, (\phi(x) \wedge \phi(y) \wedge xy = g)$. Because $G \prec \mathbb{G}$, there are $a', b' \in A$ such that $g = a'b'$.

We say that a definable $A \subseteq G$ is *generic* if $\text{RM}(A) = \text{RM}(G)$. Next, we show that finitely many translates of a generic set cover the group.

Lemma 7.2.8 *Let $A \subseteq G$ be a definable generic subset of G. There are $a_1, \ldots, a_n \in G$ such that $G = a_1 A \cup \ldots \cup a_n A$.*

Proof Because finitely many translates of G^0 cover G, we may, without loss of generality, assume that G is connected. Let $\phi(v)$ be the $\mathcal{L}_G$-formula defining A. Let $p \in S_1(G)$ be the unique generic type.
Claim For any $q \in S_1(G)$, there is $g \in G$ such that $\phi(gv) \in q$ (i.e., $\phi(v) \in gq$).
Let a and b be independent realizations of p and q. Because ab is generic, $\phi(ab)$. Let $\psi(v, w)$ be the formula $\phi(w \cdot v)$. By definability of types, there is an $\mathcal{L}_G$-formula $d_q\psi$ such that $\psi(v, g) \in q$ if and only if $G \models d_q\psi(g)$. Because b realizes the unique nonforking extension of q to $G \cup \{a\}$, $\mathbb{G} \models d_q\psi(a)$. Thus $\mathbb{G} \models \exists w \, d_q\psi(w)$. Because $G \prec \mathbb{G}$, there is $g \in G$ such that $\phi(gv) \in q$.

For each $g \in G$, let $O_g = \{q \in S_1(G) : \phi(gv) \in q\}$. This is an open subset of $S_1(G)$ and by the claim $S_1(G) = \bigcup_{g \in G} O_g$. By compactness, there are $a_1, \ldots, a_n \in G$ such that $S_1(G) = O_{a_1} \cup \ldots \cup O_{a_n}$. In particular, if $g \in G$, and q is the unique type containing the formula $v = g$, then there is an i such that $q \in O_{a_i}$. But then $\phi(a_i g)$ and $g \in a_i^{-1} A$. Thus $G = a_1^{-1} A \cup \ldots \cup a_n^{-1} A$.

When working with generic types, we frequently tacitly assume that G is somewhat saturated. We say things like "let $a \in G$ be generic over A."

By this we mean let $a \in G$ such that $\mathrm{RM}(a/A) = \mathrm{RM}(G)$. Also, if $X \subseteq G$ is definable and a is generic, we say that $a \in X$ if $a \in X(\mathbb{G})$, the elements of $\mathbb{G}$ that satisfy the formula defining X.

ω-stable Fields

We will conclude this section by using generic types to prove two important results. The first is Macintyre's theorem that any infinite ω-stable field is algebraically closed. The second is Reineke's theorem that minimal ω-stable groups are Abelian.

The proof of Macintyre's Theorem uses the following result from Galois theory (see [58] VIII §6 Theorems 10 and 11).

Theorem 7.2.9 *a) Suppose that L/K is a cyclic Galois extension of degree n, where n is relatively prime to the characteristic of K and K contains all nth roots of unity. The minimal polynomial of L/K is $X^n - a$ for some $a \in K$.*

b) Suppose that K has characteristic $p > 0$ and L/K is a Galois extension of degree p. The minimal polynomial of L/K is $X^p + X - a$ for some $a \in K$.

Theorem 7.2.10 *If $(K, +, \cdot, \ldots)$ is an infinite ω-stable field, then K is algebraically closed.*

Proof We first show that the additive group $(K, +, \ldots)$ is connected. Suppose that K^0 is the connected component of the additive group. For $a \in K \setminus \{0\}$, $x \mapsto ax$ is a definable group automorphism. By Lemma 7.1.8 K^0 is closed under multiplication by a. Thus, K^0 is an ideal of K. Because K is a field, there are no proper ideals and $K^0 = K$.

Because K is connected as an additive group, there is a unique type of maximal Morley rank. Thus, the multiplicative group $(K^\times, \cdot, \ldots)$ is also connected.

For each natural number n, the map $x \mapsto x^n$ is a multiplicative homomorphism. If a is generic, then, because a^n is interalgebraic with a, a^n is also generic. Thus, K^n, the subgroup of nth powers, contains the generic. Because the multiplicative group is connected, $K^n = K$ and every element has an nth root. In particular, if K has characteristic $p > 0$, then every element of K has a pth root. Thus, K is perfect.

Suppose that K has characteristic $p > 0$. The map $x \mapsto x^p + x$ is an additive homomorphism. If a is generic, then, because $a^p + a$ is interalgebraic with a, $a^p + a$ is also generic. Thus, as above, the homomorphism is surjective.

Claim 1 Suppose that K is an infinite ω-stable field containing all mth roots of unity for $m \leq n$. Then, K has no proper Galois extensions of degree n.

Let n be least such that there is an ω-stable field K containing all mth roots of unity for $m \leq n$ and K has a proper Galois extension L of degree n.

Let q be a prime dividing n. By Galois theory, there is $K \subseteq F \subset L$ such that L/F is Galois of degree q. The field F is a finite algebraic extension of K and thus interpretable in K (see Exercise 1.4.12). Because F is interpretable in an ω-stable structure, F is also ω-stable. Thus, by the minimality of n, $F = K$ and $n = q$.

If K has characteristic 0 or characteristic $p \neq q$, then, by Theorem 7.2.9 a), the minimal polynomial of L/K is $X^q - a$ for some $a \in K$. But every element of K has a qth root, thus $X^q - a$ is reducible, a contradiction.

If K has characteristic $p = q$, then, by Theorem 7.2.9 b), the minimal polynomial of L/K is $X^p + X - a$ for some $a \in K$. But the map $x \mapsto x^p + x - a$ is surjective; thus $X^p + X - a$ is reducible, a contradiction. This proves the claim.

Claim 2 If K is an infinite ω-stable field, then K contains all roots of unity.

Let n be least such that K does not contain all nth roots of unity. Let ξ be a primitive nth root of unity. Then $K(\xi)$ is a Galois extension of K of degree at most $n - 1$. This contradicts the previous claim.

Because K contains all roots of unity, the first claim implies that K has no proper Galois extensions. Because K is perfect, K is algebraically closed.

Minimal Groups

Next, we prove Reineke's Theorem. Recall that for G a group, the *center* of G is the group $Z(G) = \{a \in G : \forall g \in G \; ag = ga\}$.

Theorem 7.2.11 *If G is an infinite ω-stable group with no proper definable infinite subgroups, then G is Abelian.*

Proof Suppose not. Then, the center $Z(G)$ is finite and for all $a \in G \backslash Z(G)$, the centralizer $C(a) = \{g \in G : ag = ga\}$ is finite.

Let $a \in G \setminus Z(G)$, and let b be generic over a.

Claim 1 b is algebraic over $\{a, bab^{-1}\}$.

The set $\{c : cac^{-1} = bab^{-1}\}\{c : b^{-1}c \in C(a)\}$ is a finite set containing b.

Thus, $\mathrm{RM}(bab^{-1}/a) = \mathrm{RM}(b/a)$ and bab^{-1} is generic over a, so $a^G = \{gag^{-1} : g \in G\}$ is generic.

Suppose $a, b \in G \backslash Z(G)$. Because G is connected there is a unique generic type. Thus $a^G \cap b^G \neq \emptyset$ (indeed, it must be generic). If $cac^{-1} = dbd^{-1}$, then $b = d^{-1}cac^{-1}d \in a^G$. Similarly, $a \in b^G$, so $a^G = b^G$. Thus, any two elements not in $Z(G)$ are conjugate.

Let $H = G/Z(G)$. Then, H is an infinite ω-stable group, and all elements except 1 are conjugate.

Claim 2 All elements of $H \setminus \{1\}$ have the same order.

Suppose that $b^n = 1$ and $cac^{-1} = b$. Then

$$1 = b^n = (cac^{-1})^n = ca^nc^{-1}.$$

Thus $a^n = 1$.

Claim 3 Some element of H does not have order 2.

Suppose that every element of H has order 2. Let $a, b \in H$. Then

$$ab = (ab)^{-1} = b^{-1}a^{-1} = ba.$$

Thus, H is Abelian, but then for all $a \in H$, $a^H = \{a\}$, a contradiction.

Suppose that all elements of $H \setminus \{1\}$ have order $n > 0$. Clearly, n must be a prime number. Let $a \in H$, there is $b \in H$ such that $bab^{-1} = a^{-1}$. Note that $ba^{-1}b^{-1} = a$. Thus, for all $k \geq 1$

$$b^k ab^{-k} = \begin{cases} a & \text{if } k \text{ is even} \\ a^{-1} & \text{if } k \text{ is odd} \end{cases}.$$

This leads to a contradiction because n is odd, so $b^n ab^{-n} = a^{-1}$. But $b^n = 1$; thus $b^n ab^{-n} = a$.

Corollary 7.2.12 *If G is an infinite ω-stable group, then there is an infinite definable Abelian $H \leq G$.*

Proof By the Descending Chain Condition, there is an infinite definable $H \leq G$ with no infinite definable proper subgroups.

Corollary 7.2.13 *If G is a group of Morley rank 1, then G is Abelian-by-finite (i.e., there is a definable Abelian subgroup of finite index).*

Proof If $RM(G) = 1$, then G^0 is Abelian.

Corollary 7.2.14 *If G is an infinite ω-stable group with no definable infinite proper subgroups, then either G is a divisible Abelian group or every element of G has order p for some prime p.*

Proof For any prime p, $G^p = \{g^p : g \in G\}$ is a definable subgroup and hence must either be finite or all of G. If $G^p = G$, then every element is divisible by p. If G is p-divisible for all primes p, then G is divisible. If G^p is finite then $\{g \in G : g^p = 1\}$ is an infinite definable subgroup and hence must be all of G.

Note that even if G is divisible, G might have some torsion elements. For example, if K is an algebraically closed field of characteristic zero and E is an elliptic curve, then E is divisible and has n^2 elements of order n for each n.

7.3 The Indecomposability Theorem

In this section, we prove a theorem of Zil'ber's that is an important tool for studying groups of finite Morley rank. It again generalizes a result from algebraic group theory (see [14] I.2.2).

Definition 7.3.1 We say that a definable $X \subseteq G$ is *indecomposable* if and only if whenever H is a definable subgroup of G the coset space $X/H = \{xH : x \in X\}$ is either infinite or contains a unique element.

Indecomposable sets play the role of irreducible subvarieties in arbitrary finite Morley rank groups (see Exercise 7.6.13). For example, if $X \le G$ is an infinite connected definable subgroup, then X is indecomposable. If $H \le G$ is a definable group and $a, b \in X$, then $aH = bH$ if and only if $b \in a(X \cap H)$. Thus, the number of cosets in X/H is equal to the index $[X : X \cap H]$. Because X is connected, this is either one or infinite.

Theorem 7.3.2 (Zil'ber's Indecomposability Theorem) *Let G be a group of finite Morley rank and $(X_i : i \in I)$ a collection of definable indecomposable subsets of G each containing 1. Then, the subgroup of G generated by $\bigcup_{i \in I} X_i$ is definable and connected.*

Proof For each $\sigma = (i_1, \ldots, i_n) \in I^{<\omega}$, let $X^\sigma = \{x_1 \cdots x_n : x_1 \in X_{i_1}, \ldots, x_n \in X_{i_n}\}$. Because $\mathrm{RM}(G)$ is finite, there is a σ such that $\mathrm{RM}(X^\sigma) = k$ is maximal. Let $p \in S_1(G)$ be a type of Morley rank k containing the formula $v \in X^\sigma$. Let $H = \mathrm{Stab}(p)$.

Claim Each $X_i \subseteq H$.
If not, then $|X_i/H| > 1$ as $1 \in X_i \cap H$ and $X_i \not\subseteq H$. Because X_i is indecomposable, there are $a_1, a_2, \ldots$ in X_i such that $a_n H \ne a_m H$ for $n \ne m$. Because $a_m^{-1} a_n \notin H = \mathrm{Stab}(p)$, $a_n p \ne a_m p$ for $n \ne m$. Thus, $a_1 p, a_2 p, \ldots$ are infinitely many distinct types of rank k. But each of these types contains the formula $v \in X_i \cdot X^\sigma$. Thus, $X^i \cdot X^\sigma$ has rank at least $k + 1$, contradicting our choice of σ.

Thus, the group generated by $\bigcup_{i \in I} X_i$ is contained in H.
Because $H = \mathrm{Stab}(p)$, by Lemma 7.1.11, $\mathrm{RM}(H) \le \mathrm{RM}(p) = \mathrm{RM}(X^\sigma) \le \mathrm{RM}(H)$. Thus, $p \in H$ and $\mathrm{RM}(p) = \mathrm{RM}(H)$, so p is a generic type of H. Because any ω-stable group acts transitively on its generic types (see Exercise 7.6.9) and $H = \mathrm{Stab}(p)$ fixes p, H is connected. Because $X^\sigma \subseteq H$ is generic, by Lemma 7.2.7, $H = X^\sigma \cdot X^\sigma$. Thus, H is contained in the group generated by $\bigcup_{i \in I} X_i$.
We have shown that H is the group generated by $\bigcup_{i \in I} X_i$ and H is connected.

The proof of the Indecomposability Theorem shows a bit more. Namely, there are $i_1, \ldots, i_m \in I$ such that $H = X_{i_1} \cdots X_{i_m}$. If we start with a single indecomposable set X, and H is the group generated by X, then there is a number m such that every element of H is a product of m elements of X.

Let Γ be a group and S a set. An *action* of Γ on S is a map $\alpha : \Gamma \times S \to S$ such that $\alpha(1,s) = s$ for all $s \in S$ and $\alpha(\gamma, \alpha(\mu, s)) = \alpha(\gamma\mu, s)$ for all $\gamma, \mu \in \Gamma$ and $s \in S$. When no confusion arises, we write γs for $\alpha(\gamma, s)$. We say that $X \subseteq S$ is Γ-*invariant* if $\gamma X = X$ for all $\gamma \in \Gamma$. We say that Γ acts *transitively* on $X \subseteq S$ if for all $x, y \in X$ there is $\gamma \in \Gamma$ such that $\gamma x = y$. If G is a group, we say that $\alpha : \Gamma \times G \to G$ is the action of a group of automorphisms, if for each $\gamma \in \Gamma$, the function $g \mapsto \alpha(\gamma, g)$ is a group automorphism.

We say that the action is ω-stable if $(\Gamma, \cdot, S, \alpha)$ is interpretable in an ω-stable structure. Similarly, we say that the action has finite Morley rank if $(\Gamma, \cdot, S, \alpha)$ is interpretable in a finite Morley rank structure.

For example, if G is a group, G acts on itself by conjugation $\alpha(h, g) = hgh^{-1}$. If K is a field, then $K^{\times}$ acts on $(K, +)$ by $\alpha(a, x) = ax$. If G and K are ω-stable (finite Morley rank), then these actions are also ω-stable (finite Morley rank).

The next lemma shows that to test the indecomposability of a Γ-invariant X we need only show that it is indecomposable by Γ-invariant definable subgroups.

Lemma 7.3.3 *Suppose that there is an ω-stable action of Γ on a group G as a group of automorphisms, $X \subseteq G$ is Γ-invariant, and for all definable Γ-invariant subgroups H of G either $|X/H| = 1$ or X/H is infinite. Then, X is indecomposable.*

Proof Suppose that H is a definable subgroup of G and $1 < |X/H| < \aleph_0$. Suppose that $X \subseteq x_1 H \cup \ldots \cup x_n H$. If $\gamma \in \Gamma$ and $x \in X$, then $\gamma^{-1} x \in X$. Thus, $\gamma^{-1} x = x_i h$ for some $h \in H$ and $x = (\gamma x_i)(\gamma h)$. Thus $X \subseteq (\gamma x_1)(\gamma H) \cup \ldots \cup (\gamma x_n)(\gamma H)$. In particular, $X/\gamma H$ is finite.

Let $H^* = \bigcap_{\gamma \in \Gamma} \gamma H$. By the Descending Chain Condition there are $\gamma_1, \ldots, \gamma_m \in \Gamma$ such that $H^* = \gamma_1 H \cap \ldots \cap \gamma_m H$. Because X is Γ-invariant, $X/\gamma_i H$ is finite for each i and X/H^* is finite. Thus $1 < |X/H^*| < \aleph_0$. But H^* is Γ-invariant, a contradiction.

If $g, h \in G$ we let g^h denote hgh^{-1}. If $H \le G$, we let $g^H = \{g^h : h \in H\}$.

Corollary 7.3.4 *Suppose that G is an ω-stable group. If H is a definable connected subgroup of G and $g \in G$, then g^H is indecomposable.*

Proof The group H acts on G via conjugation, and g^H is invariant under this action. Thus, by the preceding lemma, it suffices to show that g^H is indecomposable for definable $N \le G$ where $hNh^{-1} = N$ for all $h \in H$. Suppose that g^H/N is finite and m is minimal such that $g^H \subseteq a_1 N \cup a_2 N \cup \ldots \cup a_m N$ for some $a_1, \ldots, a_m \in g^H$. If $h \in H$, then $ha^H h^{-1} = a^H$ and $h(a_i N)h^{-1} = a_i^h(hNh^{-1}) = a_i^h N$. Thus, for each i there is a unique j such that $a_i^h N = a_j N$. This gives a definable transitive action of H on the finite set $\{a_1, \ldots, a_m\}$. By Exercise 7.6.11, $m = 1$ as desired. Thus, g^H is indecomposable.

For $a, b \in G$, we let $[g, h]$ denote the commutator $g^{-1}h^{-1}gh$. The *commutator subgroup* G' is generated by $\{[g, h] : g, h \in G\}$.

Corollary 7.3.5 *If G is a connected group of finite Morley rank, then the commutator subgroup G' is a connected definable subgroup of G.*

Proof By Corollary 7.3.4, g^G is indecomposable. Thus, $g^{-1}(g^G)$ is indecomposable, $1 \in g^{-1}(g^G)$ and G' is the group generated by $\{g^{-1}(g^G) : g \in G\}$. By Zil'ber's Indecomposability Theorem, G' is definable and connected.

Next, we use the Indecomposability Theorem to show that for groups of finite Morley rank, simplicity is preserved under elementary equivalence.

Theorem 7.3.6 *If G is an infinite non-Abelian group of finite Morley rank and G has no nontrivial definable normal subgroups, then G is simple.*

Proof Because G^0 is a normal subgroup of G, G is connected. For $a \in G$, let $C(a)$ be the centralizer $\{g \in G : ga = ag\}$. For $g, h \in G$, $a^g = a^h$ if and only if $g \in hC(a)$. Suppose a^G is finite. Then, $C(a)$ is a finite index subgroup of G. Because G is connected, $C(a) = G$, and a is in the center, $Z(G) = \{a : ga = ag \text{ for all } g \in G\}$. Because $Z(G)$ is a definable normal subgroup and G is non-Abelian, $Z(G) = \{1\}$. Thus, we may assume a^G is infinite for all $a \in G \setminus \{1\}$.

Suppose that N is a nontrivial normal subgroup of G and $a \in N \setminus \{1\}$. Let $X = a^{-1}(a^G)$. By Corollary 7.3.4 (and Exercise 7.6.12) X is indecomposable and gXg^{-1} is indecomposable for all $g \in G$. Because N is normal, $gXg^{-1} \subseteq N$ for all $g \in G$. Let H be the subgroup of N generated by $\{gXg^{-1} : g \in G\}$. Because a^G is infinite, H is nontrivial. By Zil'ber's Indecomposability Theorem, H is definable.

We claim that H is normal. If $h = (g_1 x_1 g_1^{-1}) \cdots (g_n x_n g_n^{-1})$ where $x_1, \ldots, x_n \in X$ and $g_1, \ldots, g_n \in G$, then

$$ghg^{-1} = (gg_1 x_1 g_1^{-1} g^{-1})(gg_2 x_1 g_2^{-1} g^{-1}) \cdots (gg_n x_n g_n^{-1} g^{-1}) \in H.$$

Because G has no proper definable normal subgroups, this is a contradiction.

We leave the proof of the following corollary for Exercise 7.6.18.

Corollary 7.3.7 *If G is a simple group of finite Morley rank and $H \equiv G$, then H is simple.*

Finding a Field

If G is an algebraic group over an algebraically closed field K, then the field K is interpretable in G. Indeed, if V is any infinite variety, we can find a projection map $\pi : V \to K$ such that the image of V is a cofinite subset of

K. We can use the equivalence relation $x \sim y$ if and only if $\pi(x) = \pi(y)$ to interpret K.

To have any hope of proving the Cherlin–Zil'ber Conjecture, we would have to show that any simple group of finite Morley rank interprets an algebraically closed field. For the remainder of this section, we will show how the Indecomposability Theorem gives us ways to interpret fields in some finite Morley rank groups.

Definition 7.3.8 We say that an action $\alpha : H \times A \to A$ is *faithful* if whenever $g, h \in H$ and $g \neq h$ there is $a \in A$ such that $\alpha(g, a) \neq \alpha(h, a)$.

Theorem 7.3.9 *Let $(H, \cdot)$ and $(A, +)$ be infinite Abelian groups, and suppose that there is a faithful ω-stable action of H on A as a group of automorphisms such that no infinite definable $B \leq A$ is H-invariant. Then, we can interpret an algebraically closed field K.*

Proof Because A^0 is invariant under all definable automorphisms, A^0 is H-invariant, and hence $A = A^0$ so A is connected. Let $a \in A$ be sufficiently generic.

Claim 1 Ha is infinite.

If Ha is finite, then H^0a is finite and, by Exercise 7.6.11, $H^0a = \{a\}$. Thus, $X = \{x \in A : H^0x = \{x\}\}$ is generic and every element of A is a product of two elements of X. But then $H^0b = \{b\}$ for all $b \in A$. Because H acts faithfully, $H^0 = \{1\}$ and H is finite, a contradiction.

Claim 2 $Ha \cup \{0\}$ is indecomposable.

Because $Ha \cup \{0\}$ is H-invariant, by Lemma 7.3.3 we need only test indecomposability for H-invariant subgroups. If B is a definable proper H-invariant subgroup of A, then B is finite. Because Ha is infinite, $Ha \cup \{0\}/B$ is infinite. Thus, $Ha \cup \{0\}$ is indecomposable.

By the Indecomposability Theorem, the subgroup generated by $Ha \cup \{0\}$ is definable. Because this group is H-invariant, it must be all of A. From the proof of the Indecomposability Theorem, we see that there is an n such that every element of A is a sum of n elements in $Ha \cup \{0\}$.

Let $End(A)$ be the ring of endomorphisms of the group G. We can identify H with a subset of $End(A)$. Let R be the subring of $End(A)$ generated by H. Because H is Abelian, R is commutative. If $b \in A$, then $b = \sum_{i=1}^{m} h_i a$ for some $h_1, \ldots, h_m \in H$ and $m \leq n$. If $r \in R$, then

$$r(b) = \sum_{i=1}^{m} r h_i a = \sum_{i=1}^{m} h_i r a$$

because R is commutative. Thus if $r_1, r_2 \in R$ and $r_1 a = r_2 a$, then $r_1 = r_2$.

Suppose $ra = b$ and $b = \sum_{i=1}^{m} h_i a$. Then $h_1 + \ldots + h_m \in R$ and $ra = (h_1 + \ldots + h_m)a$. Thus, $r = h_1 + \ldots + h_m$ and every $r \in R$ is the sum of n elements of $H \cup \{0\}$.

Claim 3 The ring R is interpretable.

Define $\sim$ on $(H \cup \{0\})^n$ by

$$(h_1, \ldots, h_n) \sim (g_1, \ldots, g_n) \text{ if and only if } \sum h_i a = \sum g_i a.$$

Define $\oplus$ and $\otimes$ on $(H \cup \{0\})^n / \sim$ by

$$\overline{h} \oplus \overline{g} = \overline{l} \text{ if and only if } \sum h_i a + \sum g_i a = \sum l_i a$$

and

$$\overline{h} \otimes \overline{g} = \overline{l} \text{ if and only if } \sum_{i=1}^{n} \sum_{j=1}^{n} h_i g_j a = \sum_{k=1}^{n} l_i a.$$

Then $R \cong ((H \cup \{0\})^n / \sim, \oplus, \otimes)$.

Claim 4 R is a field.

Suppose that $r \in R$ and $r \neq 0$. If $b \in B$ and $rb = 0$, then for any $h \in H$, $r(hb) = (rh)b = (hr)b = h(rb) = 0$. Thus, the kernel of r is H-invariant. By our assumptions about A, the kernel of r is finite. Because A is connected, by Exercise 7.6.5, r is surjective. Thus, there is $c \in A$ with $rc = a$. Let $c = \sum h_i a$ and $s = \sum h_i \in R$. Then $sa = c$ and $rsa = a$. Because $1a = a$ and elements of R are determined by their actions on a, $rs = 1$.

The additive group R^+ of the interpreted field is isomorphic to A via the map $r \mapsto r(a)$, and we can view H as a subgroup of the multiplicative group of R so that the action of H on A corresponds to multiplication. In particular, the field R is infinite and, by Macintyre's Theorem, algebraically closed.

Definition 7.3.10 A group G is *solvable* if there is a chain of normal subgroups $G = G_0 \trianglerighteq \ldots \trianglerighteq G_n = \{1\}$ such that G_i / G_{i+1} is Abelian for each i.

If G is a group, we define the *derived series* by $G^{(0)} = G$, $G^{(n+1)} = G^{(n)'}$, the commutator subgroup of $G^{(n)}$.

We need two facts about solvable groups. See, for example, [89] 7.46 and 7.52.

Lemma 7.3.11 *i) A group G is solvable if and only if $G^{(n)} = \{1\}$ for some n.*

ii) If G is solvable, then all subgroups and quotients of G are solvable.

Theorem 7.3.12 *If G is an infinite connected solvable group of finite Morley rank with finite center, then G interprets an algebraically closed field.*

Proof We will prove this by induction on the rank of G. We first argue that we may, without loss of generality, assume that G is centerless. Suppose that $Z(G)$ is finite. We claim that $G/Z(G)$ is centerless. Let $a \in G$ such that $a/Z(G) \in Z(G/Z(G))$. For all $g \in G$, $a^{-1}g^{-1}ag \in Z(G)$. Thus,

$a^{-1}a^G \subseteq Z(G)$, and, hence, a^G is finite. Thus, $[G : C(a)]$ is finite. Because G is connected, $C(a) = G$ and $a \in Z(G)$. Thus, $G/Z(G)$ is solvable and centerless. By Exercise 7.6.4, $G/Z(G)$ is connected. Because $G/Z(G)$ is interpretable in G, if $G/Z(G)$ interprets an algebraically closed field, so does G.

Let $A \trianglelefteq G$ be a minimal infinite definable normal subgroup. By Lemma 7.3.11 and Corollary 7.3.5, A is solvable and A' is a proper connected normal definable subgroup of A. Moreover, any automorphism of A fixes A' setwise, thus $A' \trianglelefteq G$. By choice of A, $A' = \{1\}$ and A is Abelian.

Let $C(A) = \{g \in G : ga = ag$ for all $a \in A\}$. Because G is centerless, $C(A) \neq G$. Because A is normal, so is $C(A)$. Let $G_1 = G/C(A)$.

If $g, h \in G$ and $g = hc$ where $c \in C(A)$, then for $a \in A$

$$hah^{-1} = gcac^{-1}g^{-1} = gag^{-1}.$$

Thus, G_1 acts on A by conjugation. If $gag^{-1} = hah^{-1}$ for all $a \in A$, then $g^{-1}hah^{-1}ga = a$ and $g^{-1}h \in C(A)$. Thus, $g/C(A) = h/C(A)$ and G_1 acts faithfully on A. Moreover, because A is the smallest infinite normal definable subgroup, no infinite definable subgroup of A is G_1-invariant.

The group G_1 is solvable and, because $A \leq C(A)$, $\mathrm{RM}(G_1) < \mathrm{RM}(G)$. If $Z(G_1)$ is finite, then G_1 (and hence G) interprets an algebraically closed field by induction. Thus, we may assume that $Z(G_1)$ is infinite. Let H be a minimal definable infinite subgroup of $Z(G_1)$. Then, H is Abelian, $H \trianglelefteq G_1$, and H acts faithfully on A by conjugation.

If there are no definable infinite H-invariant proper subgroups of A, then by Theorem 7.3.9 there is an interpretable algebraically closed field. Otherwise, let $B < A$ be a minimal infinite H-invariant definable subgroup. Let $H_0 = \{h \in H : b^h = b$ for all $b \in B\}$.

Suppose that $H = H_0$ (i.e., H acts trivially on B). Because B is a minimal H-invariant subgroup, B is indecomposable by 7.3.3 and, by Lemma 7.6.12, B^g is indecomposable for all $g \in G_1$. Because $H \leq Z(G_1)$, if $h \in H$ and $b \in B$, then

$$(b^g)^h = (b^h)^g = b^g.$$

Thus, H acts trivially on B^g as well. By Zil'ber's Indecomposability Theorem, the group generated by $\langle B^g : g \in G_1 \rangle$ is a definable G_1-invariant subgroup of A. But A is a minimal G_1-invariant subgroup, thus A is generated by $\langle B^g : g \in G_1 \rangle$. Because H acts trivially on each B^g, H acts trivially on A, a contradiction.

Thus, H_0 is a proper subgroup of H and, because H is minimal, H_0 is finite. But then H/H_0 acts faithfully on B and there are no infinite definable H/H_0-invariant subgroups of B. Thus, by Theorem 7.3.9, we can interpret an algebraically closed field.

We have already seen a concrete example of Theorem 7.3.12 in Section 1.3. Let K be an algebraically closed field and G be the group of matrices

$$G = \left\{ \begin{pmatrix} a & b \\ 0 & 1 \end{pmatrix} : a, b \in K, a \neq 0 \right\};$$

then G is a connected, solvable, centerless group of finite Morley rank (see Exercise 7.6.19). The proof of Theorem 7.3.12 is an abstraction of the concrete interpretation of the field in Section 1.3.
We will give one more extension of this result.

Definition 7.3.13 A group G is *nilpotent* if there is a chain of normal subgroups $G = G_0 \trianglerighteq \ldots \trianglerighteq G_n = \{1\}$ such that $G_i/G_{i+1} \leq Z(G/G_{i+1})$ for all $i < n$.

For a group G, we define the *lower central series* as $\Gamma_0(G) = G$, $\Gamma_{n+1}(G) = [\Gamma_n(G) : G]$ the group generated by commutators $\{[a,b] : a \in \Gamma_n(G), b \in G]\}$. Then $G \trianglerighteq \Gamma_1(G) \trianglerighteq \Gamma_2(G) \ldots$. We define the *upper central series* by $Z_0(G) = \{1\}$ and $Z_n(G) = \{g \in G : g/Z_{n-1} \in Z(G/Z_{n-1})\}$.

We will use the following facts about nilpotent groups. See, for example, [89] 7.54.

Lemma 7.3.14 *A group G is nilpotent if and only if there is an n such that $\Gamma_n(G) = \{1\}$ if and only if there is an n such that $Z_n(G) = G$.*

Theorem 7.3.15 *If G is an infinite connected, solvable, nonnilpotent group of finite Morley rank, then G interprets an algebraically closed field.*

Proof Let $Z_0(G) \trianglelefteq Z_1(G) \ldots$ be the upper central series of G. Because G has finite Morley rank, there is an n such that $\mathrm{RM}(Z_n(G))$ is maximal. Then, $Z_{n+1}(G)/Z_n(G)$ is finite and, because G is nonnilpotent, $Z_n(G) \neq G$.
Consider $G/Z_n(G)$. By Lemma 7.3.11 and Exercise 7.6.4 $G/Z_n(G)$ is a connected solvable group of finite Morley rank. Because $Z(G/Z_{n+1}(G)) = Z_{n+1}(G)/Z_n(G)$ is finite, $G/Z_n(G)$ has finite center. By Theorem 7.3.12, $G/Z_n(G)$, and hence G, interprets an algebraically closed field.

7.4 Definable Groups in Algebraically Closed Fields

In this section, we will investigate groups interpretable in algebraically closed fields. Our goal is to show that any such group is definably isomorphic to an algebraic group. If G is interpretable in an algebraically closed field K, then, by elimination of imaginaries, there is a definable $X \subseteq K^n$ and a definable $f : X \times X \to X$ such that $(G, \cdot)$ is definably isomorphic

to (X, f). Thus, to study interpretable groups it suffices to study groups
where the underlying set and multiplication are definable sets. In algebraically closed fields, the definable subsets are exactly the constructible
subsets, so our goal is to show that any constructible group is definably
isomorphic to an algebraic group.

Varieties

We have already encountered two types of algebraic groups: linear algebraic
groups and elliptic curves. We define the category of algebraic groups to
include both types of examples. We begin by defining an abstract algebraic
variety. The idea is that we build abstract varieties from Zariski closed
subsets of K^n in the same way that we build manifolds from open balls in
$\mathbb{R}^n$ or $\mathbb{C}^n$.

Definition 7.4.1 A *variety*[1] is a topological space V such that V has a
finite open cover $V = V_1 \cup \ldots \cup V_n$ where for $i = 1, \ldots, n$ there is $U_i \subseteq K^{n_i}$
a Zariski closed set and a homeomorphism $f_i : V_i \to U_i$ such that:
  i) $U_{i,j} = f_i(V_i \cap V_j)$ is an open subset of U_i, and
  ii) $f_{i,j} = f_i \circ f_j^{-1} : U_{j,i} \to U_{i,j}$ is a rational map.
We call $f_1, \ldots, f_n$ *charts* for V.

Varieties arise in many natural ways. Let K be an algebraically closed
field.

Lemma 7.4.2 *i) If $V \subseteq K^n$ is Zariski closed, then V is a variety.*
  *ii) If $V \subseteq K^n$ is Zariski closed and $O \subseteq K^n$ is Zariski open, then $V \cap O$
is a variety.*
  iii) $\mathbb{P}^1(K)$ is a variety.
  *iv) If $V \subseteq \mathbb{P}^n(K)$ is Zariski closed and $O \subseteq \mathbb{P}^n(K)$ is Zariski open, then
$V \cap O$ is a variety.*

Proof
  i) Clear.
  ii) Let $O = \bigcup_{i=1}^m O_i$ where $O_i = \{x \in K^n : g_i(x) \neq 0\}$ for some $g_i \in$
$K[X_1, \ldots, X_n]$. Let $V_i = \{x \in V : g_i(x) \neq 0\}$. Then, V_i is an open subset
of V and $V \cap O = V_1 \cup \ldots \cup V_n$. Let $U_i = \{(x, y) \in K^{n+1} : x \in V$ and
$yg_i(x) = 1\}$, and let $f_i : V_i \to U_i$ be $x \mapsto (x, \frac{1}{f_i(x)})$. Then, f_i is a rational
bijection with rational inverse $(x, y) \mapsto x$, so V_i and U_i are homeomorphic.
In this case, $U_{i,j}$ is the open set $\{(x, y) \in U_i : f_j(x) \neq 0\}$ and $f_{i,j}$ is the
rational map $(x, y) \mapsto (x, \frac{1}{f_i(x)})$.

---

[1]The objects we are defining here are usually called *prevarieties* and *varieties* are
prevarieties where the diagonal $\{(x, y) : x = y\}$ is closed in $V \times V$. If a prevariety
has a group structure, the diagonal is automatically closed, so this distinction is not so
important to us. See Exercise 7.6.22.

iii) Projective 1-space $\mathbb{P}^1(K)$ is the quotient of $K^2 \setminus \{(0,0)\}$ by the equivalence relation $(x,y) \sim (u,v)$ if there is $\lambda \in K$ such that $\lambda x = u$ and $\lambda y = v$, (i.e., if $xv = yu$). Let $V_1 = \{(x,y)/ \sim: x \neq 0\}$, and let $V_2 = \{(x,y)/ \sim: x \neq 0\}$. Let $U_1 = U_2 = K$, and let $f_1((x,y)/ \sim) = y/x$, while $f_2((x,y)/ \sim) = x/y$. Then, $U_{1,2} = U_{2,1} = K \setminus \{0\}$ and $f_{i,j}(x) = f_{j,i}(x) = \frac{1}{x}$.

iv) Exercise.

A *quasiprojective* variety is the intersection of Zariski open and Zariski closed subsets of projective space. Part iv) of the preceding lemma shows that quasiprojective varieties are examples of abstract algebraic varieties.

Lemma 7.4.3 *If V is a variety, then V is interpretable in the algebraically closed field K.*

Proof Let $V = V_1 \cup \ldots \cup V_n$ with charts $f_i : V_i \to U_i$, without loss of generality, there is an m such that each $U_i \subseteq K^m$. Let $a_1, \ldots, a_n \in K$ be distinct, and let $X = \{(x,y) \in K^{m+1} : y = a_i \text{ and } x \in U_i \text{ for some } i \leq n\}$. Then, X is a Zariski closed subset of K^{m+1}. We define an equivalence relation $\sim$ on X, by $(x,a_i) \sim (y,a_j)$, if and only if $a_i = a_j$ and $x = y$ or $a_i \neq a_j$, $x \in U_{i,j}$, $y \in U_{j,i}$, and $f_{i,j}(y) = x$.

If V is a variety with charts $f_1 : V_0 \to U_0, \ldots, f_n : V_n \to U_n$, $X \subseteq U_i$ is open in U_i, and $W = f_i^{-1}(X)$, then we call W an *affine open* subset of V. Any open subset of V is a finite union of affine open subsets of V.

We will consider maps between varieties that are given locally by rational functions.

Definition 7.4.4 Suppose that V and W are varieties and $f : V \to W$. We say that f is a *morphism* if we can find $V_1, \ldots, V_n$ and $W_1, \ldots, W_m$ covers of V and W by affine open sets with homeomorphisms $f_i : V_i \to U_i$, $g_j : W_j \to U_j'$, where U_i and U_j' are open subsets of affine Zariski closed sets and $g_j \circ f \circ f_i^{-1}$ is a rational function for each $i \leq n$, $j \leq m$.

In characteristic $p > 0$, one should also consider *quasimorphisms* where the maps are locally given by quasirational functions (i.e., compositions of rational functions and $x \mapsto \sqrt[p]{x}$).

We next define the product of two varieties.

Definition 7.4.5 Suppose that V and W are varieties and $f : V \to W$. Suppose that $V = V_1 \cup \ldots \cup V_n$ and $f_i : V_i \to U_i$ and $W = W_1 \cup \ldots \cup W_m$ and $g_i : W_i \to U_i'$ are charts for V and W. We topologize $V_i \times W_j$ so that $(f_i, g_j) : V_i \times W_j \to U_i \times U_j'$ is a homeomorphism and take $\{V_i \times W_j : i \leq n, j \leq m\}$ as a finite open cover of $V \times W$.

Note that the topology on $V \times W$ is a proper refinement of the product topology. For example the line $y = x$ is a closed subset of K^2, but it is not closed in the product topology on $K \times K$.

We summarize some of the basic topological properties of varieties that we will need. These follow easily from the corresponding properties of Zariski closed sets. We leave the proofs as exercises.

Lemma 7.4.6 *Suppose that V and W are varieties.*

i) If V is a variety and $X \subseteq V$ is open, then X is a variety.

ii) There are no infinite descending chains of closed subsets of V.

iii) Any closed subset of V is a finite union of irreducible components (see Exercise 3.4.17).

iv) If $f : V \to W$ is a morphism, then f is continuous.

v) If $f : V \to W$ and for each $a \in V$ there is an open $U \subseteq V$ such that $a \in U$ and $f|U$ is a morphism, then f is a morphism.

vi) The product $V \times W$ is a variety and the topology on $V \times W$ refines the product topology.

In our proof that constructible groups are definably isomorphic to algebraic groups, we will use heavily Proposition 3.2.14, which states that if X is constructible and $f : X \to K$ is definable, then we can partition X into constructible sets $X_1, \ldots, X_m$ such that $f|X_i$ is quasirational for each $i \leq m$. The next lemma shows how we will combine Proposition 3.2.14 and Lemma 6.2.26.

Lemma 7.4.7 *Suppose that V and W are varieties, $V_0 \subseteq V$ is open, and $f : V_0 \to W$ is a definable function. There is an affine open $U \subseteq V_0$ such that $f|U$ is a quasimorphism.*

Proof Without loss of generality, we may assume that V_0 is an affine open subset of V, the closure of V_0 is irreducible, W_0 is an affine open subset of W, and $f : V_0 \to W_0$. By Proposition 3.2.14, there are quasirational functions $f_1, \ldots, f_m$ such that for each $a \in V_0$ there is $i \leq m$ such that $f(a) = f_i(a)$. Choose i such that $\{x \in V_0 : f(x) = f_i(x)\}$ has maximal rank. By Lemma 6.2.26, this set has a nonempty interior in V_0.

Algebraic Groups

Definition 7.4.8 An *algebraic group* is a group $(G, \cdot)$ where G is a variety and $\cdot$ and $x \mapsto x^{-1}$ are morphisms.

We derive some basic properties of algebraic groups.

Lemma 7.4.9 *A definable subgroup of an algebraic group is closed.*

Proof Suppose that G is an algebraic group and $H \leq G$ is definable. Let V be the closure of H in G. Suppose, for contradiction, that $a \in V \setminus H$. By Exercise 6.6.14, $\mathrm{RM}(V \setminus H) < \mathrm{RM}(H)$. Every open set containing a intersects H. If $b \in H$, then $x \mapsto bx$ is continuous. Thus, every open set containing ba intersects H and $Ha \subseteq V \setminus H$. But $x \mapsto xa$ is a definable bijection. Thus

$$\mathrm{RM}(H) = \mathrm{RM}(Ha) \leq \mathrm{RM}(V \setminus H) < \mathrm{RM}(H),$$

a contradiction.

In particular an algebraic group is connected if it has no proper algebraic subgroups of finite index.

Lemma 7.4.10 *A connected algebraic group is irreducible.*

Proof Let G be a connected algebraic group. Let $V_1, \ldots, V_m$ be the irreducible components of G. If $a \in G$, then $x \mapsto ax$ is continuous. Thus, each aV_i is irreducible and $G = aV_1 \cup \ldots \cup aV_m$. Because the decomposition into irreducible components is unique, G acts on the irreducible components. Because there are only finitely many irreducible components, $H = \{a : aV_1 = V_1\}$ is a finite index subgroup of G. Because G is connected, we must have $H = G$. Because $aV_1 = V_1$ for all $a \in G$, we must have $V_1 = G$.

For the remainder of this section, we will make the simplifying assumption that K has characteristic zero so that we do not have to deal with quasirational functions. In characteristic $p > 0$, our proof will show that every constructible group is definably isomorphic to a "quasialgebraic" group (i.e., one where multiplication and inversion are quasimorphisms). A second argument is then needed to show that quasialgebraic groups are definably isomorphic to algebraic groups. We refer the reader to [86] for details in this case.

We will be doing some model-theoretic arguments with generic types and generic elements of our field. Let $K \prec \mathbb{K}$ be a monster model. The formulas defining a group G in K will define a group $\mathbb{G}$ in $\mathbb{K}$ that will also be a monster model. If $X \subseteq G$ is definable and $a \in \mathbb{G}$, we will sometimes write "$a \in X$," to mean that a satisfies the formula defining X.

Lemma 7.4.11 *Suppose that K is an algebraically closed field of characteristic zero. Suppose that G is a variety, $(G, \cdot)$ is a group, and $\cdot$ is a morphism. Then, $(G, \cdot)$ is an algebraic group.*

Proof We must show that $x \mapsto x^{-1}$ is a morphism. Let G^0 be the connected component of G, and let $U \subseteq G^0$ be an affine open subset of G^0. Because G^0 is irreducible, it is the closure of U. Thus, U is a generic subset of G^0. Because $x \mapsto x^{-1}$ preserves the unique generic type of G^0, if a is a generic of G^0, then $a, a^{-1} \in U$.

By Lemma 7.4.7, there is an open $V \subseteq U$ such that the inverse map is a morphism on V. Let $V_1 = V \cap V^{-1}$. Then V_1 is an open subset of U containing the generic, the inverse is given by a rational function on V_1, and if $x \in V_1$, then $x^{-1} \in V_1$.

Because V contains a generic type of G, by Lemma 7.2.8, there are $a_1, \ldots, a_m \in G$ such that $G = a_1 V_1 \cup \ldots \cup a_m V_1$. On $a_i V_1$, the inverse is obtained by the composition of morphisms

$$x \mapsto a_i^{-1} x \mapsto (a_i^{-1} x)^{-1} \mapsto (a_i^{-1} x)^{-1} a_i^{-1}.$$

By Lemma 7.4.6 v), inversion is a morphism on G.

Lemma 7.4.12 *Suppose that K is an algebraically closed field of characteristic zero, G and H are algebraic groups and $f : G \to H$ is a definable group homomorphism, then f is a morphism.*

Proof Let $a \in G^0$ be generic. By Lemma 7.4.7, there is an affine open set U such that $f|U$ is a morphism. Finitely many translates of U cover G. If $x \in aU$, then $f(x)$ is given by the composition

$$x \mapsto a^{-1} x \mapsto f(a^{-1} x) \mapsto f(a) f(a^{-1} x).$$

Because $a^{-1} x \in U$, this is a composition of morphisms; thus, f is a morphism.

Constructible Groups

We now begin the proof that constructible groups are isomorphic to algebraic groups. If G is a constructible group, we must find a finite cover by definable open sets $V_1, \ldots, V_n$ and find definable charts $f_i : V_i \to U_i$ making G an algebraic variety such that multiplication becomes a morphism.

We first show that, without loss of generality, we may assume that G is connected.

Lemma 7.4.13 *Suppose that K is an algebraically closed field of characteristic zero, G is a constructible group, and G^0 is definably isomorphic to an algebraic group; then, so is G.*

Proof Let A be a set of representatives for G/G^0. For $g \in G$, let $i(g) \in A$ such that $g \in i(g) G^0$. We can choose A such that $i(1) = 1$ and $a^{-1} \in A$ for all $a \in A$. Then, G is the disjoint union $\bigcup_{a \in A} aG$.

We will use the variety structure of G^0 to topologize G. We say that $U \subseteq aG^0$ is open if and only if $\{x \in G^0 : ax \in U\}$ is open, and $X \subseteq G$ is open if and only if $X \cap aG^0$ is open for all $a \in A$. It is easy to see that this topology makes G a variety.

We must show that multiplication is a morphism. The variety $G \times G$ is covered by the disjoint open sets $aG^0 \times bG^0$ for $a, b \in A$. Let $a, b \in A$ and $x, y \in G^0$, then

$$axby = ab(b^{-1} xb)y = i(ab) i(ab)^{-1} ab(b^{-1} xb) y.$$

Because $i(ab)^{-1}ab \in G^0$ and $b^{-1}xb \in G^0$ (because G^0 is normal), $axby \in$ $i(ab)G^0$. Thus, multiplication maps $aG^0 \times bG^0$ to $i(ab)G^0$. It suffices to show that $(x, y) \mapsto i(ab)^{-1}axby$ is a morphism of G^0. Because $G^0 \trianglelefteq G$, $x \mapsto b^{-1}xb$ is a definable automorphism of G^0 and hence, by Lemma 7.4.12, a morphism. Because $i(ab)^{-1}ab \in G^0$ the map $z \mapsto i(ab)^{-1}abz$ is a morphism of G^0. Thus

$$(x, y) \mapsto i(ab)^{-1}ab(b^{-1}xb)y = i(ab)^{-1}axby$$

is a morphism of G^0, as desired.

Theorem 7.4.14 *Let K be an algebraically closed field of characteristic zero. If $G \subseteq K^n$ is a constructible group, then G is definably isomorphic to an algebraic group.*

Proof Without loss of generality, we may assume that G is connected. By quantifier elimination, $G = \bigcup_{i=1}^n F_i \cap O_i$ where F_i is Zariski closed and irreducible and O_i is open. Let V_1 be some $F_i \cap O_i$ containing the generic type of G. Note that V_1 is a variety. Because $x \mapsto x^{-1}$ is definable, we can find an open $V_2 \subseteq V_1$ containing the generic such that inversion is rational on V_2. Because $(x, y) \mapsto x \cdot y$ is definable, there is an open $W_1 \subset V_2 \times V_2$ and a rational function f such that $f(x, y) = x \cdot y$ for $(x, y) \in W_1$. Then, $V_3 = \{x \in V_2 : (y, x) \in W_1$ and $(y^{-1}, yx) \in W_1$ for all y generic over $x\}$ contains the generic of G and, by definability of types, V_3 is definable. Thus, there is $V_4 \subseteq V_3$ such that V_4 is open in V_3 and V_4 contains the generic. Let $V = V_4 \cap V_4^{-1}$ and $W = \{(x, y) \in W_1 : x, y, xy \in V\}$. Note that:

- V and W are open;
- multiplication is a morphism on W;
- if $a \in V$ and $x \in G$ is generic over a, then $(x, a) \in W$ and $(x^{-1}, xa) \in W$.

Claim 1 If $a, b \in G$, then $U_{a,b} = \{(x, y) \in V \times V : axby \in V\}$ is open in $V \times V$ and $(x, y) \mapsto axby$ is a morphism from $U_{a,b}$ to V.

Because V is generic, we can find $b', b'' \in V$ such that $b = b'b''$. Suppose that $x_0, y_0 \in V$ and $ax_0by_0 \in V$. Let c be a realization of the generic independent from a, x_0, b', b'', y_0. The following pairs are all in W: (ca, x_0), (cax_0, b'), (cax_0b', b''), $(cax_0b'b'', y_0)$. In each case, the first element of the pair is generic over the second. Because

$$X = \{(x, y) : x \in V, y \in V, (ca, x), (cax, b'), (caxb', b''), (caxb'b'', y) \in W\}$$

is the inverse image of W under a composition of morphisms, X is an open set containing (x_0, y_0). If $(x, y) \in X$, then $axby \in V$. Thus $X \subseteq U_{a,b}$ and $U_{a,b}$ is open. Because multiplication is rational on W, $(x, y) \mapsto axby$ is a composition of rational functions on X. Hence. $(x, y) \mapsto axby$ is a morphism from $U_{a,b}$ to V.

Claim 2 Let $c \in G$, then $V_c = \{y \in V : cy \in V\} = V \cap c^{-1}V$ is open and $y \mapsto cy$ is a morphism from V_c to V.

Let $x_0 \in V$. Let $a = cx_0^{-1}$ and $b = 1$. Then, $U_{a,b} = \{(x,y) \in V \times V : cx_0^{-1}xy \in V\}$ is open and $(x,y) \mapsto cx_0^{-1}xy$ is a morphism. Considering the section where $x_0 = x$, we see that V_c is open and $y \mapsto cy$ is a morphism.

Because V is generic, we can cover G by finitely many translates of V. We say that $X \subseteq cV$ is open if and only if $\{x \in V : cx \in X\}$ is open in V. By the claim 2, $V \cap cV$ is open in V; thus, $aV \cap bV$ is an open subset of aV and bV. Thus, we can topologize G by making $X \subseteq G$ open if and only if $X \cap aV$ is open for all $a \in G$. This topology makes G a variety.

To show that multiplication is a morphism, we argue that for $a, b, c \in G$ $Y = \{(x,y) \in V \times V : axby \in cV\}$ is open and $(x,y) \mapsto c^{-1}axby$ is a morphism. But $Y = U_{c^{-1}a,b}$, so this follows from the first claim.

By elimination of imaginaries, we can extend Theorem 7.4.14 to interpretable groups.

Corollary 7.4.15 *If G is a group interpretable in an algebraically closed field of characteristic zero, then G is definably isomorphic to an algebraic group.*

Corollary 7.4.16 *If K is an algebraically closed field of characteristic zero, G is an algebraic group and $H \leq G$ is an algebraic subgroup, then G/H is an algebraic group.*

Proof Because G/H is interpretable, by elimination of imaginaries it is constructible and, by Theorem 7.4.14, it is isomorphic to an algebraic group.

Both corollaries are true in finite characteristic as well.

Differential Galois Theory

Poizat [84] showed how Theorem 7.4.14 can be used to give a new proof of a result of Kolchin's in differential Galois theory.

Definition 7.4.17 Suppose that k and l are differential fields with $k \subseteq l$. The *differential Galois group* $G(l/k)$ is the group of all differential field automorphisms of l that fix the field k pointwise.

Differential Galois groups were first studied when l is obtained from k by adjoining solutions of linear differential equations. We say that $f(X) \in k\{X\}$ is a *linear differential polynomial* over k if

$$f(X) = a_n X^{(n)} + \ldots + a_2 X'' + a_1 X' + a_0 X + b$$

where $a_0, \ldots, a_n, b \in k$. If $a_n \neq 0$, then we say that f has order n. If $b = 0$, we say that $f(X) = 0$ is a *homogeneous linear differential equation*.

Definition 7.4.18 Let $x_0, \ldots, x_n$ be elements of some differential field. We define the *Wronskian* of $x_0, \ldots, x_n$ to be the determinant

$$
W(x_0, \ldots, x_n) =
\begin{vmatrix}
x_0 & x_1 & \cdots & x_n \\
x_0' & x_1' & \cdots & x_n' \\
\vdots & \vdots & \ddots & \vdots \\
x_0^{(n)} & x_1^{(n)} & \cdots & x_n^{(n)}
\end{vmatrix}
$$

The next lemma summarizes some of the basic facts we will need about linear differential equations. If k is a differential field, we let C_k denote the constant field of k.

Lemma 7.4.19 *Let k be a differential field with constants C_k. Let $f(X) \in k\{X\}$ be a homogeneous linear differential polynomial of order n.*
  i) The solutions to $f(X) = 0$ in k form a vector space over C_k of dimension at most n.
  ii) Let $x_0, \ldots, x_n \in k$; then $W(x_0, \ldots, x_n) = 0$ if and only if $x_0, \ldots, x_n$ are linearly dependent over C_k.
  iii) Let $K \supseteq k$ be differentially closed. There are $x_1, \ldots, x_n \in K$ solutions to $f(X) = 0$ such that $x_1, \ldots, x_n$ are linearly independent over C_K.

Proof i) and ii) are standard facts that can be found in any book on differential equations (for example, [39] or [65]).

iii) Given $x_1, \ldots, x_m$ with $m < n$. Because $W(x_1, \ldots, x_{m+1})$ has order $m < n$, we can find $x_{m+1} \in K$ such that $f(x_{m+1}) = 0$ and $W(x_1, \ldots, x_{m+1}) \neq 0$.

In particular, if K is differentially closed with constant field C_K, and $f(X) = 0$ is a homogeneous linear differential equation of order n, then there are $x_1, \ldots, x_n$ linearly independent over C_K such that the solution set for $f(X) = 0$ is exactly the C_K-vector space spanned by $x_1, \ldots, x_n$. In this case, we call $x_1, \ldots, x_n$ a *fundamental system of solutions*.

Definition 7.4.20 Let l/k be differential fields. We say that l is a *Picard–Vessiot extension* of k if there is a homogeneous linear differential equation $f(X) = 0$ and $x_1, \ldots, x_n \in l$ a fundamental system of solutions such that $l = k\langle x_1, \ldots, x_n \rangle$ and $C_k = C_l$. We say that l/k is a Picard–Vessiot extension for f.

Using differential closures, it is easy to show the existence of Picard–Vessiot extensions. Recall that in Exercise 6.6.32 we proved that if k is a differential field and K is the differential closure of k, then C_K is the algebraic closure of C_k. In particular, if C_k is algebraically closed, then $C_K = C_k$.

Lemma 7.4.21 *Let k be a differential field with algebraically closed constant field C_k, and let $f(X) = 0$ be a homogeneous linear differential equation over k. There is l/k a Picard–Vessiot extension for f with l contained in the differential closure of k. Moreover, if l_1 is a second Picard–Vessiot extension of k for f, then l_1 is isomorphic to l over k.*

Proof Let K be the differential closure of k. By Exercise 6.6.32, $C_K = C_k$. By Lemma 7.4.19, we can find $x_1, \ldots, x_n \in K$ a fundamental system of solutions for $f(X) = 0$. Thus, $l = k\langle x_1, \ldots, x_n \rangle$ is a Picard–Vessiot extension of k.

Suppose that l_1 is a second Picard–Vessiot extension of k for f. Let K_1 be the differential closure of l_1. By Exercise 6.6.32, $C_{K_1} = C_{l_1} = C_k$. Because K is the differential closure of k, there is a differential field embedding $\sigma : K \to K_1$ fixing k. Let $y_1 \ldots y_n \in K_1$ be a fundamental system of solutions of $f(X) = 0$ such that $l_1 = k\langle y_1 \ldots y_n \rangle$. But then each $\sigma(x_i)$ is in the span of $(y_1, \ldots, y_n)$ over C_k and each y_i is in the span of $(\sigma(x_1), \ldots, \sigma(x_n))$ over C_k. Thus, l_1 is the image of l. Thus, $f(X) = 0$ determines a unique Picard–Vessiot extension of k.

Picard–Vessiot extensions have linear algebraic differential Galois groups.

Theorem 7.4.22 *Suppose that k is a differential field with algebraically closed constant field C. Let $f(X) \in k\{X\}$ be a homogeneous linear differential equation of order n, and let l/k be a Picard–Vessiot extension for f. The differential Galois group $G(l/k)$ is isomorphic to an algebraic subgroup of $GL_n(C)$.*

Proof Suppose that $l = k\langle x_1, \ldots, x_n \rangle$, where $x_1, \ldots, x_n$ is a fundamental system of solutions to linear equation $f(X) = 0$. Because l/k is Picard–Vessiot, $C_l = C$.

Let $V = \{y \in l : f(y) = 0\}$. Then V is an n-dimensional vector space over C. We let $GL(V)$ denote the group of C-linear automorphisms of this vector space. We can identify $GL(V)$ with the algebraic group $GL_n(C)$.

Let $\mathbb{K}$ be a monster model of DCF with $l \subset \mathbb{K}$.

Claim Suppose that $y_1, \ldots, y_n \in V$. Then, $\mathrm{tp}(\overline{x}/k) = \mathrm{tp}(\overline{y}/k)$ if and only if there is $\sigma \in G(l/k)$ such that $\sigma(x_i) = y_i$ for $i = 1, \ldots, n$.

If $\overline{x}$ and $\overline{y}$ realize the same type over k, then there is an automorphism τ of $\mathbb{K}$ fixing k such that $\tau(x_i) = y_i$ for all $i = 1, dots, n$. In this case, $y_1, \ldots, y_n$ is also a fundamental system of solutions for $f = 0$ and $x_1, \ldots, x_n$ are in the C-linear span of $y_1, \ldots, y_n$. Thus, $k\langle y_1, \ldots, y_n \rangle = l$ and $\sigma = \tau|l \in G(l/k)$. On the other hand, if there is $\sigma \in G(l/k)$ with $\sigma(x_i) = y_i$, then, by quantifier elimination, $\mathrm{tp}(\overline{x}/k) = \mathrm{tp}(\overline{y}/k)$.

By Lemma 7.4.21, we can embed l into K the differential closure of k. By Theorem 6.4.10, $\mathrm{tp}(\overline{x}/k)$ is isolated by some formula $\psi(\overline{v})$. Thus, $G = \{\tau \in GL_n(V) : \psi(\tau(\overline{x}))\}$ is definable in K. Because G is a definable subgroup of the algebraic group $GL_n(C)$, by Exercise 6.6.22, G is definable in the pure field $(C, +, \cdot)$. By Lemma 7.4.9, G is an algebraic subgroup of $GL_n(C)$.

There is a very beautiful Galois theory for Picard–Vessiot extensions. For example, there is a correspondence between differential subfields of l/k and algebraic subgroups of $G(l/k)$. We refer the reader to [63] for an exposition of these results.

Kolchin generalized the notion of Picard–Vessiot extension to a class of extensions where the differential Galois groups are arbitrary algebraic groups over the constants.

Let $\mathbb{K}$ be a monster model of the theory of differentially closed fields. We consider only differential subfields of $\mathbb{K}$.

Definition 7.4.23 Let k and l be differential fields with $k \subseteq l$. We say that l/k is *strongly normal* if and only if

i) $C_l = C_k$ is algebraically closed;

ii) l/k is finitely generated;

iii) if $\sigma : \mathbb{K} \to \mathbb{K}$ is a differential field automorphism fixing k pointwise, then $\langle l, C_\mathbb{K} \rangle = \langle \sigma(l), C_\mathbb{K} \rangle$, where $\langle E, F \rangle$ denotes the differential field generated by E and F.

If C_k is algebraically closed and l/k is Picard–Vessiot, we show that l/k is strongly normal. Suppose that $l = k\langle \bar{a} \rangle$, where $\bar{a}$ is a fundamental system of solutions to a linear equation over k. For any automorphism σ fixing k pointwise, $\sigma(\bar{a}) \in \langle l, C_\mathbb{K} \rangle$, thus $\langle l, C_\mathbb{K} \rangle \supseteq \langle \sigma(l), C_\mathbb{K} \rangle$. Similarly, l is contained in $\langle \sigma(l), C_\mathbb{K} \rangle$, so equality holds.

Lemma 7.4.24 *Suppose that l/k is strongly normal and K is the differential closure of k. Then $l \subseteq K$.*

Proof Let $l = k\langle \bar{a} \rangle$ and let F be the differential closure of l. We may assume that $K \subseteq F$. Note that $C_F = C_l = C_k$. Suppose, for purposes of contradiction, that $\bar{a} \notin K$. Let $p = \operatorname{tp}(\bar{a}/K)$, let q be the unique nonforking extension of p to F and let $\bar{b}$ realize q. In particular, $\bar{b} \notin F$.

Claim $C_{F\langle \bar{b} \rangle} = C_k$.

If not, there is a K-definable function $f(\bar{x}, \bar{y})$ and $\bar{d} \in F$ such that $f(\bar{b}, \bar{d}) \in C_{F\langle \bar{b} \rangle} \setminus C_k$. Let $\phi(\bar{v}, \bar{w})$ be the formula $f(\bar{v}, \bar{w})' = 0$, and let $\psi(\bar{v}, \bar{w}, u)$ be the formula $f(\bar{v}, \bar{w}) = u$. Because p is definable and q is the unique nonforking extension of p, by definability of types, there are $\mathcal{L}_K$-formulas $d_p \phi(\bar{w})$ and $d_p \psi(\bar{w}, u)$ such that for $\bar{\alpha}, \beta \in F$

$$\phi(\bar{v}, \bar{\alpha}) \in q \Leftrightarrow F \models d_p \phi(\bar{\alpha})$$

and

$$\psi(\bar{v}, \bar{\alpha}, \beta) \in q \Leftrightarrow F \models d_p \psi(\bar{\alpha}, \beta).$$

Thus,

$$F \models d_p \phi(\bar{d}) \wedge \forall x \ (x' = 0 \to \neg d_p \psi(\bar{v}, \bar{d}, x)).$$

By model-completeness, there is $\overline{\beta} \in K$ such that

$$K \models d_p\phi(\overline{\beta}) \wedge \forall x \ (x' = 0 \to \neg d_p\psi(\overline{v}, \overline{\beta}, x)).$$

Then, because $\overline{a}$ realizes p, $f(\overline{a}, \overline{\beta})' = 0$ and $f(\overline{a}, \overline{\beta}) \neq c$ for all $c \in C_K$. Thus, $f(\overline{a}, \overline{\beta}) \in C_F \setminus C_K$, a contradiction.

Let L be the differential closure of $F\langle \overline{b} \rangle$. Then $C_L = C_F$.

Because $\overline{a}$ and $\overline{b}$ realize the same type over k, there is an automorphism of $\mathbb{K}$ fixing k and sending $\overline{a}$ to $\overline{b}$. Because l/k is strongly normal, $\overline{b} \in \langle l, C_{\mathbb{K}} \rangle$. In particular, there is a k-definable function f such that

$$\mathbb{K} \models \exists \overline{c} \ (\bigwedge c_i' = 0 \wedge f(\overline{a}, \overline{c}) = \overline{b}).$$

By model-completeness,

$$L \models \exists \overline{c} \ (\bigwedge c_i' = 0 \wedge f(\overline{a}, \overline{c}) = \overline{b}).$$

Thus $\overline{b} \in \langle l, C_L \rangle = l \subseteq F$, a contradiction.

If $l = k\langle \overline{a} \rangle / k$ is strongly normal, then, because $\overline{a}$ is in the differential closure of k, $\mathrm{tp}(\overline{a}/k)$ is isolated. Let $\psi(\overline{v})$ be an $\mathcal{L}_K$-formula isolating $\mathrm{tp}(\overline{a}/k)$.

Lemma 7.4.25 $\psi(\overline{v})$ *isolates* $\mathrm{tp}(\overline{a}/\langle k, C_{\mathbb{K}} \rangle)$.

Proof Suppose, for contradiction, that $\overline{b} \in k, \overline{c} \in C_{\mathbb{K}}$, and $\phi(\overline{v}, \overline{b}, \overline{c})$ and $\neg\phi(\overline{v}, \overline{b}, \overline{c})$ split $\psi(\overline{v})$. Then

$$\mathbb{K} \models \exists \overline{c} \ \left(\bigwedge c_i' = 0 \wedge \exists \overline{v} \exists \overline{w} \ (\psi(\overline{v}) \wedge \psi(\overline{w}) \wedge \phi(\overline{v}, \overline{b}, \overline{c}) \wedge \neg\phi(\overline{w}, \overline{b}, \overline{c})) \right).$$

By model-completeness, this is also true in the differential closure of k. But the differential closure of k has the same constants as k. Thus, ψ does not isolate $\mathrm{tp}(\overline{a}/k)$, a contradiction.

Let k be algebraically closed, and suppose that G is an algebraic group defined over k. We can view $G \subseteq \mathbb{K}^m$ as a constructible group defined over k. The k-*rational points of* G are the points in $G \cap k^m$.[2]

Theorem 7.4.26 *Suppose that l/k is strongly normal and C is the constant field of k. The differential Galois group $G(l/k)$ is isomorphic to the C-rational points of an algebraic group defined over C.*

Proof Suppose $l = k\langle \overline{a} \rangle$. Let $\psi(\overline{v})$ isolate $\mathrm{tp}(\overline{a}/k)$.

If $\psi(\overline{b})$, then $\mathrm{tp}(\overline{a}/k) = \mathrm{tp}(\overline{b}/k)$ and there is $\sigma \in G(\mathbb{K}/k)$ such that $\sigma(\overline{a}) = \overline{b}$. Because l/k is strongly normal, $\overline{b} \in \langle l, C_{\mathbb{K}} \rangle$. In particular, there is

---

[2]This is rather awkward. A more general view is that the k-rational points are the points of the algebraic group G fixed by all automorphisms of $\mathbb{K}$ fixing k pointwise.

a k-definable function $g_{\bar{b}}$ and $\bar{c} \in C_{\mathbb{K}}$ such that $g_{\bar{b}}(\bar{a}, \bar{c}) = \bar{b}$. By compactness and coding tricks, we can find a single k-definable function g such that for all $\bar{b} \in \psi(\mathbb{K})$ there is $\bar{c} \in C_{\mathbb{K}}$ such that $\bar{b} = g(\bar{a}, \bar{c})$.

Let F be the differential closure of k. Then, $l \subset F$ and $C_F = C$. If $\bar{b} \in F$, then any automorphism of l sending $\bar{a}$ to $\bar{b}$ lifts to an automorphism of $\mathbb{K}$. By model-completeness, there is $\bar{c} \in C$ such that $\bar{b} = g(\bar{a}, \bar{c})$.

It is easy to see that $\sigma \in G(l/k)$ is determined by its action on $\bar{a}$. Clearly, $\psi(\sigma(\bar{a}))$ and, if $\psi(\bar{b})$, then there is $\sigma \in G(l/k)$ with $\sigma(\bar{a}) = \bar{b}$.

Consider the relation $R(\bar{b}, \bar{d}, \bar{e})$, which asserts that if $\sigma(\bar{a}) = \bar{b}$ and $\tau(\bar{a}) = \bar{d}$, then $\sigma \circ \tau(\bar{a}) = \bar{e}$. Then $R(\bar{b}, \bar{d}, \bar{e})$ holds if and only if $\sigma(\bar{d}) = \bar{e}$. But there are constants $\bar{c} \in C$ such that $\bar{d} = g(\bar{a}, \bar{c})$. But then $\sigma(\bar{d}) = g(\bar{b}, \bar{c})$, so

$$R(\bar{b}, \bar{d}, \bar{e}) \Leftrightarrow \psi(b) \wedge \psi(d) \wedge \psi(e) \wedge \exists \bar{c} \bigwedge c_i' = 0 \wedge \bar{d} = g(\bar{a}, \bar{c}) \wedge \bar{e} = g(\bar{b}, \bar{c}).$$

Let X be the set ψ^l and define $\cdot$ on X by $\bar{b} \cdot \bar{d} = \bar{e}$ if and only if $R(\bar{b}, \bar{d}, \bar{e})$. We have shown that $(X, \cdot)$ is isomorphic to $G(l/k)$.

We can do even better. Let $Y = \{\bar{c} \in C : \psi(g(\bar{a}, \bar{c}))\}$. We define an equivalence relation E on Y by $\bar{c}_0 E \bar{c}_1$ if and only if $g(\bar{a}, \bar{c}_0) = g(\bar{a}, \bar{c}_1)$. We also define a ternary relation R^* on Y by $R^*(\bar{c}_0, \bar{c}_1, \bar{c}_2)$ if and only if $R(g(\bar{a}, \bar{c}_0), g(\bar{a}, \bar{c}_1), g(\bar{a}, \bar{c}_2))$. Clearly, R^* is E-invariant.

Because C is a pure algebraically closed field, Y, E, and R^* are definable in the language of fields. By elimination of imaginaries in algebraically closed fields, we can find a field-definable function $f : Y \to C^n$ such that $\bar{c} E \bar{c}_0$ if and only if $f(\bar{c}) = f(\bar{c}_0)$. Let G be the image of Y under f. Define on G by $x_0 \cdot x_1 = x_2$ if and only if there are $\bar{c}_0$, $\bar{c}_1$ and $\bar{c}_2 \in Y$ such that $f(\bar{c}_i) = x_i$ and $R^*(\bar{c}_0, \bar{c}_1, \bar{c}_2)$. Then, $(G, \cdot)$ is isomorphic to $G(l/k)$ and $(G, \cdot)$ is definable in the pure field structure of C. By Theorem 7.4.14, $G(l/k)$ is isomorphic to the C-rational points of an algebraic group defined over C.

7.5 Finding a Group

One of the most powerful ideas in modern model theory is that when one finds an interesting pattern of dependence it is often caused by a definable group. In this section we will prove some of the most basic results of this kind. In particular, we will prove Hrushovski's result that a "generically presented" group is a group. We begin by investigating groups and semigroups that are nearly definable.

$\bigwedge$-Definable Groups and Semigroups

Definition 7.5.1 We say that $X \subseteq \mathbb{M}$ is $\bigwedge$-*definable* if there is I with $|I| < |\mathbb{M}|$ and $\mathcal{L}_\mathbb{M}$-formulas, $\phi_i(v)$ for $i \in I$ such that

$$X = \left\{ x \in \mathbb{M} : \bigwedge_{i \in I} \phi_i(x) \right\}.$$

We will consider $G \subseteq \mathbb{M}$ that is $\bigwedge$-definable and $*$ a function definable on $D \times D$, where D is a definable set containing G. We say that $(G, *)$ is a *right-cancellation semigroup* if $*$ is associative, there is an identity element 1, and if $f * g = h * g$, then $f = h$.

Lemma 7.5.2 *If* $\mathbb{M}$ *is stable, then any* $\bigwedge$-*definable right-cancellation semigroup is a group.*

Proof Let $\bigwedge_{i \in I} \phi_i(v)$ define G.

Claim 1 We can find a definable $D \supseteq G$ such that for $x, y, z \in D$, $x * (y * z) = (x * y) * z$, $x * 1 = 1 * x = x$, and if $x * y = z * y$, then $x = z$.

Because G is a right-cancellation semigroup,

$$\mathbb{M} \models \left(\bigwedge_{i \in I} \phi_i(x) \wedge \bigwedge_{i \in I} \phi_i(y) \wedge \bigwedge_{i \in I} \phi_i(z) \right) \rightarrow$$
$$(x * (y * z) = (x * y) * z \wedge x * 1 = 1 * x = x)$$

and

$$\mathbb{M} \models \left(\bigwedge_{i \in I} \phi_i(x) \wedge \bigwedge_{i \in I} \phi_i(y) \wedge \bigwedge_{i \in I} \phi_i(z) \wedge x * y = z * y \right) \rightarrow x = z.$$

By compactness and saturation, there is a finite $I_0 \subseteq I$ such that

$$\mathbb{M} \models \left(\bigwedge_{i \in I_0} \phi_i(x) \wedge \bigwedge_{i \in I_0} \phi_i(y) \wedge \bigwedge_{i \in I_0} \phi_i(z) \right) \rightarrow$$
$$(x * (y * z) = (x * y) * z \wedge x * 1 = 1 * x = x)$$

and

$$\mathbb{M} \models \left(\bigwedge_{i \in I_0} \phi_i(x) \wedge \bigwedge_{i \in I_0} \phi_i(y) \wedge \bigwedge_{i \in I_0} \phi_i(z) \wedge x * y = z * y \right) \rightarrow x = z.$$

We define D by $\bigwedge_{i \in I_0} \phi_i(v)$.

Let $a \in G$. We must find an inverse to a in G.

Claim 2 For all definable $D \supseteq G$, there is $b \in D$ such that $a * b = b * a = 1$.

Because

$$\mathbb{M} \models \left(\bigwedge_{i \in I} \phi_i(x) \wedge \bigwedge_{i \in I} \phi_i(y) \right) \rightarrow \bigwedge_{i \in I} \phi_i(x * y)$$

there is a definable $D_1 \subseteq D$ such that $G \subseteq D_1$ and $x * y \in D$ for $x, y \in D_1$. By the claim 1, we may assume that $*$ is associative and satisfies right-cancellation on D_1.

Let $\theta(x, y)$ be $\exists u \in D_1 u * x = y$. Then, $\mathbb{M} \models \theta(a^m, a^n)$ for $m \le n$. Because $\mathbb{M}$ is stable, θ does not have the order property (see Exercise 5.5.6). Thus there is $m < n$ such that $\mathbb{M} \models \theta(a^n, a^m)$. Suppose that $c \in D_1$ and $c * a^n = a^m$. Because we have associativity and right-cancellation on D_1, $((c * a^{n-m-1}) * a) * a^m = 1 * a^m$ and $(c * a^{n-m-1}) * a = 1$. On the other hand, $a * (c * a^{n-m-1}) * a = a * 1 = 1 * a$. Thus $a * (c * a^{n-m-1}) = 1$. Thus, $c * a^{n-m-1}$ is a left and right inverse to a.

We claim that a has an inverse in G. Suppose not. Then, by a final compactness argument, we can find a definable $D \supseteq G$ such that a has no inverse in D, contradicting our second claim.

Theorem 7.5.3 *Suppose that* $\mathbb{M}$ *is* ω-*stable and* $G \subseteq \mathbb{M}$ *is an* $\bigwedge$-*definable group. Then,* G *is definable.*

Proof Let $\bigwedge_{i \in I} \phi_i(v)$ define G. Without loss of generality, we may assume that if $I_0 \subseteq I$ is finite, there is $j \in I$ such that

$$\phi_j(v) \rightarrow \bigwedge_{i \in I_0} \phi_i(v).$$

As above, by compactness we can find $l \in I$ such that if $\phi_l(x), \phi_l(y),$ and $\phi_l(z)$, then $x * (y * z) = (x * y) * z$, $x * 1 = 1 * x = x$, and if $x * y = x * z$, then $y = z$. Further, we can choose $k \in I$ such that $\phi_k(v) \rightarrow \phi_l(v)$ and $\phi_k(v)$ is of minimal Morley rank and degree. Let $\mathcal{M} \prec \mathbb{M}$ contain all parameters occurring in any ϕ_j, $j \in I$. Let $p_1, \ldots, p_m \in S_1(M)$ list the finitely many 1-types over M such that $\phi_k \in p_j$ and $\mathrm{RM}(p_j) = \mathrm{RM}(\phi_k)$ for $j = 1, \ldots, m$. Note that if y is a realization of any p_j, then $y \in G$.

Let $H = \{x \in \mathbb{M} : \phi_k(x)$ and $\phi_k(x * y)$ for all $y \in G$ such that $\mathrm{tp}(y/M) = p_j$ for some $j = 1, \ldots, m$ and $y \mathop{\smile}\limits_{M} x\}$. By definability of types (Theorem 6.3.5), H is definable. We need only show that $H = G$.

Suppose that $x \in H$ and y is a realization of p_j with $y \mathop{\smile}\limits_{M} x$. Thus $\phi_k(x * y)$. Because we have left-cancellation on $\phi_k(\mathbb{M})$, the map $g \mapsto x * g$ is one-to-one on $\phi_k(\mathbb{M})$ and $\mathrm{RM}(x * y/M \cup \{x\}) = \mathrm{RM}(y/M \cup \{x\})$. Thus

$$\mathrm{RM}(x * y/M) \ge \mathrm{RM}(x * y/M \cup \{x\}) = \mathrm{RM}(y/M \cup \{x\}) = \mathrm{RM}(y/M).$$

Because $\phi_k(x * y)$ and $\mathrm{RM}(x * y/M)$ is maximal, $x * y$ realizes some p_j. Hence $x * y \in G$. Because $\phi_k(x)$ and $y \in G$,

$$x = x * (y * y^{-1}) = (x * y) * y^{-1}.$$

But $x * y, y^{-1} \in G$, and hence $x \in G$. Thus $H \subseteq G$.

On the other hand, suppose that $x \in G$, $y \in G$ realizes some p_j and $y \underset{M}{\downarrow} x$. Because $x * y \in G$, $\phi_k(x * y)$ and $x \in H$. Thus $G \subseteq H$.

In Exercise 7.6.20, we give an example of an unstable structure where there is a $\bigwedge$-definable group that is not definable.

Generically Presented Groups

We now state and prove Hrushovski's Theorem.

Theorem 7.5.4 *Suppose that T is ω-stable and $\mathbb{M}$ is a monster model of T. Let $A \subset \mathbb{M}$, and let $p \in S_1(A)$ be a stationary type (for notational simplicity, we will assume that $A = \emptyset$, but this is no loss of generality). For $B \subset \mathbb{M}$, we let p_B denote the unique nonforking extension of p to B. Suppose that $*$ is a definable partial function such that:*

  i) for all small B, if a and b realize p_B and $a \underset{B}{\downarrow} b$, then $a * b$ is defined and $a * b$ realizes p_B and a, b and $a * b$ are pairwise-independent over B.

  ii) if a, b, c are independent realizations of p, then $a * (b * c) = (a * b) * c$.

  Then there is a group G definable in $\mathbb{M}^{eq}$ and an injective definable function σ mapping realizations of p to realizations of the generic of G such that $\sigma(a) * \sigma(b) = \sigma(a * b)$ for independent realizations of p.

Proof Suppose that f and g are definable functions. We say that f and g have the same *germ* at p if and only if whenever A is large enough so that f and g are both defined over A, and a realizes p_A, then $f(a) = g(a)$. We write $f \sim g$.

Suppose that $F(v, w)$ is a definable function. For c in some definable set A, let f_c be the function $x \mapsto F(x, a)$ and let $\mathcal{F}$ be the family of germs of functions f_a for $a \in A$. Let $\phi(v, w, u)$ be the formula "$F(v, w) = F(v, u)$". Then $f_b \sim f_c$ if and only if $\phi(v, b, c) \in p_A$ if and only if $\mathbb{M} \models d_p \phi(b, c)$. Thus the equivalence relation $\sim$ is definable on the family $\mathcal{F}$. We can thus think of the germs of functions in $\mathcal{F}$ as elements of $\mathbb{M}^{eq}$.

Suppose that g is a germ of a function in $\mathcal{F}$ and a realizes p_g. Suppose that $b_1, b_2 \in A$ such that $b_i \underset{g}{\downarrow} a$ and if $f_i(x) = F(x, b_i)$, then $f_1, f_2 \in g$. We claim that $f_1(a) = f_2(a)$. This gives a well-defined value to the germ g at a.

Choose $b_3 \in A$ such that $b_3 \underset{g}{\downarrow} a, b_1, b_2$ and if $f_3(x) = F(x, b_3)$, then $f_3 \in g$. Because $b_3 \underset{g}{\downarrow} a, b_1$ and $b_1 \underset{g}{\downarrow} a$, we have $b_1, b_3 \underset{g}{\downarrow} a$. By symmetry, $a \underset{g}{\downarrow} b_1, b_3$. Because $a \underset{\emptyset}{\downarrow} g$, we have $a \underset{\emptyset}{\downarrow} b_1, b_3$. Because f_1 and f_3 are in the same germ, $f_1(a) = f_3(a)$. Similarly $f_2(a) = f_3(a)$. Thus, $f_1(a) = f_2(a)$, as desired.

Suppose that f and g are germs at p such that if a realizes $p_{f,g}$, $f(a)$ and $g(a)$ realize p as well. One can consider the composition $f \circ g$, which is also a germ at p. Let $S(p)$ be the semigroup of all germs at p mapping realizations of p to realizations of p. In general, $S(p)$ is not definable.

We argue that $S(p)$ has right-cancellation. Suppose $g \circ h \sim f \circ h$. Let a be independent of g, h, f. Then, $h(a)$ is well-defined and $h(a) \downarrow f, g$. Because $f(h(a)) = g(h(a))$, $f \sim g$, as desired.

Consider the map $p \mapsto S(p)$, which sends a to f_a, the germ of the function $x \mapsto a * x$.

We first show that this is one-to-one. This uses the following key fact. Suppose that a, b realize p_B and $b \downarrow_B a$. Then, there is b_1 realizing p such that $b_1 \downarrow_B a$ and $b * b_1 = a$. To prove this, begin by taking c, c_1 realizations of p independent over B. Let $d = c * c_1$. Then, d, c, c_1 are pairwise-independent over B and a, b realize the same type as d, c over B.

Suppose that a_1, a_2 realize p and $f_{a_1} = f_{a_2}$. Let b realize p_{a_1, a_2}. There are c_1, c_2 such that $c_i * b = a_i$. Let d realize $p_{a_1, a_2, b, c_1, c_2}$. Then, $a_i * d = (c_i * b) * d = c_i * (b * d)$ because b, c_i, d are independent. But $f_{a_1} = f_{a_2}$ so $a_1 * d = a_2 * d$. Because $b * d$ is a realization of p independent from c_1 and c_2, f_{c_1} and f_{c_2} must be the same germ. But then $c_1 * b = c_2 * b$ and $a_1 = a_2$, as desired. Thus $a \mapsto f_a$ is one-to-one.

Next, we show that the semigroup of germs generated by the f_a is in fact $\bigwedge$-definable. Let G be the collection of germs $\{f_a \circ f_b : a, b$ realizations of $p\}$. G is $\bigwedge$-definable and is contained in the semigroup of germs generated by the f_a. If $a \in p$, then, as above, we can find b and c independent realizations of p such that $b * c = a$. If d a realization of $p_{a, b, c}$, then $a * d = (b * c) * d = b * (c * d)$ and $f_a = f_b \circ f_c$. Thus, G contains the germs f_a for a realizing p. We claim that G is closed under composition. Suppose that a, b, c realize p. It suffices to show that there are d and e realizing p such that $f_a \circ f_b \circ f_c = f_d \circ f_e$. We can, as above, find b_1 and b_2 such that $b_1 \downarrow_{a,c} b_2$, $b_i \downarrow a, b, c$, and $b_1 * b_2 = b$. Let x realize p_{a, b, c, b_1, b_2}. Using the generic associativity of $*$, we see

$$
\begin{aligned}
f_a \circ f_b \circ f_c(x) &= a * ((b_1 * b_2) * (c * x)) \\
&= a * (b_1 * (b_2 * (c * x))) \\
&= (a * b_1) * (b_2 * (c * x)) \\
&= (a * b_1) * ((b_2 * c) * x).
\end{aligned}
$$

Thus, if we let $d = a * b_1$ and $e = b_2 * c$, we see that $a * (b * (c * x)) = d * (e * x)$ so the germs $f_a \circ f_b \circ f_c = f_d \circ f_e$.

Thus, G is a $\bigwedge$-definable right-cancellation semigroup. By Lemma 7.5.2 and Theorem 7.5.3, G is a definable group.

Clearly, a generic of G will arise from $f_a \circ f_b$, where a and b are independent realizations of p. But in this case, $f_a \circ f_b = f_{a*b}$. Thus, for a realizing p, f_a is a generic of G.

We mention one corollary to Theorems 7.5.4 and 7.4.14. This result of Weil is used in the construction of Jacobians (see, for example [59] II §3).

Corollary 7.5.5 *Let K be an algebraically closed field, and let $V \subseteq K^n$ be an irreducible variety. Suppose that f is a rational function defined on* a Zariski open $U \subseteq V \times V$ such that if a, b, c are independent generic points of V, then $f(a, b)$ is a generic point of V, $a, b, f(a, b)$ are pairwise-independent, and $f(a, f(b, c)) = f(f(a, b), c)$. Then, there is an algebraic group G and a birational map $\sigma : V \to G$ such that $\sigma(a) \cdot \sigma(b) = \sigma(f(a, b))$ for $a, b \in V$ independent generic points.

The Group Configuration

Theorem 7.5.4 is the first example of a theorem where we produce a definable group once we have spotted some trace evidence of its existence. We conclude this section by stating a version of the most powerful result of this type.

Definition 7.5.6 Suppose that $\mathbb{M}$ is ω-stable. We call $a, b, c, x, y, z \in \mathbb{M}^{\mathrm{eq}}$ a *group configuration* if:

  i) $\mathrm{RM}(a) = \mathrm{RM}(b) = \mathrm{RM}(c) = \mathrm{RM}(x) = \mathrm{RM}(y) = \mathrm{RM}(z) = 1$;

  ii) any pair of elements has rank 2;

  iii) $\mathrm{RM}(a, b, c) = \mathrm{RM}(c, x, y) = \mathrm{RM}(a, y, z) = \mathrm{RM}(b, x, z) = 2$;

  iv) any other triple has rank 3;

  v) $\mathrm{RM}(a, b, c, x, y, z)$ has rank 3.

We represent the group configuration by the following diagram.

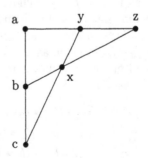

The points in the diagram have rank 1. Conditions iii) and iv) assert that each line has rank 2 while any three non-collinear points have rank 3.

There is one easy way that a group configuration arises. Suppose that G is a strongly minimal Abelian group. Let a, b, x be independent generic elements of G. Let $c = ba$, $y = cx$, and $z = bx$; then, $y = az$ and it is easy to check that conditions i)–v) hold. Remarkably, Hrushovski proved that whenever there is a group configuration there is also a definable group.

Theorem 7.5.7 *Suppose that there is a group configuration in $\mathbb{M}^{\mathrm{eq}}$. Then, there is a rank one group definable in $\mathbb{M}^{\mathrm{eq}}$.*

We give an application of the group configuration in Theorem 8.3.1. Proofs of Hrushovski's Theorem appear in [18] §4.5 and [76] §5.4.

7.6   Exercises and Remarks

Throughout the Exercises G is an ω-stable group.

Exercise 7.6.1 a) Show that the Descending Chain Condition fails for the stable group $(\mathbb{Z}, +, 0)$.

b) Suppose that G is a stable group and $\phi(x, \overline{y})$ is a formula. Show that we cannot find $\overline{a}_1, \overline{a}_2, \ldots$ such that $G_i = \{x : G \models \phi(x, \overline{a}_i)\}$ is a subgroup of G and $G_1 \supset G_2 \supset G_3 \supset \ldots$. [Hint: Suppose not and find a violation of the order property.]

Exercise 7.6.2 Prove Lemma 7.1.8.

Exercise 7.6.3 Let $p \in S_1(G)$, let G_1 be an elementary extension of G, and let p_1 be the unique nonforking extension of p to G_1. Show that the formula that defines $\text{Stab}(p)$ in G defines $\text{Stab}(p_1)$ in G_1.

Exercise 7.6.4 Show that if G is connected and $H \trianglelefteq G$ is definable, then G/H is connected.

Exercise 7.6.5 Suppose that G is a connected ω-stable group and $\sigma : G \to G$ is a definable homomorphism with finite kernel. Show that σ is surjective.

Exercise 7.6.6 Use Lemma 7.5.2 to show that any infinite stable integral domain is a field.

Exercise 7.6.7 Suppose that $(K, +, \cdot, \ldots)$ is a field of finite Morley rank. Show that K has no infinite definable subrings. [Hint: By Exercise 7.6.6, any definable subring is a subfield. Show that if there is a definable subfield, then K must have infinite rank.]

Conclude that if $\sigma : K \to K$ is a nontrivial field automorphism and $(K, +, \cdot, \sigma, 0, 1)$ has finite Morley rank, then K has characteristic $p > 0$ and the fixed field of σ is finite.

Exercise 7.6.8 Let K be a finite Morley rank field of characteristic 0.

a) Show that there are no nontrivial definable additive subgroups of K. [Hint: Let G be a definable subgroup and consider $R = \{a : aG = G\}$.] This is still an open question in characteristic p.

b) If $\sigma : K^n \to K^m$ is a definable additive homomorphism, then σ is K-linear. [Hint: Consider $\{a : \forall x\ \sigma(ax) = a\sigma(x)\}$.] In particular, the only definable homomorphisms of K^+ are $x \mapsto ax$, $a \in K$.

Exercise 7.6.9 Show that G acts transitively on the generic types of G.

Exercise 7.6.10 Show that if $\text{RM}(p) = \text{RM}(\text{Stab}(p))$, then $\text{Stab}(p)$ is connected and p is a translate of the generic of $\text{Stab}(p)$.

Exercise 7.6.11 Suppose that Γ is a connected group and there is a definable transitive action of Γ on a finite set S. Then $|S| = 1$.

Exercise 7.6.12 Show that if $X \subseteq G$ is definable and indecomposable and $g \in G$, then gX is indecomposable and gXg^{-1} is indecomposable.

Exercise 7.6.13 Suppose that G is an algebraic group and $X \subseteq G$ is an irreducible subvariety. Prove that X is indecomposable.

Exercise 7.6.14 Show that if G is an infinite group of finite Morley rank with no definable infinite proper subgroups, and $X \subseteq G$ is infinite and definable, then X generates G.

Exercise 7.6.15 Show that if G is an $\aleph_0$-saturated ω-stable group and $X \subseteq G$ is an infinite definable set, then there are $Y_1, \ldots, Y_n \subseteq X$ such that $X = Y_1 \cup \ldots \cup Y_n$ and $Y_1, \ldots, Y_n$ are indecomposable.

Exercise 7.6.16 Suppose that $(K, +, \cdot, \ldots)$ is a field of finite Morley rank and $X \subseteq K$ is an infinite definable set.

a) Show that there are $a_1, \ldots, a_n \in K$ such that $K = a_1 X + a_2 X + \ldots + a_n X$. [Hint: Without loss of generality, assume that K is $\aleph_0$-saturated and, by Exercise 7.6.15, that X is indecomposable. Let $x \in X$ and $Y = X - x$. Show that the additive subgroup A generated by $\{aY : a \in K\}$ is definable. Argue that $A = K$.]

b) Show that if the language $\mathcal{L}$ is countable, then the theory of K is categorical in all uncountable powers.

Exercise 7.6.17 Let F be an infinite field, and let G be a group of automorphisms of F such that the action of G on F has finite Morley rank. Show that $G = \{1\}$. [Hint: Without loss of generality, G is Abelian. Using Exercise 7.6.7, F has characteristic $p > 0$ and for all $\sigma \in G - \{1\}$, $Fix(\sigma)$, the fixed field of σ, is finite. Show that if $\sigma \in G - \{1\}$ then for all $n > 1$, $\sigma^n \neq 1$ and $|Fix(\sigma^n)| > |Fix(\sigma)|$. Thus, if $G \neq \{1\}$, then G is infinite. On the other hand, if G is infinite and σ is generic, so is σ^n. Derive a contradiction.]

Exercise 7.6.18 Show that if G is a simple group of finite Morley rank, and $H \equiv G$, then H is simple. [Hint: You first must show that an infinite Abelian simple group is not ω-stable.]

Exercise 7.6.19 Let K be an algebraically closed field, and let G be the affine group of matrices

$$G = \left\{ \begin{pmatrix} a & b \\ 0 & 1 \end{pmatrix} : a, b \in K, a \neq 0 \right\}.$$

a) Show that G is connected.

b) Show that

$$G' = \left\{ \begin{pmatrix} 1 & b \\ 0 & 1 \end{pmatrix} : b \in K \right\}$$

and $G'' = \{1\}$. Thus, G is solvable.

c) Show that G is centerless.

Exercise 7.6.20 Let $\mathbb{M}$ be a monster model of the theory of real closed fields. Let $I = \{x \in \mathbb{M} : |x| < \frac{1}{n} \text{ for } n = 1, 2, \ldots\}$. Show that $(I, +)$ is an $\bigwedge$-definable group that is not definable.

Exercise 7.6.21 Fill in the details in the proof of Lemma 7.4.2 iv).

Exercise 7.6.22 Let K be an algebraically closed field.
a) Show that there is a variety V with open cover $V_0 \cup V_1$ and $f_i : V_i \to K$ a homeomorphism such that $f_i(V_i \cap V_{1-i}) = K \setminus \{0\}$ and $f_i \circ f_{1-i}^{-1}$ is the identity on $K \setminus \{0\}$ for $i = 0, 1$. (The variety V looks like the line K with 0 "doubled".)
b) Show that $\Delta = \{(x, y) \in V \times V : x = y\}$ is not closed in $V \times V$.
c) Show that if G is an algebraic group, then $\Delta = \{(x, y) \in G \times G : x = y\}$ is closed in G.

Exercise 7.6.23 Prove Lemma 7.4.6.

Remarks

We know some things about the Cherlin–Zil'ber Conjecture for groups of very small rank. Of course, a connected group of rank 1 is Abelian. Cherlin proved that there are no non-Abelian simple groups of rank 2.

Theorem 7.6.24 *If G is a connected rank 2 group, then G is solvable.*

Problems arise in the analysis starting at rank 3.

Definition 7.6.25 We say that G is a *bad group* if G is connected, non-solvable, and all proper connected definable subgroups of G are nilpotent.

An algebraic group over an algebraically closed field is not a bad group. The real algebraic group $SO_3(\mathbb{R})$ is connected, nonsolvable, and all real algebraic subgroups are one-dimensional, but $SO_3(\mathbb{R})$ is unstable. It is not known whether there are any bad groups of finite Morley rank. The next result of Cherlin shows how bad groups could be obstacles to proving the Cherlin–Zil'ber Conjecture.

Theorem 7.6.26 *If G is a simple group of Morley rank 3, then*
i) G is a bad group, or
ii) G interprets an algebraically closed field K and G is definably isomorphic to $PSL_2(K)$.

For proofs of these results and more on groups of finite Morley rank, see [86] or [15].

Macintyre first noted the Descending Chain Condition for ω-stable groups. Although the Descending Chain Condition does not hold for stable groups, Baldwin and Saxl showed that there are no uniformly defined descending chains (see Exercise 7.6.1).

The notion of generic types is due to Poizat. There are important generalizations in stable and superstable groups (see [86]).

The proof we gave that an ω-stable field is algebraically closed follows the proof given by Cherlin and Shelah, who, using U-rank, generalized the result to superstable fields.

The Indecomposability Theorem and its consequences for groups of finite Morley rank are due to Zil'ber. The exercises on finite Morley rank fields are due to Poizat and Wagner [101].

Theorem 7.4.14 was first stated by van den Dries, who noticed that it follows from Weil's group chunk theorem. The proof above was given by Hrushovski. In fact, one can prove a stronger version of the theorem. Suppose that the constructible group G is defined over a subfield k; then G is definably isomorphic to an algebraic group defined over k. This requires a descent argument of Weil's. A sketch of the proof is given in [77].

The model theoretic treatment of Kolchin's differential Galois theory is due to Poizat. Surveys of differential Galois theory for linear differential equations can be found in [50], [63], and [96]. Pillay [78] developed a generalized differential Galois theory where arbitrary finite Morley rank *differential* algebraic groups arise as Galois groups.

8
Geometry of Strongly Minimal Sets

8.1 Pregeometries

In our proof of Morley's Categoricity Theorem in Chapter 6, we examined the algebraic closure relation on strongly minimal sets. In this chapter, we will return to strongly minimal sets and study more carefully the combinatorial geometry of algebraic closure. One of the great insights of modern model theory is that the local properties of the geometry of strongly minimal sets have a great influence on global properties of structures. These ideas play an important role in Hrushovski's proof of the Mordell–Lang Conjecture for function fields. In Theorem 8.3.20, we sketch how this works in one simple case.

We begin by reviewing some basic ideas from combinatorial geometry. The proofs are quite easy, and we leave them as exercises.

Definition 8.1.1 Let X be a set and let $\mathrm{cl} : \mathcal{P}(X) \to \mathcal{P}(X)$ be an operator on the power set of X. We say that (X, cl) is a *pregeometry* if the following conditions are satisfied.
  i) If $A \subseteq X$, then $A \subseteq \mathrm{cl}(A)$ and $\mathrm{cl}(\mathrm{cl}(A)) = \mathrm{cl}(A)$.
  ii) If $A \subseteq B \subseteq X$, then $\mathrm{cl}(A) \subseteq \mathrm{cl}(B)$.
  iii) (exchange) If $A \subseteq X$, $a, b \in X$, and $a \in \mathrm{cl}(A \cup \{b\})$, then $a \in \mathrm{cl}(A)$ or $b \in \mathrm{cl}(A \cup \{a\})$.
  iv) (finite character) If $A \subseteq X$ and $a \in \mathrm{cl}(A)$, then there is a finite $A_0 \subseteq A$ such that $a \in \mathrm{cl}(A_0)$.

We say that $A \subseteq X$ is *closed* if $\mathrm{cl}(A) = A$.

By Lemmas 6.1.3 and 6.1.4, if D is strongly minimal, we can associate a pregeometry by defining $\mathrm{cl}(A) = \mathrm{acl}(A) \cap D$ for $A \subseteq D$. We can generalize basic ideas about independence and dimension from strongly minimal sets to arbitrary pregeometries.

Definition 8.1.2 If (X, cl) is a pregeometry, we say that A is *independent* if $a \notin \mathrm{cl}(A \setminus \{a\})$ for all $a \in A$ and that B is a *basis* for Y if $B \subseteq Y$ is independent and $Y \subseteq \mathrm{acl}(B)$.

The natural generalization of Lemma 6.1.9 is true for all pregeometries.

Lemma 8.1.3 *If (X, cl) is a pregeometry, $Y \subseteq X$, $B_1, B_2 \subseteq Y$, and each B_i is a basis for Y, then $|B_1| = |B_2|$.*
We call $|B_i|$ the dimension of Y and write $\dim(Y) = |B_i|$.

If $A \subseteq X$, we also consider the *localization* $\mathrm{cl}_A(B) = \mathrm{cl}(A \cup B)$.

Lemma 8.1.4 *If (X, cl) is a pregeometry, then (X, cl_A) is a pregeometry.*

If (X, cl) is a pregeometry, we say that $Y \subseteq X$ is *independent over* A if Y is independent in (X, cl_A). We let $\dim(Y/A)$ be the dimension of Y in the localization (X, cl_A). We call $\dim(Y/A)$ the *dimension of Y over A*.

Definition 8.1.5 We say that a pregeometry (X, cl) is a *geometry* if $\mathrm{cl}(\emptyset) = \emptyset$ and $\mathrm{cl}(\{x\}) = \{x\}$ for any $x \in X$.

If (X, cl) is a pregeometry, then we can naturally associate a geometry. Let $X_0 = X \setminus \mathrm{cl}(\emptyset)$. Consider the relation $\sim$ on X_0 given by $a \sim b$ if and only if $\mathrm{cl}(\{a\}) = \mathrm{cl}(\{b\})$. By exchange, $\sim$ is an equivalence relation. Let $\widehat{X}$ be $X_0/\sim$. Define $\widehat{\mathrm{cl}}$ on $\widehat{X}$ by $\widehat{\mathrm{cl}}(A/\sim) = \{b/\sim : b \in \mathrm{cl}(A)\}$.

Lemma 8.1.6 *If (X, cl) is a pregeometry, then $(\widehat{X}, \widehat{\mathrm{cl}})$ is a geometry.*

We distinguish some properties of pregeometries that will play an important role.

Definition 8.1.7 Let (X, cl) be a pregeometry.
We say that (X, cl) is *trivial* if $\mathrm{cl}(A) = \bigcup_{a \in A} \mathrm{cl}(\{a\})$ for any $A \subseteq X$.
We say that (X, cl) is *modular* if for any finite-dimensional closed $A, B \subseteq X$

$$\dim(A \cup B) = \dim A + \dim B - \dim(A \cap B).$$

We say that (X, cl) is *locally modular* if (X, cl_a) is modular for some $a \in X$.

In Exercise 8.4.8, we show that a localization of a modular geometry is modular.
We give several illustrative examples.

Example 8.1.8 *Pure Sets*

Let D be a set with no structure. Then for all $a \in D$, $\operatorname{acl}(a) = \{a\}$ and $\operatorname{acl}(\emptyset) = \emptyset$. Thus, (D, acl) is a trivial geometry.

Example 8.1.9 *Successor*

Let $D \models \operatorname{Th}(\mathbb{Z}, s)$, where $s(x) = x + 1$. Then, $\operatorname{acl}(\emptyset) = \emptyset$ and $\operatorname{acl}(A) = \{s^n(a) : a \in A, n \in \mathbb{Z}\}$ for any $A \subseteq D$. Thus, (D, acl) is a trivial pregeometry that is not a geometry.

Example 8.1.10 *Projective Geometry*

Let F be a division ring and V be an infinite vector space over F. We view V as a structure in the language $\mathcal{L} = \{+, 0, \lambda_a : a \in F\}$ where $\lambda_a(x) = ax$. Then, V is a strongly minimal set, and for any set $A \subseteq V$ the algebraic closure of A is equal to the smallest F-subspace spanned by A. The usual dimension theorem for intersections of linear subspaces shows that this pregeometry is modular. This is not a geometry because $\operatorname{cl}(\emptyset) = \{0\}$ and for any $a \in V \setminus \{0\}$, $\operatorname{cl}(a)$ is the line through a and 0. To form the associated geometry, we take as points the lines through 0. The closure of a set of lines is the set of all lines in their linear span. Thus, the associated geometry is just the projective space associated to V. If $\dim V = n$, then the projective space has dimension $n - 1$.

Example 8.1.11 *Affine Geometry*

Let V and F be as above. We define a second geometry on V where the closure of a set A is the smallest affine space containing it and $\operatorname{cl}(\emptyset) = \emptyset$. (An *affine space* is any translate of a linear space). Here $\operatorname{cl}(\{a\}) = \{a\}$, so this is a geometry. Let $a, b, c \in V$ be noncollinear. Then, $\dim(a, b, c, c + b - a) = 3$, whereas $\dim(a, b) = \dim(c, c + b - a) = 2$ and $\operatorname{cl}(a, b) \cap \operatorname{cl}(c, c + b - a) = \emptyset$ because these are parallel lines. Thus, the geometry is not modular. If we localize at zero, then the pregeometry is exactly projective geometry, so this is locally modular.

For $F = \mathbb{Q}$, we can view this as the algebraic closure geometry of a strongly minimal set by viewing V as a structure in the language $\{\tau\}$ where $\tau(x, y, z) = x + y - z$. (For arbitrary F, add function symbols for $ax + y - az$ for each $a \in F$.)

Example 8.1.12 *Algebraically Closed Fields*

Let K be an algebraically closed field of infinite transcendence degree. We claim that (K, acl) is not locally modular. Let k be an algebraically closed subfield of finite transcendence degree. We will show that even localizing at k the pregeometry is not modular. Let a, b, x be algebraically independent over k. Let $y = ax + b$. Then, $\dim(k(x, y, a, b)/k) = 3$ and

$\dim(k(x,y)/k) = \dim(k(a,b)/k) = 2$. We contradict modularity by showing that $\mathrm{acl}(k(x,y)) \cap \mathrm{acl}(k(a,b)) = k$. To see this, suppose for purposes of contradiction that $d \in (\mathrm{acl}(k(a,b)) \cap \mathrm{acl}(k(x,y))) \setminus k$. Because $k(x,y)$ has transcendence degree 2 over k, we may, without loss of generality, assume that y is algebraic over $k(d,x)$. Let $k_1 = \mathrm{acl}(k(d))$. Then, there is $p(X,Y) \in k_1[X,Y]$ an irreducible polynomial such that $p(x,y) = 0$. By model-completeness, $p(X,Y)$ is still irreducible over $\mathrm{acl}(k(a,b))$. Thus, $p(X,Y)$ is $\alpha(Y - aX - b)$ for some $\alpha \in \mathrm{acl}(k(a,b))$ which is impossible as then $\alpha \in k_1$ and $a,b \in k_1$.

Algebraically closed fields are the only known naturally arising examples of nonlocally modular strongly minimal sets. Zil'ber conjectured that every non-locally modular strongly minimal set interprets an algebraically closed field. In [41], Hrushovski gave a general method of constructing non-locally modular strongly minimal sets and showed that many of these sets do not even interpret groups. Hrushovski's method has been generalized to construct many interesting pathological structures. For example, in [42] Hrushovski showed that there is a strongly minimal structure $(D, +, \cdot, \oplus, \otimes)$ where $(D, +, \cdot)$ is an algebraically closed field of characteristic $p \geq 0$ and $(D, \oplus, \otimes)$ is an algebraically closed field of characteristic $q \neq p$.

Next, we give a useful characterization of modularity.

Lemma 8.1.13 *Let* (X, cl) *be a pregeometry. The following are equivalent.*

i) (X, cl) *is modular.*

ii) If $A \subseteq X$ *is closed and nonempty,* $b \in X$, *and* $x \in \mathrm{cl}(A, b)$, *then there is* $a \in A$ *such that* $x \in \mathrm{cl}(a, b)$.

iii) If $A, B \subseteq X$ *are closed and nonempty, and* $x \in \mathrm{cl}(A, B)$, *then there are* $a \in A$ *and* $b \in B$ *such that* $x \in \mathrm{cl}(a, b)$.

Proof

i)$\Rightarrow$ ii) By the finite nature of closure, we may assume that $\dim A$ is finite. If $x \in \mathrm{cl}(b)$, we are done, so we may assume $x \notin \mathrm{cl}(b)$. By modularity,

$$\dim(A, b, x) = \dim A + \dim(b, x) - \dim(A \cap \mathrm{cl}(b, x))$$

and

$$\dim(A, b, x) = \dim(A, b) = \dim A + \dim b - \dim(A \cap \mathrm{cl}(b)).$$

Because $\dim(b, x) = \dim(b) + 1$, there is $a \in A$ such that $a \in \mathrm{cl}(b, x) \setminus \mathrm{cl}(b)$. By exchange, $x \in \mathrm{cl}(b, a)$.

ii)$\Rightarrow$ iii) We may suppose that A and B are finite-dimensional. We proceed by induction on $\dim A$. If $\dim A$ is zero then iii) holds. Suppose that $A = \mathrm{cl}(A_0, a)$, where $\dim A_0 = \dim A - 1$. Then $x \in \mathrm{cl}(A_0, B, a)$. By ii), there is $c \in \mathrm{cl}(A_0, B)$ such that $x \in \mathrm{cl}(c, a)$. By induction, there is $a_0 \in A_0$ and $b \in B$ such that $c \in \mathrm{cl}(a_0, b)$. Again by ii), there is $a^* \in \mathrm{cl}(a_0, a) \subseteq A$ such that $x \in \mathrm{cl}(a^*, b)$.

iii)$\Rightarrow$ i) Suppose that $A, B \subseteq X$ are finite-dimensional and closed. We prove i) by induction on $\dim A$. If $\dim A = 0$, then we are done. Suppose that $A = \mathrm{cl}(A_0, a)$, where $\dim A_0 = \dim A - 1$ and we assume, by induction, that

$$\dim(A_0, B) = \dim A_0 + \dim B - \dim(A_0 \cap B).$$

First, assume that $a \in \mathrm{cl}(A_0, B)$. Then $\dim(A_0, B) = \dim(A, B)$ and, because $a \notin A_0$, $\dim A = \dim A_0 + 1$. Because $a \in \mathrm{cl}(A_0, B)$, by iii) there is $a_0 \in A_0$ and $b \in B$ such that $a \in \mathrm{cl}(a_0, b)$. Because $a \notin \mathrm{cl}(a_0)$, by exchange, $b \in \mathrm{cl}(a, a_0)$. Thus $b \in A$. But $b \notin A_0$, because otherwise $a \in A_0$. Therefore, $\dim(A \cap B) = \dim(A_0 \cap B) + 1$, as desired.

Next, suppose that $a \notin \mathrm{cl}(A_0, B)$. In this case, we need to show that $A \cap B = A_0 \cap B$. Suppose that $b \in B$ and $b \in \mathrm{cl}(A_0, a) \setminus \mathrm{cl}(A_0)$. Then, by exchange, $a \in \mathrm{cl}(A_0, b)$, a contradiction.

If V is a vector space, $a_1, \ldots, a_n$ is a basis for $A \subseteq V$, and $b_1, \ldots, b_m$ is a basis for B, then any x in the span of $A \cup B$ is the sum of a linear combination of the a_i and a linear combination of the b_j. Thus, condition iii) above holds.

Lemma 8.1.13 can be used to give another proof that algebraically closed fields are not locally modular (see Exercise 8.4.9).

8.2 Canonical Bases and Families of Plane Curves

Let D be a strongly minimal set. In this section, we will consider families of strongly minimal subsets of D^2.

Definition 8.2.1 Suppose that $D \subseteq \mathbb{M}^n$ is strongly minimal and $A \subseteq \mathbb{M}^{\mathrm{eq}}$ is definable. We say that a definable $C \subseteq D^2 \times A$ is a *family of plane curves* if for all $a \in A$ the set $C_a = \{(x, y) \in D : (x, y, a) \in C\}$ is a strongly minimal subset of D^2.

Consider the following two examples. Suppose that V is a $\mathbb{Q}$ vector space. Let $E = \{(x, y, z) \in V^3 : y = mx + z\}$ where $m \in \mathbb{Q}$. For $a \in V$, let $E_a = \{(x, y) : (x, y, a) \in E\}$. We think of E as describing the *family of plane curves* $\{E_a : a \in V\}$. We call V the *parameter space* for the family E. Note that in this case the parameter space is rank 1. Indeed, if E is a family of plane curves in V^2 and the parameter space has rank greater than 1, then there is a one-dimensional family C such that for any E_a there is C_b such that $E_a \triangle C_b$ is finite. This says that every family of plane curves is "essentially one-dimensional."

On the other hand, suppose that K is an algebraically closed field. Fix $n \in N$ and consider

$$E = \{(x, y, z_0, \ldots, z_{n-1}) : y = x^n + z_{n-1}x^{n-1} + \ldots + z_1 x + z_0\}.$$

If

$$E_a = \{(x,y) : y = x^n + \sum_{i=0}^{n-1} a_i x^i\}$$

for $a = (a_0, \ldots, a_{n-1}) \in K^n$, then $\left\{ E_a : a \in K^n \right\}$ is an n-dimensional family of strongly minimal sets.

In this section, we will examine families of plane curves and show that a strongly minimal set is locally modular if and only if any family of plane curves is "essentially one-dimensional." To make these notions precise, we must first digress and discuss canonical bases.

Canonical Bases

We work in a monster model $\mathbb{M}$. To find canonical bases, we will usually have to work in $\mathbb{M}^{\text{eq}}$. We recall from Lemma 1.3.10 that every automorphism of $\mathbb{M}$ lifts to an automorphism of $\mathbb{M}^{\text{eq}}$ and every automorphism of $\mathbb{M}^{\text{eq}}$ restricts to an automorphism of $\mathbb{M}$. Thus, we can naturally identify the automorphism groups $Aut(\mathbb{M})$ and $Aut(\mathbb{M}^{\text{eq}})$. Recall that $\text{acl}^{\text{eq}}(A)$ is the algebraic closure of A in $\mathbb{M}^{\text{eq}}$ and $\text{dcl}^{\text{eq}}(A)$ is the definable closure of A in $\mathbb{M}^{\text{eq}}$.

Definition 8.2.2 Let $X \subseteq \mathbb{M}^n$ be definable. We say that $A \subset \mathbb{M}^{\text{eq}}$ is a *canonical base* for X if σ fixes X setwise if and only if σ fixes A pointwise for all $\sigma \in Aut(\mathbb{M})$.

If $p \in S_n(\mathbb{M})$, then $A \subseteq \mathbb{M}^{\text{eq}}$ is a canonical base for p if $\sigma p = p$ if and only if σ fixes A pointwise for all $\sigma \in Aut(\mathbb{M})$.

For any theory, we can find canonical bases for definable sets in $\mathbb{M}^{\text{eq}}$.

Lemma 8.2.3 *Suppose that $X \subseteq \mathbb{M}$ is definable. There is $\alpha \in \mathbb{M}^{\text{eq}}$ such that α is a canonical base for X. Indeed, if X is A-definable, we can find a canonical base in $\text{dcl}^{\text{eq}}(A)$.*

Proof Suppose that X is defined by the formula $\phi(\overline{x}, \overline{a})$. Let E be the equivalence relation

$$\overline{a} \, E \, \overline{b} \Leftrightarrow (\phi(\overline{x}, \overline{a}) \leftrightarrow \phi(\overline{x}, \overline{b})).$$

Let $\alpha = \overline{a}/E \in \mathbb{M}^{\text{eq}}$. Then, α is a canonical base for X.

Next, we consider canonical bases for types. We first note that canonical bases are determined up to definable closure in $\mathbb{M}^{\text{eq}}$. This works equally well for definable sets instead of types.

Lemma 8.2.4 *If A is a canonical base for $p \in S_n(\mathbb{M})$, then B is a canonical base for p if and only if $\text{dcl}^{\text{eq}}(A) = \text{dcl}^{\text{eq}}(B)$.*

Proof Suppose that $C \subset M$ and $|C| < |M|$; let $Aut(M/C)$ denote the automorphisms of M fixing C pointwise. The proof of Proposition 4.3.25 generalized to M^{eq} shows that

$$\mathrm{dcl}^{eq}(C) = \{x \in M^{eq} : \sigma(x) = x \text{ for all } \sigma \in Aut(M/C)\}.$$

Suppose that B is a canonical base for p. If σ is an automorphism fixing B pointwise, then $\sigma p = p$ and σ fixes A pointwise. Thus $A \subseteq \mathrm{dcl}^{eq}(B)$. Similarly, if τ is an automorphism fixing A pointwise, $\tau p = p$ and τ fixes B pointwise. Thus $B \subseteq \mathrm{dcl}^{eq}(A)$. Hence $\mathrm{dcl}^{eq}(A) = \mathrm{dcl}^{eq}(B)$.

Conversely, suppose that $\mathrm{dcl}^{eq}(A) = \mathrm{dcl}^{eq}(B)$. If $\sigma \in Aut(M)$, then σ fixes A pointwise if and only if σ fixes B pointwise. Because A is a canonical base for p, so is B.

Definition 8.2.5 If A is any canonical base for p, let $\mathrm{cb}(p) = \mathrm{dcl}^{eq}(A)$.

By Lemma 8.2.4, this definition of $\mathrm{cb}(p)$ does not depend on the choice of A. It is easy to see that $\mathrm{cb}(p)$ is the largest possible choice of canonical base for p.

Using definability of types, it is easy to find canonical bases in ω-stable theories.

Lemma 8.2.6 *Suppose that M is ω-stable and $p \in S_n(M)$. Then, p has a canonical base in M^{eq}.*

Proof For each $\mathcal{L}$-formula $\phi(\overline{v}, \overline{w})$, let $X_\phi = \{\overline{a} \in M : \phi(\overline{v}, \overline{a}) \in p\}$. By definability of types, X_ϕ is definable. If σ is an automorphism of M, then $\sigma p = p$ if and only if σ fixes each X_ϕ setwise. Let $\alpha_\phi \in M^{eq}$ be a canonical base for X_ϕ, and let $A = \{\alpha_\phi : \phi$ an $\mathcal{L}$-formula$\}$. Then, $\sigma p = p$ if and only if σ fixes A pointwise. Thus, A is a canonical base for p.

A more careful analysis shows that we can always find a finite canonical base in $\mathrm{acl}^{eq}(A)$ for any A over which p does not fork.

Theorem 8.2.7 *Suppose that M is ω-stable and $p \in S_n(M)$ does not fork over $A \subseteq M$. There is $\alpha \in \mathrm{acl}^{eq}(A)$, a canonical base for p. If $p|A$ is stationary, then we can find a canonical base $\alpha \in \mathrm{dcl}^{eq}(A)$.*

Proof Suppose that $\phi(\overline{v}, \overline{w})$ is an $\mathcal{L}$-formula such that $\phi(\overline{v}, \overline{a}) \in p$ and $\mathrm{RM}(\phi(\overline{v}, \overline{a})) = \mathrm{RM}(p)$. Let $X = \{\overline{b} : \phi(\overline{v}, \overline{b}) \in p\}$. By definability of types, X is definable. Indeed, by Theorem 6.3.9, X is definable over $\mathrm{acl}^{eq}(A)$ and, if $p|A$ is stationary, X is definable over A.

Claim If σ is an automorphism of M, then $\sigma p = p$ if and only if $\sigma X = X$.
If $\sigma p = p$, then

$$\overline{c} \in X \Leftrightarrow \phi(\overline{v}, \overline{c}) \in p \Leftrightarrow \phi(\overline{v}, \overline{c}) \in \sigma p \Leftrightarrow \overline{c} \in \sigma X.$$

Thus $\sigma X = X$.

Conversely, suppose $\sigma X = X$. Then, $\bar{a} \in \sigma X$ and $\phi(\bar{v}, \bar{a}) \in \sigma p$. Because $\text{RM}(p) = \text{RM}(\sigma(p))$ and $\deg_{\text{M}}(\phi(\bar{v}, \bar{a})) = 1$, $\sigma p = p$.

Thus, $B \subset \text{M}^{\text{eq}}$ is a canonical base for p if and only if B is a canonical base for X. Because X is $\text{acl}^{\text{eq}}(A)$-definable, by Lemma 8.2.4, we can find a canonical base in $\text{acl}^{\text{eq}}(A)$. If $p|A$ is stationary, then X is A-definable and we can find a canonical base in $\text{dcl}^{\text{eq}}(A)$.

If $p \in S_n(\text{M})$, then p is definable over any canonical base. Thus, the converse to Theorem 8.2.7 holds as well.

Corollary 8.2.8 *Suppose* M *is* ω-*stable, then* $p \in S_n(\text{M})$ *does not fork over* A *if and only if* $\text{cb}(p) \subseteq \text{acl}^{\text{eq}}(A)$.

We say that the canonical base for a set X (or a type p) has rank α if α is least such that there is a canonical base $b \in \text{M}^{\text{eq}}$ with $\text{RM}(b) = \alpha$.

For strongly minimal theories, we can compute ranks in M^{eq} using the following elimination of imaginaries result of Lascar and Pillay. We have already proved a special case in Lemma 3.2.19, and the proof of the general case is a straightforward generalization, which we leave as an exercise.

Lemma 8.2.9 *Let* M *be a strongly minimal set and let* $X \subset \text{M}$ *be infinite. Suppose that* E *is an* $\emptyset$-*definable equivalence relation on* M^m. *Let* $\bar{a} \in \text{M}^m$ *and* $\alpha = \bar{a}/E$. *There is a finite* $C \subset \text{M}^k$ *(for some* k) *such that any automorphism of* M *fixing* X *fixes* α *if and only if it fixes* C *setwise.*

In particular, if M_X *is* M *viewed as an* $\mathcal{L}_X$-*structure, then for every* $\alpha \in \text{M}_X^{\text{eq}}$ *there is* $\bar{d} \in \text{M}$ *such that* $\text{acl}^{\text{eq}}(\alpha, X) = \text{acl}^{\text{eq}}(\bar{d}, X)$.

Suppose that M is strongly minimal and $\alpha \in \text{M}^{\text{eq}}$. Let $X \subset \text{M}$ be infinite such that $\alpha \underset{\smile}{\mid} X$. Because $\text{RM}(\alpha) = \text{RM}(\alpha/X)$, it suffices to calculate $\text{RM}(\alpha/X)$. By Lemma 8.2.9, there is $\bar{d} \in \text{M}$ such that the set α is interalgebraic with $\bar{d}$ over X. Then, $\text{RM}(\alpha)$ is equal to $\text{RM}(\bar{d}/X)$.

Families of Plane Curves

Using canonical bases, we can make precise the idea that in locally modular strongly minimal sets families of plane curves are "essentially one-dimensional."

Definition 8.2.10 Suppose that $D \subset \text{M}^n$ is strongly minimal and ϕ is the strongly minimal formula defining D. We say that D is *linear* if for all $p \in S_2(D)$, if $\phi(v_1) \wedge \phi(v_2) \in p$ and $\text{RM}(p) = 1$, then the canonical base for p has rank at most 1.

Suppose that $\phi(v_1, v_2, \bar{b})$ is strongly minimal. Let $\mathcal{F}$ be the family of sets $C_{\bar{a}} = \{(x, y) : \phi(x, y, \bar{a})\}$ where $\bar{a}$ and $\bar{b}$ realize the same type. There is a natural equivalence relation on $\mathcal{F}$, $C_{\bar{a}} \sim C_{\bar{c}}$ if and only if $C_{\bar{a}} \triangle C_{\bar{c}}$ is finite. In Exercise 8.4.11, we show that $\sim$ is definable. If p is the generic type of

ϕ, then the Morley rank of the canonical base of p intuitively corresponds to the dimension of $\mathcal{F}/\sim$. Thus, D is linear if and only if there is no family of plane curves of dimension greater than 1. Algebraically closed fields are nonlinear because we have the family of curves $C_{(a,b)} = \{(x,y) : y = ax+b\}$, whereas vector spaces are linear.

Next, we show that the linear strongly minimal sets are exactly the locally modular ones.

Theorem 8.2.11 *Let $D \subseteq \mathbb{M}^n$ be a strongly minimal set. The following are equivalent:*
  i) for some small $B \subset D$, the pregeometry D_B is modular;
  ii) D is linear;
  iii) for any $b \in D \setminus \mathrm{acl}(\emptyset)$, D_b is modular;
  iv) D is locally modular.

Proof Often when we want to prove things about arbitrary strongly minimal sets $D \subseteq \mathbb{M}^n$, we instead assume that $\mathbb{M}$ is strongly minimal. This is no great loss of generality. By extending the language, we may assume that D is $\emptyset$-definable. By Corollary 6.3.7, any subset of D^n that is definable is definable using parameters from D. Thus, to study definability in D, we can ignore all of $\mathbb{M}$ outside of D. For this reason, we may, without loss of generality, assume that D is the universe of our structure.

i)$\Rightarrow$ ii) We first claim that if D is nonlinear, then D_B is also nonlinear. Suppose that $p \in S_2(D)$, $\mathrm{RM}(p) = 1$, α is a canonical base for p, and $\mathrm{RM}(\alpha) \geq 2$. If α' realizes a nonforking extension of $\mathrm{tp}(\alpha)$ to B, then α' is a canonical base for a rank 1 type and D_B is nonlinear. Thus, adding the parameters B to the language, we assume that $B = \emptyset$ and D is modular.

Let $p \in S_2(D)$ with $\mathrm{RM}(p) = 1$. Let $\phi(v_1, v_2, \bar{a})$ be a strongly minimal formula in p. Let b_1, b_2 realize p. Let $X = \mathrm{acl}(\bar{a}) \cap \mathrm{acl}(b_1, b_2)$. By modularity,

$$\dim X = \dim(\bar{a}) + \dim(b_1, b_2) - \dim(\bar{a}, b_1, b_2).$$

Because $\dim(\bar{a}, b_1, b_2) = \dim(\bar{a}) + 1$ and $1 \leq \dim(b_1, b_2) \leq 2$, $\dim X \leq 1$. Thus,

$$\begin{aligned}
\dim(b_1, b_2/\bar{a}) &= \dim(b_1, b_2, \bar{a}) - \dim(\bar{a}) \\
&= \dim(b_1, b_2) - \dim X \\
&= \dim(b_1, b_2/X).
\end{aligned}$$

Thus $\dim(b_1, b_2/X) = 1$ and p does not fork over X. By Theorem 8.2.7, $\mathrm{cb}(p) \subseteq \mathrm{acl}^{\mathrm{eq}}(X)$, so $\mathrm{RM}(\alpha) \leq 1$.

ii)$\Rightarrow$ iii) Let $b \in D \setminus \mathrm{acl}(\emptyset)$. We will use the equivalence from Lemma 8.1.13. Suppose that B is a finite-dimensional closed set. Suppose that $a_1 \in \mathrm{acl}(a_2, B, b)$. We must find $d \in \mathrm{acl}(B, b)$ such that $a_1 \in \mathrm{acl}(a_2, d, b)$. Clearly, we may assume that $a_1 \notin \mathrm{acl}(B, b)$, $a_2 \notin \mathrm{acl}(B, b)$, and $a_1 \notin \mathrm{acl}(a_2, b)$ (otherwise we are done). Thus, $\dim(a_1, a_2/b) = 2$ and $\dim(a_1, a_2/Bb) = 1$.

Let $\alpha \in \mathrm{acl}^{\mathrm{eq}}(B, b)$ be a canonical base for the type of a_1, a_2 over $\mathrm{acl}^{\mathrm{eq}}(B, b)$.

Claim $\alpha \in \mathrm{acl}^{\mathrm{eq}}(a_1, a_2)$.

Because α is a canonical base, $\mathrm{RM}(a_1, a_2\alpha) = 1$. Because $\mathrm{RM}(a_1, a_2/\alpha) <$ $\mathrm{RM}(a_1, a_2)$, $\mathrm{RM}(\alpha/a_1, a_2) < \mathrm{RM}(\alpha)$. By ii), $\mathrm{RM}(\alpha) = 1$; thus $\alpha \in$ $\mathrm{acl}^{\mathrm{eq}}(a_1, a_2)$.

Because $\mathrm{RM}(a_1, a_2/b) = 2$ and $\mathrm{RM}(a_1, a_2/\alpha) = 1$, $\alpha \notin \mathrm{acl}(b)$. Thus, because $b \notin \mathrm{acl}(\emptyset)$, $b \notin \mathrm{acl}(\alpha)$. Thus, a_1 and b realize the same type over α and, by saturation, there is $d \in D$ such that $\mathrm{tp}(a_1, a_2/\alpha) = \mathrm{tp}(b, d/\alpha)$. Then

$$d \in \mathrm{acl}(b, \alpha) \subseteq \Big(\mathrm{acl}(a_1, a_2, b) \cap \mathrm{acl}(Bb)\Big)$$

and $d \notin \mathrm{acl}(b)$.

We claim that $d \notin \mathrm{acl}(a_2, b)$. If $d \in \mathrm{acl}(a_2, b)$, then, because $d \notin \mathrm{acl}(b)$, $a_2 \in \mathrm{acl}(d, b) \subseteq \mathrm{acl}(B, b)$, a contradiction.

Thus, because $d \in \mathrm{acl}(a_1, a_2, b) \setminus \mathrm{acl}(a_2, b)$, $a_1 \in \mathrm{acl}(a_2, b, d)$, as desired.

iii)$\Rightarrow$ iv) and iv)$\Rightarrow$ i) are clear.

One-Based Theories

We give one further characterization of locally modular strongly minimal sets. In Exercise 8.4.6, we show that a pregeometry is modular if and only if any two closed sets A and B are independent over $A \cap B$. It turns out to be interesting to look at theories where any two sets A and B that are algebraically closed in $\mathbb{M}^{\mathrm{eq}}$ are independent (in the sense of forking) over $A \cap B$.

Definition 8.2.12 Suppose T is an ω-stable theory with monster model $\mathbb{M}$. We say that T is *one-based* if whenever $A, B \subseteq \mathbb{M}^{\mathrm{eq}}$, $A = \mathrm{acl}^{\mathrm{eq}}(A)$, and $B = \mathrm{acl}^{\mathrm{eq}}(B)$, then $A \underset{A \cap B}{\downarrow} B$.

The next lemma explains why we call these theories one-based.[1]

Lemma 8.2.13 *Suppose that T is ω-stable. The following are equivalent.*

i) T is one-based.

ii) For all $\bar{a} \in \mathbb{M}^{\mathrm{eq}}$ and $B \subseteq \mathbb{M}^{\mathrm{eq}}$, if $\mathrm{tp}(\bar{a}/B)$ is stationary, then $\mathrm{cb}(\mathrm{tp}(\bar{a}/B)) \subseteq \mathrm{acl}^{\mathrm{eq}}(\bar{a})$.

Proof

i) $\Rightarrow$ ii) Let $A = \mathrm{acl}^{\mathrm{eq}}(\bar{a})$. Because $\mathrm{tp}(\bar{a}/\mathrm{acl}^{\mathrm{eq}}(B))$ does not fork over B, we may without loss of generality assume that $B = \mathrm{acl}^{\mathrm{eq}}(B)$. Because T is one-based, $\bar{a} \underset{A \cap B}{\downarrow} B$. Thus $\mathrm{cb}(\mathrm{tp}(\bar{a}/B)) \subseteq A \cap B \subseteq A$.

---

[1]Compare this to Exercise 8.4.12 for arbitrary ω-stable theories.

ii) $\Rightarrow$ i) Let $A, B \subseteq \mathbb{M}^{\mathrm{eq}}$ with $\mathrm{acl}^{\mathrm{eq}}(A) = A$, $\mathrm{acl}^{\mathrm{eq}}(B) = B$, and $\bar{a} \in A$. For any ω-stable theory, $\mathrm{cb}(\mathrm{tp}(\bar{a}/B))$ is contained in $\mathrm{acl}^{\mathrm{eq}}(B) = B$. By ii), $\mathrm{cb}(\mathrm{tp}(\bar{a}/B)) \subseteq \mathrm{acl}^{\mathrm{eq}}(\bar{a}) \subseteq A$. Thus, $\mathrm{cb}(\mathrm{tp}(\bar{a}/B)) \subseteq A \cap B$ and $\mathrm{tp}(\bar{a}/B)$ does not fork over $A \cap B$.

Theorem 8.2.14 *Suppose that* $\mathbb{M}$ *is strongly minimal. Then,* T *is one-based if and only if* D *is locally modular.*

Proof We first assume

(*) For every $\alpha \in \mathbb{M}^{\mathrm{eq}}$, there is $\bar{d} \in \mathbb{M}$ such that $\mathrm{acl}^{\mathrm{eq}}(\alpha) = \mathrm{acl}^{\mathrm{eq}}(\bar{d})$.

Under this assumption we claim that $\mathbb{M}$ is one-based if and only if $\mathbb{M}$ is modular. Suppose that $\mathbb{M}$ is one-based. If $A, B \subset \mathbb{M}$ are algebraically closed, then $A \underset{\mathrm{acl}^{\mathrm{eq}}(A) \cap \mathrm{acl}^{\mathrm{eq}}(B)}{\downarrow} B$. By (*), we can find $\bar{d} \in A \cap B$ such that $A \underset{\bar{d}}{\downarrow} B$. By monotonicity, $A \underset{A \cap B}{\downarrow} B$, as desired.

Suppose, on the other hand, that $\mathbb{M}$ is modular. If $A, B \subseteq \mathbb{M}^{\mathrm{eq}}$ are algebraically closed in $\mathbb{M}^{\mathrm{eq}}$, we can find $A_0, B_0 \subseteq \mathbb{M}$ such that $\mathrm{acl}^{\mathrm{eq}}(A_0) = A$ and $\mathrm{acl}^{\mathrm{eq}}(B_0) = B$. By modularity, $A_0 \underset{\mathrm{acl}(A_0) \cap \mathrm{acl}(B_0)}{\downarrow} B_0$. Thus, $A \underset{A \cap B}{\downarrow} B$ by Corollary 6.3.21.

We need to show that one-basedness is preserved by localization. Suppose that $X \subset \mathbb{M}$. Let $\mathbb{M}_X$ denote $\mathbb{M}$ viewed as an $\mathcal{L}_X$-structure.

Claim $\mathbb{M}$ is one-based if and only if $\mathbb{M}_X$ is one-based.

Suppose $\mathbb{M}$ is one-based. If $A, B \subseteq \mathbb{M}^{\mathrm{eq}}$, then

$$\mathrm{acl}^{\mathrm{eq}}(AX) \underset{\mathrm{acl}^{\mathrm{eq}}(AX) \cap \mathrm{acl}^{\mathrm{eq}}(AX)}{\downarrow} \mathrm{acl}^{\mathrm{eq}}(BX).$$

Thus, $\mathbb{M}_X$ is one-based.

Suppose $\mathbb{M}_X$ is one-based. Let $B \subseteq \mathbb{M}$ and $\bar{a} \in \mathbb{M}$. We want to show that $\mathrm{cb}(\mathrm{tp}(\bar{a}/B)) \subseteq \mathrm{acl}^{\mathrm{eq}}(\bar{a})$. Because $\mathrm{tp}(\bar{a}/B)$ does not fork over some finite $B_0 \subseteq B$, we may, without loss of generality, assume that B is finite. Also, without loss of generality, we may assume that $\bar{a}, B \underset{\emptyset}{\downarrow} X$. Otherwise, we replace $\bar{a}$ and $\overline{B}$ by $\bar{a}'$ and $\overline{B}'$, realizing a nonforking extension of $\mathrm{tp}(\bar{a}, B)$ over X.

Because $\bar{a}, B \underset{}{\downarrow} X$, $\bar{a} \underset{B}{\downarrow} X$ by transitivity. Let $\bar{c}$ be a canonical base for $\mathrm{tp}(\bar{a}/B, X)$. Because $\bar{a} \underset{B}{\downarrow} X$, $\bar{c} \in \mathrm{acl}^{\mathrm{eq}}(B)$ and, because $\mathbb{M}_X$ is one-based, $\bar{c} \in \mathrm{acl}^{\mathrm{eq}}(\bar{a}, X)$. But $\bar{c} \underset{\bar{a}}{\downarrow} X$ because $\bar{a}, B \underset{\emptyset}{\downarrow} X$. Thus, $\bar{c} \in \mathrm{acl}^{\mathrm{eq}}(\bar{a})$.

We can now finish the proof. Suppose that $\mathbb{M}$ is locally modular. We can find $d \in \mathbb{M}$ such that $\mathbb{M}_d$ is modular. By Lemma 8.2.9, if $X \subset \mathbb{M}$ is infinite, then $\mathbb{M}_{X,d}$ satisfies (*). By Exercise 8.4.8, $\mathbb{M}_{X,d}$ is also modular. Thus, $\mathbb{M}_{X,d}$ is one-based and $\mathbb{M}$ is one-based. On the other hand, if $\mathbb{M}$ is one-based and $X \subset \mathbb{M}$ is infinite, then $\mathbb{M}_X$ is one-based and satisfies (*). Thus, $\mathbb{M}_X$ is modular. By Theorem 8.2.11, $\mathbb{M}$ is locally modular.

There are much stronger versions of Theorem 8.2.14.

Theorem 8.2.15 *Suppose that T is uncountably categorical and $\mathbb{M}$ is the monster model of T. The following are equivalent.*
i) T is one-based.
ii) Every strongly minimal $D \subseteq \mathbb{M}^n$ is locally modular.
iii) Some strongly minimal $D \subseteq \mathbb{M}^n$ is locally modular.

For a proof, see Theorem 4.3.1 in [18].

8.3   Geometry and Algebra

In this section, we will sketch some important results showing the relationship between the geometry of strongly minimal sets and the presence of definable algebraic structure. We conclude with a sketch of how these ideas come together in Hrushovski's proof of the Mordell–Lang Conjecture for function fields.

Nontrivial Locally Modular Strongly Minimal Sets

So far, the only examples we have given of nontrivial locally modular strongly minimal sets are affine and projective geometries. In both cases, there is a group present. The following remarkable theorem of Hrushovski shows that this is always the case.

Theorem 8.3.1 *Suppose that $\mathbb{M}$ is strongly minimal, nontrivial, and locally modular; then there is an infinite group definable in $\mathbb{M}^{eq}$.*

Proof We will deduce this result using the group configuration theorem. A direct proof of this result would be an easier special case of the group configuration theorem. The reader can find direct proofs in [18] or [76].

Because $\mathbb{M}$ is nontrivial, we can find a finite $A \subset \mathbb{M}$ and $b, c \in \mathbb{M} \setminus \mathrm{acl}(A)$ such that $c \in \mathrm{acl}(A, b) \setminus (\mathrm{acl}(A) \cup \mathrm{acl}(b))$. Choose $d \in \mathbb{M}$ independent from A, b, c. By Theorem 8.2.11, $\mathbb{M}_d$ is modular. Adding d to the language, we may assume that $\mathbb{M}$ is modular. Because d is independent from A, b, c, we still have $c \in \mathrm{acl}(A, b) \setminus (\mathrm{acl}(A) \cup \mathrm{acl}(b))$.

Let $C = \mathrm{acl}(A) \cap \mathrm{acl}(b, c)$. By modularity (see Exercise 8.4.6),

$$\dim(b, c, A) = \dim(b, c) + \dim(A) - \dim C.$$

Because $\dim(b, c, A) = \dim A + 1$ and $\dim(b, c) = 2$, $\dim C = 1$. Thus, there is $a \in C$ with $\dim(a) = 1$. Note that

$$\dim(a, c) = \dim(a, b) = \dim(b, c) = \dim(a, b, c) = 2.$$

Choose $y, z \in \mathbb{M}$ such that (b, c) and (y, z) realize the same type over $\mathrm{acl}^{eq}(a)$ and (y, z) are independent from (b, c) over a (i.e., $\dim(y, z/a, b, c) =$

$\dim(y, z/a) = \dim(b, c/a) = 1)$. Thus, $\dim(a, b, c, x, y) = 3$. Because (y, z) and (b, c) realize the same type over a,

$$\dim(a, z) = \dim(a, y) = \dim(y, z) = \dim(a, y, z) = 2.$$

Because $a \in \mathrm{acl}(b, c)$, $z \in \mathrm{acl}(b, c, y)$. Thus, $\dim(b, c, y) = 3$ and $\dim(b, y) = 2$. Symmetric arguments show that

$$\dim(b, z) = \dim(c, y) = \dim(c, z) = 2$$

and

$$\dim(b, c, z) = \dim(b, y, z) = \dim(c, y, z) = 3.$$

Let $X = \mathrm{acl}(c, y) \cap \mathrm{acl}(b, z)$. By modularity,

$$\dim(c, y, b, z) = \dim(c, y) + \dim(b, z) - \dim X.$$

Thus, $\dim X = 1$ and there is $x \in X$ with $\dim(x) = 1$. Then

$$\dim(c, x, y) = \dim(c, x) = \dim(c, y) = 2$$

and

$$\dim(b, x, z) = \dim(b, x) = \dim(b, z) = 2.$$

If $u = y$ or z and $v = b$ or c, then $a, b, c, x, y, z \in \mathrm{acl}(a, u, v)$, thus; $\dim(a, u, v) = 3$. Similarly, $\dim(b, x, y) = \dim(c, x, z) = 3$.

Using the fact that dimension and Morley rank are the same in strongly minimal sets, we have:

i) $\mathrm{RM}(a) = \mathrm{RM}(b) = \mathrm{RM}(c) = \mathrm{RM}(x) = \mathrm{RM}(y) = \mathrm{RM}(z) = 1$;

ii) any pair of elements has rank 2;

iii) $\mathrm{RM}(a, b, c) = \mathrm{RM}(c, x, y) = \mathrm{RM}(a, y, z) = \mathrm{RM}(b, x, z) = 2$;

iv) all other triples have rank 3;

v) $\mathrm{RM}(a, b, c, x, y, z) = 3$.

Thus, a, b, c, x, y, z is a group configuration and, by Theorem 7.5.7, there is a definable rank one group in $\mathbb{M}^{\mathrm{eq}}$.

Theorem 8.3.1 is just the beginning of the story. In fact, the group that we define tells us a great deal about the structure of $\mathbb{M}$. For example, Hrushovski proved the following.

Theorem 8.3.2 *Let $\mathbb{M}$ be a nontrivial locally modular strongly minimal set. Then, there is a rank 1 Abelian group G definable in $\mathbb{M}^{\mathrm{eq}}$ such that G acts definably as a group of automorphisms of the unique rank 1 type of $\mathbb{M}$.*

One-Based Groups

Theorem 8.3.1 shows that to understand nontrivial locally modular strongly minimal sets we must understand locally modular groups. In this section we will go one step further and analyze ω-stable one-based groups.

Suppose that $(G, \cdot, \ldots)$ is an ω-stable, one-based group. We will prove that G^0 is Abelian and any definable $X \subseteq G^n$ is a Boolean combination of cosets of definable subgroups. This analysis of one-based groups is due to Hrushovski and Pillay.

Let $G \prec \mathbb{G}$ be a monster model.

Lemma 8.3.3 *If $H \le G^n$ is a connected definable subgroup, then $\mathrm{cb}(H) \subseteq \mathrm{acl}^{\mathrm{eq}}(\emptyset)$.*

Proof We will assume that $n = 1$. If $n > 1$, then we can replace G by the group $G^* = (G^n, \cdot, G, \pi_1, \ldots, \pi_n, \ldots)$, where G is a predicate that picks out the elements of the form $(g, 1, \ldots, 1)$ and $\pi_1, \ldots, \pi_n$ are the coordinate maps. It is easy to see that G^* is one-based (see Exercise 8.4.13) and every $X \subseteq G^m$ is definable in G if and only if it is definable in G^*.

Let $g \in \mathbb{G}$ be generic over G. Let p be the generic type of H, and let a realize p with $a \underset{G}{\downarrow} g$. Let q be the nonforking extension of $\mathrm{tp}(ga/G, g)$ to $\mathbb{G}$. Note that

$$\mathrm{RM}(q) = \mathrm{RM}(ga/G, g) = \mathrm{RM}(a/G, g) = \mathrm{RM}(H).$$

Let α be a canonical base for H, and let β be a canonical base for q.
Claim 1 $\alpha \in \mathrm{dcl}^{\mathrm{eq}}(\beta)$.
Let $\phi(\overline{v}, \overline{b})$ be the formula defining H. Let $\mathbb{H} = \{g \in \mathbb{G} : \mathbb{G} \models \phi(\overline{v}, \overline{b})\}$. It suffices to show that if σ is an automorphism of $\mathbb{G}$ and $\sigma q = q$, then σ fixes $\mathbb{H}$ setwise. Because $\phi(g^{-1}v, \overline{b}) \in q$, $\phi(\sigma(g)^{-1}v, \sigma(\overline{b})) \in \sigma q$. If $\sigma q = q$, then $\phi(g^{-1}v, \overline{b}) \wedge \phi(\sigma(g)^{-1}v, \sigma(\overline{b})) \in q$. The formula $\phi(\overline{v}, \sigma(\overline{b}))$ defines the group $\sigma \mathbb{H}$. Thus, q asserts that v is in $g\mathbb{H} \cap \sigma(g)\sigma\mathbb{H}$. Hence, $\mathrm{RM}(g\mathbb{H} \cap \sigma(g)\mathbb{H}) \ge \mathrm{RM}(q)$. Because $g\mathbb{H} \cap \sigma(g)\sigma\mathbb{H}$ is a coset of $\mathbb{H} \cap \sigma\mathbb{H}$,

$$\mathrm{RM}(q) \ge \mathrm{RM}(\mathbb{H}) \ge \mathrm{RM}(\mathbb{H} \cap \sigma\mathbb{H}) \ge \mathrm{RM}(q).$$

Thus, $\mathbb{H} \cap \sigma\mathbb{H}$ is a definable subgroup of $\mathbb{H}$ of finite index. Because $\mathbb{H}$ is connected, $\mathbb{H} \cap \sigma\mathbb{H} = \mathbb{H}$. Thus $\mathbb{H} \subseteq \sigma\mathbb{H}$. A symmetric argument shows that $\sigma\mathbb{H} \subseteq \mathbb{H}$. Thus $\sigma\mathbb{H} = \mathbb{H}$, as desired.

Because $\mathbb{G}$ is one-based, $\beta \in \mathrm{acl}^{\mathrm{eq}}(ga)$, and, by the claim, $\alpha \in \mathrm{acl}^{\mathrm{eq}}(ga)$.
Claim 2 $\alpha \underset{\emptyset}{\downarrow} ga$.
Because $a \underset{G}{\downarrow} g$, by symmetry $g \underset{G}{\downarrow} a$. Because g is generic over G, ga is generic over G, a. In particular

$$\mathrm{RM}(G) \ge \mathrm{RM}(ga/\alpha) \ge \mathrm{RM}(ga/G) = \mathrm{RM}(G).$$

Thus $\alpha \underset{\emptyset}{\downarrow} ga$.
Because $\alpha \in \mathrm{acl}^{\mathrm{eq}}(ga)$ and $\alpha \underset{\emptyset}{\downarrow} ga$, $\alpha \in \mathrm{acl}^{\mathrm{eq}}(\emptyset)$, as desired.

One-based groups have very few definable subgroups.

Corollary 8.3.4 *If $\mathbb{G}$ is an ω-stable, one-based group, then there are at most countably many definable subgroups of $\mathbb{G}^n$.*

Proof Any definable subgroup H has a canonical base in $\operatorname{acl}^{\operatorname{eq}}(\emptyset)$. Because our language is countable, $\operatorname{acl}^{\operatorname{eq}}(\emptyset)$ is countable and there are only countably many definable subgroups.

Next, we prove that one-based groups are Abelian-by-finite.

Lemma 8.3.5 *If $\mathbb{G}$ is a connected one-based ω-stable group, then $\mathbb{G}$ is Abelian. Thus every one-based ω-stable group is Abelian-by-finite.*

Proof For $g \in \mathbb{G}$, let $H_g = \{(h, g^{-1}hg) : h \in \mathbb{G}\} \subseteq \mathbb{G} \times \mathbb{G}$. Then, $H_g = H_h$ if and only if $g/Z(\mathbb{G}) = h/Z(\mathbb{G})$. If $\mathbb{G}$ is non-Abelian, then $Z(\mathbb{G})$ has infinite index in $\mathbb{G}$; thus $\{H_g : g \in \mathbb{G}\}$ is an infinite collection of definable subgroups of $\mathbb{G} \times \mathbb{G}$. Because $\mathbb{G}$ is saturated, there must be $|\mathbb{G}|$ subgroups, a contradiction.

Lemma 8.3.6 *If $p \in S_1(G)$, then there is $b \in G$ such that "$v \in \operatorname{Stab}(p)b$" $\in p$. In, particular, every 1-type is the translate of the generic type of its stabilizer.*

Proof By adding the canonical base for p to the language, we may, without loss of generality, assume that $\operatorname{cb}(p) = \emptyset$. If p' is the unique nonforking extension of p to $\mathbb{G}$, then the formula defining the stabilizer of p defines the stabilizer of p'. Because any automorphism of $\mathbb{G}$ fixes p', it also fixes the stabilizer of p'. Thus, $\operatorname{Stab}(p)$ is defined over $\emptyset$.

Let G_1 be a $|G|^+$-saturated elementary extension of G. Let $g \in G_1$ be generic over G. Let a be a realization of p such that $a \underset{G}{\downarrow} G_1$. Let q be a nonforking extension of $\operatorname{tp}(ga/G_1)$ to $\mathbb{G}$. Let α be a canonical base for $g\operatorname{Stab}(p')$ and let β be a canonical base for q.

Claim 1 α and β are interdefinable.

Let σ be an automorphism of $\mathbb{G}$. We must show that $\sigma q = q$ if and only if $\sigma(g\operatorname{Stab}(p')) = g\operatorname{Stab}(p')$. Because $q = gp'$ and $\sigma(p') = p'$,

$$\sigma q = \sigma(g)\sigma(p') = \sigma(g)p'.$$

Thus

$$
\begin{aligned}
\sigma q = q \ &\Leftrightarrow \sigma(g)p' = gp' \\
&\Leftrightarrow g^{-1}\sigma(g)p' = p' \\
&\Leftrightarrow g^{-1}\sigma(g) \in \operatorname{Stab}(p') \\
&\Leftrightarrow \sigma(g)\operatorname{Stab}(p') = g\operatorname{Stab}(p') \\
&\Leftrightarrow \sigma(g)\operatorname{Stab}(p') = g\operatorname{Stab}(p').
\end{aligned}
$$

The last equivalence follows because $\sigma(g\operatorname{Stab}(p')) = \sigma(g)\sigma(\operatorname{Stab}(p')) = \sigma(g)\operatorname{Stab}(p')$.

Claim 2 $a \underset{G}{\downarrow} ga, \alpha$.

First, note that

$$\mathrm{RM}(G) \geq \mathrm{RM}(ga/G) \geq \mathrm{RM}(ga/G, a) \geq \mathrm{RM}(g/G, a) = \mathrm{RM}(g/G)$$

because $a \underset{G}{\downarrow} G_1$. But $\mathrm{RM}(g/G) = \mathrm{RM}(G)$. Thus, $\mathrm{RM}(ga/G) = \mathrm{RM}(ga/G, a)$ and $a \underset{G}{\downarrow} ga$.

Because G is one-based, $\beta \in \mathrm{acl}^{\mathrm{eq}}(ga)$. But α and β are interdefinable. Thus, $\alpha \in \mathrm{acl}^{\mathrm{eq}}(ga)$ and $a \underset{G}{\downarrow} ga, \alpha$.

Let $c \in G_1$ realize p with $c \underset{G}{\downarrow} \alpha$. Because $c \in G_1$, β is a canonical base for $\mathrm{tp}(ga/G_1)$, and β and α are interdefinable,

$$c \underset{G,\alpha}{\downarrow} ga.$$

Because $c \underset{G}{\downarrow} \alpha$, by transitivity, $c \underset{G}{\downarrow} ga, \alpha$.

Because a and c are both realizations of p, $a \underset{G}{\downarrow} ga, \alpha$ and $c \underset{G}{\downarrow} ga, \alpha$, a and c both realize the unique nonforking extension of p to G, ga, α. Thus $\mathrm{tp}(a/G, ga, \alpha) = \mathrm{tp}(c/G, ga, \alpha)$.

Clearly, $ga \in g\mathrm{Stab}(p')a$. Because $g\mathrm{Stab}(p')$ is definable from α, we must have $ga \in g\mathrm{Stab}(p')c$ and $a \in \mathrm{Stab}(p')c$.

Let $\phi(v, w)$ be the formula "$v \in \mathrm{Stab}(p)w$." By definability of types, there is an $\mathcal{L}_G$-formula $d_p\phi$ defining ϕ. Because $c \in G_1$ and $G \prec G_1$, there is $b \in G$ with $G \models d_p\phi(b)$. Thus "$v \in \mathrm{Stab}(p)b$" $\in p$.

The lemma above also works for n-types.

Corollary 8.3.7 *Suppose that $p \in S_n(G)$ then, there is $\bar{b} \in G^n$ such that* "$v \in \mathrm{Stab}(p)\bar{b}$"$\in p$.

Proof As we argued above, the group G^n is also one-based, and we can view p as a 1-type over G^n and apply the previous lemma.

Next, we show that, for $\bar{a} \in G^n$, $\mathrm{tp}(\bar{a}/G)$ is determined by the cosets of definable subgroups to which $\bar{a}$ belongs.

Corollary 8.3.8 *Suppose that $p, q \in S_n(G)$ and for all definable subgroups H of G^n and all $\bar{a} \in G^n$,*

$$\text{"}\bar{v} \in H\bar{a}\text{"} \in p \text{ if and only if "}\bar{v} \in H\bar{a}\text{"} \in q.$$

Then $p = q$.

Proof Choose $\bar{b}$ and $\bar{c}$ realizing p and q with $\bar{b} \underset{G}{\downarrow} \bar{c}$. Suppose, without loss of generality, that $\mathrm{RM}(p) \geq \mathrm{RM}(q)$. We will show that $\bar{b}$ also realizes q.

By Corollary 8.3.7, there is $\bar{a} \in G^n$ such that "$\bar{v} \in \mathrm{Stab}(q)\bar{a}$" $\in q$. By assumption, "$\bar{v} \in \mathrm{Stab}(q)\bar{a}$" $\in p$ as well. Let q' be the nonforking extension of q to $\mathcal{G}$. The same formula defines $\mathrm{Stab}(q)$ and $\mathrm{Stab}(q')$. Thus, $\bar{b}, \bar{c} \in \mathrm{Stab}(q')\bar{a}$. Thus $\bar{b}\bar{c}^{-1} \in \mathrm{Stab}(q')$.

Claim $\bar{c} \underset{G}{\downarrow} \bar{b}\bar{c}^{-1}$.

Because $\bar{b} \underset{G}{\downarrow} \bar{c}$, $\mathrm{RM}(\bar{b}\bar{c}^{-1}/G, \bar{c}) = \mathrm{RM}(\bar{b}/G, \bar{c}) = \mathrm{RM}(p) \geq \mathrm{RM}(q) \geq \mathrm{RM}(\mathrm{Stab}(q)) \geq \mathrm{RM}(\bar{b}\bar{c}^{-1}/G) \geq \mathrm{RM}(\bar{b}\bar{c}^{-1}/G, \bar{c})$. Thus, $\mathrm{RM}(\bar{b}\bar{c}^{-1}/G, \bar{c}) = \mathrm{RM}(\bar{b}\bar{c}^{-1}/G)$, as desired.

Let $\bar{d}$ realize q'. Because $\bar{b}\bar{c}^{-1} \in \mathrm{Stab}(q')$, $\bar{b}\bar{c}^{-1}\bar{d}$ realizes q'. But $\bar{c}$ and $\bar{d}$ both realize the unique nonforking extension of q to $G \cup \{\bar{b}\bar{c}^{-1}\}$. Thus, $\bar{b} = \bar{b}\bar{c}^{-1}\bar{c}$ realizes q.

Theorem 8.3.9 *If G is an ω-stable one-based group and $X \subseteq G^n$ is definable, then X is a finite Boolean combination of cosets of definable subgroups $H \leq G^n$.*

Proof This follows from Corollary 8.3.8 and Exercise 4.5.13.

Zariski Geometries

In Theorem 8.3.1, we saw that nontrivial locally modular, strongly minimal sets interpret groups. Zil'ber conjectured that if a strongly minimal set is nonlocally modular, it is because it interprets an algebraically closed field. As we noted above, Hrushovski refuted Zil'ber's conjecture, but Hrushovski and Zil'ber found an important class of strongly minimal sets where Zil'ber's conjecture is true. This work also answers the interesting metamathematical question: Can one characterize the topological spaces that arise from the Zariski topology on an algebraic curve?[2] We will describe their work but refer the reader to [44] and [45] for the proofs and [67] for a more lengthy survey.

We say that a topological space is *Noetherian* if there are no infinite descending chains of closed sets. If K is a field, then the Zariski topology on K^n is Noetherian. If K is a differential field, the Kolchin topology on K^n is Noetherian (see Exercise 4.5.43).

A closed set X is *irreducible* if there are no proper closed subsets X_0 and X_1 such that $X = X_0 \cup X_1$. The following lemma is an easy application of König's Lemma, we leave the proof as an exercise.

Lemma 8.3.10 *Suppose that X is a Noetherian topology. If $Y \subseteq X$ is closed, then there are irreducible closed sets $Y_1, \ldots, Y_n$ such that $Y = Y_1 \cup \ldots \cup Y_n$ and Y is not the union of any proper subset of $\{Y_1, \ldots, Y_n\}$. Moreover, if $W_1, \ldots, W_m$ are irreducible closed sets, $Y = W_1 \cup \ldots \cup W_m$, and Y is not the union of any proper subset of $\{W_1, \ldots, W_m\}$, then $n = m$ and $W_1, \ldots, W_n$ is a permutation of $Y_1, \ldots, Y_n$. We call $Y_1, \ldots, Y_n$ the irreducible components of Y.*

---

[2]This work could be thought of in the same spirit as the result that a Pappian projective plane is coordinatized by a field.

In Noetherian topological spaces, we can inductively assign an ordinal *dimension* to any nonempty irreducible closed set X by

$$\dim X = \sup\{\dim Y + 1 : Y \subset X, Y \neq \emptyset, Y \text{ irreducible}\}.$$

Note that if $a \in X$, then $\dim\{a\} = \sup \emptyset = 0$. The dimension of a reducible closed set is the maximum dimension of an irreducible component.

Definition 8.3.11 A *Zariski geometry* is an infinite set D and a sequence of Noetherian topologies on $D, D^2, D^3, \ldots$ such that the following axioms hold.

(Z0) i) If $f : D^n \to D^m$ is defined by $f(x) = (f_1(x), \ldots, f_m(x))$ where each $f_i : D^n \to D$ is either constant or a coordinate projection, then f is continuous.

ii) Each diagonal $\Delta_{i,j}^n = \{x \in D^n : x_i = x_j\}$ is closed.

(Z1) (Weak QE): If $C \subseteq D^n$ is closed and irreducible, and $\pi : D^n \to D^m$ is a projection, then there is a closed $F \subset \overline{\pi(C)}$ such that $\pi(C) \supseteq \overline{\pi(C)} \setminus F$.

(Z2) (Uniform one-dimensionality): i) D is irreducible.

ii) Let $C \subseteq D^n \times D$ be closed and irreducible. For $a \in D^n$, let $C(a) = \{x \in D : (a, x) \in C\}$. There is a number N such that, for all $a \in D^n$, either $|C(a)| \leq N$ or $C(a) = D$. In particular, any proper closed subset of D is finite.

(Z3) (Dimension theorem): Let $C \subseteq D^n$ be closed and irreducible. Let W be a nonempty irreducible component of $C \cap \Delta_{i,j}^n$. Then $\dim C \leq \dim W + 1$.

The basic example of a Zariski geometry is a smooth algebraic curve C over an algebraically closed field where C^n is equipped with the Zariski topology. In this case, Z0 is clear, Z1 follows from quantifier elimination, and Z2 follows from the fact that C is strongly minimal. The verification of Z3 uses the smoothness of C (although a weaker condition suffices).

Zariski geometries may be locally modular. If X is an infinite set, we can topologize X^n by taking the positive quantifier-free definable sets in the language of equality (allowing parameters) as the closed sets. This determines a trivial Zariski geometry on X. If K is a field, we could topologize K^n by taking the affine subsets (i.e., cosets of subspaces). This is a nontrivial, locally modular Zariski geometry. Clearly, a locally modular Zariski geometry will not interpret a field, but Hrushovski and Zil'ber showed that this is the only restriction.

First, one must see how model theory enters the picture. Given a Zariski geometry D, let $\mathcal{L}$ be the language with an n-ary relation symbol for each closed subset of D^n. Let $\mathcal{D}$ be D viewed in the natural way as an $\mathcal{L}$-structure.

Lemma 8.3.12 *The theory of $\mathcal{D}$ admits quantifier elimination, and D is a strongly minimal set. Moreover, Morley rank in $\mathcal{D}$ is exactly dimension.*

Theorem 8.3.13 *Suppose that D is a nonlocally modular Zariski geometry; then D interprets an algebraically closed field K. If $X \subseteq K^n$ is definable in D, then X is definable using only the field structure of K (we say that K is a pure field).*

Hrushovski and Zil'ber give a more refined version of Theorem 8.3.13 with additional geometric information. By a *family of plane curves in D* we mean closed sets $X \subseteq D^m$ and $C \subset D^2 \times X$ such that, for all $a \in X$, if we let C_a denote $\{(x, y) \in D^2 : (x, y, a) \in C\}$, then C_a is a one-dimensional irreducible closed subset of D^2.

Definition 8.3.14 We say that a family of plane curves is *ample* if whenever x and y are independent generic points of D^2 there is a plane curve C_a with $x, y \in C_a$.

An ample family is called *very ample* if for x, y (not necessarily independent) generic points in D^2 there is a curve $C(a)$ with $x \in C(a)$ and $y \notin C(a)$. In this case, we say that the family *separates points*.

We say that D is (very) ample if there is a (very) ample family of plane curves.

Using Theorem 8.2.11 it is not hard to show that a Zariski geometry is nonlocally modular if and only if it is ample.

Corollary 8.3.15 *If D is an ample Zariski geometry, then D interprets an algebraically closed field.*

For most model-theoretic applications, we only need the results for ample Zariski geometries, but the following result shows that very ample geometries are intimately related to smooth algebraic curves.

Theorem 8.3.16 *If D is a very ample Zariski geometry, then there is an interpretable algebraically closed field K, C a smooth quasiprojective curve defined over K, and a definable bijection $f : D \to C$ such that the induced maps $f^n : D^n \to C^n$ are homeomorphisms for all n.*

Hrushovski and Sokolović showed that Zariski geometries arise naturally in differentially closed fields (see [67] for a proof).

Theorem 8.3.17 *If K is a differentially closed field and D is a strongly minimal set, there is a finite $X \subset D$ such that if we topologize $(D \backslash X)^n$ with the restriction of the Kolchin topology, then $(D \setminus X)$ is a Zariski geometry.*

This shows that, in differentially closed fields, any nonlocally modular, strongly minimal set interprets an algebraically closed field. This field must have finite rank. Sokolović showed that any finite rank field interpretable in a differentially closed field is definably isomorphic to the field of constants.

Thus, in differentially closed fields, any nonlocally modular strongly minimal set interprets the constants.[3]

Applications to Diophantine Geometry

Theorem 8.3.9 plays an important role in Hrushovski's applications of model theory to Diophantine geometry. We will try to briefly give the flavor of these applications.

Definition 8.3.18 Suppose that K is an algebraically closed field. An algebraic group A that is embedded as an irreducible closed set in projective n-space over K is called an *Abelian variety*. We say that A is *simple* if A has no proper infinite algebraic subgroups.

For example, an elliptic curve is a simple Abelian variety. The next lemma summarizes some of the group-theoretic properties of Abelian varieties. See [38] for details.

Lemma 8.3.19 *Suppose that A is an Abelian variety.*

i) A is a commutative divisible group.

ii) If K has characteristic zero, the subgroup of n-torsion points of A has size n^{2d}, where d is the dimension of A.

iii) The torsion points of A are Zariski dense in A.

If G is an algebraic group and we consider the structure $(G, \cdot, \ldots)$ where we have predicates for all constructible subsets of G^n, then G is not one-based. Indeed, we can always find a definable subset $X \subseteq G$ and a definable map $X \mapsto K$ such that the image is a cofinite subset of K. Using this map, we can interpret the field, a nonlocally modular, strongly minimal set, contradicting Exercise 8.4.14.

Surprisingly, this is not true for differential algebraic groups. Let K be a differentially closed field, and let C be the field of constants of K. If A is an Abelian variety defined over K, then we can still view A as a group definable in the differentially closed field K. In Exercise 4.5.43, we introduced the Kolchin topology, the extension of the Zariski topology where we also consider solution sets of algebraic differential equations, and showed that there are no infinite descending chains of Kolchin closed sets. Even if there are no interesting algebraic subgroups of A, there will be new Kolchin closed subgroups. The following result is due to Hrushovski and Sokolović, drawing on earlier work of [64] and Buium[20].

Theorem 8.3.20 *Suppose that K is a differentially closed field and A is a simple Abelian variety defined over K that is not isomorphic to an Abelian variety defined over C. Let $A^{\#}$ be the closure in the Kolchin topology of*

---

[3]Recently, Pillay and Ziegler [82] have given a direct proof of this avoiding the Zariski Geometry machinery.

the torsion points of A. Then, $A^{\#}$ is a one-based group and there are no proper infinite definable subgroups of $A^{\#}$.

The results of Manin and Buium show that $A^{\#}$ is a finite Morley rank group with no infinite definable subgroups. There are two cases to consider. If all strongly minimal subsets of $A^{\#}$ are locally modular, then $A^{\#}$ is one-based. If there is any nonlocally modular strongly minimal subset of $A^{\#}$, then, using Zariski geometries, we can interpret the constants in $A^{\#}$. In this case, one can, by some additional model-theoretic arguments, show that A is isomorphic to an Abelian variety defined over the constants. A more detailed sketch is given in [66].

We argue that in fact $A^{\#}$ is strongly minimal. Suppose that $X \subseteq A^{\#}$ is strongly minimal. Because $A^{\#}$ is one-based, X is a Boolean combination of cosets of definable subgroups of $A^{\#}$. Because $A^{\#}$ has no infinite definable proper subgroups, $A^{\#} \setminus X$ is finite and $A^{\#}$ is strongly minimal.

The following simple application is a special case of Hrushovski's proof of the Mordell–Lang Conjecture for function fields.

Theorem 8.3.21 *Let K and k be algebraically closed fields of characteristic zero with $k \subseteq K$. Let A be a simple Abelian variety defined over K that is not isomorphic to an Abelian variety defined over the algebraic closure of k. If $V \subset A$ is a proper subvariety of A, then V contains only finitely many torsion points of A.*

Proof We can define a derivation δ on K such that the constant field is k (see [58] X §7). If $\widehat{K}$ is the differential closure of (K, δ), then, by Exercise 6.6.32, the constant field of $\widehat{K}$ is k. Thus, replacing K by $\widehat{K}$ if necessary, we may assume that K is differentially closed. Let $A^{\#}$ be the Kolchin closure of the torsion points of A. We will show that $A^{\#} \cap V$ is finite.

If $A^{\#} \cap V$ is infinite, then, because $A^{\#}$ is strongly minimal, $A^{\#} \setminus V$ is finite. Because $A^{\#}$ contains the torsion points of A, $A^{\#}$ is Zariski dense in A. Thus, V is Zariski dense in A, contradicting the fact that V is a proper subvariety.

8.4 Exercises and Remarks

Throughout the Exercises assume that we are working in an ω-stable theory.

Exercise 8.4.1 Prove Lemma 8.1.3.

Exercise 8.4.2 Prove Lemma 8.1.4.

Exercise 8.4.3 Suppose that (X, cl) is a pregeometry, $A, Y \subseteq X$, $\dim(A) < \aleph_0$ and $\dim(Y) < \aleph_0$. Then, $\dim(Y/A) = \dim(A \cup Y) - \dim(A)$.

Exercise 8.4.4 Prove Lemma 8.1.6.

Exercise 8.4.5 Suppose that $\mathcal{M} \models \text{Th}(\mathbb{Z}, s)$. Show that the associated geometry $(\widehat{M}, \widehat{\text{acl}})$ has $\text{acl}(A) = A$ for all $A \subseteq \widehat{M}$.

Exercise 8.4.6 Show that a pregeometry is modular if and only if any two closed sets A and B are independent over $A \cap B$.

Exercise 8.4.7 Prove Lemma 8.2.9.

Exercise 8.4.8 Suppose that (X, cl) is a modular pregeometry. Show that any localization of (X, cl) is also modular.

Exercise 8.4.9 a) Suppose that K is an algebraically closed field, $x, a_0, \ldots, a_n \in K$ are algebraically independent, and

$$y = \sum_{i=0}^{n} a_i x^i,$$

then y is not algebraic over $k(x)$ for k a subfield of $\text{acl}(a_0, \ldots, a_n)$ of dimension less than $n + 1$.

b) Use Lemma 8.1.13 to conclude that algebraically closed fields are not locally modular.

Exercise 8.4.10 Suppose that A is a canonical base for X and $\text{dcl}^{\text{eq}}(A) = \text{dcl}^{\text{eq}}(B)$, then B is also a canonical base for X.

Exercise 8.4.11 Suppose that D is a strongly minimal set and $C \subseteq D \times A$ is a family of strongly minimal subsets of $D \times D$. Let $\sim$ be the equivalence relation $a \sim b$ if and only if $C_a \triangle C_b$ is finite. Show that $\sim$ is definable.

Exercise 8.4.12 Suppose that $p \in S_n(A)$ is stationary and $p' \in S_n(\mathbb{M})$ is the unique nonforking extension of p. Let $\overline{a}_1, \overline{a}_2, \ldots$ be a Morley sequence for p. Show that there is an n such that $\text{cb}(p) \subseteq \text{dcl}^{\text{eq}}(\overline{a}_1, \ldots, \overline{a}_n)$. In other words for any stationary type p, we can find a finite set of realizations of p over which p' does not fork. [Hint: Use Exercise 6.6.35 and its generalization to n-types.]

Thus, if $p = \text{tp}(\overline{a}/B)$ is stationary, then a canonical base for p can be found in the definable closure of a sufficiently large finite set of independent realizations of p.

Exercise 8.4.13 Suppose that $\mathbb{M}$ is one-based. Consider the structure $\mathcal{M}^*$ with universe $\mathbb{M}^n$ where, for some $a \in \mathbb{M}$, we have a predicate picking out $\{(x, a, a, \ldots, a) : x \in \mathbb{M}\}$ which we identify with $\mathbb{M}$. We also have projection maps $\pi_1, \ldots, \pi_n$ where $\pi_i(x_1, \ldots, x_n) = (x_i, a, a, \ldots, a)$. For each function (relation) symbol of $\mathcal{L}$, we have a function (relation) symbol in $\mathcal{L}^*$, which we interpret in the natural way. Show that $\mathbb{M}^*$ is one-based.

Exercise 8.4.14 Show that if $\mathbb{M}$ is an ω-stable one-based structure, then any strongly minimal set D interpretable in $\mathbb{M}$ is locally modular. [Hint: Prove that D is one-based.]

Exercise 8.4.15 a) Prove Lemma 8.3.10.

b) Let α be an ordinal. Topologize α by taking closed sets of the form $\{\gamma : \gamma \leq \beta\}$ for $\beta < \alpha + 1$. Show that α is a Noetherian topology, and calculate the dimension of the space.

Exercise 8.4.16 Prove that a Zariski geometry is nonlocally modular if and only if it is ample.

Remarks

Zil'ber was the first to recognize the importance of studying the combinatorial geometry of algebraic closure. Pillay's *Geometric Stability Theory* is an excellent reference for the material in this chapter.

Although Zil'ber's conjecture is false, the analog for o-minimal theories is true. Peterzil and Starchenko [74] proved that, in an o-minimal theory, if there are two-dimensional families of plane curves, then one can interpret a real closed field.

Zariski geometries also arise when studying compact complex manifolds. Suppose that M is a complex manifold. We say that $X \subseteq M$ is *analytic* if for all $a \in M$ there is an open neighborhood U of a and $f_1, \ldots, f_m$ holomorphic functions on U such that

$$X \cap U = \{x \in U : f_1(x) = \ldots = f_m(x) = 0\}.$$

If M is compact, then there are no infinite descending chains of analytic subsets. Thus, we can give M^n a Noetherian topology where the closed sets are exactly the analytic subsets.

If M is a complex manifold we consider the structure $\mathcal{M}$, where the underlying set is M and we have relation symbols for each analytic subset of $M, M^2, M^3, \ldots$. If $a \in M$, then $\{a\}$ is analytic so this language will have size $2^{\aleph_0}$.

Zil'ber proved the following result (see [80]).

Theorem 8.4.17 *If M is a compact complex manifold, then $\mathcal{M}$ has elimination of quantifiers and is totally transcendental. Indeed, if $A \subseteq M^n$ is analytic, then $\mathrm{RM}(A)$ is at most the complex dimension of A.*

If $\mathcal{M}$ is strongly minimal, then M is a Zariski geometry.

One way to get a compact complex manifold is by taking a smooth projective variety. By Chow's Theorem (see [37] B.2.2), every analytic subset of projective space is already defined algebraically. Thus, these examples are interpretable in the field $\mathbb{C}$ and hence nothing new. In particular, a smooth projective variety is strongly minimal if and only if it is an algebraic curve. Riemann's Existence Theorem (see [37] B.3.1) says that any one-dimensional compact complex manifold is a smooth projective algebraic curve. Thus, to look for new examples, we need to consider manifolds of dimension greater than one.

Suppose that $\alpha_1, \ldots, \alpha_{2n} \in \mathbb{C}^n$ and $\alpha_1, \ldots, \alpha_{2n}$ are linearly independent over $\mathbb{R}$. Let Λ be the lattice $\mathbb{Z}\alpha_1 \oplus \ldots \oplus \mathbb{Z}\alpha_{2n}$. We can form a compact complex manifold M, by taking the quotient structure $\mathbb{C}^n/\Lambda$. We call M a *complex torus*. Because Λ is an additive subgroup of $\mathbb{C}^n$, M is a complex Lie group.

If $\alpha_1, \ldots, \alpha_{2n}$ are algebraically independent over $\mathbb{Q}$, we call M a *generic torus*. Classical results from complex manifolds (see [91] VIII 1.4) imply that M is strongly minimal. Pillay [81] showed that M is a locally modular group.

The most striking recent result in model theory is Hrushovski's [43] proof of the Mordell–Lang Conjecture for function fields. We conclude our remarks by explaining the conjecture and stating Hrushovski's result.

Let K be an algebraically closed field of characteristic zero and A an Abelian variety defined over K. Suppose that Γ is a subgroup of A. We say that Γ is *finite rank* if there is a finitely generated subgroup Γ_0 such that $\Gamma \subseteq \{g \in A : ng \in \Gamma_0 \text{ for some } n = 1, 2, \ldots\}$. For example, taking $\Gamma_0 = \{0\}$, the torsion subgroup of A is of finite rank.

Mordell–Lang Conjecture (characteristic zero) Suppose that K has characteristic zero, A is an Abelian variety, Γ is a finite rank subgroup of A and X is a proper subvariety of A. Then, $X \cap \Gamma$ is a finite union of cosets of subgroups of A.

Next we show how the Mordell-Lang conjecture implies the Mordell Conjecture. Suppose that C is a curve of genus $g > 1$ defined over a number field k. The Mordell Conjecture asserts that C has only finitely many k-rational points (i.e., points with coordinates in k).

To any curve C of genus $g \geq 1$, we can associate a g-dimensional Abelian variety $J(C)$ defined over k called the Jacobian of C (see [38]). The curve C is a subvariety of $J(C)$, and $J(C)$ is the smallest Abelian variety in which C embeds. If C has genus 1, then C is an elliptic curve and $J(C) = C$.

Let C have genus $g > 1$. Let Γ be the k-rational points of $J(C)$. The Mordell–Weil theorem (see [38]) asserts that Γ is a finitely generated group. Thus, $C \cap \Gamma$ is a finite union of cosets of subgroups of Γ. If any of these subgroups is infinite, then the Zariski closure of that coset is also a coset and the Zariski closure must be the entire curve C. But then there would be a group structure defined on C and C would be an Abelian variety contradicting the fact that $J(C) \supset C$ and $J(C)$ is the smallest Abelian variety in which C embeds. Thus, $C \cap \Gamma$ is finite and C contains only finitely many k-rational points.

The Mordell Conjecture was proved by Faltings, who also proved the Mordell–Lang Conjecture in case Γ is finitely generated. The full characteristic zero Mordell–Lang Conjecture was proved by McQuillen, building on Faltings' work as well as work of Raynaud, Hindry, and Vojta.

In number theory, when studying question about number fields, it is often insightful to ask the same question about finitely generated extensions of algebraically closed fields (we call these *function fields*). Long before Faltings' work, Manin proved the function field case of the Mordell Conjecture.

Theorem 8.4.18 *Let k be algebraically closed of characteristic zero, and let $K \supset k$. Let C be a curve of genus $g > 1$ defined over K. Then, either C has only finitely many K-rational points or C is isomorphic to a curve defined over k.*

So far, we have only considered characteristic zero. What about characteristic $p > 0$? The obvious generalization of the Mordell–Lang Conjecture to characteristic p is false. Suppose that C is a curve of genus $g > 1$ defined over $\mathbb{F}_p$ and let $J(C)$ be its Jacobian. Let α be a generic point of C. Let Γ be the group of $\mathbb{F}_p(\alpha)$-rational points of $J(C)$. By the Lang–Néron Theorem (the function field version of the Mordell–Weil Theorem (see [38])), Γ is finitely generated. If σ is the Frobenius automorphism $x \mapsto x^p$, then $\sigma^n(\alpha) \in C$ for all n. Thus, $C \cap \Gamma$ is infinite, but C is not a coset of a subgroup. In this case, our curve C is defined over the prime field. Abramovich and Voloch [1] stated a plausible version of the Mordell–Lang Conjecture for function fields in characteristic p and proved it in many important cases. Their conjecture was proved by Hrushovski, who also gave a new proof for the function field case in characteristic zero.

Theorem 8.4.19 (Mordell–Lang Conjecture for function fields)
Let k be an algebraically closed field with $K \supset k$. Let A be an Abelian variety defined over K, X a subvariety of A, and Γ a finite rank subgroup of A. Suppose that $X \cap \Gamma$ is Zariski dense in X. Then, there is a sub-Abelian variety $A_1 \subseteq A$, an Abelian variety B defined over k, a surjective homomorphism $g : A_1 \to B$, and a subvariety X_0 of B defined over k such that $g^{-1}(X_0)$ is a translate of X.

For more on Hrushovski's result, see [16], [17], and [79].

Appendix A
Set Theory

In this Appendix, we will survey some of the elementary results from set theory that we use in the text. We give very few proofs and refer the reader to set theory texts such as [26], [47], or [57] for further details.

We will work in ZFC, Zermelo–Fraenkel set theory with the Axiom of Choice. The Axiom of Choice asserts that if $(A_i : i \in I)$ is a family of nonempty sets, then there is a function f with domain I such that $f(i) \in A_i$ for all $i \in I$.

Zorn's Lemma and Well-Orderings

If X is a set and $<$ is a binary relation on X, we say that $(X, <)$ is a *partial order* if $(X, <) \models \forall x \, \neg (x < x)$ and $(X, <) \models \forall x \forall y \forall z \, ((x < y \wedge y < z) \rightarrow x < z)$.

We say that $(X, <)$ is a *linear order* if in addition
$(X, <) \models \forall x \forall y \, (x < y \vee x = y \vee y < x)$.

If $(X, <)$ is a partial order, then we say that $C \subseteq X$ is a *chain* in X if C is linearly ordered by $<$.

Theorem A.1 (Zorn's Lemma) *If $(X, <)$ is a partial order such that for every chain $C \subseteq X$ there is $x \in X$ such that $c \leq x$ for all $c \in C$, then there is $y \in X$ such that there is no $z \in X$ with $z > x$. In other words, if every chain has an upper bound, then there is a maximal element of X.*

We give one application of Zorn's Lemma. We say that a linear order $(A, <)$ is a *well-order* if for any nonempty $C \subseteq A$, there is $a \in C$ such that $a \leq b$ for all $b \in C$. The following characterization is also useful.

Lemma A.2 $(A, <)$ *is a well-order if and only if there is no infinite descending chain* $a_0 > a_1 > a_2 > \ldots$ *in* A.

Theorem A.3 (Well-Ordering Principle) *If A is any set, then there is a well-ordering of A.*

Proof Let $X = \{(Y, R) : Y \subseteq A$ and R is a well-ordering of $A\}$. We say that $(Y, R) < (Y_1, R_1)$ if $Y \subset Y_1$, $R \subset R_1$, and if $a \in Y_1 \setminus Y$ and $b \in Y$; then bRa (i.e. every new element is greater than every old element). Suppose that $C \subset X$ is a chain. Let

$$\widehat{Y} = \bigcup_{(Y,R) \in C} Y \text{ and } \widehat{R} = \bigcup_{(Y,R) \in C} R.$$

We claim that $\widehat{R}$ is a well-ordering of $\widehat{Y}$. We first show that $\widehat{R}$ is a linear order. Clearly, $\neg(a \widehat{R} a)$ for all $a \in \widehat{Y}$. If $a_1, a_2, a_3 \in \widehat{Y}$ such that $a_1 \widehat{R} a_2$ and $a_2 \widehat{R} a_3$, then we can find $(Y_i, R_i) \in C$ such that $a_i \in Y_i$ for $i = 1, 2, 3$. Because C is a chain, there is j such that $(Y_i, R_i) \leq (Y_j, R_j)$ for each $i = 1, 2, 3$. Because (Y_i, R_i) is transitive, $a_1 R_j a_3$ and $a_1 \widehat{R} a_3$.

If $a_0 > a_1 > \ldots$ is a decreasing chain in $\widehat{R}$, we can find $(Y, R) \in C$ such that $a_0 \in Y$. Because of the way we order X, all of the $a_i \in Y$. In this case, R would not be a well-order, a contradiction. Thus $(\widehat{Y}, \widehat{R}) \in X$. Clearly, $(\widehat{Y}, \widehat{R}) \geq (Y, R)$ for all $(Y, R) \in C$. Thus, every chain has an upper bound.

By Zorn's Lemma, there is $(Y, R) \in X$ maximal. We claim that $Y = A$. Suppose that $a \in A \setminus Y$. Let $Y' = A \cup \{a\}$, and let $R' = R \cup (Y \times \{a\})$ (i.e., we order Y' by making a the largest element). Then, R' is a well-ordering and we have contradicted the maximality of (Y, R). Thus $,R$ is a well-ordering of A.

Zorn's Lemma and the Well-Ordering Principle are equivalent forms of the Axiom of Choice.

Ordinals

Definition A.4 We say that X is *transitive* if, whenever $x \in X$ and $y \in x$, then $y \in X$. We say that a set X is an *ordinal* if X is transitive and well-ordered by $\in$. Let On be the class of all ordinals.

Lemma A.5 i) On *is transitive and well-ordered by* $\in$.
ii) *If α and β are ordinals, then the orderings $(\alpha, \in)$ and $(\beta, \in)$ are isomorphic if and only if $\alpha = \beta$.*

It follows from i) that On is not a set. If On were a set, then On is itself an ordinal and $On \in On$. This gives rise to an infinite descending chain contradicting the fact that On is well-ordered by $\in$.

Because $\in$ is an ordering of On we often write $\alpha < \beta$ instead of $\alpha \in \beta$. Note that $\alpha = \{\beta \in On : \beta < \alpha\}$.

Every well-ordering is isomorphic to an ordinal.

Proposition A.6 *If $(X, <)$ is a well-ordering, then there is an ordinal α such that $(X, <)$ is isomorphic to $(\alpha, \in)$. We call α the order type of $(X, <)$.*

Lemma A.7 *i) $\emptyset$ is an ordinal and if $\alpha \in On$, and $\alpha \neq \emptyset$ then $\emptyset \in \alpha$. Thus, $\emptyset$ is the least ordinal.*

ii) If α is an ordinal, then $suc(\alpha) = \alpha \cup \{\alpha\}$ is an ordinal, and if $\beta \in On$, then $\beta \leq \alpha$ or $suc(\alpha) \leq \beta$.

iii) If C is a set of ordinals, then $\delta = \bigcup_{\alpha \in C} \alpha$ is an ordinal, and δ is the least upper bound of the ordinals in C.

Lemma A.7 gives us a description of the first ordinals. By i), $0 = \emptyset$ is the least ordinal. The next ordinals are $1 = \{\emptyset\}$, $2 = \{\emptyset, \{\emptyset\}\}, \ldots$. In general, we let $n + 1 = suc(n)$. Note that $n = \{0, 1, \ldots, n - 1\}$. Thus, the natural numbers are an initial segment of the ordinals. The next ordinal is $\omega = \{0, 1, 2, 3, \ldots\}$.

If $\alpha \in On$, we say that α is a *successor ordinal* if $\alpha = suc(\beta)$ for some ordinal β. If $\alpha \neq 0$ and α is not a successor ordinal then we can say α is a *limit ordinal*. The next proposition is the main tool for proving things about ordinals.

Theorem A.8 (Transfinite Induction) *Suppose that C is a subclass of the ordinals such that*

i) $0 \in C$,

ii) if $\alpha \in C$, then $suc(\alpha) \in C$, and

iii) if α is a limit ordinal and $\beta \in C$ for all $\beta < \alpha$, then $\alpha \in C$.

Then $C = On$.

We can define addition, multiplication, and exponentiation of ordinals. If $\alpha, \beta \in On$, let X be the well-order obtained by putting a copy of β after a copy of α. More precisely, $X = (\{0\} \times \alpha) \cup (\{1\} \times \beta)$ with the lexicographic order. Then $\alpha + \beta$ is the order type of X. Let Y be the well-order obtained by taking the lexicographic order on $\beta \times \alpha$. Then $\alpha \cdot \beta$ is the order type of $\beta \times \alpha$. We define α^β by transfinite recursion as follows:

i) $\alpha^0 = 1$;

ii) $\alpha^{suc(\beta)} = \alpha^\beta \alpha$;

iii) if β is a limit ordinal, then $\alpha^\beta = \sup\{\alpha^\gamma : \gamma < \beta\} = \bigcup_{\gamma < \beta} \alpha^\gamma$.

Addition and multiplication are not commutative, but we do have the following properties.

Lemma A.9 *i)* $\mathrm{suc}(\alpha) = \alpha + 1$.

ii) $\mathrm{suc}(\alpha + \beta) = \alpha + \mathrm{suc}(\beta)$.

iii) $\alpha + (\beta + \gamma) = (\alpha + \beta) + \gamma$.

iv) $\alpha(\beta\gamma) = (\alpha\beta)\gamma$.

v) $\alpha(\beta + \gamma) = \alpha\beta + \alpha\gamma$.

vi) If $\beta = \sup_{\gamma \in C} \gamma$, then $\alpha + \beta = \sup_{\gamma \in C} \alpha + \gamma$.

We can start building the ordinals above ω:

$\omega, \omega + 1, \omega + 2, \ldots, \sup\{\omega + n : n < \omega\} = \omega + \omega = \omega 2, \omega 2 + 1, \omega 2 + 2, \ldots,$
$\sup\{\omega 2 + n : n < \omega\} = \omega 2 + \omega = \omega 3, \ldots, \omega 3, \ldots, \omega 4, \ldots, \omega 5 \ldots, \sup\{\omega n : n < \omega\} = \omega \times \omega = \omega^2, \omega^2 + 1, \ldots, \omega^3, \ldots, \omega^4, \ldots, \sup\{\omega^n : n < \omega\} = \omega^\omega$.

Continuing this way: $\ldots, \omega^{\omega+1}, \ldots, \omega^{\omega+2}, \ldots, \omega^{\omega 2}, \ldots, \omega^{\omega 3}, \ldots, \omega^{\omega^2}, \ldots,$
$\omega^{\omega^n}, \ldots, \omega^{\omega^\omega}, \ldots, \omega^{\omega^{\omega^\omega}}, \ldots$

This is the limit of the ordinals we can easily describe. The next ordinal is

$$\epsilon_0 = \sup\{\omega, \omega^\omega, \omega^{\omega^\omega}, \ldots\}.$$

We could now continue as before. Indeed, all of the ordinals we have described so far are still quite small.

Cardinals

We need a method of comparing sizes of sets. Let A be any set. By the Well-Ordering Principle, there is a well-ordering $<$ of A and, by Proposition A.6, there is an ordinal α such that $(A, <)$ is isomorphic to α. We let $|A|$ be the least ordinal α such that there is a well-ordering of A isomorphic to α.

Proposition A.10 *The following are equivalent.*

i) $|A| = |B|$.

ii) There is a bijection $f : A \to B$.

iii) There are one-to-one functions $f : A \to B$ and $g : B \to A$.

We say that A is *countable* if $|A| \le \omega$. All of the ordinals $\alpha \le \epsilon_0$ that we described above are countable. Let $\omega_1 = \{\alpha \in On : \alpha \text{ is countable}\}$. It is easy to see that ω_1 is transitive and well-ordered by $\in$. If ω_1 is countable, then $\omega_1 \in \omega_1$ and we get a contradiction. Thus, ω_1 is the first uncountable ordinal. Note that $|\omega_1| = \omega_1$.

We say that an ordinal α is a *cardinal* if $|\alpha| = \alpha$. We recursively define ω_α for $\alpha \in On$ as follows:

$\omega_0 = \omega$;

$\omega_{\alpha+1} = \{\delta \in On : |\delta| = \omega_\alpha\}$;

if α is a limit ordinal, then $\omega_\alpha = \sup_{\beta < \alpha} \omega_\beta$.

Proposition A.11 *i) Each ω_α is a cardinal and $\omega_\alpha < \omega_\beta$ if $\alpha < \beta$.*

ii) If κ is a cardinal, then either $\kappa < \omega$ or $\kappa = \omega_\alpha$ for some $\alpha \in On$.

We also use the notation $\aleph_\alpha = \omega_\alpha$. When we are thinking of it as an ordinal, we use ω_α and when we are thinking of it as a cardinal, we use $\aleph_\alpha$.

If κ is a cardinal, there is a least cardinal greater than κ, which we call κ^+. We say that κ is a *successor cardinal* if $\kappa = \lambda^+$ for some λ, otherwise (if κ is nonzero), we say that κ is a *limit cardinal*. Note that infinite successor cardinals are limit ordinals.

For any limit ordinal $\alpha \geq \omega$, the *cofinality* of α is the least cardinal λ such that there is a function $f : \lambda \to \alpha$ and the image of f is unbounded in α. We let $\operatorname{cof}(\alpha)$ denote the cofinality of α.

For example, $\operatorname{cof}(\omega) = \aleph_0$ because a finite function cannot be unbounded in ω. On the other hand, $\operatorname{cof}(\omega_\omega) = \omega$ because the function $n \mapsto \omega_n$ has unbounded image.

If $\kappa \geq \aleph_0$ is a cardinal, we say that κ is *regular* if $\operatorname{cof}(\kappa) = \kappa$; otherwise, we say that κ is *singular*.

Proposition A.12 *If $\kappa \geq \aleph_0$ is a cardinal, then κ^+ is regular.*

$\aleph_0$ is a regular limit cardinal. It may be the only cardinal with both properties. We say that $\kappa > \aleph_0$ is *inaccessible* if κ is a regular limit cardinal. Although we cannot prove that inaccessible cardinals exist, it seems likely that we also cannot prove that they do not exist. Inaccessible cardinals are quite large.

Proposition A.13 *If $\kappa > \aleph_0$ is inaccessible, then $\kappa = \aleph_\kappa$.*

Proof An induction shows that $\omega_\alpha \geq \alpha$ for all α. If $\kappa = \aleph_\alpha$ where $\alpha < \kappa$, then $\beta \mapsto \omega_\beta$ is an unbounded map from α into κ, a contradiction.

Cardinal Arithmetic

We define addition and multiplication of cardinals. If $|X| = \kappa$ and $|Y| = \lambda$, then $\kappa + \lambda = |(\{0\} \times X) \cup (\{1\} \times Y)|$ and $\kappa\lambda = |X \times Y|$. These operations are commutative but not very interesting.

Lemma A.14 *Let κ and λ be cardinals. If κ and λ are both finite, then these operations agree with the usual arithmetic operations. If either κ or λ is infinite, then*

$$\kappa + \lambda = \kappa\lambda = \max\{\kappa, \lambda\}.$$

Corollary A.15 *i) If $|I| = \kappa$ and $|A_i| \leq \kappa$ for all $i \in I$, then $|\bigcup A_i| \leq \kappa$.*
ii) If κ is regular, $|I| < \kappa$, and $|A_i| < \kappa$ for all $i \in I$, then $|\bigcup A_i| < \kappa$.
iii) Let κ be an infinite cardinal. Let X be a set and $\mathcal{F}$ a set of functions $f : X^{n_f} \to X$. Suppose that $|\mathcal{F}| \leq \kappa$ and $A \subseteq X$ with $|A| \leq \kappa$. Let $cl(A)$ be the smallest subset of X containing A closed under the functions in $\mathcal{F}$. Then $|cl(A)| \leq \kappa$.

Exponentiation is much more interesting. If A and B are sets, then A^B is the set of functions from A to B and $|A|^{|B|} = |A^B|$.

Lemma A.16 *Let κ, λ, and μ be cardinals.*
  i) $(\kappa^\lambda)^\mu = \kappa^{\lambda\mu}$.
  ii) If $\lambda \geq \aleph_0$ and $2 \leq \kappa < \lambda$, then $2^\lambda = \kappa^\lambda = \lambda^\lambda$.
  iii) If κ is regular and $\lambda < \kappa$, then $\kappa^\lambda = \sup\{\kappa, \mu^\lambda : \mu < \kappa\}$.

Proof iii) If $f : \lambda \to \kappa$, then, because κ is regular, there is $\alpha < \kappa$ such that $f : \lambda \to \alpha$. Thus $\kappa^\lambda = \bigcup_{\alpha < \kappa} \alpha^\lambda$. The right-hand side is the union of κ sets each of size μ^λ for some $\mu < \kappa$.

We say that an inaccessible cardinal κ is *strongly inaccessible* if $2^\lambda < \kappa$ for all $\lambda < \kappa$.

Corollary A.17 *If κ is strongly inaccessible and $\lambda < \kappa$, then $\kappa^\lambda = \kappa$.*

We know by Cantor that $2^\kappa > \kappa$ for all cardinals κ. The next theorem is a slight generalization.

Proposition A.18 (König's Theorem) *If $\kappa \geq \aleph_0$, then $\kappa^{\mathrm{cof}(\kappa)} > \kappa$.*

This gives us Cantor's theorem because $2^\kappa = \kappa^\kappa > \kappa$ but also gives us, for example, that $\aleph_\omega^{\aleph_0} > \aleph_\omega$.

ZFC is too weak to answer basic questions about cardinal exponentiation. The most interesting is the Continuum Hypothesis.

Continuum Hypothesis (CH) $2^{\aleph_0} = \aleph_1$.

Generalized Continuum Hypothesis (GCH) $2^{\aleph_\alpha} = \aleph_{\alpha+1}$.

The Continuum Hypothesis is unprovable in ZFC, but GCH is consistent with ZFC.[1] Assuming the Generalized Continuum Hypothesis, we get a complete picture of cardinal exponentiation.

Proposition A.19 *Assume the Generalized Continuum Hypothesis. Let $\kappa, \lambda \geq 2$ with at least one infinite.*
  i) If $\lambda \leq \kappa$, then $\lambda^\kappa = \kappa^+$.
  ii) If $\lambda < \mathrm{cof}(\kappa)$, then $\kappa^\lambda = \kappa$.
  iii) If $\mathrm{cof}(\kappa) \leq \lambda < \kappa$, then $\kappa^\lambda = \kappa^+$.

Finite Branching Trees

Definition A.20 A *finite branching tree* is a partial order $(T, <)$ such that:
  i) there is $r \in T$ such that $r \leq x$ for all $x \in T$;
  ii) if $x \in T$, then $\{y : y < x\}$ is finite and linearly ordered by $<$;

---

[1] Provided ZFC itself is consistent.

iii) if $x \in T$, then there is a finite (possibly empty) set $\{y_1, \ldots, y_m\}$ of incomparable elements such that each $y_i > x$ and, if $z > x$, then $z \geq y_i$ for some i.

A *path* through T is a function $f : \omega \to T$ such that $f(n) < f(n+1)$ for all n.

Lemma A.21 (König's Lemma) *If T is an infinite finite branching tree, then there is a path through T.*

Proof Let $S(x) = \{y : y \geq x\}$ for $x \in T$. We inductively define $f(n)$ such that $S(f(n))$ is infinite for all n. Let r be the minimal element of T, then $S(r)$ is infinite. Let $f(0) = r$. Given $f(n)$, let $\{y_1, \ldots, y_m\}$ be the immediate successors of $f(n)$. Because $S(f(n)) = S(y_1) \cup \ldots \cup S(y_n)$, $S(y_i)$ is infinite for some i. Let $f(n+1) = y_i$.

Forcing Constructions

Definition A.22 Let $(P, <)$ be a partial order. We say that $F \subseteq P$ is a *filter* if:
i) if $p \in F$, $q \in P$, and $p < q$, then $q \in F$;
ii) if $p, q \in F$, there is $r \in F$ such that $r \leq p$ and $r \leq q$.
We say that $D \subseteq P$ is *dense* if for all $p \in P$ there is $q \in D$ such that $q \leq p$. If $\mathcal{D}$ is a collection of dense subsets of P, we say that $G \subseteq P$ is a $\mathcal{D}$-*generic* filter if $D \cap G \neq \emptyset$ for all $D \in \mathcal{D}$.

Lemma A.23 *For any partial order P, if $\mathcal{D}$ is a countable collection of dense subsets of P, then there is a $\mathcal{D}$-generic filter G.*

Proof Let $D_0, D_1, \ldots$, list $\mathcal{D}$. Choose $p_0 \in P$. Given p_n, we can find $p_{n+1} \leq p_n$ with $p_{n+1} \in D_n$. Let $G = \{q : q \geq p_n \text{ for some } n\}$.

Lemma A.23 is the best we can do without extra assumptions. Let P be the set of all finite sequences of zeros and ones ordered by $p < q$ if $p \supset q$. The following sets are dense:
$E_n = \{p \in P : n \in \operatorname{dom}(p)\}$ for $n \in \omega$;
$D_f = \{p \in P : \exists n \in \operatorname{dom}(p) \ p(n) \neq f(n)\}$ for $f \in 2^\omega$.
If G is a filter meeting all of the E_n then $g = \bigcup_{p \in G} p$. Then $g : \omega \to 2$. If G meets D_f, then $g \neq f$. Thus if $\mathcal{D} = \{E_n, D_f : n \in \omega, f \in 2^\omega\}$, then there is no $\mathcal{D}$-generic filter.

We say that p and $q \in P$ are *compatible* if there is $r \leq p, q$ and say that $(P, <)$ satisfies the *countable chain condition* if whenever $A \subset P$ and any two elements of A are incomparable, then $|A| \leq \aleph_0$.

Martin's Axiom If $(P, <)$ is a partial order satisfying the countable chain condition, and $\mathcal{D}$ is a collection of dense subsets of P with $|\mathcal{D}| < 2^{\aleph_0}$, then there is a $\mathcal{D}$-generic filter on P.

Of course, if the Continuum Hypothesis is true, then Martin's Axiom is a trivial consequence of Lemma A.23. On the other hand, Martin's Axiom is consistent with, but not provable from, ZFC $+\neg$CH.

Appendix B
Real Algebra

We prove some of the algebraic facts needed in Section 3.3. All of these results are due to Artin and Schreier. See [58] XI for more details. All fields are assumed to be of characteristic 0.

Definition B.1 A field K is *real* if -1 can not be expressed as a sum of squares of elements of K. In general, we let $\sum K^2$ be the sums of squares from K.

If F is orderable, then F is real because squares are nonnegative with respect to any ordering.

Lemma B.2 *Suppose that F is real and $a \in F \setminus \{0\}$. Then, at most one of a and $-a$ is a sum of squares.*

Proof If a and b are both sums of squares, then $\frac{a}{b} = \frac{a}{b^2} b$ is a sum of squares. Thus, if F is real, at least one of a and $-a$ is not in $\sum F^2$.

Lemma B.3 *If F is real and $-a \in F \setminus \sum F^2$, then $F(\sqrt{a})$ is real. Thus, if F is real and $a \in F$, then $F(\sqrt{a})$ is real or $F(\sqrt{-a})$ is real.*

Proof We may assume that $\sqrt{a} \notin F$. If $F(\sqrt{a})$ is not real, then there are $b_i, c_i \in F$ such that

$$-1 = \sum (b_i + c_i \sqrt{a})^2 = \sum (b_i^2 + 2c_i b_i \sqrt{a} + c_i^2 a).$$

Because $\sqrt{a}$ and 1 are a vector space basis for $F(\sqrt{a})$ over F,

$$-1 = \sum b_i^2 + a \sum c_i^2.$$

Thus

$$-a = \frac{1 + \sum b_i^2}{\sum c_i^2} = \frac{\left(\sum b_i^2\right)\left(\sum c_i^2\right) + \left(\sum c_i^2\right)}{\left(\sum c_i^2\right)^2}$$

and $-a \in \sum F^2$, a contradiction.

Lemma B.4 *If F is real, $f(X) \in F[X]$ is irreducible of odd degree n, and $f(\alpha) = 0$, then $F(\alpha)$ is real.*

Proof We proceed by induction on n. If $n = 1$, this is clear. Suppose, for purposes of contradiction, that $n > 1$ is odd, $f(X) \in F[X]$ is irreducible of degree n, $f(\alpha) = 0$, and $F(\alpha)$ is not real. There are polynomials g_i of degree at most $n - 1$ such that $-1 = \sum g_i(\alpha)^2$. Because F is real, some g_i is nonconstant. Because $F(\alpha) \cong F[X]/(f)$, there is a polynomial $q(X) \in F[X]$ such that

$$1 = \sum g_i^2(X) + q(X)f(X).$$

The polynomial $\sum g_i^2(X)$ has a positive even degree at most $2n - 2$. Thus, q has odd degree at most $n - 2$. Let β be the root of an irreducible factor of q. By induction, $F(\beta)$ is real, but $-1 = \sum g_i^2(\beta)$, a contradiction.

Definition B.5 We say that a field R is *real closed* if and only if R is real and has no proper real algebraic extensions.

If R is real closed and $a \in R$, then, by Lemmas B.2 and B.3, either $a \in R^2$ or $-a \in R^2$. Thus, we can define an order on R by

$$a \geq 0 \Leftrightarrow a \in R^2.$$

Moreover, this is the only way to define an order on R because the squares must be nonnegative. Also, if R is real closed, every polynomial of odd degree has a root in R.

Lemma B.6 *Let F be a real field. There is $R \supseteq F$ a real closed algebraic extension. We call R a* real closure *of F.*

Proof Let $I = \{K \supseteq F : K \text{ real}, K/F \text{ algebraic}\}$. The union of any chain of real fields is real; thus, by Zorn's Lemma, there is a maximal $R \in I$. Clearly, R has no proper real algebraic extensions; thus, R is real closed.

Corollary B.7 *If F is any real field, then F is orderable. Indeed, if $a \in F$ and $-a \notin \sum F^2$, then there is an ordering of F, where $a > 0$.*

Proof By Lemma B.3, $F(\sqrt{a})$ is real. Let R be a real closure of F. We order F by restricting the ordering of R because a is a square in R, $a > 0$.

The following theorem is a version of the Fundamental Theorem of Algebra.

Theorem B.8 *Let R be a real field such that*
i) for all $a \in R$, either $\sqrt{a}$ or $\sqrt{-a} \in R$ and
ii) if $f(X) \in R[X]$ has odd degree, then f has a root in R.
If $i = \sqrt{-1}$, then $K = R(i)$ is algebraically closed.

Proof

Claim 1 Every element of K has a square root in K.

Let $a + bi \in K$. Note that $\frac{a + \sqrt{a^2 + b^2}}{2}$ is nonnegative for any ordering of R. Thus, by i), there is $c \in R$ with

$$c^2 = \frac{a + \sqrt{a^2 + b^2}}{2}.$$

If $d = \frac{b}{2c}$, then $(c + di)^2 = a + bi$.

Let $L \supseteq K$ be a finite Galois extension of R. We must show that $L = K$. Let $G = Gal(L/R)$ be the Galois group of L/R. Let H be the 2-Sylow subgroup of G.

Claim 2 $G = H$.

Let F be the fixed field of H. Then F/R must have odd degree. If $F = R(x)$, then the minimal polynomial of x over R has odd degree, but the only irreducible polynomials of odd degree are linear. Thus, $F = R$ and $G = H$.

Let $G_1 = Gal(L/K)$. If G_1 is nontrivial, then there is G_2 a subgroup of G_1 of index 2. Let F be the fixed field of G_2. Then, F/K has degree 2. But by Claim 1, K has no extensions of degree 2. Thus, G_1 is trivial and $L = K$.

Corollary B.9 *Suppose that R is real. Then R is real closed if and only if $R(i)$ is algebraically closed.*

Proof
($\Rightarrow$) By Theorem B.8.
($\Leftarrow$) $R(i)$ is the only algebraic extension of R, and it is not real.

Let $(R, <)$ be an ordered field. We say that R has the *intermediate value property* if for any polynomial $p(X) \in R[X]$ if $a < b$ and $p(a) < 0 < p(b)$, then there is $c \in (a, b)$ with $p(c) = 0$.

Lemma B.10 *If $(R, <)$ is an ordered field with the intermediate value property, then R is real closed.*

Proof Let $a > 0$ and let $p(X) = X^2 - a$. Then $p(0) < 0$, and $p(1 + a) > 0$; thus, there is $c \in R$ with $c^2 = a$.

Let

$$f(X) = X^n + \sum_{i=0}^{n-1} a_i X^i$$

where n is odd. For M large enough, $f(M) > 0$ and $f(-M) < 0$; thus, there is a c such that $f(c) = 0$.

By Theorem B.8, $R(i)$ is algebraically closed. Because R is real, it must be real closed.

Lemma B.11 *Suppose that R is real closed and $<$ is the unique ordering, then $(R, <)$ has the intermediate value property.*

Proof Suppose $f(X) \in R[X]$, $a < b$, and $f(a) < 0 < f(b)$. We may assume that $f(X)$ is irreducible (for some factor of f must change signs). Because $R(i)$ is algebraically closed, either $f(X)$ is linear, and hence has a root in (a, b), or

$$f(X) = X^2 + cX + d,$$

where $c^2 - 4d < 0$. But then

$$f(X) = \left(X + \frac{c}{2}\right)^2 + \left(d - \frac{c^2}{4}\right)$$

and $f(x) > 0$ for all x.

We summarize as follows.

Theorem B.12 *The following are equivalent.*
i) R is real closed.
ii) For all $a \in R$, either a or $-a$ has a square root in R and every polynomial of odd degree has a root in R.
iii) We can order R by $a \geq 0$ if and only if a is a square and, with respect to this ordering, R has the intermediate value property.

Finally, we consider the question of uniqueness of real closures. We first note that there are some subtleties. For example, there are nonisomorphic real closures of $F = \mathbf{Q}(\sqrt{2})$. The field of real algebraic numbers is one real closure of F. Because $a + b\sqrt{2} \mapsto a - b\sqrt{2}$ is an automorphism of F, $\sqrt{2}$ is not in $\sum F^2$. Thus, by Corollary B.5, $F(\sqrt{-2})$ is real. Let R be a real closure of F containing $F(\sqrt{-2})$. Then, R is not isomorphic to the real algebraic numbers over F.

This is an example of a more general phenomenon. It is proved by successive applications of Lemmas B.2 and B.3.

Lemma B.13 *If $(F, <)$ is an ordered field, then there is a real closure of F in which every positive element of F is a square.*

Because $\mathbb{Q}(\sqrt{2})$ has two distinct orderings, it has two nonisomorphic real closures. The field $\mathbb{Q}(t)$ of rational functions over $\mathbb{Q}$ has $2^{\aleph_0}$ orderings and hence $2^{\aleph_0}$ nonisomorphic real closures.

The next theorem shows that once we fix an ordering of F, there is a unique real closure that induces the ordering.

Theorem B.14 *Let $(F, <)$ be an ordered field. Let R_0 and R_1 be real closures of F such that $(R_i, <)$ is an ordered field extension of $(F, <)$. Then, R_0 is isomorphic to R_1 over F and the isomorphism is unique.*

The proof of Theorem B.14 uses Sturm's algorithm.

Definition B.15 Let R be a real closed field. A *Sturm sequence* is a finite sequence of polynomials $f_0, \ldots, f_n$ such that:

i) $f_1 = f_0'$;

ii) for all x and $0 \le i \le n - 1$, it is not the case that $f_i(x) = f_{i+1}(x) = 0$;

iii) for all x and $1 \le i \le n - 1$, if $f_i(x) = 0$, then $f_{i-1}(x)$ and $f_{i+1}(x)$ have opposite signs;

iv) f_n is a nonzero constant.

If $f_0, \ldots, f_n$ is a Sturm sequence and $x \in \mathbb{R}$, define $v(x)$ to be the number of sign changes in the sequence $f_0(x), \ldots, f_n(x)$.

Suppose that $f \in R[X]$ is nonconstant and does not have multiple roots. We define a Sturm sequence as follows:

$f_0 = f$;

$f_1 = f'$.

Given f_i nonconstant, use the Euclidean algorithm to write

$$f_i = g_i f_{i-1} - f_{i+1}$$

where the degree of f_{i+1} is less than the degree of f_{i-1}. We eventually reach a constant function f_n.

Lemma B.16 *If f has no multiple roots, then $f_0, \ldots, f_n$ is a Sturm sequence.*

Proof

iv) If $f_n = 0$, then $f_{n-1} | f_i$ for all i. But f has no multiple roots; thus f and f' have no common factors, a contradiction.

ii) If $f_i(x) = f_{i+1}(x) = 0$, then by induction $f_n(x) = 0$, contradicting iv).

iii) If $1 \le i \le n - 1$ and $f_i(x) = 0$, then $f_{i-1}(x) = -f_{i+1}(x)$. Thus, $f_{i-1}(x)$ and $f_{i+1}(x)$ have opposite signs.

Theorem B.17 (Sturm's Algorithm) *Suppose that R is a real closed field, $a, b \in R$, and $a < b$. Let f be a polynomial without multiple roots. Let $f = f_0, \ldots, f_n$ be a Sturm sequence such that $f_i(a) \ne 0$ and $f_i(b) \ne 0$ for all i. Then, the number of roots of f in (a, b) is equal to $v(a) - v(b)$.*

Proof Let $z_1 < \ldots < z_m$ be all the roots of the polynomials $f_0, \ldots, f_n$ that are in the interval (a, b). Choose $c_1, \ldots, c_{m-1}$ with $z_i < c_i < z_{i+1}$. Let $a = c_0$ and $b = c_m$. For $0 \le i \le m - 1$, let r_i be the number of roots of f in the interval (c_i, c_{i+1}). Clearly, $\sum r_i$ is the number of roots of f in the interval (a, b). On the other hand,

$$v(a) - v(b) = \sum_{i=0}^{m-1} (v(c_i) - v(c_{i+1})).$$

Thus, it suffices to show that if $c < z < d$ and z is the only root of any f_i in (c, d), then

$$v(d) = \begin{cases} v(c) - 1 & z \text{ is a root of } f \\ v(c) & \text{otherwise} \end{cases}.$$

If $f_i(b)$ and $f_i(c)$ have different signs, then $f_i(z) = 0$. We need only see what happens at those places.

If z is a root of f_i, $i > 0$, then $f_{i+1}(z)$ and $f_{i-1}(z)$ have opposite signs and f_{i+1} and f_{i-1} do not change signs on $[c, d]$. Thus, the sequences $f_{i-1}(c), f_i(c), f_{i+1}(c)$ and $f_{i-1}(d), f_i(d), f_{i+1}(d)$ each have one sign change. For example, if $f_{i-1}(z) > 0$ and $f_{i-1}(z) < 0$, then these sequences are either $+, +, -$ or $+, -, +$, and in either case both sequences have one sign change.

If z is a root of f_0, then, because $f'(z) \neq 0$, f is monotonic on (c, d). If f is increasing on (c, d), the sequence at c starts $-, +, \ldots$ and the sequence at d starts $+, +, \ldots$. Similarly, if f is decreasing, the sequence at c starts $+, -, \ldots$, and the sequence at b starts $-, -, \ldots$. In either case, the sequence at c has one more sign change than the sequence at d. Thus, $v(c) - v(d) = 1$, as desired.

Corollary B.18 *Suppose that $(F, <)$ is an ordered field. Let f be a nonconstant irreducible polynomial over F. If R_0 and R_1 are real closures of F compatible with the ordering, then f has the same number of roots in both R_0 and R_1.*

Proof Let $f_0, \ldots, f_n$ be the Sturm sequence from Lemma B.16. Note that each $f_i \in F[X]$. We can find $M \in F$ such that any root of f_i is in $(-M, M)$ (if $g(X) = X^n + \sum a_i X^i$, then any root of g has absolute value at most $1 + \sum |a_i|$, for example). Then, the number of roots of f in R_i is equal to $v(-M) - v(M)$, but $v(M)$ depends only on F.

Proof of Theorem B.14 Let $K \subset R_0$ be a finite extension of F. Say $K = F(\alpha)$. Let $f(X)$ be the irreducible polynomial of α over F. Let $\alpha_1 < \ldots < \alpha_n$ be the roots of f in R_0, and let $\beta_1 < \ldots < \beta_n$ be the roots of f in R_1. We can map $F(\alpha_1, \ldots, \alpha_n)$ into R_1 by $\alpha_i \mapsto \beta_i$, and this is the only field embedding that could possibly extend to an isomorphism.

Using this idea and Zorn's Lemma, build a maximal embedding of subfields, this must be an isomorphism.

Uniqueness follows because the ith root of $f(X)$ in R_0 must be sent to the ith root of $f(X)$ in R_1.

References

[1] D. Abramovich and F. Voloch, Towards a proof of the Mordell–Lang conjecture in characteristic p, Int. Math. Res. Not. 2 (1992), 103–115.

[2] J. Ax, The elementary theory of finite fields, Ann. Math. 88 (1968), 239–271.

[3] J. Ax and S. Kochen, Diophantine problems over local fields I, Am. J. Math. 87 (1965), 605–630.

[4] J. Ax and S. Kochen, Diophantine problems over local fields II, Am. J. Math. 87 (1965), 631–648.

[5] J. Ax and S. Kochen, Diophantine problems over local fields III, Ann. Math. 83 (1966), 437–456.

[6] M. Atiyah and I. Macdonald, *Introduction to Commutative Algebra*, Addison-Wesley, Reading, MA, 1996.

[7] J. Baldwin, *Fundamentals of Stability Theory*, Springer-Verlag, New York, 1988.

[8] J. Barwise, *Handbook of Mathematical Logic*, North-Holland, Amsterdam, 1977.

[9] H. Becker and A. Kechris, *The Descriptive Set Theory of Polish Group Actions*, Cambridge University Press, Cambridge, UK 1996.

[10] K. Binmore, *Fun and Games*, Heath, Boston, 1992.

[11] L. Blum, Differentially closed fields: a model theoretic tour, in *Contributions to Algebra*, H. Bass, Phyllis J. Cassidy and Jerald Kovacic, eds., Academic Press, New York, 1977.

[12] J. Bochnak, M. Coste, and M-F. Roy, *Real Algebraic Geometry*, Springer-Verlag, New York, 1998.

[13] A. Borel, Injective endomorphisms of algebraic varieties, Arch. Math. 20 (1969), 531-537.

[14] A. Borel, *Linear Algebraic Groups*, Springer-Verlag, New York, 1991.

[15] A. Borovik and A. Nesin, *Groups of Finite Morley Rank*, Oxford Science Publications, Oxford, UK, 1994.

[16] E. Bouscaren, ed., *Model Theory and Algebraic Geometry: An Introduction to E. Hrushovski's Proof of the Geometric Mordell–Lang Conjecture*, Springer-Verlag, New York, 1998.

[17] E. Bouscaren, Proof of the Mordell–Lang conjecture for function fields, in [16].

[18] S. Buechler, *Essential Stability Theory*, Springer-Verlag, New York, 1996.

[19] S. Buechler, Vaught's conjecture for superstable theories of finite rank, Ann. Pure Appl. Logic, to appear.

[20] A. Buium, *Differential Algebra and Diophantine Geometry*, Hermann, Paris, 1994.

[21] B. F. Caviness and J. R. Johnson, eds., *Quantifier Elimination and Cylindric Algebraic Decomposition*, Springer-Verlag, New York, 1998.

[22] C. C. Chang and H. J. Keisler, *Model Theory*, North-Holland, Amsterdam, 1990.

[23] Z. Chatzidakis, L. van den Dries, and A. Macintyre, Definable sets over finite fields, J. Reine Angew. Math. 427 (1992), 107–135.

[24] M. Davis, J. Matijasevič, and J. Robinson, Hilbert's 10th Problem. Diophantine equations: Positive aspects of a negative solution, in *Mathematical Developments from Hilbert's Problems*, F. Browder, ed., American Mathematical Societ, Providence, RI, 1976.

[25] J. Denef, The rationality of the Poincare series associated to the p-adic points on a variety, Invent. Math. 77 (1984), 1–23.

[26] K. Devlin, *Fundamentals of Contemporary Set Theory*, Springer-Verlag, New York, 1979.

[27] K. Devlin, *Constructibility*, Springer-Verlag, New York, 1984.

[28] L. van den Dries, Some applications of a model theoretic fact to (semi-) algebraic geometry, Indag. Math. 4 (1982), 397–401.

[29] L. van den Dries, *Tame Topology and o-minimal Structures*, Cambridge University Press, Cambridge, UK, 1998.

[30] L. van den Dries, A. Macintyre, and D. Marker, The elementary theory of restricted analytic fields with exponentiation, Ann. Math. 140 (1994), 183–205.

[31] H. D. Ebbinghaus, J. Flum, and W. Thomas, *Mathematical Logic*, Springer-Verlag, New York, 1989.

[32] H. D. Ebbinghaus and J. Flum, *Finite Model Theory*, Springer-Verlag, New York, 1995.

[33] M. Fairtlough and S. Wainer, Hierarchies of provably recursive functions, in *Handbook of Proof Theory*, S. Buss, ed., Elsevier Amsterdam, 1998.

[34] D. Flath and S. Wagon, How to pick out the integers in the rationals: An application of number theory to logic, Am. Math. Mon. 98 (1991), 812–823.

[35] K. Gödel, Review of T. Skolem's "On the imposibility of a complete characterization of the number sequence by means of a finite axiom system", in *Collected Works v.1 1929–1936*, S. Feferman, J. Dawson, W. Goldfarb, C. Parsons and R. Solovay, eds., Oxford University Press, Oxford, UK, 1986, 376–380.

[36] R. Graham, B. Rothschild, and J. Spencer, *Ramsey Theory*, Wiley, New York, 1980.

[37] R. Hartshorne, *Algebraic Geometry*, Springer-Verlag, New York, 1977.

[38] M. Hindry and J. Silverman, *Diophantine Geometry: An Introduction*, Springer, New York, 2000.

[39] M. Hirsch and S. Smale, *Differential Equations, Dynamical Systems and Linear Algebra*, Academic Press, New York, 1974.

[40] W. Hodges, *Model Theory*, Cambridge University Press, Cambridge, UK, 1993.

[41] E. Hrushovski, A new strongly minimal set, Ann. Pure Appl. Logic 62 (1993), 147–166.

[42] E. Hrushovski, Strongly minimal expansions of algebraically closed fields. Isr. J. Math. 79 (1992), 129–151.

[43] E. Hrushovski, The Mordell–Lang conjecture for function fields, J. Am. Math. Soc. 9 (1996), 667–690.

[44] E. Hrushovski and B. Zil'ber, Zariski geometries, Bull. Am. Math. Soc. 28 (1993), 315–323.

[45] E. Hrushovski and B. Zil'ber, Zariski geometries, J. Am. Math. Soc. 9 (1996), 1–56.

[46] N. Immerman, Descriptive Complexity, Springer, New York, 1999.

[47] T. Jech, Set Theory, Academic Press, New York, 1978.

[48] A. Kanamori, The Higher Infinite, Springer-Verlag, New York, 1994.

[49] A. Kanamori and K. McAloon, On Gödel incompleteness and finite combinatorics, Ann. Pure Appl. Logic 33 (1987), 23–41.

[50] I. Kaplansky, An Introduction to Differential Algebra, Hermann, Paris, 1957.

[51] R. Kaye, Models of Peano Arithmetic, Oxford Science Publications, Oxford, UK, 1991.

[52] A. Kechris, Classical Descriptive Set Theory, Springer-Verlag, New York, 1995.

[53] H. J. Keisler, Model Theory, in [8].

[54] B. Kim and A. Pillay, From stability to simplicity, Bull. Symbolic Logic 4 (1998), 17–36.

[55] S. Kochen, Model theory of local fields, in Proceedings of the International Summer Institute and Logic Colloquium, Kiel, 1974, G. Müller, A. Oberschelp, and K. Potthoff, eds., Springer-Verlag, Berlin, 1975.

[56] E. Kolchin, Constrained extensions of differential fields, Adv. Math. 12 (1974), 141–170.

[57] K. Kunen, Set Theory: An Introduction to Independence Proofs, North-Holland, Amsterdam, 1980.

[58] S. Lang, Algebra, Addison-Wesley, Reading, MA, 1971.

[59] S. Lang, Abelian Varieties, Springer-Verlag, New York, 1983.

[60] A. Macintyre, On definable dubsets of the p-adic field, J. Symbolic Logic 41 (1976), 605–610.

[61] A. Macintyre and D. Marker, Primes and their residue rings in models of open induction, Ann. Pure Appl. Logic 43 (1989), 57–77.

[62] A. Macintyre and A. Wilkie, On the decidability of the real exponential field, in *Kreiseliana*, P. Odifreddi, eds., A. K. Peters, Wellesley, MA, 1996.

[63] A. Magid, *Lectures on Differential Galois Theory*, American Mathematical Society, Providence, RI, 1994.

[64] Y. Manin, Rational points on an algebraic curve over a function field, Am. Math. Soc. Transl. 50 (1966), 189–234.

[65] D. Marker, Model theory of differential fields, in *Model Theory of Fields*, D. Marker, M. Messmer, and A. Pillay, eds., Springer-Verlag, New York, 1995.

[66] D. Marker, Manin kernels, in *Connections between Model Theory and Algebraic and Analytic Geometry*, A. Macintyre, ed., Quaderni di mathematica, Naples, 2001.

[67] D. Marker, Zariski geometries, in [16].

[68] H. Matsamura, *Commutative Ring Theory*, Cambridge University Press, Cambridge, UK 1980.

[69] L. Mayer, Vaught's conjecture for o-minimal theories, J. Symbolic Logic 53 (1988), 146–159.

[70] A. Mekler, Stability of nilpotent groups of class 2 and prime exponent, J. Symbolic Logic 46 (1981) 781–788.

[71] M. Morley, Categoricity in power, Trans. Am. Math. Soc. 114 (1965), 514–538.

[72] M. Nadel, $\mathcal{L}_{\omega_1,\omega}$ and admissible fragments, in *Model-Theoretic Logics*, J. Barwise and S. Feferman eds., Springer-Verlag, New York, 1985.

[73] J. Paris and L. Harrington, A mathematical incompleteness in Peano Arithmetic, in [8].

[74] Y. Peterzil and S. Starchenko, A trichotomy theorem for o-minimal structures, Proc. London Math. Soc. 77 (1998), 481–523.

[75] A. Pillay, *An Introduction to Stability Theory*, Oxford Science Publications, Oxford, UK, 1983.

[76] A. Pillay, *Geometric Stability Theory*, Oxford Science Publications, Oxford, UK, 1996.

334   References

[77] A. Pillay, Model theory of algebraically closed fields, in [16].

[78] A. Pillay, Differential Galois theory I, Illinois J. Math. 42 (1998), 678–699.

[79] A. Pillay, Model theory and Diophantine geometry, Bull. Am. Math. Soc. 34 (1997), 405–422.

[80] A. Pillay, Some model theory of compact complex spaces, in *Hilbert's tenth problem: Relations with Arithmetic and Algebraic geometry (Ghent, 1999)*, Contemporary Mathematics vol 270, American Mathematical Societ, Providence, RI, 2000.

[81] A. Pillay, Definable sets in generic complex tori, Ann. Pure Appl. Logic 77 (1996), 75–80.

[82] A. Pillay and M. Ziegler, Jet spaces of varieties over differential and difference fields, preprint, 2001.

[83] A. Pillay and C. Steinhorn, Definable sets in ordered structures I, Trans. Am. Math. Soc. 295 (1986), 565–593.

[84] B. Poizat, Une théorie de Galois imaginaire, J. Symbolic Logic 48 (1983), 1151–1171.

[85] B. Poizat, Review of S. Shelah's "The number of non-isomorphic models of an unstable first-order theory," J. Symbolic Logic 47 (1982), 436–438.

[86] B. Poizat, *Groupes Stables*, Nur al-Mantiq wal-Ma'riah 1987. Translated as *Stable Groups*, American Mathematical Society, Providence, RI, 2001.

[87] M. Prest, *Model Theory and Modules*, Cambridge University Press, Cambridge, UK, 1988.

[88] H. Putnam, Models and reality, *Philosophy of Mathematics: Selected Readings*, P. Benacerraf and H. Putnam eds., Cambridge University Press, Cambridge, UK, 1983.

[89] J. Rose, *A Course in Group Theory*, Cambridge University Press, Cambridge, UK, 1978.

[90] W. Rudin, Injective polynomial maps are automorphisms, Am. Math. Mon. 102 (1995), 540–543.

[91] I. Shafarevich, *Basic Algebraic Geometry 1 and 2*, Springer-Verlag, New York, 1994.

[92] S. Shelah, *Classification Theory and the Number of Non-isomorphic Models*, North-Holland, Amsterdam, 1978.

[93] S. Shelah, L. Harrington, and M. Makkai, A proof of Vaught's conjecture for ω-stable theories, Isr. J. Math. 49 (1984), 259–280.

[94] J. Shoenfield, *Mathematical Logic*, A. K. Peters Ltd., Natick, MA, 2001.

[95] J. Silverman, *The Arithmetic of Ellptic Curves*, Springer-Verlag, New York, 1986.

[96] M. Singer, Direct and inverse problems in differential Galois theory, in *Selected Works of Ellis Kolchin*, Hyman Bass, A. Buium, and P. Cassidy, eds., American Mathematical Society, Providence, RI, 1999.

[97] J. Spencer, *The Strange Logic of Random Graphs*, Springer, New York, 2001.

[98] J. Steel, On Vaught's Conjecture, in *Cabal Seminar 76–77*, A. Kechris and Y. Moschovakis, eds., Springer-Verlag, New York, 1978.

[99] R. Vaught, Denumerable models of complete theories, *Infinitistic Methods*, Pergamon, New York 1961.

[100] C. Wagner, On Martin's conjecture, Ann. Math. Logic 22 (1982), 47-67.

[101] F. Wagner, Fields of finite Morley rank, J. Symbolic Logic 66 (2001), 703–706.

[102] R. Walker, *Algebraic Curves*, Springer-Verlag, New York, 1978.

[103] A. Wilkie, Model completeness results for expansions of the real field by restricted Pfaffian functions and the exponential function, J. Am. Math. Soc. 9 (1996), 1051–1094.

[104] M. Ziegler, Introduction to stability theory and Morley rank, in [16].

Index

Graduate Texts in Mathematics

(continued from page ii)